Food Intake in Fish

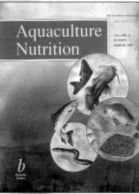

Food Intake in Fish

Edited by

Dominic Houlihan
University of Aberdeen, UK

Thierry Boujard
Unité mixte INRA-IFREMER, Sainte Pee sur Nivelle, France

and

Malcolm Jobling
NFH, University of Tromsø, Norway

Blackwell
Science

© 2001 by
Blackwell Science Ltd
Editorial Offices:
Osney Mead, Oxford OX2 0EL
25 John Street, London WC1N 2BS
23 Ainslie Place, Edinburgh EH3 6AJ
350 Main Street, Malden
  MA 02148 5018, USA
54 University Street, Carlton
  Victoria 3053, Australia
10, rue Casimir Delavigne
  75006 Paris, France

Other Editorial Offices:

Blackwell Wissenschafts-Verlag GmbH
Kurfürstendamm 57
10707 Berlin, Germany

Blackwell Science KK
MG Kodenmacho Building
7–10 Kodenmacho Nihombashi
Chuo-ku, Tokyo 104, Japan

Iowa State University Press
A Blackwell Science Company
2121 S. State Avenue
Ames, Iowa 50014–8300, USA

The right of the Authors to be identified as
the Authors of this Work has been asserted in
accordance with the Copyright, Designs and
Patents Act 1988.

The authors have endeavoured to contact any known
copyright holders for any previously published
illustrations and apologises if any formal permission
to use any of the illustrations in this book has not
been forthcoming by the time of publication.

First published 2001

Set in 10/13 pt Times
by Sparks Computer Solutions Ltd, Oxford
http://www.sparks.co.uk
Printed and bound in Great Britain by
MPG Books Ltd, Bodmin, Cornwall

DISTRIBUTORS

Marston Book Services Ltd
PO Box 269
Abingdon
Oxon OX14 4YN
(*Orders:* Tel: 01235 465500
        Fax: 01235 465555)

USA and Canada
Iowa State University Press
A Blackwell Science Company
2121 S. State Avenue
Ames, Iowa 50014-8300
(*Orders:* Tel: 800-862-6657
        Fax: 515-292-3348
        Web www.isupress.com
        email: orders@isupress.com)

Australia
Blackwell Science Pty Ltd
54 University Street
Carlton, Victoria 3053
(*Orders:* Tel: 03 9347 0300
        Fax: 03 9347 5001)

A catalogue record for this title
is available from the British Library

ISBN 0-632-05576-6

Library of Congress
Cataloging-in-Publication Data
is available

For further information on
Blackwell Science, visit our website:
www.blackwell-science.com

List of Contributors

Alanärä, Anders   Department of Aquaculture, Swedish University of Agricultural Sciences, S-901 83 Umea, Sweden.

Azzaydi, Mezian   Department of Physiology and Pharmacology, Faculty of Biology, University of Murcia, 3010 Murcia, Spain.

Baras, Etienne   Laboratoire de Démographie des Poissons et d'Aquaculture, Université de Liège, 10 Chemin de la Justice, 4500 Tihange, Belgium.

Beauchaud, Marilyn   Université Jean Monnet de St Etienne, 23 rue Pierre Michelon, 42023 St Etienne Cedex 2, France.

Bégout Anras, Marie-Laure   CREMA L'Houmeau, BP 5, 17137 L'Houmeau, France.

Björnsson, Björn Thrandur   Fish Endocrinology Laboratory, Department of Zoology, Göteborg University, PO Box 463, 40530 Göteborg, Sweden.

Bolliet, Valérie   Unité de recherches en Hydrobiologie, Unité mixte INRA/IFREMER, Station d'Hydrobiologie INRA, BP. 3, 64310 Saint Pée sur Nivelle, France.

Boujard, Thierry   Unité de recherches en Hydrobiologie, Unité mixte INRA/IFREMER, Station d'Hydrobiologie INRA, BP. 3, 64310 Saint Pée sur Nivelle, France.

Carter, Chris   Tasmanian Aquaculture and Fisheries Institute, University of Tasmania, P.O. Box 1214, 7250 Launceston, Tasmania, Australia.

Covès, Denis   Laboratoire de Recherche en Pisciculture de Méditerranée, Station Expérimentale d'Aquaculture Ifremer, 34250 Palavas, France.

Damsgård, Børge   Norwegian Institute of Fisheries and Aquaculture (FISKERIFORSKNING), 9005 Tromsø, Norway.

de la Higuera, Manuel   Departmento de Biologia Animal y Ecologia, Facultad de Ciencias, Universidad de Granada, Avenida Fuenta Nueva, 18001 Granada, Spain.

de Pedro, Nuria   Departamento de Biología Animal II (Fisiología Animal), Facultad de Ciencias Biológicas, Universidad Complutense, 28040 Madrid, Spain.

Dias, Jorges   Instituto de Ciencias biomedicas da Universidade do Porto, Aquatic Sciences Department, Largo Prof. Abel Salazar 2, 4050 Porto, Portugal.

Gomes, Emidio   Instituto de Ciencias biomedicas da Universidade do Porto, Aquatic Sciences Department, Largo Prof. Abel Salazar 2, 4050 Porto, Portugal.

Gudmundsson, Olafur   Agricultural Research Institute, Rala Building, Keldnaholt, 112 Reykjavik, Iceland

Houlihan, Dominic   Faculty of Science and Engineering, University of Aberdeen, AB24 3FX Aberdeen, United Kingdom.

Jobling, Malcolm   Norwegian College of Fishery Science (NFH), University of Tromsø, 9037 Tromsø, Norway.

Juell, Jon-Erik   Agricultural University of Norway, P.O.Box N-5065, 1432 Aas, Norway.

Kadri, Sunil   Division of Environmental & Evolutionary Biology, IBLS, University of Glasgow, G12 8QQ Glasgow, United Kingdom.

Kestemont, Patrick   Unité de Recherches en Biologie des Organismes, Facultés Universitaires N.D. de la Paix, 61, rue de Bruxelles, 5000 Namur, Belgium.

Kettunen, Juhani   Finnish Game and Fisheries Research Institute, PO Box 6, 00721 Helsinki, Finland.

Kiessling, Anders   Department of Aquaculture, Swedish University of Agricultural Sciences, 901 83 Umea, Sweden.

King, Jonathan   The Centre for Environment, Fisheries and Aquaculture Science, Conwy Laboratory, Benarth Toad, LL32 8UB Conwy, North Wales, United Kingdom.

Koskela, Juha   Finish Game and Fisheries Research Institute, Laukaa Fisheries Research and Aquaculture, 41360 Valkola, Finland.

Kristiansen, Henrik   Mobile Nutrients ApS, Forskerparken Foulum, P.O. Box 10, 8830 Tjele, Denmark.

Lagardère, Jean-Paul   CREMA L'Houmeau, BP 5, 17137 L'Houmeau, France.

Lamb, Charles F. Department of Biology Black Hills State University 1200 University Spearfish, SD 57799-9095 U.S.A.

Madrid, Juan Antonio Department of Physiology and Pharmacology, Faculty of Biology, University of Murcia, 3010 Murcia, Spain.

Médale, Françoise Unité de recherches en Hydrobiologie, Unité mixte INRA/IFREMER, Station d'Hydrobiologie INRA, BP. 3, 64310 Saint Pée sur Nivelle, France.

Paspatis, Mihaelis Aquaculture Department, Institute of Marine Biology of Crete, PO Box 2214, 71003 Heraclion, Crete, Greece.

Petursdottir, Thuridur E Agricultural Research Institute, Rala Building, Keldnaholt, 112 Reykjavik, Iceland.

Ruohonen, Kari Finnish Game and Fisheries Research Institute, Evo Fisheries Research Station, 16970 Evo, Finland.

Sánchez-Vázquez, F. Javier Department of Physiology and Pharmacology, Faculty of Biology, University of Murcia, 3010 Murcia, Spain.

Contents

3 Techniques for Measuring Feed Intake   **49**
Malcolm Jobling, Denis Covès, Børge Damsgård, Henrik R. Kristiansen,
Juha Koskela, Thuridur E. Petursdottir, Sunil Kadri and Olafur Gudmundsson

4 Experimental Design in Feeding Experiments   **88**
Kari Ruohonen, Juhani Kettunen and Jonathan King

5 Gustation and Feeding Behaviour   108
Charles F. Lamb

6 Environmental Factors and Feed Intake: Mechanisms and Interactions   131
Patrick Kestemont and Etienne Baras

10 Effects of Feeding Time on Feed Intake and Growth — 233
Valérie Bolliet, Mezian Azzaydi and Thierry Boujard

11 Effects of Nutritional Factors and Feed Characteristics on Feed Intake — 250
Manuel de la Higuera

12 Regulation of Food Intake by Neuropeptides and Hormones — 269
Nuria de Pedro and Björn Thrandur Björnsson

13 Physiological Effects of Feeding   297
*Chris Carter, Dominic Houlihan, Anders Kiessling, Francoise Médale
and Malcolm Jobling*

14 Feeding Management   332
Anders Alanärä, Sunil Kadri and Mihalis Paspatis

Preface

This book arose from collaboration among participants in a European Union COST programme: regulation of voluntary feed intake in fish. COST is a framework for research and development co-operation in Europe and the objective of the programme was to promote, co-ordinate and harmonise European research on feed intake in fish at a crucial moment for the development of new management techniques in fish feeding. Indeed, food intake in fish is an area of study that is of great importance to the applied science of fisheries and aquaculture, as well as being of basic biological interest. For example, the more we know about the various factors that influence the ingestion of feed the better able we will be to manipulate the rearing environment to ensure improved growth performance and feed utilisation of cultured fish, and also to decrease the amount of waste per unit of fish produced.

The book seeks to illustrate how insights into the biological and environmental factors that underlie the feeding responses of fish may be used to address practical issues of feed management. The main focus is on the feeding of fish held in captivity, and there is an emphasis on the application of the results of feeding studies in the practical context of fish culture. Feeding in fish, as in all animals, encompasses a wide range of factors. The animals encounter the food and capture it, make decisions concerning palatability and nutritional value, and then either ingest or reject it. Ingested food is then digested and absorbed, after which the animal's physiological systems deal with the influx of nutrients.

Palatability of feed is a major factor determining feed acceptance. Animals of several taxonomic groups produce, or concentrate, noxious substances, which reduce their attractiveness as prey, and many plants produce and store secondary metabolites that act as feeding deterrents. Feeds based on plant, rather than on animal, materials may lead to palatability problems in aquaculture, particularly when carnivorous fish species are being reared. In addition to containing substances that influence whether or not a fish will ingest them, plants may contain a range of antinutritional factors that have effects on digestion, absorption and nutrient metabolism. However, there will be a need to make greater use of plant materials in feed formulations as aquacultural production increases. The inclusion of plant proteins in formulated feeds reduces production costs, so there are important economic incentives to improve the palatability of plant feedstuffs, and for the investigation of potential feeding stimulants. The first two chapters of the book present basic information about feed composition and manufacture.

Chapter 1 deals with feed composition and the methods used in the analysis of feeds. As the farming of fish is often reliant upon formulated feeds we need to have a good knowledge of the nutrients present, and how we should analyse for them. Although chemical analyses provide useful information about the nutrients present in feeds and feedstuffs, it is essential to carry out growth trials on live animals to obtain a complete assessment of the biological value of feeds. Such trials will usually incorporate digestibility studies to provide an assessment of nutrient bioavailabilities and feed evaluation will also involve an assessment of how efficiently the various nutrients have been used to support growth. The chapter describes the methods used for carrying out digestibility studies, and provides information about the biological evaluation of feeds and feed ingredients.

In Chapter 2, the question of feed formulation is addressed. Feed fed to aquatic organisms may be subject to leaching of nutrients while it remains in the water column or at the bottom of a tank. This creates a special set of problems for the manufacturer of feeds for fish and other aquatic animals. Dry feeds are most commonly used in present-day fish culture, and both the manufacturer and the farmer need to know how ingredient selection and processing can affect the availability of the nutrients present in the finished feed. These are topics considered in this chapter, as is the important question of the transition from feed formulations based upon fish meal to those based upon plant proteins.

Measuring feed consumption of animals held in water is not easy, and this is probably the major reason for the shortage of information on feed intake in fish, relative to terrestrial animals. In the 1950s some researchers trained fish to hit a rod to obtain food; this approach resulted in several short-term studies into feed preferences and feeding behaviour of fish. These first self-feeding devices were rather crude in construction, but the results obtained were very promising. The devices became gradually more refined as time progressed, and the introduction of computerised systems added a new dimension of sophistication. Self-feeders represent one type of on-demand feeding system, of which there are currently two main forms: self-feeders and interactive feedback feeders. When the former are used the fish must act to trigger the feed delivery system and feed themselves, whereas the latter type of system monitors how much feed is being consumed and controls delivery in response; the fish feed and the delivery of additional food is automatically regulated accordingly.

In addition to these methods, where the fish determine the amount of food delivered, a number of techniques have also been developed to measure the amount of food consumed when the experimenter or fish farmer controls the delivery of feed. The various techniques are considered in Chapter 3. There is no single method that is suitable for use under all conditions; different techniques are best for individual fish, small groups of fish held under experimental conditions, and larger groups in a cultured context. In addition to considering the measurement of feed intake of fish held in captivity the authors also present some information about the measurement of food consumption of wild fish.

Chapter 4 considers how we do experiments involving the measurement of food consumption, and how we analyse the data obtained. In a novel approach the authors pose the question: how should we design experiments to obtain results that reduce the amount of uncertainty? This question is of central importance because a lower level of uncertainty means increased reliability of scientific results. In the context of the aquaculture industry, reductions in uncertainty mean fewer risks of making mistakes when estimating economic returns for investment. The authors emphasise the need to decide upon the method to be used for statistical analysis in advance, as the method chosen will determine the best experimental approach.

There are lessons for all in this chapter; adoption of the principles would improve experimentation, not only in fish biology but also in the life sciences generally.

Fish may use inputs from various combinations of sensory systems during the different phases of feeding behaviour, but the ultimate acceptance or rejection of a food item is almost invariably dependent upon inputs from chemoreceptors. Fish possess several chemosensory systems, including gustation (taste), olfaction (smell) and a common chemical sense, but attributing specific roles to each system is difficult because each system responds to aqueous chemical stimuli, some of which may be common across the different systems. Nevertheless, it seems that the gustatory sense is the most important in the acquisition and ingestion of food, and in the rejection of potentially harmful or toxic substances. The gustatory system, which is very highly developed in some fish species, is the subject of Chapter 5. The emphasis is on the adaptations shown by the 'taste specialists', species that possess taste buds in epithelia outwith the oropharynx and have substantial enlargements of the brain regions responsible for processing gustatory information.

Chapters 6 and 7 consider the environmental factors that can influence fish feeding. Light, temperature, water velocity, social factors, predators and disturbance by humans can all influence the distribution of the fish in the water and their chances of gaining access to food. Thus there may be numerous combinations of biotic and abiotic factors that can have an important influence on feeding by fish. Most people who have fed fish have experienced that the fish may suddenly stop feeding for no apparent reason, and have observed that rates of feeding may decrease or increase over a period of days. If we have a better understanding of the factors that can influence feeding we may be better able to manage the allocation of food, and to reduce the variability in feeding that is usually evident both among individuals and among groups of fish.

Behavioural factors influence feed intake, and may even lead to feed refusal. Examples of factors that can influence feeding behaviour are stocking density, the sex ratio and the reproductive status of the fish, and biological rhythms. Endogenous clock mechanisms seem to control some of these rhythms, but environmental factors, such as day length or temperature, may either control others, or act as time setters or zeitgebers. There are three chapters that consider the rhythmic activity of feeding. Chapter 8 concentrates upon a discussion of whether the rhythms are based around 24 hours, days, weeks or months. There are many questions concerning whether these rhythms are generated within the fish and the extent to which external factors can influence the rhythms. Chapter 9 deals with anticipatory behaviour. When faced with the periodic delivery of food animals may develop some sort of activity in anticipation of the forthcoming meal. For example, the animals may become active some time prior to feeding, and such behaviour may be coupled to the cycles of feed delivery. The anticipatory behaviour is often seen to persist in fasting animals, so the suggestion is that there is also an internal timing mechanism in fish, as there seems to be in birds and mammals. The practical implications are that fish may adjust to the arrival of meals at set times, and that there may be optimal feeding times for each species. One very important topic concerns whether the presence of rhythms can influence not only the amount of food consumed but also the efficiency with which the food is converted into growth. This is discussed in Chapter 10. It is an intriguing possibility that if we knew more about the physiological rhythms then we could time feeding in order to maximise growth, improve efficiency, reduce waste and decrease pollution.

The diet of an animal should meet requirements for essential nutrients, and food should also be consumed in sufficient quantity to enable growth and reproduction to be maintained. Thus, nutrient composition is expected to influence whether or not a feed is ingested preferentially, but selection of feeds is also influenced by the sensitivity of the fish to other components, such as chemical deterrents and antinutritional factors. The topic of feed selection is dealt with in Chapter 11, in which it is suggested that fish recognise and discriminate between feeds having different chemical patterns. It is proposed that this enables them to create a diet that meets their needs, via the acceptance of the feeds that most closely match their nutritional preference and by rejection of others.

Chapter 12 deals with the hormones that are involved in the regulation of feeding in fish. Even in mammals the interactions of the factors that influence satiety are not completely understood, although it appears that feeding is under the control of a central feeding system in combination with a peripheral satiation system. The same general principles also probably apply to fish, and many of the chemical factors involved in the regulation of feeding appear to have similar structures, and roles, in the different groups of vertebrates.

Chapter 13 is concerned with how animals deal with a nutrient influx in terms of their physiology. Feed intake represents a challenge to the animal's physiology in that, following digestion and absorption, the nutrients provide the energy for maintenance and are incorporated into tissues as growth. After the ingestion of a meal there is an increase in oxygen consumption. This is partly a reflection of the energy demand for the synthesis of proteins. The proportion of ingested amino acids synthesised into proteins is central to the efficiency with which food is converted into growth. The more we understand about the balance between amino acid oxidation and protein retention, the better will be our ability to optimise feed composition and the timing of meal delivery.

Chapter 14 discusses ways of calculating the amount of feed to be distributed to farmed fish. The efficiency with which fish utilise the feed supply is a major factor that determines the economic returns of a fish farm, so the farmer should aim to feed the fish in a manner that ensures both rapid growth and minimal waste. The approach adopted by the authors involves a consideration of how knowledge about energy requirements can be converted into practical advice about feeding under given sets of conditions. The chapter also contains examples illustrating the use of on-demand feeding systems in a practical setting. These devices allow a continuous record of feed demand to be kept, so may reveal both day-to-day and seasonal variations in feeding, and also give information about the diel timing of feeding at different times of the year.

Fish are an important part of the diet of human populations in many countries. They are an excellent source of protein, contain unsaturated fatty acids that have a beneficial effect on human health, and are also valuable sources of some vitamins and minerals. Ultimately, the value of cultured fish as a source of nutrients for humans will depend upon the conditions under which they have been reared, and the types of feed with which they have been fed. Usually it is fish muscle, or fillet, that is used as a source of food by humans, but internal organs, such as gonads and livers, may also be consumed. The final chapter of the book addresses the question of the partitioning of nutrients among the different organs and tissues of fish species, and also considers how fish body composition may be influenced by the chemical composition of the feed.

Although much of the research into feeding behaviour and food intake in fish incorporates studies of basic principles, its purpose is often applied. For this reason we conclude this preface with a statement of the hope that this book will prove to be of interest not only to students and researchers within fish biology, but also to those directly involved in the aquaculture industry. It is certain that aquaculture will continue to expand, and will provide an increasing proportion of the world supply of fish products. This expansion will be dependent upon continued developments within feed formulation, further studies of feeding behaviour and feed management, and research aimed at the improvement of feed efficiency and the reduction of environmental loads from fish farms.

Dominic Houlihan
Thierry Boujard
Malcolm Jobling

Chapter 1
Feed Composition and Analysis

Malcolm Jobling

1.1 Introduction

Farmed fish and livestock do not usually consume natural prey organisms and forages, but are provided with feed that has been treated or preserved in some way prior to presentation. The feed is usually manufactured from a range of raw ingredients, or feedstuffs, and is formulated to contain an array of essential and non-essential nutrients. A nutrient can be considered to be a chemical element or compound that an animal can utilise in metabolism or growth, an essential nutrient being one that the animal cannot synthesise *de novo.* Thus, an animal's requirements for essential nutrients must be met via the diet. Feedstuffs may vary from simple compounds, such as common salt (NaCl), to meals of plant and animal origin that are made up of complex mixtures of compounds from several nutrient classes.

This chapter provides a brief introduction to the major nutrients found in feeds and feedstuffs, and methods for the evaluation of feeds and feedstuffs are also considered. Methods described include those for analysis of the chemical composition of feeds, and the biological methods used to assess the utilisation of feeds by animals.

1.2 Nutrient classes

Feeds and feedstuffs contain a range of chemical substances that can be grouped into classes according to constitution, properties and function (Fig. 1.1). All feeds will contain a certain proportion of water, so an initial division of feed components into moisture and dry matter can be made. The dry matter is conveniently divided into organic and inorganic material.

Carbohydrates, lipids and proteins make up the major part of the dry matter of feeds; these organic macronutrients can be used directly as metabolic fuels, can be stored within the body for utilisation at a later date, or be deposited in the structural materials that represent the somatic growth of the animal. The main component of the dry matter of feedstuffs of plant origin is carbohydrate, exceptions being oilseeds (e.g. peanut, sunflower, soybean) which contain large amounts of protein and lipid. In contrast to plants, the carbohydrate content of the animal body is low. One of the main reasons for the difference in carbohydrate content between plants and animals is that plant cell walls are made up of carbohydrate, largely cellulose, whereas the walls of animal cells are composed of protein and lipid. Furthermore,

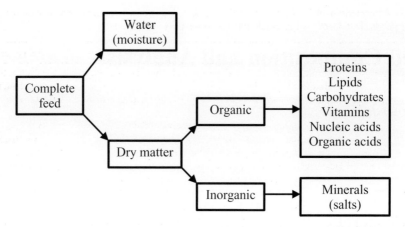

Fig. 1.1   Hierarchical subdivision of a feed showing the major chemical components.

plants store most of their energy as carbohydrates, such as starch, whereas lipids are an animal's main energy store.

Proteins are the major nitrogen-containing compounds present in both plant and animal feedstuffs. Like proteins, nucleic acids are also nitrogen-containing compounds, but these are present in small quantity. Organic acids, such as citric, malic, succinic and pyruvic, will also be present in most feedstuffs in small amounts, as will members of the heterogeneous group of essential organic micronutrients known as vitamins.

Inorganic matter will usually make up a small proportion of a feed or feedstuff. Calcium and phosphorus are generally the major mineral components in feedstuffs of animal origin, whereas potassium and silicon are the main inorganic elements in plants.

1.2.1 Proteins

Proteins are large organic, nitrogen-containing compounds comprising long chains of amino acids. The elemental compositions of proteins tend to be similar, with approximate percentages being $C = 50–55\%$, $H = 6–8\%$, $O = 20–23\%$, $N = 15–18\%$ and $S = 0–4\%$. The majority of the amino acids, of which 20 or so may be incorporated into proteins, have a chemical structure $RCH(NH_2)COOH$, with a carboxyl (-COOH) and an amino ($-NH_2$) group attached to the α carbon atom. R represents the side chain, which differs in configuration depending upon the amino acid. The amino acids can be divided into a number of series depending upon their structure and side chain configurations (Table 1.1):

(1)   *Aliphatic series* – glycine, alanine, serine, threonine, valine, leucine, isoleucine.
(2)   *Aromatic series* – phenylalanine, tyrosine.
(3)   *Sulphur amino acid series* – cysteine, cystine, methionine.
(4)   *Heterocyclic series* – tryptophan, proline, hydroxyproline.
(5)   *Acidic series* – aspartic acid, glutamic acid.
(6)   *Basic series* – arginine, histidine, lysine.

Two amide derivatives of amino acids are also found as constituents of proteins; these are the

Table 1.1 Characteristics of the different series of amino acids. Indispensable (essential) amino acids are indicated in bold type, and conditionally indispensable amino acids are shown in italics.

Series	Characteristics	Amino acid
Aliphatic	Aliphatic amino acids containing one carboxyl group and one amino group	Gly, Ala, Ser, Val, **Thr, Leu, Ile**
Sulphur amino acids	Aliphatic, monoamino-monocarboxylic amino acids containing sulphur	**Met,** *Cys*
Acidic	Aliphatic, dicarboxylic amino acids; aqueous solutions are acidic	Asp, Glu
Basic	Aliphatic amino acids giving basic aqueous solutions	**Arg, His, Lys**
Aromatic	Amino acids with the aromatic, or benzenoid, ring	**Phe,** *Tyr*
Heterocyclic	Heterocyclic structure incorporating nitrogen; proline and hydroxyproline have an imino (NH) group, but no amino group	**Trp,** Pro, Hyp

amides of aspartic and glutamic acids. Asparagine, the β-amide of aspartic acid, is a constituent of many proteins, and it is hydrolysed to aspartic acid and NH_3 by acid. Glutamine, the γ-amide of glutamic acid, occurs in biological materials both in free form and as a constituent of proteins. β-Alanine, an isomer of alanine, is not a constituent of proteins, but it occurs as the free amino acid, and as a component of several biological molecules: the vitamin pantothenic acid, coenzyme A, and the peptides carnosine and anserine found in muscle. The two amino acids citrulline and ornithine are part of the metabolic sequence leading to the formation of urea; they are not constituents of proteins, but do occur in the free form in animal tissues.

The structure of a protein is dependent upon the amino acid sequence (the primary structure) which determines the molecular conformation (secondary and tertiary structures). Proteins sometimes occur as molecular aggregates which are arranged in an orderly geometric fashion (quaternary structure). Constituents other than amino acids may be covalently bound and incorporated into proteins. For example, phosphoproteins such as milk casein and egg phosvitin contain phosphoric esters of serine, and glycoproteins such as collagen and some fish serum proteins contain monosaccharide or oligosaccharide units bound to serine, threonine or asparagine.

Proteins – or more correctly some of the amino acids they contain – are an essential component of the diet for all animals. The amino acids may, therefore, be classified according to whether or not protein synthesis and growth can proceed in the absence of a dietary supply (Macrae *et al.* 1993; Friedman 1996). Essential amino acids are those that animals cannot synthesise, or cannot synthesise in sufficient quantity to enable the maintenance of good rates of growth, whereas non-essential, or dispensable, amino acids can be synthesised *de novo* from other compounds (Table 1.1):

- *essential or indispensable* – arginine, histidine, isoleucine, leucine, lysine, methionine, phenylalanine, threonine, tryptophan, valine.
- *conditionally indispensable* – cystine, tyrosine.
- *non-essential or dispensable* – alanine, asparagine, aspartic acid, glutamine, glutamic acid, glycine, proline, serine.

Some of the 10 essential amino acids can be replaced by their α-hydroxy or β-keto analogues, illustrating that it is the carbon skeleton of the essential amino acid that the animal is unable to synthesise. Cystine and tyrosine are classified as being conditionally indispensable because

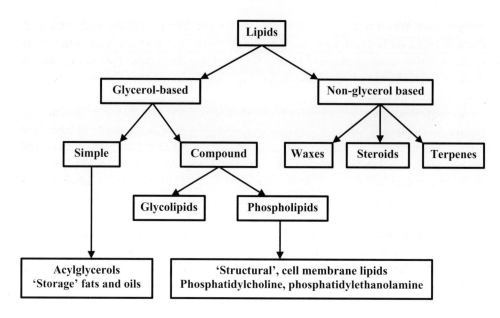

Fig. 1.2   Hierarchical diagram illustrating the classification of lipids.

they can be synthesised from methionine and phenylalanine, respectively. Thus, a dietary supply of cystine and tyrosine is not required if sufficient quantities of their respective precursors are available. In addition to being components of proteins, several of the amino acids are precursors for biologically active molecules: for example, histidine is decarboxylated to form histamine, tyrosine is the precursor for thyroxine and catecholamines, and tryptophan is the precursor of serotonin (5-HT) and melatonin. Estimates of quantitative amino acid requirements have been made for a range of fish species, and the data have been summarised in several overviews (Wilson 1989, 1991; NRC 1993; Cowey 1994; Jobling 1994; Mambrini & Kaushik 1995; Akiyama *et al.* 1998; Kaushik 1998).

1.2.2 Lipids

The lipids are a heterogeneous class of water-insoluble organic compounds that are readily soluble in organic solvents such as chloroform, hexane and diethyl ether (Fig. 1.2). More specifically, the term lipids may be used to describe substances which conform to these solubility characteristics and which contain esterified fatty acids. The glycerol-based lipids can be broadly subdivided into simple, neutral lipids, which are involved in storage, and more complex polar lipids, several of which form the structural components of cell membranes (Sargent *et al.* 1989; Macrae *et al.* 1993; Bell 1998).

Acylglycerols, the major neutral lipids, are formed by esterification of fatty acids with the trihydric alcohol, glycerol. Fatty acids are long-chain organic acids having the general formula $CH_3(C_XH_Y)COOH$. The hydrocarbon chain is either saturated ($Y = 2X$), or there are double bonds between some of the adjacent carbon atoms (unsaturated fatty acid). Triacylglycerols make up the great bulk of neutral lipids found in nature, but mono- and diacylglycerols are also found. The wax esters are a group of compounds closely related to the acylglycerols. Wax esters, which are formed by the esterification of fatty acids with long-chain alcohols, are found as water-proofing agents on the outer surfaces of fruits and leaves of plants, and on the exoskeleton of insects. Wax esters are also a major lipid storage reserve in some zooplanktonic crustacea and, hence, may provide an important energy source for a number of fish species.

Although the majority of the lipid extracted from animal tissues will be made up of neutral, storage triacylglycerols, the total lipids will also contain smaller quantities of other lipid classes. Among the polar lipids it is the phospholipids that are important components of cell membranes, where they form lipoprotein complexes. Phospholipids are esters of glycerol, in which two of the alcohol groups are esterified with fatty acids. The third alcohol group is esterified with phosphoric acid which, in turn, is esterified by a nitrogenous base, the amino acid serine, the sugar alcohol inositol or by glycerol phosphate. The nature of the esterification compound provides the basis of the name of the specific families of phospholipids, e.g. phosphatidylcholine, phosphatidylethanolamine, phosphatidylserine and phosphatidyl-inositol. The major cell membrane phospholipids are phosphatidylcholine and phosphatidylethanolamine, whereas phosphatidylinositol appears to have a number of important roles in signal transduction over biomembranes.

The fatty acids that are esterified to the glycerol backbone of the acylglycerols and phospholipids differ in the numbers of carbon atoms in the molecule, and in the number and positioning of the double bonds between the carbon atoms. These differences impart different physical and chemical properties to the various lipids. The fatty acids form distinct series, and shorthand systems have been devised for classifying these fatty acid series (Table 1.2). For example, palmitic acid may be represented by the shorthand formula 16:0, indicating that palmitic acid is a fatty acid with 16 carbon atoms and no double bonds. A fatty acid that lacks double bonds between the carbons is known as a saturated fatty acid. The monounsaturated

Table 1.2 Classification and naming of some representative fatty acids. Note that in the scientific designation anoic refers to a fatty acid without double bonds in the carbon chain, enoic to a fatty acid with one double bond, dienoic to two double bonds, trienoic to three, tetraenoic to four, etc. The shorthand notation gives the number of carbons, the number of double bonds, and the position of the first double bond counting from the methyl end.

Trivial name (scientific designation)	Number of carbon atoms	Number of double bonds	Fatty acid series	Shorthand notation
Saturated fatty acids (SFAs)				
Lauric (dodecanoic)	12	0		12:0
Palmitic (hexadecanoic)	16	0		16:0
Stearic (octadecanoic)	18	0		18:0
Monounsaturated fatty acids (MUFAs)				
Palmitoleic (hexadecenoic)	16	1	(n-7)	16:1 (n-7)
Oleic (octadecenoic)	18	1	(n-9)	18:1 (n-9)
Erucic (docosenoic)	22	1	(n-9)	22:1 (n-9)
Polyunsaturated fatty acids (PUFAs)				
Linoleic (octadecadienoic)	18	2	(n-6)	18:2 (n-6)
γ-Linolenic (octadecatrienoic)	18	3	(n-6)	18:3 (n-6)
α-Linolenic (octadecatrienoic)	18	3	(n-3)	18:3 (n-3)
Highly-unsaturated fatty acids (HUFAs)				
Arachidonic (eicosatetraenoic)	20	4	(n-6)	20:4 (n-6)
EPA (eicosapentaenoic)	20	5	(n-3)	20:5 (n-3)
DHA (docosahexaenoic)	22	6	(n-3)	22:6 (n-3)

fatty acid oleic acid is designated as 18:1(n-9), showing that oleic acid is an 18 carbon fatty acid with one double bond, the (n-9) denoting that the double bond occurs between the ninth and tenth carbon atoms from the methyl end. The shorthand formula 18:3(n-3) represents α-linolenic acid, an 18 carbon fatty acid with three double bonds, the first double bond being found in the link between the third and fourth carbon atoms from the methyl end. Fatty acids with one or more double bonds are unsaturated fatty acids. Those having two to four double bonds in the molecule are often termed polyunsaturated fatty acids (PUFAs), and those with more than four double bonds may be described as being highly-unsaturated fatty acids (HUFAs).

Lipid nutrition and metabolism have been extensively studied in those fish species, such as salmonids, which are farmed intensively (Henderson & Tocher 1987; Sargent *et al.* 1989, 1995; Kanazawa 1993; Bell 1998). Fish are not able to synthesise PUFAs of the (n-3) and (n-6) series *de novo*, so fatty acids of these two series are essential nutrients that must be supplied in the diet. The fatty acids may be supplied as 18:2(n-6) and 18:3(n-3) provided that the fish are able to convert and chain-elongate these fatty acids into the longer-chain HUFAs, e.g. 20:5(n-3), 22:6(n-3), 20:4(n-6), that are essential components of cell membranes. In those fish that are incapable of converting the 18C fatty acids to their 20C and 22C homologues, 18:2(n-6) and 18:3(n-3) will not be effective as essential fatty acids (EFAs). In species that are able to perform the conversion reactions, the 18C, 20C and 22C fatty acids will all have EFA properties, although the 20C and 22C fatty acids are the most efficacious. It appears that the EFA requirements of many freshwater fish species can be met via a dietary supply of 18:2(n-6) and 18:3(n-3), although the pike, *Esox lucius*, a piscivorous freshwater species, seems incapable of converting 18C fatty acids to the longer-chain HUFAs. Consequently, the pike has a dietary requirement for long-chain HUFAs. The marine fish species investigated thus far also seem to have very limited ability to convert 18C fatty acids to their long-chain HUFA metabolites, so these fish also require a dietary supply of the HUFAs. The broad differences in EFA requirements observed for marine and freshwater fish may depend upon the composition of the natural diet, i.e. whether the species is herbivorous, omnivorous or carnivorous, and upon the relative abundance of the particular fatty acids within freshwater and marine food chains. The (n-6) series fatty acids are typically found in terrestrial and freshwater environments, whilst in marine ecosystems fatty acids of the (n-3) series tend to dominate. The food chain in fresh water is characterised by 18:2(n-6), 18:3(n-3) and 20:5(n-3) fatty acids, whereas the fatty acid profiles of marine phyto- and zooplankton tend to be dominated by 18:3(n-3), 20:5(n-3) and 22:6(n-3) fatty acids, and this pattern is generally reflected at higher levels in the food chain.

1.2.3 Carbohydrates

The term carbohydrate expresses the originally determined empirical formula $C_x(H_2O)_y$, but some compounds not showing the 2:1 ratio of hydrogen to oxygen also have many of the chemical properties of carbohydrates, e.g. deoxyribose ($C_5H_{10}O_4$). Further, a number of compounds that contain small proportions of nitrogen and sulphur in addition to carbon, hydrogen and oxygen have characteristics considered typical for this class of nutrients (Macrae *et al.* 1993). Classification of the carbohydrates may be made according to the size of the molecule (Fig. 1.3). Monosaccharides are simple sugars that cannot be hydrolysed into smaller units,

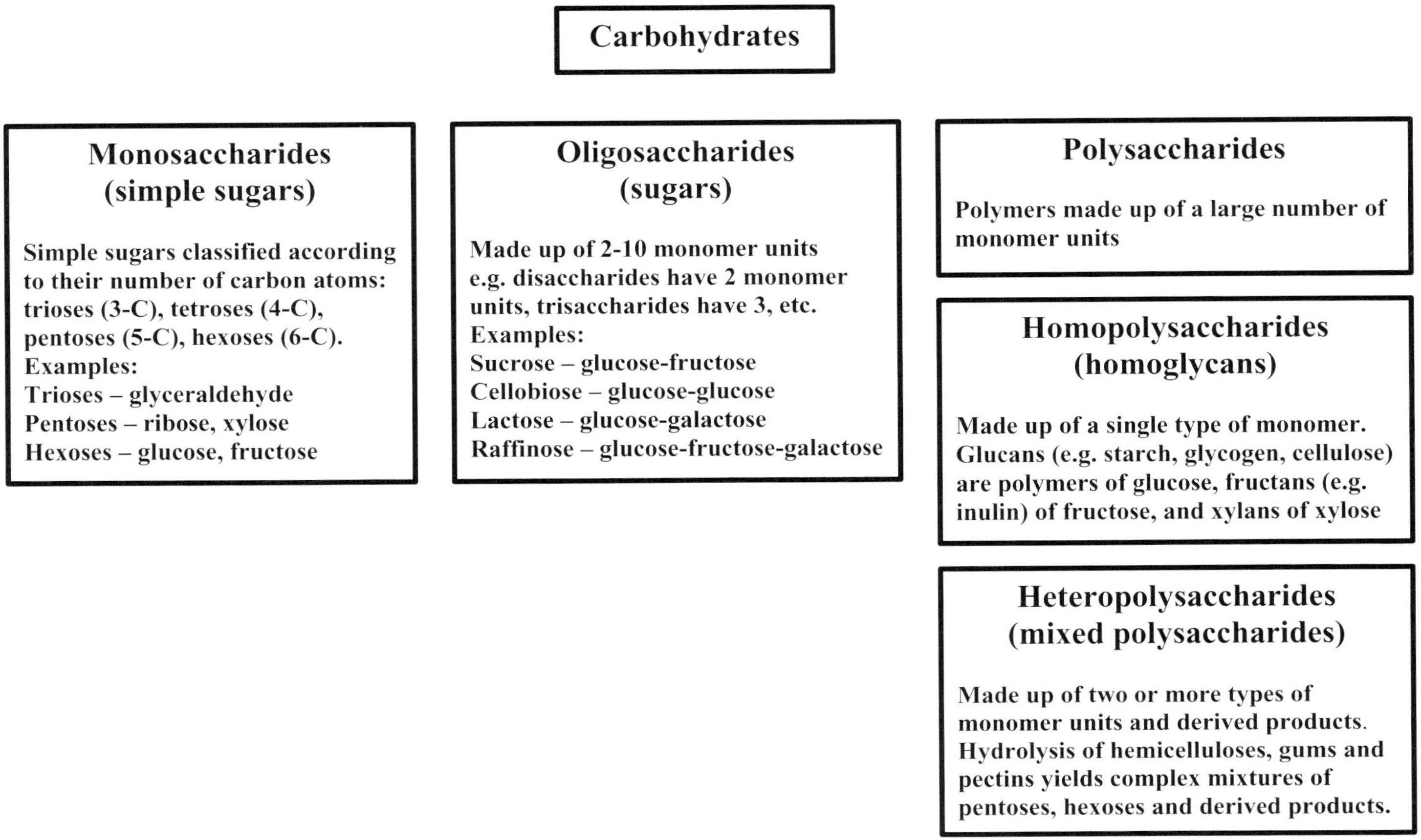

Carbohydrates

**Monosaccharides
(simple sugars)**

Simple sugars classified according to their number of carbon atoms: trioses (3-C), tetroses (4-C), pentoses (5-C), hexoses (6-C).
Examples:
Trioses – glyceraldehyde
Pentoses – ribose, xylose
Hexoses – glucose, fructose

**Oligosaccharides
(sugars)**

Made up of 2-10 monomer units e.g. disaccharides have 2 monomer units, trisaccharides have 3, etc.
Examples:
Sucrose – glucose-fructose
Cellobiose – glucose-glucose
Lactose – glucose-galactose
Raffinose – glucose-fructose-galactose

Polysaccharides

Polymers made up of a large number of monomer units

**Homopolysaccharides
(homoglycans)**

Made up of a single type of monomer. Glucans (e.g. starch, glycogen, cellulose) are polymers of glucose, fructans (e.g. inulin) of fructose, and xylans of xylose

**Heteropolysaccharides
(mixed polysaccharides)**

Made up of two or more types of monomer units and derived products. Hydrolysis of hemicelluloses, gums and pectins yields complex mixtures of pentoses, hexoses and derived products.

Fig. 1.3   Classification of carbohydrates based upon division according to molecular size.

and the monosaccharides are usually classified according to the number of carbon atoms they possess, e.g. trioses ($C_3H_6O_3$) have three carbon atoms, tetroses have four carbon atoms, pentoses have five and hexoses have six. Monosaccharides occur naturally in only very small amounts, the simple sugars usually being polymerised into larger oligosaccharide or polysaccharide molecules. When two to ten monosaccharide units are linked together they form an oligosaccharide, and when more than ten monosaccharide units are joined together they produce a polysaccharide. Most polysaccharide molecules contain several hundred to several thousand monosaccharide residues. In nature, the carbohydrates are usually present as long-chain polysaccharides, the polysaccharides having either a structural or energy storage function.

Some polysaccharides – the homopolysaccharides or homoglycans – contain a single monomer monosaccharide, whereas heteropolysaccharides are mixed polysaccharides which, on hydrolysis, yield mixtures of monosaccharides and derived products. For example, both cellulose and starch are homoglycans that yield only glucose on hydrolysis. The main difference between these two polysaccharides is the nature of the linkages between the adjacent glucose monomer units. There are α linkages in starch and β linkages in cellulose, and this gives rise to differences in physical and chemical properties between the two, including the ease with which they can be digested by animals. Animals produce a digestive enzyme, amylase, which attacks the α linkages in starch, but the β linkages in cellulose are more resistant to digestion. Cellulose is usually considered to be highly indigestible for monogastric animals. Microbes present in the gut of ruminants, and some other animals, produce cellulase, which means that cellulose has nutritional value for animals having the appropriate gastrointestinal microflora.

The polysaccharides present in feeds and feedstuffs include starch, cellulose, hemicellulose, pectic substances and seed gums, glycogen, chitin, and polysaccharides of seaweed and microbial origin. Starch, the principal carbohydrate reserve of terrestrial plants, occurs in storage granules. Two forms of starch may be found in the storage granules – amylose and amylopectin. Amylose is a straight-chained α-1,4 glucan, whereas amylopectin is a more complex molecule with side branches. Amylopectin consists of an α-1,4 glucan chain with α-1,6 branch points. When starch is heated in water the granules swell – a process known as gelatinisation – and this imparts thickening and gel-forming properties. Gelatinisation also improves the ease with which the starch can be digested. Cellulose, a β-1,4 glucan, is the principal structural component of the cell walls of terrestrial plants. Much of the texture of plant feedstuffs is due to cellulose, along with lignaceous substances.

Hemicelluloses are present in plant cell walls. On hydrolysis they yield the five-carbon sugars (pentoses) arabinose and xylose, along with glucuronic acid and some deoxy sugars. Pectic substances, which occur along with lignin in the middle lamella of plant cell walls, form gels in the presence of calcium ions due to the cross-linking of pectin chains. Many other plant polysaccharides are soluble in water and find commercial uses as thickeners or gelling agents. These include gums derived from seeds, plant exudates and seaweeds, and gums produced by microbial fermentation or by chemical modification of cellulose. One example is alginic acid, a component of algal cell walls that can comprise up to 40% of the dry weight of the brown seaweeds (Phaeophyceae). Alginic acid, a linear copolymer of two uronic acids, mannuronic acid and guluronic acid, is extracted from kelps, such as *Macrocystis pyrifera*, and is sold as the sodium alginate salt. Solutions gel at room temperature in the presence of calcium ions, which means that alginates can be used as effective binding agents in feeds. Many species of red algae (Rhodophyceae) contain agar and carrageenans as structural elements in their cell walls. These polysaccharides are galactans made up of repeating sequences of 1,3- and 1,4-linked units of sulphated or methylated sugar residues. All of these algal polysaccharides are soluble in warm water and, on cooling, they form firm gels. For example, *K*-carrageenan has an affinity for protein and forms strong gels in the presence of potassium ions. Derivatives of cellulose are also used as thickening or gelling agents: treatment of cellulose with chloroacetic acid under strongly alkaline conditions yields carboxymethylcellulose (CMC), which can be used to increase the viscosity of feed mixtures.

Animal products usually contain much lower proportions of carbohydrates than do plants. The main storage carbohydrate of animals is glycogen, which is found in the liver and in muscles. Glycogen is similar in structure to amylopectin, the major difference being that glycogen has relatively large numbers of α-1,6 linkages in the molecule. The polysaccharide chitin is found in the exoskeletons of insects and crustaceans where it acts as a strengthening agent. Chitin is composed of long chains of an amino-sugar, *N*-acetylglucosamine. This means that the chitin molecule contains atoms of nitrogen in addition to carbon, hydrogen and oxygen. Structurally, chitin is a close relative of cellulose; the hydroxyl (-OH) group at position 2 in the glucose residues of cellulose is replaced by an *N*-acetylamino (-NHCOCH$_3$) group in the *N*-acetylglucosamine units that make up chitin.

1.2.4 Vitamins

Vitamins are defined as complex organic compounds required for normal metabolism, but that cannot be synthesised by the animal (Halver 1989; Chan 1993; Macrae *et al.* 1993; Chew 1996). Some of the vitamins can be synthesised in small amounts by the microflora of the gastrointestinal tract. These compounds were initially termed vitamines, from 'vital amines', because it was thought that they contained amino-nitrogen. It is now known that only a few of them contain amino-nitrogen, but the amended group name vitamins is still used. The vitamins are a mixed group of compounds that are not closely related to each other chemically, and it has become practice to divide the vitamins into two groups on the basis of their solubility characteristics: lipid-soluble vitamins and water-soluble vitamins. The lipid-soluble vitamins are usually found, and extracted from feeds, in association with lipids. They are absorbed from the gastrointestinal tract along with lipids, are not normally excreted and tend to be stored in the body. In contrast, the water-soluble vitamins are not normally stored in the body in appreciable amounts, and any excess is excreted.

Eleven water-soluble and four lipid-soluble vitamins are known to be required by fish (Halver 1989; Wilson 1991; NRC 1993). Eight of the water-soluble vitamins, the B complex, have coenzyme functions and are required in small quantities. The other water-soluble vitamins – ascorbic acid (vitamin C), myo-inositol and choline – are required in larger amounts, and these compounds are sometimes referred to as the macrovitamins. The lipid-soluble vitamins are vitamins A, D, E and K. During the early years, the chemical structures of the vitamins were unknown and they were assigned letters of the alphabet for convenience. Once vitamins had been isolated and their chemical structures determined, letters were sometimes replaced by names, e.g. thiamin (vitamin B$_1$), riboflavin (vitamin B$_2$), ascorbic acid (vitamin C) and biotin (vitamin H), based upon chemical structure, function or source. There are also substances (provitamins) present in feeds which are not themselves vitamins, but which are capable of being converted into vitamins in the body. For example, some carotenes are pro-vitamins of vitamin A, and the amino acid tryptophan can be considered a provitamin of nicotinic acid.

1.2.5 Minerals

A mineral may be defined as being an homogeneous inorganic substance. When a feed or feedstuff is burned, the organic portion oxidises, and the ashes that remain are minerals. The animal body requires seven minerals, i.e. calcium, sodium, magnesium, potassium, phosphorus, chlorine and sulphur, in relatively large amounts, and several others, e.g. cobalt, copper, iron, iodine, selenium, zinc, manganese and molybdenum, in trace amounts (Osborne & Voogt 1978; Macrae *et al.* 1993). The major minerals serve as structural components of tissues, function in cellular metabolism, and have important roles in osmoregulation and acid–base balance. The functions of the trace minerals vary considerably, but they are generally involved in the regulation of various cellular metabolic functions. Minerals may act as cofactors in biochemical reactions because they have the ability to chelate with proteins. This can stabilise a protein's structure or change its configuration, thereby influencing function via the binding of substrates, activation of an enzyme–substrate complex, or by influencing an enzyme's affinity for its substrate. Copper, manganese, cobalt, zinc and selenium are all trace elements with important metabolic functions, iron is a component of the respiratory pigment haemoglobin, and iodine is required for the production of thyroid hormones.

Not all of the minerals found in the fish body are obtained exclusively via the diet. Fish are surrounded by water that contains mineral salts in solution, and they are capable of absorbing some dissolved minerals from the water (Lall 1989). As a consequence, the roles of the inorganic elements as nutrients for fish are difficult to ascertain, and the determination of quantitative dietary requirements is complicated by the exchange of ions between the fish and the external environment. Nevertheless, much of the requirement for many of the mineral elements is probably met from dietary sources rather than by direct absorption from the water; this is almost certainly the case for freshwater fish species.

1.3 Analysis of feeds and feedstuffs

The assessment of the energy value of feeds and feedstuffs is central to the study of energy metabolism and energy balance. The major energy sources in feedstuffs are the organic constituents lipids, proteins and carbohydrates. These can be completely oxidised in a bomb calorimeter to yield carbon dioxide, water, oxides of nitrogen and sulphur, and heat. Bomb calorimetry involves the rapid combustion of a sample in oxygen at increased pressure, and the heat produced is measured. The calorimeter is calibrated with benzoic acid, a thermochemical standard, allowing the *heat of combustion* of the sample to be calculated. This is the maximum potential energy present, which in studies of animal nutrition is usually referred to as the gross energy of a feed or feedstuff.

Chemical composition is measured using a series of standard laboratory methods; modern methods of analysis, including chromatographic and spectrochemical techniques, enable de-

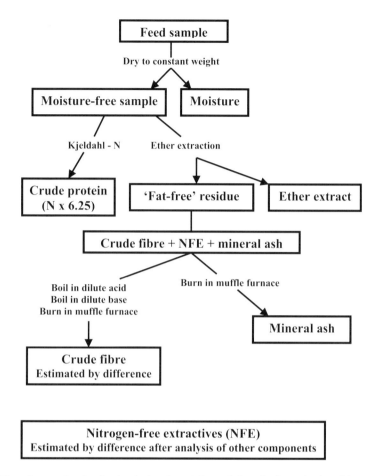

Fig. 1.4 Flow diagram showing the chemical analyses to be carried out for determination of the proximate composition of feeds and feedstuffs.

terminations of individual amino acids, fatty acids, vitamins and mineral elements to be made (Osborne & Voogt 1978; Christie 1982; AOAC 1990; Macrae *et al.* 1993). Such detailed information is rather tedious to obtain, and expensive, automatic analytical equipment is also required. Consequently, analysis is not usually carried out in this detail, and the composition of feeds and feedstuffs (and animal tissue samples) is most frequently given as *proximate composition*. The system of proximate analysis divides the feed components into six fractions: moisture, crude protein, ether extract, ash, crude fibre and nitrogen-free extractives (Fig. 1.4).

Moisture content is determined as the weight loss resulting from drying a known weight of sample at 105°C to constant weight. This method is usually satisfactory, but in some cases the use of such high temperature will lead to the loss of other volatile components. Consequently, modifications employing lower drying temperatures (e.g. 60°C) or freeze-drying are sometimes used for the determination of moisture content.

The crude protein content of the sample is calculated from its nitrogen content. The total nitrogen present in a sample is usually determined by the Kjeldahl method (Osborne & Voogt

1978; AOAC 1990). Protein is then estimated from nitrogen using a conversion factor: the fact that most proteins contain approximately 16% nitrogen forms the basis for the commonly used conversion factor 6.25. Many variations on the Kjeldahl method exist, but the method basically involves heating the sample with sulphuric acid, a boiling aid (e.g. potassium sulphate) and a catalyst (e.g. copper, mercury or selenium) to convert the nitrogen present to ammonium ions. Upon addition of alkali, ammonia is released: this may be measured titrimetrically following distillation, or the ammonia can be reacted with phenol and sodium hypochlorite to give a coloured derivative which can be measured spectrophotometrically. This method does not give the 'true' protein content of the sample because of interference due to the presence of non-protein, nitrogenous compounds in the sample being analysed, e.g. amino acids, nucleic acids, urea. Consequently, this fraction is designated crude protein.

The ether extract fraction is determined by subjecting the sample to extraction with petroleum ether for a defined period (usually 4 h or longer). The residue obtained after evaporation of the solvent is the ether extract. The ether-soluble materials encompass a variety of organic compounds, including neutral lipids, fatty acid esters, some of the compound lipids, lipid-soluble vitamins, provitamins and some pigments. The primary reason for determining the ether extract fraction is to obtain information about the feed components with high caloric value. Provided that the ether extract comprises primarily neutral lipids and fatty acid esters, this will be a valid approach, but the designation of this fraction as 'lipid', 'fat' or 'oil' is not strictly correct.

The ash content is determined by ignition of a known weight of dry sample in a muffle furnace, and ash weight is the weight of mineral residues left after the burning of organic material. In the analysis of feeds and feedstuffs burning is usually carried out at 500°C, but when determining the ash content of biological material combustion temperatures of 450–500°C are usually used. Temperatures above 550°C are avoided because they may cause decomposition of some of the calcium carbonate present in skeletal structures.

The carbohydrate present in a feed or feedstuff is contained in two fractions, the crude fibre and the nitrogen-free extractives (Osborne & Voogt 1978; AOAC 1990). Crude fibre is assessed by boiling an ether-extracted sample in dilute acid, then boiling it in a dilute base, followed by filtering, drying and burning in a furnace. The difference in weight before and after burning is the crude fibre fraction, primarily plant structural carbohydrates such as cellulose and hemicellulose. In addition, detergent extraction methods may be used. Neutral-detergent extraction involves boiling samples in a solution containing sodium lauryl sulphate. The non-soluble material remaining after extraction is referred to as neutral-detergent fibre (NDF), and this contains the major plant cell wall components cellulose, hemicellulose and lignin, and several other minor components. Acid-detergent extraction is carried out by boiling samples in a solution containing cetyl trimethylammonium bromide in sulphuric acid. The residue remaining after extraction, termed acid-detergent fibre (ADF), contains cellulose, lignin, some pectins and other minor constituents. When the sum of the percentages of moisture, crude protein, ether extract, ash and crude fibre is subtracted from 100, the difference is designated the nitrogen-free extractives (NFE). The NFE represent a complex mixture of compounds, including some of the cellulose and hemicelluloses, lignin, sugars, pectins, resins and tannins, organic acids and pigments, and some of the water-soluble vitamins.

Although it is not completely adequate as a measure of the chemical composition of feeds, feedstuffs and animal tissues, a proximate analysis provides useful quantitative information about the nutrients and their distributions. Proximate analysis is the most widely used method for the analysis of feeds and feedstuffs, but the data are often supplemented with more detailed analyses of specific nutrients. In studies of fish feeding and growth the chemical components most often studied in greater detail are the proteins and amino acids, and the lipids and fatty acids.

1.3.1 Protein analysis

The important nutritional role of proteins has led to the development of many analytical techniques. These vary in their selectivity and sensitivity, and are semiquantitative because there is no universal protein against which calibration can be made. Amino acid analysis of protein hydrolysates is an important technique used in the evaluation of protein quality.

In addition to the determination of total nitrogen (Kjeldahl method) the main methods of protein analysis are ultraviolet (UV) absorption, and techniques involving the reaction of the protein with a chromophore. All the techniques may be subject to interference from non-protein compounds, so the degree of accuracy depends upon the purity of the protein sample being analysed: accuracy can be improved by precipitation of the protein prior to analysis. UV absorption can be used to estimate protein because most proteins have an absorption maximum at 280 nm due to the presence of tyrosine and tryptophan residues in their primary structure. This is a quick method for protein estimation, but may be inaccurate due to interference from nucleic acids and polyphenols. Methods involving visible-region spectrophotometry are also widely used to measure proteins (Macrae *et al.* 1993). These methods rely upon the reaction of the protein with a chromophore. One widely employed colour reaction of peptides and proteins, that is not given by free amino acids, is the biuret reaction. Treatment of a peptide or protein with copper sulphate and alkali yields a purple complex as the copper chelates with the peptide linkages within the peptide or protein backbone. Colour development in the Lowry, bicinchoninic acid (BCA) and Bradford techniques is reliant upon the reduction of copper(II) to copper(I). The extent to which this reaction occurs depends upon the aromatic amino acid composition of the protein being analysed.

The determination of amino nitrogen following protein hydrolysis is a more accurate method for measuring protein than those listed above, but the hydrolysis procedure may destroy some of the amino acids, e.g. tryptophan. Hydrolysis of the proteins will be required for determination of the amino acid composition of a feed or feedstuff (AOAC 1990; Macrae *et al.* 1993). In some cases it may be necessary to remove non-protein substances from samples prior to hydrolysis since these may affect the accuracy of the analysis. Protein hydrolysis can be achieved enzymatically, or by treatment with alkali or acid. Alkaline hydrolysis is generally only used in the determination of tryptophan, and enzymatic hydrolysis is rarely used except for the determination of glutamine and asparagine. The usual procedure involves heating the protein with excess 6 M HCl under reflux at 110°C for 24 h, and the HCl is then removed. The problem of tryptophan has already been mentioned, but cystine, cysteine and methionine also suffer some oxidative degradation during acid hydrolysis. Systems that employ either standard liquid-phase or vapour-phase hydrolysis are available commercially, and systems using microwave irradiation of samples can substantially reduce the time required for hydrolysis. Following hydrolysis, the amino acids must be separated from each other, and chromatography is almost invariably the method of choice. High-performance liquid chromatography (HPLC) separation followed by post-column derivatisation and detection (by spectrophotometer or fluorometer) is the traditional method used for amino acid analysis.

Protein and amino acid analysis provides useful information, but the results may not be used directly to assess protein quality because the 'quality' of a protein depends upon its amino acid composition, the availability of the amino acids to the animal, and upon their physiological utilisation following digestion and absorption. Some of the amino acids within a feed or feedstuff can combine with other substances to form bonds that are not hydrolysed by digestive enzymes, thereby reducing the biological availability of the amino acids in question. Such bonds are, however, hydrolysed by the acid treatment that precedes a chemical analysis, leading to the possibility that the information obtained may be incorrectly interpreted.

1.3.2 Lipid analysis

Lipids are readily separated from proteins and other cellular components due to their solubility in non-polar solvents such as diethyl ether, chloroform and chloroform–methanol mixtures, but the solvents vary in the efficiency with which they extract the different components of the total lipid fraction of a sample (Osborne & Voogt 1978; Christie 1982; AOAC 1990). If lipid class and fatty acid composition analyses are to be undertaken, mild extraction methods are required. The extracted lipids should also be stored at very low temperature, and preferably oxygen-free, to avoid possible degradation of fatty acids.

Some lipid classes can be assayed by enzymatic or colorimetric methods, and kits for these types of assay are commercially available, but chromatography is the lipid analyst's key tool (Christie 1982; Macrae *et al.* 1993). Lipid classes may be separated using thin-layer chromatography (TLC) and then identified using a flame ionisation detector. Alternatively, the total lipids can be separated into component classes by high-performance TLC and then detected by densitometry following treatment with a visualising reagent, e.g. 2',7'-dichloro-fluorescein or copper acetate reagent. Fatty acids may be determined specifically using chromatographic methods, the quickest method being to use gas chromatography with an added internal standard. There are several problems in the direct measurement of fatty acids, so they are usually derivatised to their methyl esters prior to analysis. Methyl ester derivatives of the fatty acids of acylglycerols and polar lipids may be prepared by methanolysis using methanolic potassium hydroxide or by acid-catalysed transesterification, and the methyl esters then identified by gas chromatography via comparison with internal standards.

1.3.3 Automated methods of analysis

Proximate analysis is time-consuming and expensive, large amounts of sample may be required, and the analyses are destructive. Consequently, effort has been directed towards the search for methods that can provide quick, non-destructive, quantitative assessments of the constituents of biological materials, and a number of alternative methods are now available (Foster *et al.* 1988; Burns & Ciurczak 1992; Macrae *et al.* 1993; Osborne *et al.* 1993; Deurenberg & Schutz 1995). These newer techniques include X-ray computed tomography (CT), total body electrical conductivity (TOBEC), near-infrared reflectance spectroscopy (NIRS), and near-infrared transmittance (NIT). These techniques have the advantage that they are very rapid, once the initial calibration has been carried out, enabling large numbers of samples to be handled within relatively short periods of time.

CT is a non-invasive, non-destructive technique that was initially developed as a diagnostic tool in human medicine. The technique has, however, also been applied to the estimation of body composition of a range of domestic animals. The CT technique can be used to gain quantitative information about both the chemical composition of the body tissues and the distribution of the various lipid depots. However, the high investment and maintenance costs of running a CT machine generally preclude the use of this technique for making routine measurements. TOBEC is another non-invasive, non-destructive method that has been used to estimate a variety of composition parameters in biological materials, e.g. lipid content and water content. The technique involves placing a sample within a low-frequency electromagnetic field, measuring the conductivity of the material, and then predicting composition based upon the principle that the different chemical constituents differ in electrical properties. The equations developed for prediction of chemical composition from TOBEC are, however, specific for each material and calibration is, therefore, required for each type of biological material examined. TOBEC readings are also sensitive to a variety of extraneous factors, so measurements need to be made under standardised conditions if accurate and repeatable predictions of chemical composition are to be obtained.

Spectroscopy is being increasingly used for the rapid analysis of the chemical composition of feeds and feedstuffs. Spectrophotometers have been developed that measure transmission, absorption or reflection of light at several wavelengths. Both near- and mid-infrared spectroscopy have been used to assess the composition of feeds, but near-infrared techniques are used most frequently due to the simpler sample preparation. The composition of biological materials is reflected in the types of bonds between the atoms or functional groups that make up the material, and this determines the wavelengths and amounts of light that are absorbed and transmitted. Consequently, the spectrum of light that is reflected from, or transmitted through, a sample contains detail about its chemical composition. In practical application, near-infrared spectra are obtained for a number of samples, and proximate, or more detailed chemical, analyses are then performed on the same samples. A multivariate statistical model is then developed to describe the relationship between the spectral characteristics and the chemical components of interest. The model can then be used for prediction of the chemical composition of additional samples from their spectra.

1.4 Nutrient availability and feed evaluation

Proximate analyses and bomb calorimetry provide information about the chemical compositions and gross energy contents of feeds and feedstuffs, but digestibility studies are required to reveal the availabilities of the various nutrients to fish. Further, growth trials remain fundamental to feed assessment, and for the determination of nutrient requirements (Zeitoun *et al.* 1976; Cho *et al.* 1982, 1994; Mercer 1982, 1992; Baker 1986; Macrae *et al.* 1993; Shearer 1995; Friedman 1996).

1.4.1 Nutrient absorption and digestibility

An estimate of nutrient availability, the proportion of the nutrients that are absorbed, can be obtained by assessing *digestibility* (*A*), also known as absorption efficiency. Methods developed for digestibility determination can be divided into *direct* and *indirect*. The direct method requires both knowledge of the total amount of feed consumed and the collection of all the faeces produced. Feed consumption is frequently unknown and total collection of faeces is usually impracticable, so most estimations of digestibility in fish have been made using the indirect method (Fig. 1.5). This involves the inclusion of an inert marker in the feed, with digestibility being calculated as:

$$A = 100 - 100 \left([X_A/X_B] \times [Y_B/Y_A] \right)$$

where X_A and X_B are the concentrations of the inert marker in the feed and faeces, respectively, Y_A is the nutrient concentration in the feed, and Y_B is the concentration of the nutrient in the faeces. A number of conditions must be fulfilled for the indirect method to yield accurate results. The marker must be non-toxic. It should not interfere with feeding, digestion and absorption, nor should it be absorbed or metabolised. The marker should also pass through the gut at the same rate as the other digesta, i.e. there should not be any separation of marker and other feed components during passage through the gut. Accurate analysis of the marker in both the feed and faeces must be possible, and it is advantageous if analyses can be performed cheaply and effectively. Finally, the feed containing the marker must be fed over a sufficiently long time period to enable representative sampling of faeces to be made, i.e. samples collected should be free of contamination by faeces produced during consumption of previously unmarked feed.

Several marker substances have been used in digestibility trials with livestock, the markers, e.g. chromic oxide, titanium oxide, rare earth elements, celite or acid-insoluble ash, lignin and chromogens, being added to the feed at low concentration (McCarthy *et al.* 1977; Wainman 1977; Kennelly *et al.* 1980). Chromic oxide has been commonly used in digestibility studies with fish, but it is suspected of violating some of the prerequisites of an inert marker: it may cause disturbance to digestive function, it has carcinogenic properties, and it may separate from the other digesta during passage through the gut. Acid-insoluble ash (AIA), barium carbonate, yttrium and ytterbium oxides, and ferro-nickel microtracers have all been suggested as being viable alternatives to chromic oxide (Atkinson *et al.* 1984; Riche *et al.* 1995; Refstie *et al.* 1997; Kabir *et al.* 1998). AIA is a natural component of feeds, but is usually present in too low a concentration to be used as a natural internal marker, so the concentration of AIA is increased by the addition of celite during feed manufacture (Atkinson *et al.* 1984).

The way in which the faeces are collected (Fig. 1.5) will influence the estimate of digestibility. When faeces are collected by siphoning or netting from the water there may be leaching of some faecal components, and this will lead to digestibility being overestimated: digestibility estimates obtained from analysis of faeces collected 1 h or more after defaecation may be ca. 10% higher than those obtained when faeces are allowed only limited contact with the water (Windell *et al.* 1978; Smith *et al.* 1980; Spyridakis *et al.* 1989; Anderson *et al.* 1995). Faecal collection systems have been developed that ensure that faecal contact with water

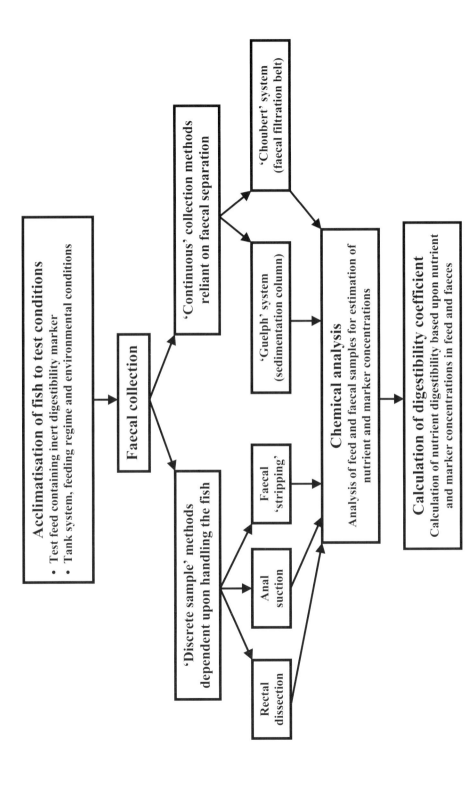

Fig. 1.5   Flow diagram showing the procedures to be carried out in a digestibility trial when using the inert marker method for the estimation of digestibility of feeds and feedstuffs.

is kept to a minimum (Choubert *et al.* 1979, 1982; Vens-Cappell 1984), and there are also systems that rely upon separation of the faeces from the water in a sedimentation column (Cho *et al.* 1982; Satoh *et al.* 1992; Hajen *et al.* 1993; Allan *et al.* 1999). The latter are simple, but they have the disadvantage that the faeces may be in contact with the water for some time, thereby increasing the risk of some dissolution and loss of nutrients due to leaching.

Faecal contact with the water can be excluded by collecting the faeces by anal suction, stripping or intestinal dissection. These methods require handling of the fish, are unsuitable for use on small fish, may require sampling from large numbers of fish, and the faeces may become contaminated with urine, mucus and incompletely digested feed. Contamination leads to an underestimation of digestibility, so estimates obtained from analysis of faeces collected by these methods are almost inevitably lower than those obtained when the faeces have been collected by filtration and decantation (Windell *et al.* 1978; Smith *et al.* 1980; Spyridakis *et al.* 1989; Hajen *et al.* 1993; Fernández *et al.* 1998; Storebakken *et al.* 1998a; Allan *et al.* 1999).

Animal feeds are now often formulated on the basis of digestible energy and/or amino acid availabilities, rather than gross energy or protein contents. This necessitates that information about the digestibilities of the different feed ingredients be available (Cho *et al.* 1994). Information about the digestibility of a feedstuff is usually obtained by substituting part of a 'reference' feed (for which nutrient digestibilities are known) with the feedstuff in question, estimating nutrient digestibilities of the 'new' feed, and then calculating the digestibilities of the nutrients in the feedstuff 'by difference' (Cho *et al.* 1982). An assumption behind this calculation is that digestibilities of feed ingredients are independent and additive, i.e. digestibilities of the macronutrients and energy estimated for a given feedstuff are not influenced by interactions with components derived from other feed ingredients (Storebakken *et al.* 1998b; Allan *et al.* 1999). It is known that nutrient interactions occur, so the assumption of independent, additive digestibilities will rarely be fulfilled. Consequently, it is advisable to estimate the error introduced by adopting this method of calculation. This can be done by comparing digestibilities of complete feeds obtained by direct measurement, with those calculated from the combination of the proportional contributions of the various feed components and their nutrient digestibilities estimated using the 'substitution' method (Cho *et al.* 1982; Allan *et al.* 1999).

1.4.2 Growth trials and biological evaluation

Feeds and feed ingredients are usually evaluated on the basis of their ability to support growth, and 'quality' is assessed using some form of expression that indicates how efficiently a feed or feed ingredient has been retained as growth. Calculation of an efficiency involves collection of both growth and consumption data. The importance of having accurate information about feed consumption cannot be stated too strongly: efficiencies are expressed as gain (growth) divided by feed consumption, so errors in measurements of feed consumption may have serious consequences for estimates of feed efficiency. There are compelling arguments for calculating efficiency as output divided by input, i.e. gain:feed, although in many studies with fish the data have been presented as a feed:gain ratio, i.e. input divided by output. The expression of the data as gain:feed is logical from the standpoint of cause-and-effect, and also leads to clarity in the description of the response, i.e. if gain:feed increases, efficiency increases.

The simplest way to express feed efficiency is to calculate gain:feed as the wet weight gain per weight of feed consumed. A similar expression, the protein efficiency ratio (PER), is often used to describe how well a protein source is used to support growth: PER is calculated as weight gain per unit weight protein consumed. Both of these gain:feed ratios give rather crude estimations of efficiencies because they do not account for possible differences in the chemical composition of the gain deposited by animals given different feeds. The animal body can be considered as two compartments, 'lean body mass' (LBM) and lipid energy stores. The composition of LBM is similar across many animal species and life history stages, being approximately 19 moisture:5 protein:1 ash, and smaller amounts of carbohydrates and structural lipids. Thus, the bulk of the weight gain in growing animals comprises moisture, protein and lipid, with lipid and protein being the main forms in which energy is stored in the animal body. The quantities of protein and lipid stored can be estimated by carrying out carbon and nitrogen balance trials; that is, by measuring the amounts of these elements entering and leaving the body and so, by difference, the amounts retained. Such trials are, however, quite difficult to carry out because elaborate apparatus is required for making accurate measurements. Consequently, assessments of nutrient and energy retention are usually made in other ways.

Energy, or nutrient, retention can be measured if the chemical composition of the animal is estimated at the beginning and end of the trial. In the *comparative slaughter method* this is done by taking a sample of the test animals at the start of the trial, slaughtering them, carrying out a proximate analysis to determine chemical composition, and determining energy content by bomb calorimetry. These data are then used to predict the initial composition of the animals used in the growth trial. At the end of the trial the test animals are slaughtered, their chemical composition and energy content determined, and the amounts of each component deposited during the course of the trial estimated by difference. The comparative slaughter method requires no elaborate apparatus, other than that needed for making the chemical analyses, but is obviously expensive and laborious when applied to many groups of animals used for testing several feeds or feed ingredients. The method becomes less laborious if body composition can be measured non-destructively on the living animal; non-destructive assessment of body composition has the added advantage that repeated measurements can be made on individuals over time, so the measurements can be undertaken on the same individuals both at the beginning and end of the growth trial. Assuming that feed intake has also been measured satisfactorily, the data can then be used to provide detailed assessments of nutrient retention efficiencies.

If both energy consumption and energy retention are known, it is possible to calculate a feed efficiency in terms of feed energy – a more satisfactory assessment than that based upon weights alone. Comparative slaughter also allows for better measures of the utilisation of protein than PER; net protein utilisation (NPU) – also known as protein productive value (PPV) – may be calculated as the weight of protein gained per unit weight of protein consumed. The best method of evaluating 'protein quality' is, however, usually considered to be the calculation of the biological value (BV). This is defined as the proportion of the nitrogen absorbed from the diet that is retained in the body. Thus, BV is often determined by carrying out a nitrogen balance trial in which data for nitrogen intake, faecal losses, and the losses of nitrogen in excretory products are collected. Alternatively, BV can be estimated using the comparative slaughter method provided that information about protein intake and protein digestibility has been obtained.

Calculations of feed efficiency are subject to influence by a range of extraneous factors, so it is of utmost importance that trials be carried out under strictly controlled conditions, and that adequate information be given about test conditions when reporting results. For example, the efficiency with which an animal utilises a feed is influenced by the amount consumed: if feed supply is extremely restricted the animal will lose weight, if the animal is fed at maintenance level, and weight gain is zero, then gain:feed is zero, but as feed supplies are increased above maintenance levels, and more feed is consumed, the animal will gain weight and gain:feed will have a positive value. Thus, as feed consumption increases, gain:feed increases, and calculated values of feed efficiency increase. This increase is not usually monotonic, and there may be a decline in feed efficiency at very high levels of feed supply (Jobling 1994, 1997).

Feed efficiency may also be influenced by the size and sex of the test animals, their reproductive status, and the time of year at which the growth trial is carried out. For example, small fish tend to use feed more efficiently for growth than do larger conspecifics (Jobling 1994). Further, the fact that fish of some species reduce their feed intake, or even become anorectic, just prior to the spawning season will be reflected in feed efficiencies calculated at these times. Temperature may also have profound effects upon the interactions between feeding, metabolism and growth in fish, so the temperature at which a growth trial is carried out will have consequences for how efficiently the feed is utilised for growth. The interactions are complex, but as a generalisation it can be said that when feed supply is limited the best growth and most efficient utilisation of the feed will be achieved at a low temperature, but as feed supply is increased both growth and gain:feed are best at a higher temperature (Jobling 1994, 1997).

1.5 Concluding comments

Most, if not all, of the nutrients required in the diets of animals for the maintenance of normal health and growth have probably been identified, although it is possible that one or more essential nutrients may still be added to the list. However, if there are unidentified essential nutrients they must be required in extremely small amounts. Nevertheless, although qualitative nutrient requirements may be known with reasonable certainty, many nutritionally related problems remain to be solved. There are, for example, few – if any – species for which there is complete information about quantitative essential nutrient requirements. The determination of quantitative requirements is likely to remain a Sisyphean task because requirements may differ between life history stages, and genetic selection that results in animals with novel growth characteristics will almost certainly lead to changes in requirements for essential nutrients. Adequate nutrition is, however, an essential component of animal well-being, and most feeds are formulated to meet the known nutritional requirements of the species, or life history stage, in question. However, when allowed a free choice animals will usually select feedstuffs in proportions that provide them with a reasonably well-balanced diet: this probably also applies to fish, but very few dietary selection studies have been carried out with representatives from this group of animals.

The potential value of a feed for supplying a particular nutrient can be determined by chemical analysis, but the real value of the feed to the animal can only be arrived at after making allowances for the losses that occur during digestion, absorption and metabolism. Thus, complete evaluations of feeds and feedstuffs require combinations of proximate analyses, feed intake and digestibility studies, and examination of nutrient retention in a growth trial incorporating slaughter and chemical analysis of test subjects. The LBM of animals is reasonably constant in composition, and any relative changes in the chemical composition of an animal body that occur during the course of a growth trial will be a reflection of differences in the rates of accretion of LBM and lipid reserves: percentage moisture decreases as the proportions of dry matter and lipid increase, whereas the proportions of protein and ash show smaller decreases as percentage lipid increases.

Examination of the changes in body protein and lipid allow assessments of protein and energy retention to be made in terms of feed efficiencies, these being most widely used for feed evaluation. Assessments of feed protein and feed energy are usually given priority, and there are good reasons for this. Protein sources are the most expensive of the feed ingredients, and the energy-supplying nutrients are those present in the feed in greatest quantity. This means that if a feed has been formulated to meet other nutrient requirements first, and is then found to be deficient in protein or energy, a major revision of constituents will probably be needed. In contrast, a deficiency of a vitamin or mineral can often be rectified by simply adding a small quantity in concentrated form. A further feature of the energy-supplying nutrients which distinguishes them from the others is the manner in which they influence growth performance. In other words, the growth of animals tends to show a continuous response to changes in the quantities of feed energy supplied, so the intake of feed energy is the pacemaker of growth and production.

1.6 References

Akiyama, T., Oohara, I & Yamamoto, T. (1998) Comparison of essential amino acid requirements with A/E ratio among fish species. *Fisheries Science*, **63**, 963–970.

Allan, G.L., Rowland, S.J., Parkinson, S., Stone, D.A.J. & Jantrarotai, W. (1999) Nutrient digestibility for juvenile silver perch *Bidyanus bidyanus*: development of methods. *Aquaculture*, **170**, 131–145.

Anderson, J.S., Lall, S.P., Anderson, D.M. & McNiven, M.A. (1995) Availability of amino acids from various fish meals fed to Atlantic salmon (*Salmo salar*). *Aquaculture*, **138**, 291–301.

AOAC (1990) *Official Methods of Analysis*. Association of Official Analytical Chemists, Arlington.

Atkinson, J.L., Hilton, J.W. & Slinger, S.J. (1984) Evaluation of acid-insoluble ash as an indicator of feed digestibility in rainbow trout (*Salmo gairdneri*). *Canadian Journal of Fisheries and Aquatic Science*, **41**, 1384–1386.

Baker, D.H. (1986) Problems and pitfalls in animal experiments designed to establish dietary requirements for essential nutrients. *Journal of Nutrition*, **116**, 2339–2349.

Bell, J.G. (1998) Current aspects of lipid nutrition in fish farming. In: *Biology of Farmed Fish* (eds K.D. Black & A.D. Pickering), pp. 114–145. Sheffield Academic Press, Sheffield.

Burns, D.A. & Ciurczak, E.W. (eds) (1992) *Handbook of near-infrared analysis.* Marcel Dekker, New York.

Chan, A.C. (1993) Partners in defense, vitamin E and vitamin C. *Canadian Journal of Physiology and Pharmacology*, **71**, 725–731.

Chew, B.P. (1996) Importance of antioxidant vitamins in immunity and health in animals. *Animal Feed Science and Technology*, **59**, 103–114.

Cho, C.Y., Hynes, J.D., Wood, K.R. & Yoshida, H.K. (1994) Development of high-nutrient-dense, low-pollution diets and prediction of aquaculture wastes using biological approaches. *Aquaculture*, **124**, 293–305.

Cho, C.Y., Slinger, S.J. & Bayley, H.S. (1982) Bioenergetics of salmonid fishes: energy intake, expenditure and productivity. *Comparative Biochemistry and Physiology*, **73B**, 25–41.

Choubert, G., de la Noüe, J. & Luquet, P. (1979) Continuous quantitative automatic collector for fish feces. *The Progressive Fish-Culturist*, **41**, 64–67.

Choubert, G., de la Noüe, J. & Luquet, P. (1982) Digestibility in fish: improved device for the automatic collection of feces. *Aquaculture*, **29**, 185–189.

Christie, W.W. (1982) *Lipid Analysis.* Pergamon Press, Oxford.

Cowey, C.B. (1994) Amino acid requirements of fish: a critical appraisal of present values. *Aquaculture*, **124**, 1–11.

Deurenberg, P. & Schutz, Y. (1995) Body composition: overview of methods and future directions of research. *Annals of Nutrition and Metabolism*, **39**, 325–333.

Fernández, F., Miquel, A.G., Guinea, J. & Martinez, R. (1998) Digestion and digestibility in gilthead sea bream (*Sparus aurata*): the effect of diet composition and ration size. *Aquaculture*, **166**, 67–84.

Foster, M.A., Fowler, P.A., Fuller, M.F. & Knight, C.H. (1988) Non-invasive methods for assessment of body composition. *Proceedings of the Nutrition Society*, **47**, 375–385.

Friedman, M. (1996) Nutritional value of proteins from different food sources. A review. *Journal of Agricultural & Food Chemistry*, **44**, 6–29.

Hajen, W.E., Beames, R.M., Higgs, D.A. & Dosanjh, B.S. (1993) Digestibility of various feedstuffs by post-juvenile chinook salmon (*Oncorhynchus tshawytscha*) in sea water. 1. Validation of technique. *Aquaculture*, **112**, 321–332.

Halver, J.E. (1989) The vitamins. In: *Fish Nutrition* (ed. J.E. Halver), pp. 31–109. Academic Press, London.

Henderson, R.J. & Tocher, D.R. (1987) The lipid composition and biochemistry of freshwater fish. *Progress in Lipid Research*, **26**, 281–347.

Jobling, M. (1994) *Fish Bioenergetics.* Chapman & Hall, London.

Jobling, M. (1997) Temperature and growth: modulation of growth rate via temperature change. In: *Global Warming: Implications for freshwater and marine fish* (eds C.M. Wood & D.G. McDonald), pp. 225–253. Cambridge University Press, Cambridge.

Kabir, N.M.J., Wee, K.L. & Maguire, G. (1998) Estimation of apparent digestibility coefficients in rainbow trout (*Oncorhynchus mykiss*) using different markers 1. Validation of microtracer F-Ni as a marker. *Aquaculture*, **167**, 259–272.

Kanazawa, A. (1993) Essential phospholipids of fish and crustaceans. In: *Fish Nutrition in Practice* (eds S.J. Kaushik & P. Luquet), pp. 519–530. INRA, Paris.

Kaushik, S.J. (1998) Whole body amino acid composition of European seabass (*Dicentrarchus labrax*), gilthead seabream (*Sparus aurata*) and turbot (*Psetta maxima*) with an estimation of their IAA requirement profiles. *Aquatic Living Resources*, **11**, 355–358.

Kennelly, J.J., Aherne, F.X. & Apps, M.J. (1980) Dysprosium as an inert marker for swine digestibility studies. *Canadian Journal of Animal Science*, **60**, 441–446.

Lall, S.P. (1989) The Minerals. In: *Fish Nutrition* (ed. J.E. Halver), pp. 219–257. Academic Press, London.

Macrae, R., Robinson, R.K. & Sadler, M.J. (eds) (1993) *Encyclopaedia of Food Science, Food Technology and Nutrition.* Academic Press, London.

Mambrini, M. & Kaushik, S.J. (1995) Indispensable amino acid requirements of fish: correspondence between quantitative data and amino acid profiles of tissue proteins. *Journal of Applied Ichthyology*, **11**, 240–247.

McCarthy, J.F., Bowland, J.P. & Aherne, F.X. (1977) Influence of method upon the determination of apparent digestibility in the pig. *Canadian Journal of Animal Science*, **57**, 131–135.

Mercer, L.P. (1982) The quantitative nutrient-response relationship. *Journal of Nutrition*, **112**, 560–566.

Mercer, L.P. (1992) The determination of nutritional requirements: mathematical modeling of nutrient-response curves. *Journal of Nutrition*, **122**, 706–708.

NRC (National Research Council, USA) (1993) *Nutrient Requirements of Fish.* National Academy of Sciences, Washington

Osborne, B.G., Fearn, T. & Hindle, P.H. (1993) *Practical NIR Spectroscopy with Applications in Food and Beverage Analysis.* Longman, Singapore.

Osborne, D.R. & Voogt, P. (1978) *The Analysis of Nutrients in Foods.* Academic Press, London.

Refstie, S., Helland, S.J. & Storebakken, T. (1997) Adaptation to soybean meal in diets for rainbow trout, *Oncorhynchus mykiss. Aquaculture*, **153**, 263–272.

Riche, M., White, M.R. & Brown, P.B. (1995) Barium carbonate as an alternative indicator to chromic oxide for use in digestibility experiments with rainbow trout. *Nutrition Research*, **15**, 1323–1331.

Sargent, J., Bell, J.G., Bell, M.V., Henderson, R.J. & Tocher, D.R. (1995) Requirement criteria for essential fatty acids. *Journal of Applied Ichthyology*, **11**, 183–198.

Sargent, J., Henderson, R.J. & Tocher, D.R. (1989) The Lipids. In: *Fish Nutrition* (ed. J.E. Halver), pp. 153–218. Academic Press, London.

Satoh, S., Cho, C.Y. & Watanabe, T. (1992) Effect of fecal retrieval timing on digestibility of nutrients in rainbow trout diet with the Guelph and TUF feces collection systems. *Nippon Suisan Gakkaishi*, **58**, 1123–1127.

Shearer, K.D. (1995) The use of factorial modeling to determine the dietary requirements for essential elements in fishes. *Aquaculture*, **133**, 57–72.

Smith, R.R., Peterson, M.C. & Allred, A.C. (1980) Effect of leaching on apparent digestion coefficients of feedstuffs for salmonids. *The Progressive Fish-Culturist*, **42**, 195–199.

Spyridakis, P., Metailler, R., Gabaudan, J. & Riaza, A. (1989) Studies on nutrient digestibility in European sea bass (*Dicentrarchus labrax*) 1. Methodological aspects concerning faeces collection. *Aquaculture*, **77**, 61–70.

Storebakken, T., Kvien, I.S., Shearer, K.D., Grisdale-Helland, B., Helland, S.J. & Berge, G.M. (1998a) The apparent digestibility of diets containing fish meal, soybean meal or bacterial meal fed to Atlantic salmon (*Salmo salar*): evaluation of different faecal collection methods. *Aquaculture*, **169**, 195–210.

Storebakken, T., Shearer, K.D., Refstie, S., Lagocki, S. & McCool, J. (1998b) Interactions between salinity, dietary carbohydrate source and carbohydrate concentration on the digestibility of macronutrients and energy in rainbow trout (*Oncorhynchus mykiss*). *Aquaculture*, **163**, 347–359.

Vens-Cappell, B. (1984) The effects of extrusion and pelleting of feed for trout on the digestibility of protein, amino acids and energy, and on feed conversion. *Aquacultural Engineering*, **3**, 71–89.

Wainman, F.W. (1977) Digestibility and balance in ruminants. *Proceedings of the Nutrition Society*, **36**, 195–202.

Wilson, R.P. (1989) Amino acids and proteins. In: *Fish Nutrition* (ed. J.E. Halver), pp. 111–151. Academic Press, London.

Wilson, R.P. (ed.) (1991) *Handbook of Nutrient Requirements of Finfish.* CRC Press, Boca Raton.

Windell, J.T., Foltz, J.W. & Sarokon, J.A. (1978) Methods of fecal collection and nutrient leaching in digestibility studies. *The Progressive Fish-Culturist*, **40**, 51–55.

Zeitoun, I.H., Ullrey, D.E., Magee, W.T., Gill, J.L. & Bergen, W.G. (1976) Quantifying nutrient requirements of fish. *Journal of the Fisheries Research Board of Canada*, **33**, 167–172.

Chapter 2
Feed Types, Manufacture and Ingredients

Malcolm Jobling, Emidio Gomes and Jorges Dias

2.1 Introduction

A complete diet meets the needs of an animal for nutrients and energy for maintenance, growth and reproduction, and this must be kept in mind when feeds are being formulated for the rearing of animals held in captivity. Thus, important aspects of feed formulation relate to the bioavailability of essential nutrients and energy, but the acceptability of a feed may depend not only on its chemical composition but also on the sensitivity of the animal towards its components. Even when readily accepted, nutritionally balanced feeds have been formulated, their physical qualities may have important influences upon stability and nutrient bioavailability.

The feeding of aquatic animals, such as fish, poses some problems not encountered in the feeding of terrestrial animals; fish feeds need to be water-stable so that they remain intact until ingested, and there should also be minimal losses of water-soluble nutrients due to leaching. If adequate attention is not given to these problems during the course of feed manufacture, there may be a substantial loss of nutrients before the feed has been consumed by the fish. Thus, a combination of poor physical quality of feeds and suboptimal feeding practices can lead to significant nutrient and feed losses. Such losses result in adverse effects on water quality and decrease the efficiency of production.

This chapter provides a brief introduction to the feed types used in the farming of aquatic animals, discusses aspects of feed formulation and manufacture, and gives a survey of the major ingredients used in fish feeds.

2.2 Feed types

Feeds used in intensive fish farming may be classified as wet, moist or dry depending upon their moisture content (Hardy 1989; Jobling 1994; Goddard 1996). Wet feeds generally contain 50–70% moisture, moist pellets contain 35–40% moisture, and dry feeds contain less than 10% moisture. Wet and moist feeds are made by combining wet and dry meal mixes, proportions ranging from 90:10 to 50:50 (wet:dry). The wet ingredients will usually consist of ground fresh or frozen fish, fish-processing waste and/or acid-preserved hydrolysed fish waste (silage). In general, the dry mixes will contain fish meal, cooked starch, vitamin and mineral premixes and alginate binder. Although wet and moist feeds are used on some commercial fish farms, the bulk of the feed used in intensive fish farming is commercially produced dry feed. The dry feeds can be broadly divided into compressed, expanded and extruded pellet types.

2.3 Manufacture of dry feeds

Feed formulation and ingredient selection will usually be made to meet the known nutritional needs of the target species, but the choice of equipment and processing conditions used in manufacture can have a major influence on the physical characteristics of the finished feed (Fig. 2.1) (Thomas & van der Poel 1996; Thomas *et al.* 1997, 1998). Three processing techniques currently dominate the production of dry, pelleted fish feeds, and their use usually – but not always – results in improved bioavailability of the nutrients in the raw materials. The three processing techniques of grinding, steam conditioning and extrusion invariably affect the physical and chemical characteristics of a feed. These characteristics include water stability and durability, pellet hardness, nutrient bioavailability and organoleptic properties. The technology used for the manufacture of dry feeds for fish is based on that developed for the manufacture of feeds for terrestrial animals. The basic steps in feed manufacture

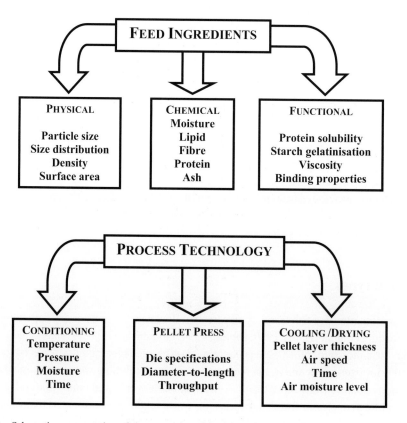

Fig. 2.1 Schematic representation of the properties of feed ingredients and processing conditions that may influence the characteristics of the finished feed.

are grinding and mixing of ingredients, pelleting (with a pellet mill, expander or extruder), cutting, drying, 'top-dressing' (coating), cooling and bagging (Fig. 2.2) (Hardy 1989). In the manufacture of fish feeds account needs to be taken of water stability (i.e. resistance to breakdown in water) and buoyancy (i.e. sinking rate) of the finished pellet, and to constraints imposed by formulation and lipid incorporation.

As an initial step in fish feed manufacture most ingredients are subjected to particle size reduction by grinding. The degree of particle size reduction depends both on the size of the fish to be fed and upon the technological process to be used for pelleting, but particle sizes of dry ingredients should be below 500 μm (NRC 1993). Particle sizes should be lower when feeds are destined for larval or juvenile fish. Different types of grinders, such as plate mills or roller mills, can be used, but particle size reduction by impact using hammer mills is most

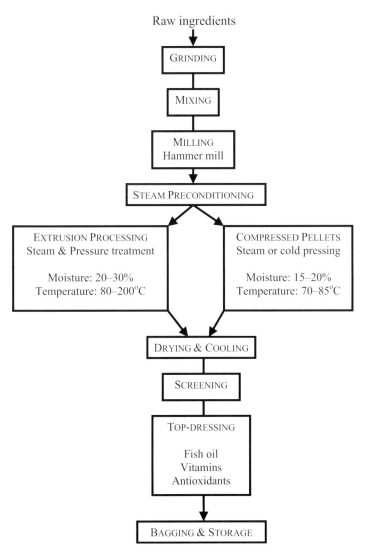

Fig. 2.2 Schematic representation of the steps involved in the manufacture of dry, pelleted fish feeds.

common. Grinding improves the mixing properties of ingredients, increases the bulk density of the feedstuff, and facilitates the penetration of steam into a feed mixture. This will influence hardness, density and starch gelatinisation, and have effects on the durability and water stability of pellets. A suitable particle size distribution among ingredients will also promote homogeneity within the feed mixture, and this is generally associated with an improvement in pellet quality.

From a nutritional point of view, grinding of feedstuffs may improve nutrient digestibility via increases in surface area, and it facilitates the destruction of some of the antinutritional factors that may be present in ingredients of plant origin (Tacon & Jackson 1985). For example, fine-grinding of peas improves the digestibility of starch and protein in terrestrial animals such as pigs and chickens (Daveby *et al.* 1998; Hess *et al.* 1998). Studies on the effects of ingredient particle size on nutrient digestibility and overall growth in fish are scarce. Booth *et al.* (2000) reported that fine-grinding of feed ingredients had minor effects on digestibility and growth characteristics of silver perch, *Bidyanus bidyanus*, whereas both steam conditioning and extrusion influenced feed digestibility. Similarly, Sveier *et al.* (1999) reported that fish meal particle size did not affect apparent digestibility of protein in Atlantic salmon, *Salmo salar*, but fish fed feeds containing coarse-ground fish meal had reduced feed intake and slower gastric evacuation than those fed finer-ground fish meal. Berge *et al.* (1994) suggested that a more rapid release of amino acids from digestion of the proteins in finely ground particles may result in a high rate of amino acid deamination in the liver, thereby leading to a high proportion of the protein being used as an energy substrate rather than for protein accretion.

Correct mixing of ingredients is essential to ensure that each pellet is nutritionally complete. The type of mixer used depends on production capacity requirements, the bulk density of the ingredients or mixed feed, and the amount of liquid to be added. A horizontal mixer is most commonly used in the production of fish feeds. Several factors, including the physical properties of the ingredients (particle size and shape, density, hygroscopicity, static charge and adhesiveness) and mixing parameters (addition sequence of the ingredients, mixing speed and time) may influence the efficacy of the mixing.

Pelleting is the compaction of ingredients by forcing them through a die opening using a mechanical process. The objective is to produce nutritionally complete feed pellets, that can withstand the rigours of transport and handling, at low manufacturing cost. The conditioning of the feed mixture prior to pelleting is of great importance. In animal feed manufacture, conditioning involves treating the feed mixture with heat, water and pressure to convert it into a physical state that facilitates compaction. Therefore, conditioning will have an influence on the physical, nutritional and hygienic quality of the finished feed (Thomas *et al.* 1997). Treatment with heat and water (or steam) induces a wide range of physical and chemical changes in the feed ingredients, including thermal softening, denaturation of proteins and gelatinisation of starch. These physicochemical changes enhance the binding properties of the particles leading to improved pellet quality characteristics relating to hardness and durability (i.e. reduced risk of losses due to fragmentation and abrasion). Pressurised conditioners may be used to de-aerate the feed mixture. De-aeration results in a decrease in energy consumption during subsequent compaction and to a reduction in pellet porosity. Pellets may be produced by steam compression, expansion or extrusion techniques (Fig. 2.2).

2.3.1 Compressed pellets

Steam pelleting, during which the feed ingredients are compressed together, produces a dense pellet that sinks quite rapidly. Another characteristic of compressed pellets is the relatively high proportion of dust and fines in the finished feed, which affects feed wastage. The steam pelleting process involves the use of moisture, heat and pressure to cause agglomeration of the feed ingredients into larger homogeneous particles. The feed ingredients are finely ground prior to mixing and pelleting, and some grain starch is included in the mix to ensure good pellet binding. Fibrous materials and lipids are detrimental to binding, and the amount of lipid included in the pellet mix does not usually exceed 10%. Additional lipid can be sprayed on to the feed after pelleting, and lipid levels of 16–20% can be achieved.

During pellet manufacture steam is added to the dry feed ingredients, increasing the moisture content of the mixture to 15–20%, and raising the temperature (usually to 70–85°C). This increase in temperature causes partial gelatinisation of the starch, which aids binding of the feed ingredients. The mixture is then passed through a pellet die, and the moisture content reduced by forcing air over the warm pellets immediately after they leave the pelleting machine. A double pelleting system has been developed to improve the physical quality of the finished feed. This method consists of pelleting the feed mixture and then repelleting the warm soft pellet by passing it immediately through a second die.

2.3.2 Expanded pellets

Expansion techniques rely on high-pressure conditioning of the feed mixture within an angular expander. The configuration of the expander is similar to that of a single-screw extruder, consisting of a thick-walled mixing tube, a heavy screw and attachments for addition of steam. However, the angular expander has a conical discharge valve, which can be adjusted to provide a variable annular gap. Steam is added before the feed mixture enters the expander chamber where the temperature can exceed 120°C. The extruder screw exerts a shearing and mixing action, and this alters the feed mix structure to such an extent that binding between feed particles is enhanced. The discharge valve and pressure (20–30 bar/cm^2) are adjusted hydraulically during operation depending on the desired conditioning. The degree of starch gelatinisation obtained by expansion can exceed 60%, microbial content of the mixture can be significantly reduced, and there is the possibility of adding liquid such as oils and molasses. In comparison to compressed pellets, the pellets produced by the angular expander have improved hardness and durability, and the production capacity is higher. The lipid content of expanded pellets can be increased to 20–22% by top-dressing with oil.

2.3.3 Extruded pellets

Extrusion techniques were introduced into fish feed production in an attempt to increase the digestible energy content of the feeds via increased lipid incorporation. Extrusion has all the advantages of the expansion technique. Further, due to the porosity of the extruded pellet there is the possibility of increasing lipid incorporation by top-dressing (coating) with oil. Although more expensive and technically demanding than the conventional compression process, extrusion pelleting technology is flexible and versatile. It has the advantages of improving the physical quality of the feeds by decreasing the levels of dust and fines, and of improving the water stability of the pellet. Extruded pellets are water-stable for ca. 4 h, but may be stable for 24 h or more if additional binders are included in the feed mix (Kearns 1993).

The processing temperature, in the pressurised conditioning chamber, can range from 80°C to 200°C, and the moisture content of the feed mix can be increased to 20–30%. Thus, extrusion pelleting involves processing of ingredients at greater temperatures, moisture levels and pressures than those used in conventional steam pelleting. This can lead to reduced availability of some nutrients, such as lysine, via non-enzymatic browning (Maillard reaction), and increase the risk of destruction of others, such as heat-sensitive vitamins (Phillips 1989; Camire *et al.* 1990; Macrae *et al.* 1993). The development of stable forms of some of the vitamins, such as the phosphate derivatives of ascorbic acid (Dabrowski *et al.* 1994), together with post-processing supplementation, has eliminated some of these problems.

In extrusion processing the starch present in the feed mixture is almost completely gelatinised, and it is this that produces a firmly bound, water-stable pellet. Gelatinisation of the starch also improves the ease with which it is digested by the fish. During extrusion the feed ingredients are first treated with steam to bring the moisture content up to 20–30%, and the temperature to 65–95°C. The mixture is then passed to the pressurised extrusion barrel, where the temperature increases (usually to 110–150°C), and the feed mix is then extruded through a die plate. When the hot mixture emerges from the extrusion barrel, some of the water vaporises, leading to a rapid increase in volume and porosity. The pellets are then cooled and dried, and additional ingredients, such as oils and antioxidants, are sprayed on to the absorbent surface of the pellets. This ability of the pellets to absorb oil enables feed lipid concentrations to be increased to much higher levels than is possible with compressed steam pellets.

By making adjustments to the temperature and pressure profiles within the extruder, and by altering the starch content of the feed mixture, it is possible to produce extruded pellets of different density and buoyancy characteristics. Thus, there is the possibility to manufacture floating, slow-sinking or sinking pellets with the same extrusion equipment. This is of importance as it allows the production of feeds tailored to the feeding behaviour of a given species. For example, the extruded feeds employed in the farming of salmonids sink slowly through the water column, and have increased availability relative to denser compressed steam pellets, and extruded floating pellets are used in channel catfish, *Ictalurus punctatus*, and turbot, *Psetta maxima*, farming. However, silver perch seemed to be reluctant to consume floating, extruded pellets, and fish given floating or slowly sinking pellets ate less than those fed on sinking feed (Booth *et al.* 2000).

2.3.4 Special types of feeds

Feed production technology must be adaptable to provide feeds suited to different sizes of animals, ranging from first-feeding larvae (weighing milligrams) to large broodstock (perhaps weighing many kilograms) (Pigott & Tucker 1989). Thus, for most cultivated species of fish or crustaceans, feed particle size can vary considerably during the life cycle, perhaps ranging from less than 50 μm to 20 mm in diameter. The development of microparticulate feeds to replace live prey in fish and crustacean larviculture is a specific domain. Starter feeds must be small, yet nutritionally complete particles. This requires specialised formulation and manufacturing technology. Formulated feeds for larval fish are usually classified as microbound, microcoated or microencapsulated. The efficacy of such feeds varies widely among species, the main problems seeming to be low acceptability, i.e. low rates of feed intake, and poor digestion of the formulated feeds by small larvae (Watanabe & Kiron 1994). The materials used for binding, coating or encapsulating the feeds should be easily broken down in the intestinal lumen, to release the nutrients, but should also have properties that make the microparticle water-stable (Férnandez-Díaz & Yúfera 1995). Despite the problems with formulation and water stability, such feeds may be utilised effectively by the larvae of some species, even though the compound feeds may be inferior to live prey as start-feeds (Cahu *et al.* 1998).

2.4 Feed ingredients

A wide range of raw materials can be used for the manufacture of feeds for aquatic animals (Cho *et al.* 1982, 1994; Tacon & Jackson 1985; Hardy 1989, 1996; Pigott & Tucker 1989; Macrae *et al.* 1993; Jobling 1994, 1998; Pond *et al.* 1995; Friedman 1996). In addition to nutritional value, ingredients also have functional properties (e.g. water absorption properties and pellet binding capacity) that have important influences on the production and physical quality of feed pellets (Fig. 2.1). Feeds for fish in intensive aquaculture are formulated to be nutrient- and energy-dense, and many are based on ingredients of marine origin. Since most cultivated teleosts seem to utilise carbohydrates poorly, the ingredients chosen for feed manufacture are generally protein- and energy-rich. Proteinaceous ingredients are important constituents of the feeds, and will usually comprise a significant part of the formulation.

2.4.1 Protein sources of animal origin

Fish meals are generally considered to be among the best of the protein sources, because they have an essential amino acid profile that seems to meet the requirements of most teleost species, and nutrient bioavailability is also usually high. Fish meals are also good sources of essential fatty acids, minerals and certain vitamins. Fish meals contain a high percentage of protein (usually 60–75%), an appreciable amount of mineral ash (10–20%), and a proportion of lipid (5–10%). Most fish meal is produced from small, pelagic fish species such as sardines, anchovies, capelin, herring and menhaden, but meals are also produced from the smaller members of the codfish family, and from the wastes arising from the fish processing industry. The nutritional value, or quality, of a fish meal will depend upon the species of fish from which it has been prepared, the degree of freshness (extent of spoilage) of raw materials, 'cooking' and drying temperatures used during meal production, and antioxidant additives, storage and transport conditions (Fig. 2.3).

Following capture, fish undergo deteriorative changes; protein is broken down to peptides, free amino acids and ammonia, and biogenic amines arise from the decarboxylation of certain amino acids. The total volatile nitrogen (TVN), as a proportion of total nitrogen, rises as fish spoils, and TVN and biogenic amine concentrations are often used as indicators of raw material freshness. High concentrations of biogenic amines, such as cadaverine, histamine,

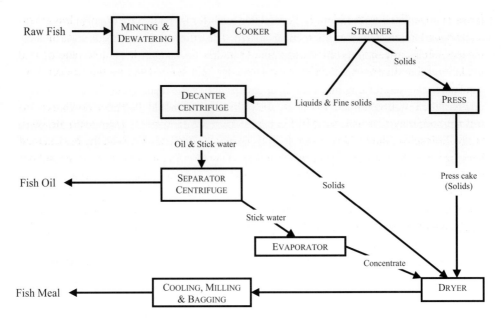

Fig. 2.3 Schematic representation of the processes involved in the production of fish meals and fish oil.

putrescine and spermidine, in feeds have been accompanied by reductions in growth and feed efficiency (Aksnes & Mundheim 1997; Aksnes *et al.* 1997).

Processing conditions also influence the quality of fish meal, with high processing temperatures (100–150°C) giving meals of lower digestibility than milder conditions (60–70°C) (Anderson *et al.* 1995; Aksnes & Mundheim 1997). The differences may arise because of the formation of enzyme-resistant disulphide bridges, or other cross-linkages, between protein chains, the oxidation of amino acids, such as methionine and tryptophan, or the reaction of lipid oxidation products with amino acids (Camire *et al.* 1990; Macrae *et al.* 1993).

Thus, the best quality fish meals are produced by processing fresh whole fish under low-temperature conditions, and the quality of the meals produced decreases if 'stale' fish and/or higher cooking and drying temperatures are used (Pike *et al.* 1990; Anderson *et al.* 1993, 1997; Jobling 1994, 1998; Hardy 1996; Aksnes & Mundheim 1997). The meals produced from filleting wastes, and other fish processing by-products, are considered to be of poor quality due to their relatively high ash, and low protein, content.

Some promising results have been obtained using acid-preserved fish silage as a protein source, but both beneficial and negative effects have been reported when silages containing free amino acids and partially hydrolysed proteins have been used in fish feeds (Tacon & Jackson 1985; Hardy 1996). The silages (which are liquid) can be used as is, and may be used to produce a protein concentrate, or even co-dried with other protein ingredients to produce a meal (Hardy 1996). Fish silages are usually prepared by exploiting lactic acid bacterial fermentation, but the ensiling process can also involve formic, hydrochloric or sulphuric acid preservation (Arason 1994; Kristinsson & Rasco 2000). Use of fish silages prepared using organic or inorganic acids may be associated with a reduction of feed intake in salmonids due to the low pH imparted to the feed by the acids (Rungruangsak & Utne 1981). There may be beneficial effects when silage concentrates are used to replace part of the fish meal in feeds for salmonids (Espe *et al.* 1999). Higher inclusion levels appear, however, to be less efficacious. There are some additional problems with fish silage, because any deficiencies in ensiling technique may result in the development of bacteria, tryptophan degradation and/or lipid oxidation, all of which would limit the value of using this raw material for the production of fish feeds.

Fish protein concentrates (FPCs) are usually obtained by enzymatic hydrolysis, but chemical methods may also be used. FPCs can be prepared by relying on autolytic hydrolysis, but hydrolysis via the addition of exogenous enzymes is considered to be the best method because it allows good control over the hydrolysis, and thereby the properties of the resulting product (Kristinsson & Rasco 2000). FPCs are generally of high biological value, and may also contain free amino acids and peptides that act as feeding attractants, stimulants or flavour enhancers and potentiators. However, the hydrolysis may also give rise to 'bitterness', resulting from hydrophobic peptides containing basic amino acids. For example, basic tripeptides containing asparagine and lysine, along with leucine or glycine may impart a bitter taste to FPCs. More extensive hydrolysis decreases the bitterness because the hydrophobic peptides are far more bitter than mixtures of their constituent amino acids. The flavour-potentiating properties seem to reside within the acidic peptides, including some dipeptides (Glu-Asp, Glu-Glu, Glu-Ser, Thr-Ser) and tripeptides (Asp-Glu-Ser, Glu-Asp-Glu, Glu-Gly-Ser, Ser-Glu-Glu). FPCs may also be good sources of essential fatty acids, but most lipids are removed during manufacture to prevent the negative effects of oxidation during storage (Kristinsson & Rasco 2000). FPCs are more expensive than fish meals, but a combination of high digestibility, and the presence of potential feeding attractants and stimulants, means that they may be a useful component of feeds prepared for fish larvae, where price is less critical than performance (Tacon & Jackson 1985). For example, the inclusion of small amounts of hydrolysed proteins in salmon starter feeds has been reported to give positive effects (Berge & Storebakken 1996). Similarly, krill meal has also been reported to have chemical properties that promote increased feed intake in several fish species (Arndt *et al.* 1999).

Several protein sources derived from terrestrial animals have been used in the formulation of fish feeds. These include meat meal, meat and bone meal, poultry by-product meal, blood meal and worm meal. All have a relatively high protein content, and may have promise as partial substitutes for fish meal, but there may be negative effects at high levels of inclusion in feeds. The composition and freshness of the raw material, as well as cooking and drying conditions, influence the nutritional quality of meals prepared from rendered animal products. Such products may vary widely in terms of nutrient and energy digestibility, but modern handling and processing techniques seem to have resulted in the production of meals of significantly improved quality (Bureau *et al.* 1999).

2.4.2 Single-cell proteins (SCPs)

The term single-cell protein is applied to a wide range of products of microbial (microorganism) origin (Tusé 1984; Tacon & Jackson 1985). These may be algal products, or fungal (including yeasts) and bacterial products resulting from fermentation processes. The yeasts or bacteria can be grown on soluble wastes from breweries, sewage processing, wood pulping operations, and 'cracking products' from the oil industry; these are substrates which are available in relatively large quantities and/or are waste products which may cause environmental deterioration. Thus, considerable effort has been expended upon developing fermentation processes for the large-scale production of microbial protein. Production of SCP via fermentation of yeast or bacterial cultures is considered desirable because the organisms have rapid rates of growth, they can be grown on relatively inexpensive media, they make efficient use of the energy source, and finished products with a high protein content can be prepared relatively easily (Tusé 1984).

Traditionally, the largest producers of SCP have been the brewing and distilling industries; the yeast by-products are sold as protein and vitamin supplements for inclusion in animal feeds (Hardy 1989). Several of the SCPs are too expensive to form the main protein source in fish feeds, and concerns have been expressed about the presence of undesirable compounds, such as heavy metals, in others. Interest in industrial production of SCP using oil and petroleum fractions as the growth medium waned during the late-1970s due partly to increasing costs of the petrochemical energy source, but there is currently interest in the production of SCP via bacterial fermentation using natural gas as the energy source and ammonia as the source of nitrogen for protein synthesis.

2.4.3 Antinutritional factors (ANFs)

Many plant ingredients have been evaluated as protein sources in fish feeds, but their value may be reduced due to the presence of ANFs (Table 2.1). The term ANF encompasses a variety of compounds, including protease inhibitors, phyto-oestrogens, lectins (haemagglutinins), goitrogens, antivitamins, phytates, saponins (haemolytic agents), various oligosaccharides and antigenic proteins (allergens) (Tacon & Jackson 1985; Hendricks & Bailey 1989; Macrae *et al.* 1993; Liener 1994; Anderson & Wolf 1995; Friedman 1996; Alarcón *et al.* 1999). The ANFs can be classified according to their heat sensitivity, defined as the amount of heat necessary to inactivate the ANF without altering the biological value of other components. Thus, ANFs may be heat labile (lectins and protease inhibitors) or heat stable (e.g. tannins, oligosaccharides, saponins, phyto-oestrogens, phytate and alkaloids) (Melcion & van der Poel 1993). The deleterious effects of ANFs include reduced feed intake and nutrient bioavailability, and depressed growth (Kaushik *et al.* 1995; van den Ingh *et al.* 1996; Refstie *et al.* 1998; Alarcón *et al.* 1999; Arndt *et al.* 1999).

Most protein-rich plant ingredients are characterised by having the majority of their phosphorus bound in phytate (Ca-Mg salt of phytic acid). The phosphorus in phytate is limited in bioavailability to most fish because they lack the digestive enzyme phytase. However, when fish are fed on feeds containing relatively high levels of plant protein sources, phosphorus requirements can be met by the addition of an inorganic P source. The inclusion of microbial phytase in the feed is an approach to increase phytic phosphorus bioavailability and thereby reduce the use of inorganic phosphorus supplements (Rodehutscord & Pfeffer 1995; Oliva-Teles *et al.* 1998). This may have additional benefits by reducing the risk of phytate binding to essential minerals, and it also alleviates any potentially negative effects of phytate on digestive enzyme function.

Techniques to reduce, remove or inactive ANFs include selective plant breeding (e.g. low-glucosinolate rape, or canola) and various extraction and processing techniques (Table 2.1) (Melcion & van der Poel 1993). Due to the heat sensitivity of some ANFs, the use of expansion or extrusion in fish feed manufacture may lead to significant reductions in levels of some

Table 2.1 Examples of antinutritional factors (ANFs) and toxins present in plant ingredients commonly used in the formulation of fish feeds.

ANF or toxicant	Major sources	Remedial actions for reduction, removal or destruction
Enzyme inhibitors		
Protease (trypsin)	Legumes (e.g. soybean)	Heat treatment
Amylase	Legumes and cereals	Heat treatment
Lectins (haemagglutinins)	Legumes	Heat treatment
Alkaloids		
Glycoalkaloids	Potato	Selective breeding
Quinolizidine alkaloids	Lupin	Selective breeding
Goitrogens		
Glucosinolates (glycosides)	Brassicas (e.g. rapeseed)	Selective breeding
		Solvent extraction
Glycopeptides	Some legumes (e.g. soybean, peanut)	Moist heat treatment
Phytate (phytic acid)	Most plants (e.g. cereals, legumes, rapeseed)	Enzymatic degradation
Phytoestrogens (e.g. coumestans, isoflavones)	Legumes, cereals and rapeseed	Solvent extraction
Tannins (polyphenols)	Many plants (e.g. legumes, oilseeds)	Dehulling and milling
		Soaking/solvent extraction
Saponins (triterpenoids)	Many plants, especially legumes (e.g. soybean)	Solvent extraction
Oligosaccharides (e.g. raffinose, stachyose)	Most legumes and oilseeds	Enzymatic degradation
		Protein concentrates
Gossypol	Cottonseed	Moist heat treatment
		Selective breeding
Fatty acids		
Cyclopropene fatty acids (sterculic and malvalic)	Cottonseed	
Erucic acid	Rapeseed	Solvent extraction
		Selective breeding
Aflatoxins	'Mouldy' feed ingredients	Correct handling and storage

of these adverse compounds. Further, as some ANFs are present in seed hulls, the dehulling and extrusion of some plant ingredients, such as lupins and other legumes, prior to incorporation into fish feeds will reduce ANF levels. Such treatments may reduce ANFs to levels where they have few, if any, negative effects on feed intake, nutrient utilisation and growth (Robaina *et al.* 1995; Burel *et al.* 1998; Refstie *et al.* 1998).

2.4.4 *Plant protein sources*

The most important protein sources of plant origin are the oilseed meals, prepared from the material remaining after oil has been extracted from soybeans, cottonseed, rape/canola, peanuts, sunflower and safflower seeds (Tacon & Jackson 1985; Hardy 1989; Friedman 1996). It is also usual to include maize (corn) or cereal meals in feed formulations, but these are included as a cheap energy source and for the binding properties provided by the starch they contain rather than for their protein content.

    Although ingredients of plant origin are being increasingly used in fish feeds, the total replacement of fish meal by plant protein sources has rarely been successful, the high inclusion levels of plant protein sources usually resulting in reduced growth and less efficient feed utilisation (de la Higuera *et al.* 1988; Robaina *et al.* 1995; Nengas *et al.* 1996; Bureau *et al.* 1998; Burel *et al.* 1998). The poorer growth commonly observed in fish fed feeds containing high proportions of plant proteins may be related to a decrease in feed intake resulting from reduced feed palatability (Reigh & Ellis 1992; Gomes *et al.* 1995; Bureau *et al.* 1998; Refstie *et al.* 1998; Arndt *et al.* 1999). However, the possible influences of an essential amino acid imbalance (e.g. sulphur amino acids) (Médale *et al.* 1998), low phosphorus availability, and the metabolic effects of ANFs cannot be ignored (Alarcón *et al.* 1999). Many plant proteins are deficient in one or more of the essential amino acids, so the use of plant protein sources as major ingredients in fish feeds may necessitate amino acid supplementation to provide an amino acid profile that matches the essential amino acid (EAA) requirements of the given fish species (Murai *et al.* 1986; Kaushik *et al.* 1995).

Soybean meal is considered to be the best widely available plant protein source, but soybeans contain several ANFs. Some of the ANFs, such as the trypsin and chymotrypsin inhibitors, are partially destroyed or inactivated by heating and drying, but others – such as phytate and the oligosaccharides raffinose and stachyose – are less affected by the normal processing procedures used in meal production. This incomplete destruction of the ANFs may reduce the potential for using conventional soybean meal in feed formulations, and much effort has been expended in devising processing techniques for improving the nutritional value of soybeans. Modern processing techniques may employ a range of chemical, enzymatic and physical treatments (Phillips 1989; Anderson & Wolf 1995); once the trypsin inhibitors and allergens have been removed it appears that the nutritional value of soybean meal is similar to that of high-quality fish meal (Médale *et al.* 1998).

Other plant protein sources, such as rape/canola and lupin, may hold promise for inclusion in fish feeds. Both plants grow well in cooler climates than soybeans, and they are being cultivated in Europe and North America for inclusion in animal feeds. Lupin seeds have a high protein content (ca. 35%), although they are deficient in methionine. Unlike soybeans, lupins do not contain high concentrations of many of the ANFs, but they do contain quinolizidine alkaloids, which make the seeds bitter and potentially toxic. Lupin cultivars with a low alkaloid content have been developed, and it is these which seem to hold most promise as animal feedstuffs (Burel *et al.* 1998; van Barneveld 1999).

Meals prepared from older varieties of rape had limited use in animal feeds because of their content of ANFs, but rape cultivars – canola – with lower levels of ANFs are now available. Canola is low in glucosinolates and erucic acid, but canola meals may have quite a high fibre content and also contain raffinose, stachyose, phytate and tannins. Feeds containing low-glucosinolate meals may, nevertheless, have a negative impact on growth in a number of ways. Despite being low in glucosinolates, such meals have been reported to affect thyroid function in juvenile rainbow trout, *Oncorhynchus mykiss*, and the content of other ANFs may have resulted in reduced digestibility and poorer nutrient utilisation (Burel *et al.* 2000). Protein concentrates have been prepared from canola to reduce the levels of fibre and other undesirable compounds in the finished products: these concentrates comprise 60–65% protein and have an amino acid profile that resembles that of fish meal.

2.4.5 Lipid sources

Many different lipid sources have been used in fish feeds, including lipids of both plant and animal origin, and those from both terrestrial and aquatic environments (Hardy 1989; Sargent *et al*. 1989; Jobling 1994; Bell 1998). Terrestrial lipids of animal origin, such as pork fat and beef tallow, contain high levels of saturated fatty acids, and are poor sources of essential fatty acids (EFAs) of the (n-3) and (n-6) series. Oils derived from the seeds of a number of plant species, such as soya and sunflower, contain quite high levels of unsaturated fatty acids of the (n-6) series. The best sources of the long-chain highly unsaturated fatty acids (HUFAs) of the (n-3) series are marine fish oils. Consequently, marine fish oils have traditionally been used as a lipid source in feeds for most species of intensively farmed fish. However, fish of many species can chain-elongate 18C fatty acids of the (n-3) and (n-6) series, so can utilise plant oils provided that they contain sufficient 18:3(n-3) to meet the EFA requirement for fatty acids of this series. Linseed oil contains appreciable quantities of both 18:3(n-3) and 18:2(n-6), and so may be a useful source of EFAs for many fish species. The EFA requirements may also be met by providing the fish with mixtures of oils derived from marine fish, linseed, rapeseed and soya. By using mixtures of oils from several sources it may be possible to create formulated feeds having similar fatty acid profiles to the natural prey of the fish.

2.4.6 *Other ingredients*

In addition to the protein and lipid sources, complete feeds will also contain an array of other ingredients (Hardy 1989). These will include vitamin and mineral mixes. These mixes are usually formulated so that, at the dosages included in the feeds, they meet the known requirements of the species in question for each of the vitamins and minerals (Halver 1989; Lall 1989; Wilson 1991; NRC 1993; Jobling 1994). Complete feeds may also contain additional binding agents, antioxidants and antifungal agents, carotenoid pigments (e.g. in feeds used for on-growing salmonids), feeding attractants and flavour enhancers, and drugs or antibiotics (in medicated feeds). These latter ingredients will usually be included in relatively small quantities, and although they have little major nutritional value they are of practical importance in a number of ways.

2.5 Feed characteristics and feed acceptability

Fish feeds must not only be formulated in accordance with the nutritional requirements of the species in question, but must also be ingested in sufficient quantity if the fish is to survive and grow. Feed acceptance depends upon a variety of chemical, nutritional and physical characteristics, all of which can be influenced by the choice of feed ingredients and processing conditions used in the manufacture of the feed. Criteria for acceptance of feeds by fish are availability and appearance, particle size, and organoleptic properties relating to smell, taste and texture. First, the fish must be able to locate the feed, then it must be attracted to it and be able to ingest it, and finally the fish must be willing to ingest and swallow the feed.

The ability of fish to detect and ingest a feed can be affected by physical characteristics such as pellet density (sinking rate), size (shape, diameter and length), colour (contrast), and texture (hardness). All of these properties may be influenced by the processing conditions used during feed manufacture. Many species feed in the water column or on the bottom, and may be reluctant to eat floating pellets. Further, feed availability may be increased when pellets sink slowly through the water column, and slow-sinking dry pellets are probably the feed type of choice for the rearing of species that naturally feed in the water column (Tucker 1998). Prey movement may be important in the initiation of feeding responses by visual feeders (e.g. Wright & O'Brien 1982; Holmes & Gibson 1986), and successful feeding will also depend upon there being sufficient contrast between the feed and the background (see Chapter 6). Thus, feed colour preferences may be expected to differ depending upon environmental setting, e.g. tank colour and light intensity (Ginetz & Larkin 1973; Clarke & Sutterlin 1985). Feed movement and a high degree of contrast with the background may be of particular importance during the rearing of larval and juvenile fish. Further, the importance of the previous history of the fish should not be ignored, because there may be a significant learning component in the feeding behaviour of fish: individuals often develop preferences for feed types with which they are familiar (e.g. Bryan 1973; Clarke & Sutterlin 1985; Cox & Pankhurst 2000).

The size, shape and texture of the feed may also affect the amount eaten. For example, the optimal feed size for a range of fish species seems to be 25–50% of the mouth width (Wankowski & Thorpe 1979; Knights 1985; Tabachek 1988; Linnér & Brännäs 1994; Tucker 1998), with fish that naturally consume large prey probably accepting the largest pellets. Fish may respond more rapidly following presentation of large pellets than small, but pellets of a size that initially appear to be most attractive may not be those that are ingested most rapidly once captured (Stradmeyer *et al.* 1988; Linnér & Brännäs 1994; Smith *et al.* 1995). For example, Arctic charr, *Salvelinus alpinus*, were reported to respond more rapidly to large pellets than to small, but handling time increased with increasing pellet size. Consequently, the shortest sum of reaction and handling times was recorded for pellets of intermediate size, and pellets of intermediate size were those that were least often missed or rejected (Linnér & Brännäs 1994). The texture of pellets also influences acceptability, with soft pellets often proving to be more acceptable to the fish (Lemm 1983; Stradmeyer *et al.* 1988; Stradmeyer 1989). There may also be an interaction between pellet size and texture, in that fish may be capable of handling larger feed particles when they are soft (Knights 1985).

Chemical, in addition to physical, properties of pellets will also have a major influence on attractiveness and acceptability. The chemical properties of the pellets will largely depend upon the range of ingredients used in feed manufacture. It is known that many plants contain chemicals that defend them against attack from herbivores; defence chemicals are found in both terrestrial and aquatic plants, although the nature of the chemicals may differ in plants from different environments (Forsyth 1968; Dreisbach 1971; Hay & Fenical 1988; Hay 1996). The defence chemicals are often termed 'secondary compounds or metabolites' because they are metabolites that do not generally participate in primary physiological processes. The presence of secondary metabolites in an organism is usually characterised by specialised synthesis, transport and storage, and concentrations are often subject to environmental or ontogenetic regulation. Further, the defence chemicals are often compartmentalised, i.e. concentrations differ among organs and tissues. There may also be specialised systems for discharge or activation, to ensure increased efficacy of defensive action and reduce the risk of autotoxicity (Berenbaum 1995; Zimmer & Butman 2000). The range of secondary compounds produced by plants includes terpenes, polyphenolics, alkaloids, and a range of aromatic compounds and amino acid derivatives (Forsyth 1968; Dreisbach 1971; Hay & Fenical 1988; see also section on ANFs). The amino acid derivatives include coumarins, cyanogenic glycosides and glucosinolates, whereas polyphenols are derived from compounds that participate in the tricarboxylic acid cycle.

Animals of several phyla also contain defensive chemicals which serve to reduce the risk of bodily harm. Chemical defences are rare in organisms at the top of the food chain, i.e. in organisms that are at low risk of being consumed, but are widespread amongst sponges (Porifera), ascidians (Chordata: Tunicata), soft corals (Coelenterata), bryozoans, polychaetes and oligochaete worms (Annelida), insects (Arthropoda) and molluscs (Dreisbach 1971; Gerhart *et al.* 1991; Connaughton *et al.* 1994; Berenbaum 1995; Hay 1996; Lindquist & Hay 1996; Zimmer & Butman 2000). Some species obtain defensive chemicals via sequestration from their food, i.e. they specialise in feeding on plants or animals that are potentially toxic to other species, including their own predators. For example, the Spanish dancer nudibranch, *Hexabranchus sanguineus*, feeds on sponges, *Halichondria* spp., which contain chemicals that deter feeding by fishes. The nudibranch sequesters these chemicals and concentrates them in its mantle tissue and egg masses, where they serve as a defence against potential consumers. Other species synthesise their defensive chemicals *de novo*. For example, aquatic dytiscid beetles synthesise and secrete a range of steroids, and many of the defensive secretions of insects are products of amino acid metabolism; quinones derive from tyrosine, formic acid from serine, isobutyric acid from isoleucine and valine, and alkyl sulphides from methionine (Gerhart *et al.* 1991; Berenbaum 1995).

The fact that many plants contain secondary metabolites with a defensive function means that their inclusion in fish feeds may give rise to palatability problems (see also sections on ANFs and Plant protein sources). For example, rainbow trout were reported to find feed less palatable when fish meal was partially substituted with a leaf protein concentrate, and there may also be acceptability problems when feeds are formulated to include high concentrations of soybean meal (Fowler 1980; Tacon & Jackson 1985; Mackie & Mitchell 1985; Dias *et al.* 1997; Kubitza *et al.* 1997; Papatryphon & Soares 2000). Some animal protein sources with potential for inclusion in fish feeds, e.g. lumbricid worms, may also contain feeding deterrents (Tacon *et al.* 1983; Mackie & Mitchell 1985). Bitter-tasting compounds, such as quinine and caffeine, result in decreased feed intake of several fish species (Mackie & Mitchell 1985; Lamb & Finger 1995), and certain antibiotics also seem to possess feeding deterrent properties (Schreck & Moffitt 1987; Poe & Wilson 1989; Toften *et al.* 1995).

The negative effects of the feeding deterrents may, at least to some extent, be ameliorated by formulating feeds to include chemical mixtures with stimulant properties. This may be efficacious both when feeds are formulated to contain large proportions of a plant protein source or when the feed contains an antibiotic (Toften *et al.* 1995; Dias *et al.* 1997; Kubitza *et al.* 1997; Papatryphon & Soares 2000). There are four main types of low-molecular-weight compounds, which seem to serve as stimulants of feeding behaviour in fish. They may act alone, but appear to be more effective as part of a more complex mixture. The four types of compounds most often reported to have stimulant properties are free amino acids, quaternary ammonium compounds, nucleotides and nucleosides, and organic acids, although other substances may evoke feeding behaviour in some fish species (Mackie & Mitchell 1985; Carr & Derby 1986; Jones 1992; Takeda & Takii 1992; Carr *et al.* 1996; see also Chapter 5). Identification of the natural feeding stimulants and stimulant mixtures may become more important as fish feeds are formulated to contain increasing proportions of ingredients that either lack such chemicals, or contain compounds with deterrent properties.

2.6 Concluding comments

Formulated feeds prepared for intensively reared carnivorous fish typically contain 40–50% protein. Traditionally, most of this protein has been supplied by including large proportions of fish meals in the feeds, and dietary lipids have usually been provided by the inclusion of marine fish oils. Fish meal production has stabilised at about 6–7 million tonnes, annually, and there would seem to be little prospect of any increase because fish resources appear to be fully exploited. This realisation has stimulated research into alternative protein sources for inclusion in formulated feeds for farmed fish, and increased use is being made of plant protein sources. These are not completely novel ingredients, but are materials that have been shown to have the potential of being improved in quality by additional processing and/or nutrient supplementation.

Extrusion pelleting technology is becoming increasingly popular for the production of fish feeds due to its versatility and flexibility. Extrusion technology enables production of feeds with a high energy content (i.e. high lipid concentrations), the heat processing involved can reduce levels of some ANFs in plant ingredients, and there is an almost complete gelatinisation of the starch in the feed mix. The latter effects may result in increased nutrient and energy digestibility (Jeong *et al.* 1991; Pfeffer *et al.* 1991; Bergot 1993). For example, although neither Vens-Cappell (1984) nor Akimoto and co-workers (1992) found any effect of feed production method (compressed pelleting versus extrusion) on protein and lipid digestibility in rainbow trout, starch digestibility was improved by extrusion. Similar results were reported in a study with European sea bass, *Dicentrarchus labrax*, provided with feeds containing 30% wheat gluten, but differing in the manufacture process (Robaina *et al.* 1999).

The bioavailability of nutrients and energy present in feed ingredients is a major factor determining their suitability for inclusion in fish feed formulations. The digestibility of protein (Table 2.2) and energy (Table 2.3) in feed ingredients has been widely studied in salmonids, especially the rainbow trout (e.g. Cho *et al.* 1982; Cho & Kaushik 1990; Wilson 1991; Gomes *et al.* 1995; Pfeffer *et al.* 1995; Bureau *et al.* 1999), and in the channel catfish (e.g. Tucker & Robinson 1990; Robinson 1991; Wilson 1991), but information for many other species is still fragmentary. Such information is likely to increase as the farming of 'novel' species becomes better established and attempts are made to develop feeds tailored to meet the requirements of each particular species.

Finally, it must be recalled that a formulated feed must not only meet the nutrient needs of the species, but it must also be consumed in sufficient quantities to support survival, growth and reproduction. Many properties of a feed influence whether or not it will be detected, captured and ingested. These properties include physical characteristics relating to size, shape and colour, and organoleptic properties such as texture, taste and smell. All of these properties may be influenced by feed manufacturing processes, such as heat and pressure treatment. It is, however, the choice of ingredients used in the formulation that has the greatest influence on organoleptic properties, because this choice has an influence upon the chemical composition of the feed with respect to nutrient and non-nutrient components, the balance between the nutrients, and whether or not attractant or deterrent chemicals are present in the finished feed.

Table 2.2 Bioavailability (digestibility) of protein in some ingredients commonly used in the manufacture of fish feeds. Values are given as Apparent Digestibility Coefficients (ADCs) in percentages.

Feed ingredient	Protein digestibility (ADC, %)						
	CC	RT	CS	ESB	HSB	GSB	RD
Animal products							
Herring meal		92–95	91	95–96		83–96	
Anchovy meal	85–90	94	92				
Menhaden meal	70–87	90	83		88		77–96
Poultry by-product	65	68–91	74–85			80–82	49
Feather meal	63–74	58–87	71			25–58	
Blood meal							
(flame-dried)	23	16	29			46	
(spray-dried)	74	82–99		90–93	86	90	100
Meat-and-bone meal	61–82	83–88		91–92	73	36–79	74–79
Fish protein concentrate (FPC)		95		96–98			
Plant products							
Cottonseed meal	81–83				84	75	76–85
Wheat middlings		92	86		92		87
Soybean meal	77–97	96	74–77	88–90	80	86–91	80–86
Corn (maize) gluten	80–92	96				90	

Abbreviations: CC, channel catfish, *Ictalurus punctatus* (sources Tucker & Robinson 1990; Robinson 1991); RT, rainbow trout, *Oncorhynchus mykiss* (sources Cho *et al.* 1982; Bureau *et al.* 1999); CS, chinook salmon, *Oncorhynchus tshawytscha* (source Hajen *et al.* 1993); ESB, European sea bass, *Dicentrarchus labrax* (source Da Silva & Oliva-Teles 1998); HSB, hybrid striped bass, *Morone saxatilis × Morone chrysops* (source Sullivan & Reigh 1995); GSB, gilthead sea bream, *Sparus aurata* (sources Nengas *et al.* 1995; Lupatsch *et al.* 1997); RD, red drum, *Sciaenops ocellatus* (sources Gaylord & Gatlin 1996; McGoogan & Reigh 1996).

Table 2.3 Bioavailability (digestibility) of energy in some ingredients commonly used in the manufacture of fish feeds. Values are given as Apparent Digestibility Coefficients (ADCs) in percentages.

Feed ingredient	Energy digestibility (ADC, %)						
	CC	RT	CS	ESB	HSB	GSB	RD
Animal products							
Herring meal		91	93	94–97		80–94	
Anchovy meal			92				
Menhaden meal	85–92		84		96		60–95
Poultry by-product		71–87	65–72			78–80	72
Feather meal	67	70–80	57			7–64	
Blood meal							
(flame-dried)		50	32			58	
(spray-dried)		82–94		92–94		83	58
Meat-and-bone meal	76–81	68–85		84–86	80	15–69	54–86
Fish protein concentrate (FPC)		94		94–96			
Plant products							
Cottonseed meal	56–80				73	39	22–70
Wheat middlings		46	45		61		34
Soybean meal	56–72	75	66–70	69–70	55	45–72	38–63
Corn (maize) gluten		83				80	

Abbreviations: CC, channel catfish, *Ictalurus punctatus* (sources Tucker & Robinson 1990; Robinson 1991); RT, rainbow trout, *Oncorhynchus mykiss* (sources Cho *et al.* 1982; Bureau *et al.* 1999); CS, chinook salmon, *Oncorhynchus tshawytscha* (source Hajen *et al.* 1993); ESB, European sea bass, *Dicentrarchus labrax* (source Da Silva & Oliva-Teles 1998); HSB, hybrid striped bass, *Morone saxatilis × Morone chrysops* (source Sullivan & Reigh 1995); GSB, gilthead sea bream, *Sparus aurata* (sources Nengas *et al.* 1995; Lupatsch *et al.* 1997); RD, red drum, *Sciaenops ocellatus* (sources Gaylord & Gatlin 1996; McGoogan & Reigh 1996).

2.7 References

Akimoto, A., Takeuchi, T., Satoh, S. & Watanabe, T. (1992) Effect of extrusion processing on nutritional value of brown fish meal diets for rainbow trout. *Nippon Suisan Gakkaishi*, **58**, 1477–1482.

Aksnes, A. & Mundheim, H. (1997) The impact of raw material freshness and processing temperature for fish meal on growth, feed efficiency and chemical composition of Atlantic halibut (*Hippoglossus hippoglossus*). *Aquaculture*, **149**, 87–106.

Aksnes, A., Izquierdo, M.S., Robaina, L., Vergara, J.M. & Montero, D. (1997) Influence of fish meal quality and feed pellet on growth, feed efficiency and muscle composition in gilthead seabream (*Sparus aurata*). *Aquaculture*, **153**, 251–261.

Alarcón, F.J., Moyano, F.J. & Diaz, M. (1999) Effect of inhibitors present in protein sources on digestive proteases of juvenile sea bream (*Sparus aurata*). *Aquatic Living Resources*, **12**, 233–238.

Anderson, J.S., Higgs, D.A., Beames, R.M. & Rowshandeli, M. (1997) Fish meal quality assessment for Atlantic salmon (*Salmo salar* L.) reared in sea water. *Aquaculture Nutrition*, **3**, 25–38.

Anderson, J.S., Lall, S.P., Anderson, D.M. & McNiven, M.A. (1993) Evaluation of protein quality in fish meals by chemical and biological assays. *Aquaculture*, **115**, 305–325.

Anderson, J.S., Lall, S.P., Anderson, D.M. & McNiven, M.A. (1995) Availability of amino acids from various fish meals fed to Atlantic salmon (*Salmo salar*). *Aquaculture*, **138**, 291–301.

Anderson, R.L. & Wolf, W.J. (1995) Compositional changes in trypsin inhibitors, phytic acid, saponins and isoflavones related to soybean processing. *Journal of Nutrition*, **125**, 581S–588S.

Arason, S. (1994) Production of fish silage. In: *Fisheries Processing: Biotechnological applications* (ed. A.M. Martin). pp. 244–272. Chapman & Hall, London.

Arndt, R.E., Hardy, R.W., Sugiura, S.H. & Dong, F.M. (1999) Effects of heat treatment and substitution level on palatability and nutritional value of soy defatted flour in feeds for coho salmon, *Oncorhynchus kisutch*. *Aquaculture*, **180**, 129–145.

Bell, J.G. (1998) Current aspects of lipid nutrition in fish farming. In: *Biology of Farmed Fish* (eds K.D. Black & A.D. Pickering). pp. 114–145. Sheffield Academic Press, Sheffield.

Berenbaum, M.R. (1995) The chemistry of defense: theory and practice. *Proceedings of the National Academy of Sciences, USA*, **92**, 2–8.

Berge, G.E. & Storebakken, T. (1996) Fish protein hydrolyzate in starter diets for Atlantic salmon (*Salmo salar*) fry. *Aquaculture*, **145**, 205–212.

Berge, G.E., Lied, E. & Espe, M. (1994) Absorption and incorporation of dietary free and protein bound (U^{14}C)-lysine in Atlantic cod (*Gadus morhua*). *Comparative Biochemistry and Physiology*, **109A**, 681–688.

Bergot, F. (1993) Digestibility of native starches of various botanical origins by rainbow trout (*Oncorhynchus mykiss*). In: *Fish Nutrition in Practice* (eds S.J. Kaushik & P. Luquet). pp. 857–865. INRA Editions, Paris.

Booth, M.A., Allan, G.L. & Warner-Smith, R. (2000) Effects of grinding, steam conditioning and extrusion of a practical diet on digestibility and weight gain of silver perch, *Bidyanus bidyanus*. *Aquaculture*, **182**, 287–299.

Bryan, J.E. (1973) Feeding history, parental stock, and food selection in rainbow trout. *Behaviour*, **45**, 123–153.

Bureau, D.P., Harris, A.M. & Cho, C.Y. (1998) The effects of purified alcohol extracts from soy products on feed intake and growth of chinook salmon (*Oncorhynchus tshawytscha*) and rainbow trout (*Oncorhynchus mykiss*). *Aquaculture*, **161**, 27–43.

Bureau, D.P., Harris, A.M. & Cho, C.Y. (1999) Apparent digestibility of rendered animal protein ingredients for rainbow trout (*Oncorhynchus mykiss*). *Aquaculture*, **180**, 345–358.

Burel, C., Boujard, T., Corraze, G., *et al.* (1998) Incorporation of high levels of extruded lupin in diets for rainbow trout (*Oncorhynchus mykiss*): nutritional value and effect on thyroid status. *Aquaculture*, **163**, 325–345.

Burel, C., Boujard, T., Escaffre, A-M., *et al.* (2000) Dietary low-glucosinolate rapeseed meal affects thyroid status and nutrient utilisation in rainbow trout. *British Journal of Nutrition*, **83**, 653–664.

Cahu, C., Infante, J.Z., Escaffre, A-M., Bergot, P. & Kaushik, S. (1998) Preliminary results on sea bass (*Dicentrarchus labrax*) larvae rearing with compound diet from first feeding. Comparison with carp (*Cyprinus carpio*) larvae. *Aquaculture*, **169**, 1–7.

Camire, M.E., Camire, A & Krumhar, K. (1990) Chemical and nutritional changes in foods during extrusion. *Critical Reviews in Food Science and Nutrition*, **29**, 35–57.

Carr, W.E.S. & Derby, C.D. (1986) Chemically stimulated feeding behavior in marine animals. Importance of chemical mixtures and involvement of mixture interactions. *Journal of Chemical Ecology*, **12**, 989–1011.

Carr, W.E.S., Netherton, J.C., III, Gleeson, R.A. & Derby, C.D. (1996) Stimulants of feeding behavior in fish: analyses of tissues of diverse marine organisms. *Biological Bulletin*, **190**, 149–160.

Cho, C.Y. & Kaushik, S.J. (1990) Nutritional energetics in fish: energy and protein utilization in rainbow trout (*Salmo gairdneri*). *World Review of Nutrition and Dietetics*, **61**, 132–172.

Cho, C.Y., Hynes, J.D., Wood, K.R. & Yoshida, H.K. (1994) Development of high-nutrient-dense, low-pollution diets and prediction of aquaculture wastes using biological approaches. *Aquaculture*, **124**, 293–305.

Cho, C.Y., Slinger, S.J. & Bayley, H.S. (1982) Bioenergetics of salmonid fishes: energy intake, expenditure and productivity. *Comparative Biochemistry and Physiology*, **73B**, 25–41.

Clarke, L.A. & Sutterlin, A.M. (1985) Associative learning, short-term memory, and colour preference during first feeding by juvenile Atlantic salmon. *Canadian Journal of Zoology*, **63**, 9–14.

Connaughton, V.P., Schuur, A., Targett, N.M. & Epifanio, C.E. (1994) Chemical suppression of feeding in larval weakfish (*Cynoscion regalis*) by trochophores of the serpulid polychaete *Hydroides dianthus*. *Journal of Chemical Ecology*, **20**, 1763–1771.

Cox, E.S. & Pankhurst, P.M. (2000) Feeding behaviour of greenback flounder larvae, *Rhombosolea taprina* (Günther) with differing exposure histories to live prey. *Aquaculture*, **183**, 285–297.

Da Silva, J.G. & Oliva-Teles, A (1998) Apparent digestibility coefficients of feedstuffs in seabass (*Dicentrarchus labrax*) juveniles. *Aquatic Living Resources*, **11**, 187–191.

Dabrowski, K., Matusiewicz, M. & Blom, J.H. (1994) Hydrolysis, absorption and bioavailability of ascorbic acid esters in fish. *Aquaculture*, **124**, 169–192.

Daveby, Y.D., Razdan, A. & Aman, P. (1998) Effect of particle size and enzyme supplementation of diets based on dehulled peas on the nutritive value for broiler chickens. *Animal Feed Science and Technology*, **74**, 227–239.

De la Higuera, M., Garciá-Gallego, M., Sanz, A., Cardenete, G., Suárez, M.D. & Moyano, F.J. (1988) Evaluation of lupin seed meal as an alternative protein source in feeding of rainbow trout (*Salmo gairdneri*). *Aquaculture*, **71**, 37–50.

Dias, J., Gomes, E.F. & Kaushik, S.J. (1997) Improvement of feed intake through supplementation with an attractant mix in European seabass fed plant-protein rich diets. *Aquatic Living Resources*, **10**, 385–389.

Dreisbach, R.H. (1971) *Handbook of Poisoning: Diagnosis and Treatment.* Lange Medical Publications, Los Altos.

Espe, M., Sveier, H., Høgøy, I. & Lied, E. (1999) Nutrient absorption and growth of Atlantic salmon (*Salmo salar* L.) fed fish protein concentrate. *Aquaculture*, **174**, 119–137.

Fernández-Díaz, C. & Yúfera, M. (1995) Capacity of gilthead seabream, *Sparus aurata* L., larvae to break down dietary microcapsules. *Aquaculture*, **134**, 269–278.

Forsyth, A.A. (1968) *British Poisonous Plants.* HMSO, London.

Fowler, L.G. (1980) Substitution of soybean and cottonseed products for fish meal in diets fed to chinook salmon *Oncorhynchus tshawytscha* and coho salmon *Oncorhynchus kisutch. The Progressive Fish-Culturist*, **42**, 87–91.

Friedman, M. (1996) Nutritional value of proteins from different food sources. A review. *Journal of Agricultural & Food Chemistry*, **44**, 6–29.

Gaylord, T.G. & Gatlin, D.M. (1996) Determination of digestibility coefficients of various feedstuffs for red drum (*Sciaenops ocellatus*). *Aquaculture*, **139**, 303–314.

Gerhart, D.J., Bondura, M.E. & Commito, J.A. (1991) Inhibition of sunfish feeding by defensive steroids from aquatic beetles: structure–activity relationships. *Journal of Chemical Ecology*, **17**, 1363–1370.

Ginetz, R.M. & Larkin, P.A. (1973) Choice of colors of food items by rainbow trout (*Salmo gairdneri*). *Journal of the Fisheries Research Board of Canada*, **30**, 229–234.

Goddard, S. (1996) *Feed Management in Intensive Aquaculture.* Chapman & Hall, London.

Gomes, E.F., Rema, P. & Kaushik, S.J. (1995) Replacement of fish meal by plant proteins in the diet of rainbow trout (*Oncorhynchus mykiss*): digestibility and growth performance. *Aquaculture*, **130**, 177–186.

Hajen, W.E., Higgs, D.A., Beames, R.M. & Dosanjh, B.S. (1993) Digestibility of various feedstuffs by post-juvenile chinook salmon (*Oncorhynchus tshawytscha*) in sea water. 2. Measurement of digestibility. *Aquaculture*, **112**, 333–348.

Halver, J.E. (1989) The Vitamins. In: *Fish Nutrition* (ed. J.E. Halver). pp. 31–109. Academic Press, London.

Hardy, R.W. (1989) Diet Preparation. In: *Fish Nutrition* (ed. J.E. Halver). pp. 475–548. Academic Press, London.

Hardy, R.W. (1996) Alternate protein sources for salmon and trout diets. *Animal Feed Science and Technology*, **59**, 71–80.

Hay, M.E. (1996) Marine chemical ecology: what's known and what's next? *Journal of Experimental Marine Biology and Ecology*, **200**, 103–134.

Hay, M.E. & Fenical, W. (1988) Marine plant-herbivore interactions: the ecology of chemical defense. *Annual Review of Ecology and Systematics*, **19**, 111–145.

Hendricks, J.D. & Bailey, G.S. (1989) Adventitious toxins. In: *Fish Nutrition* (ed. J.E. Halver), pp. 605–651. Academic Press, London.

Hess, V., Thibault, J.N., Melcion, J.P., van Eyes, J. & Sève, B. (1998) Influence de la variété et du microbroyage sur la digestibilité iléale de l'azote et des acides aminés du pois: digestibilité réelle de l'azote et pertes endogènes spécifiques. *Journées de Recherche Porcine*, **30**, 223–229.

Holmes, R.A. & Gibson, R.N. (1986) Visual cues determining prey selection by the turbot, *Scophthalmus maximus* L. *Journal of Fish Biology*, **29** (Supplement A), 49–58.

Jeong, K.S., Takeuchi, T. & Watanabe, T. (1991) Improvement of nutritional quality of carbohydrate ingredients by extrusion process in diets of red sea bream. *Nippon Suisan Gakkaishi*, **57**, 1543–1549.

Jobling, M. (1994) *Fish Bioenergetics.* Chapman & Hall, London.

Jobling, M. (1998) Feeding and nutrition in intensive fish farming. In: *Biology of Farmed Fish* (eds K.D. Black & A.D. Pickering). pp. 67–113. Sheffield Academic Press, Sheffield.

Jones, K.A. (1992) Food search behaviour in fish and the use of chemical lures in commercial and sports fishing. In: *Fish Chemoreception* (ed. T.J. Hara). pp. 288–320. Chapman & Hall, London.

Kaushik, S.J., Cravedi, J.P., Lalles, J.P., Sumpter, J., Fauconneau, B. & Laroche, M. (1995) Partial or total replacement of fish meal by soybean protein on growth, protein utilization, potential estrogenic or antigenic effects, cholesterolemia and flesh quality in rainbow trout, *Oncorhynchus mykiss. Aquaculture*, **133**, 257–274.

Kearns, J.P. (1993) Extrusion of aquatic feeds. *Technical Bulletin of the American Soybean Association* **AQ40**, 16–34.

Knights, B. (1985) Feeding behaviour and fish culture. In: *Nutrition and Feeding in Fish* (eds C.B. Cowey, A.M. Mackie & J.G. Bell). pp. 223–241. Academic Press, London.

Kristinsson, H.G. & Rasco, B.A. (2000) Fish protein hydrolysates: production, biochemical and functional properties. *Critical Reviews in Food Science and Nutrition*, **40**, 43–81.

Kubitza, F., Lovshin, L.L. & Lovell, R.T. (1997) Identification of feed enhancers for juvenile largemouth bass *Micropterus salmoides. Aquaculture*, **148**, 191–200.

Lall, S.P. (1989) The minerals. In: *Fish Nutrition* (ed. J.E. Halver). pp. 219–257. Academic Press, London.

Lamb, C.F. & Finger, T.E. (1995) Gustatory control of feeding behavior in goldfish. *Physiology & Behavior*, **57**, 483–488.

Lemm, C.A. (1983) Growth and survival of Atlantic salmon fed semi-moist or dry starter diets. *The Progressive Fish-Culturist*, **45**, 72–74.

Liener, I.E. (1994) Implications of antinutritional components in soybean foods. *Critical Reviews in Food Science and Nutrition*, **34**, 31–67.

Lindquist, N. & Hay, M.E. (1996) Palatability and chemical defense of marine invertebrate larvae. *Ecological Monographs*, **66**, 431–450.

Linnér, J. & Brännäs, E. (1994) Behavioral response to commercial food of different sizes and self-initiated food size selection by Arctic char. *Transactions of the American Fisheries Society*, **123**, 416–422.

Lupatsch, I., Kissil, G.W., Sklan, D. & Pfeffer, E. (1997) Apparent digestibility coefficients of feed ingredients and their predictability in compound diets for gilthead seabream, *Sparus aurata* L. *Aquaculture Nutrition*, **3**, 81–89.

Mackie, A.M. & Mitchell, A.I. (1985) Identification of gustatory feeding stimulants for fish – applications in aquaculture. In: *Nutrition and Feeding in Fish* (eds C.B. Cowey, A.M. Mackie & J.G. Bell). pp. 177–189. Academic Press, London.

Macrae, R., Robinson, R.K. & Sadler, M.J. (eds) (1993) *Encyclopaedia of Food Science, Food Technology and Nutrition.* Academic Press, London.

McGoogan, B.B. & Reigh, R.C. (1996) Apparent digestibility of selected ingredients in red drum (*Sciaenops ocellatus*) diets. *Aquaculture*, **141**, 233–244.

Médale, F., Boujard, T., Vallée, F., *et al.* (1998) Voluntary feed intake, nitrogen and phosphorus losses in rainbow trout (*Oncorhynchus mykiss*) fed increasing dietary levels of soy protein concentrate. *Aquatic Living Resources*, **11**, 239–246.

Melcion, J. P., & Van der Poel, A.F.B. (1993) Process technology and antinutritional factors: principles, adequacy and process optimization. In: *Recent Advances in Antinutritional Factors in Legume Seeds* (eds A.F.B. Van der Poel, J. Huisman & H.S. Saini). pp. 419–434. EAAP Publication, Wageningen, the Netherlands.

Murai, T., Ogata, H., Kosutarak, P. & Arai, S. (1986) Effects of amino acid supplementation and methanol treatment on utilisation of soy flour by fingerling carp. *Aquaculture*, **56**, 197–206.

NRC (National Research Council, USA) (1993) *Nutrient Requirements of Fish.* National Academy of Sciences, Washington.

Nengas, I., Alexis, M.N., Davies, S.J. & Petichakis, G. (1995) Investigation to determine digestibility coefficients of various raw materials in diets for gilthead sea bream, *Sparus auratus* L. *Aquaculture Research*, **26**, 185–194.

Nengas, I., Alexis, M. & Davies, S. (1996) Partial substitution of fishmeal with soybean meal products and derivatives in diets for the gilthead seabream *Sparus aurata* (L.). *Aquaculture Research*, **27**, 147–156.

Oliva-Teles, A., Pereira, J.P., Gouveia, A. & Gomes, E. (1998) Utilisation of diets supplemented with microbial phytase by seabass (*Dicentrarchus labrax*) juveniles. *Aquatic Living Resources*, **11**, 255–259.

Papatryphon, E. & Soares, J.H. Jr. (2000) The effect of dietary feeding stimulants on growth performance of striped bass, *Morone saxatilis*, fed a plant feedstuff-based diet. *Aquaculture*, **185**, 329–338.

Pfeffer, E., Beckmann-Toussaint, J., Henrichfreise, B. & Jansen, H.D. (1991) Effect of extrusion on efficiency of utilization of maize starch by rainbow trout (*Oncorhynchus mykiss*). *Aquaculture*, **96**, 293–303.

Pfeffer, E., Kinzinger, S. & Rodehutscord, M. (1995) Influence of the proportion of poultry slaughter by-products and of untreated or hydrothermically treated legume seeds in diets for rainbow trout, *Oncorhynchus mykiss* (Walbaum), on apparent digestibilities of their energy and organic compounds. *Aquaculture Nutrition*, **1**, 111–117.

Phillips, R.D. (1989) Effect of extrusion cooking on the nutritional quality of plant proteins. In: *Protein Quality and the Effects of Processing* (eds R.D. Phillips & J.W. Finley). pp. 219–246. Marcel Dekker, New York.

Pigott, G.M. & Tucker, B.W. (1989) Special feeds. In: *Fish Nutrition* (ed. J.E. Halver). pp. 653–679. Academic Press, London.

Pike, I.H., Andorsdottir, G. & Mundheim, H. (1990) The role of fish meal in diets for salmonids. *IAFMM Technical Bulletin*, **24**, 35pp.

Poe, W.E. & Wilson, R.P. (1989) Palatability of diets containing sulfadimethoxine, ormetoprim, and Romet 30 to channel catfish fingerlings. *The Progressive Fish-Culturist*, **51**, 226–228.

Pond, W.G., Church, D.C. & Pond, K.R. (1995) *Basic Animal Nutrition and Feeding.* John Wiley & Sons, Chichester.

Refstie, S., Storebakken, T. & Roem, A. J. (1998) Feed consumption and conversion in Atlantic salmon (*Salmo salar*) fed diets with fish meal, extracted soybean meal or soybean meal with reduced content of oligosaccharides, trypsin inhibitors, lectins and soya antigens. *Aquaculture*, **162**, 301–312.

Reigh, R.C. & Ellis, S.C. (1992) Effects of dietary soybean and fish-protein ratios on growth and body composition of red drum (*Sciaenops ocellatus*) fed isonitrogenous diets. *Aquaculture*, **104**, 279–292.

Robaina, L., Corraze, G., Aguirre, P., Blanc, D., Melcion, J.P. & Kaushik, S.J. (1999) Digestibility, post-prandial ammonia excretion and selected plasma metabolites in European sea bass (*Dicentrarchus labrax*) fed pelleted or extruded diets with or without wheat gluten. *Aquaculture*, **179**, 45–56.

Robaina, L., Izquierdo, M.S., Moyano, F.J., *et al.* (1995) Soybean and lupin seed meals as protein sources in diets for gilthead seabream (*Sparus aurata*): nutritional and histological implications. *Aquaculture*, **130**, 219–233.

Robinson, E.H. (1991) A Practical Guide to Nutrition, Feeds and Feeding of Catfish. *MAFES Bulletin* **979**, 18pp.

Rodehutscord, M. & Pfeffer, E. (1995) Effects of supplemental microbial phytase on phosphorus digestibility and utilization in rainbow trout (*Oncorhynchus mykiss*). *Water Science and Technology*, **31**, 143–147.

Rungruangsak, K. & Utne, F. (1981) Effects of different acidified wet feeds on protease activities in the digestive tract and on growth rate of rainbow trout (*Salmo gairdneri* Richardson). *Aquaculture*, **22**, 67–79.

Sargent, J., Henderson, R.J. & Tocher, D.R. (1989) The lipids. In: *Fish Nutrition* (ed. J.E. Halver). pp. 153–218. Academic Press, London.

Schreck, J.A. & Moffitt, C.M. (1987) Palatability of feed containing different concentrations of erythromycin thiocyanate to chinook salmon. *The Progressive Fish-Culturist*, **49**, 241–247.

Smith, I.P., Metcalfe, N.B. & Huntingford, F.A. (1995) The effects of food pellet dimensions on feeding responses by Atlantic salmon (*Salmo salar* L.) in a marine net pen. *Aquaculture*, **130**, 167–175.

Stradmeyer, L. (1989) A behavioural method to test feeding responses of fish to pelleted diets. *Aquaculture*, **79**, 303–310.

Stradmeyer, L., Metcalfe, N.B. & Thorpe, J.E. (1988) Effect of food pellet shape and texture on the feeding response of juvenile Atlantic salmon. *Aquaculture*, **73**, 217–228.

Sullivan, J.A. & Reigh, R.C. (1995) Apparent digestibility of selected feedstuffs in diets for hybrid striped bass (*Morone saxatilis* × *Morone chrysops*). *Aquaculture*, **138**, 313–322.

Sveier, H., Wathne, E. & Lied, E. (1999) Growth, feed and nutrient utilisation and gastrointestinal evacuation time in Atlantic salmon (*Salmo salar* L.): the effect of dietary fish meal particle size and protein concentration. *Aquaculture*, **180**, 265–282.

Tabachek, J-A. L. (1988) The effect of feed particle size on the growth and feed efficiency of Arctic charr [*Salvelinus alpinus* (L.)]. *Aquaculture*, **71**, 319–330.

Tacon, A.G.J. & Jackson, A.J. (1985) Utilisation of conventional and unconventional protein sources in practical fish feeds. In: *Nutrition and Feeding in Fish* (eds C.B. Cowey, A.M. Mackie & J.G. Bell). pp. 119–145. Academic Press, London.

Tacon, A.G.J., Stafford, E.A. & Edwards, C.A. (1983) A preliminary investigation of the nutritive value of three terrestrial lumbricid worms for rainbow trout. *Aquaculture*, **35**, 187–199.

Takeda, M. & Takii, K. (1992) Gustation and nutrition in fishes: application to aquaculture. In: *Fish Chemoreception* (ed. T.J. Hara). pp. 271–287. Chapman & Hall, London.

Thomas, M. & van der Poel, A.F.B. (1996) Physical quality of pelleted animal feed 1. Criteria for pellet quality. *Animal Feed Science and Technology*, **61**, 89–112.

Thomas, M., van Zuilichem, D.J. & van der Poel, A.F.B. (1997) Physical quality of pelleted animal feed. 2. Contribution of processes and its conditions. *Animal Feed Science and Technology*, **64**, 173–192.

Thomas, M., van Vliet, T. & van der Poel, A.F.B. (1998) Physical quality of pelleted animal feed. 3. Contribution of feedstuff components. *Animal Feed Science and Technology*, **70**, 59–78.

Toften, H., Jørgensen, E.H. & Jobling, M. (1995) The study of feeding preferences using radiography: oxytetracycline as a feeding deterrent and squid extract as a stimulant in diets for Atlantic salmon. *Aquaculture Nutrition*, **1**, 145–149.

Tucker, C.S. & Robinson, E.H. (1990) *Channel Catfish Farming Handbook*. Van Nostrand Reinhold, New York.

Tucker, J.W. Jr. (1998) *Marine Fish Culture*. Kluwer Academic Publishers, Boston.

Tusé, D. (1984) Single-cell protein: current status and future prospects. *CRC Critical Reviews in Food Science and Nutrition*, **19**, 273–325.

Van Barneveld, R.J. (1999) Understanding the nutritional chemistry of lupin (*Lupinus* spp.) seed to improve livestock production efficiency. *Nutrition Research Reviews*, **12**, 203–230.

Van den Ingh, T.S.G.A.M., Olli, J. & Krogdahl, Å. (1996) Alcohol-soluble components in soybeans cause morphological changes in the distal intestine of Atlantic salmon, *Salmo salar* L. *Journal of Fish Diseases*, **19**, 47–53.

Vens-Cappell, B. (1984) The effects of extrusion and pelleting of feed for trout on the digestibility of protein, amino acids and energy, and on feed conversion. *Aquacultural Engineering*, **3**, 71–89.

Wankowski, J.W.J. & Thorpe, J.E. (1979) The role of food particle size in the growth of juvenile Atlantic salmon (*Salmo salar* L.). *Journal of Fish Biology*, **14**, 351–370.

Watanabe, T. & Kiron, V. (1994) Prospects in larval fish dietetics. *Aquaculture*, **124**, 223–251.

Wilson, R.P. (ed.) (1991) *Handbook of Nutrient Requirements of Finfish*. CRC Press, Boca Raton.

Wright, D.I. & O'Brien, W.J. (1982) Differential location of *Chaoborus* larvae and *Daphnia* by fish: the importance of motion and visible size. *The American Midland Naturalist*, **108**, 68–73.

Zimmer, R.K. & Butman, C.A. (2000) Chemical signaling processes in the marine environment. *Biological Bulletin*, **198**, 168–187.

Chapter 3
Techniques for Measuring Feed Intake

Malcolm Jobling, Denis Covès, Børge Damsgård, Henrik R. Kristiansen,
Juha Koskela, Thuridur E. Petursdottir, Sunil Kadri and
Olafur Gudmundsson

3.1 Introduction

The measurement of feed intake in fish can be considered at different levels depending upon the aims and constraints of particular studies. In some cases data acquisition may concern a fish group, whereas in others the focus may be on individuals within a group, or even on isolated fish. The investigator may wish to measure feed intake over a short period, such as in a single meal, the study may require measurements of daily feed intake to be made, or the investigator may wish to collect temporal series to investigate feeding rhythms in chronobiology studies. Test conditions, holding facilities (tanks or aquaria, pens or cages, ponds and lakes), and ease of access to the fish (captive or wild populations) will also differ between studies. All of these factors will influence both the method to be used for the measurement of feed intake and the accuracy of the data collected.

Whether or not it is possible to collect quantitative, or only qualitative, data depends upon the food type, which may range from the live prey consumed by wild fish to the different types of formulated feeds used in fish culture, and the sampling protocols employed. Some of the techniques used in the study of feeding in fish are invasive, and some are also destructive; so they are limited to a research application. Other techniques are neither invasive nor destructive, and may be used under commercial farming conditions as a feed management tool to enhance farm productivity.

In this chapter consideration is given to the technical aspects of some commonly used methods employed in the investigation of feed intake in fish. The chapter opens with a discussion of stomach contents analysis, a method that is commonly used to gain insights into the feeding habits of wild fish, and then progresses to the examination of methods that are more widely used in laboratory studies of feeding behaviour and feed intake. The applications and limitations of the different methods are discussed in an attempt to guide potential users towards the choice of methods appropriate for the examination of the various facets of feeding in fish.

3.2 Stomach contents analysis

Stomach contents analysis is primarily a method for qualitative estimation of dietary composition by investigation of prey items in the fish stomach. This type of analysis is often used in field studies of fish ecology, and summaries of the methods are given by Hynes (1950), Windell and Bowen (1978), Hyslop (1980), Talbot (1985) and Bowen (1996).

3.2.1 Technical aspects

Stomach contents analysis may be carried out on either live or dead fish. The most common practices are outlined in Fig. 3.1. When using live fish, the stomach content may be removed by stomach pumping or flushing. Stomach pumping is carried out by palpating the abdominal region of the fish, often after introducing water into the stomach. Alternatively, a tube may be inserted into the stomach, and lavage performed using one or two tubes to flush the stomach (Twomey & Giller 1990; dos Santos & Jobling 1992; Bromley 1994; Andersen 1998). Emetics (arsenous acid, tartar emetics, apomorphine) may be used to cause the fish to vomit (Talbot 1985). Food can also be removed by inserting a spoon into the oesophagus and stomach, but this method is rarely used. Dubets (1954) described a metal gastroscope that enabled direct observation of stomach contents to be made.

The analysis may be carried out after sampling in the field or in tanks, sometimes after feeding tagged pre-weighed prey in order to estimate individual feed intake (Amundsen *et al.* 1995; Bagge *et al.* 1995). After sampling, the fish are commonly killed by a blow to the head or via lethal anaesthesia, and fish length and weight recorded. An abdominal incision is made from the gills to the anal opening, and the stomach removed by cutting at the upper end of the oesophagus and behind the pyloric sphincter. A larger part of the gut may be sampled in those fish which lack a well-defined stomach. The mouth cavity should be checked for food items, and these included in the stomach contents sample. This is especially important when fish are sampled by sport fishing and trawling.

Before further analysis, the stomachs may be kept frozen or preserved in buffered formalin, 70% or 96% ethanol, or ethanol and glycerine. Caution should be taken when storing

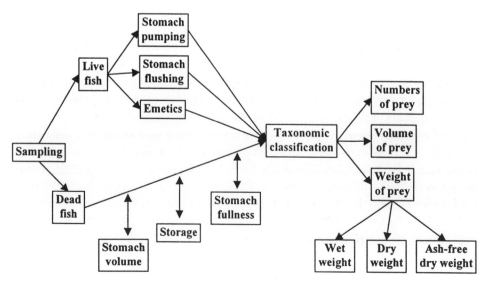

Fig. 3.1   Flow diagram illustrating the most common practices for stomach contents analysis in fish.

the samples, because formalin preservation may cause an increase in weight (Parker 1963; Hyslop 1980), and mass loss may occur in alcohol (Levings *et al.* 1994). Formalin storage may lead to dissolution of bones and otoliths used to estimate the age of prey found in the stomach.

Stomach volume may be estimated by ligaturing the pyloric sphincter, binding the oesophagus to a burette and determining the volume of water required to dilate a stomach subjected to a known pressure head (usually 50 cm) (Jobling *et al.* 1977). Alternatively, stomach capacity may be determined by measuring the amount of water that can be injected into an empty stomach prior to bursting (Kimball & Helm 1971). If the stomach contains food items the displacement volume may be determined in a graduated cylinder (Hyslop 1980), or calculated from geometrical algorithms (Ruttner-Kolisko 1977).

Stomach fullness may be estimated as the proportion of the maximum stomach volume occupied by prey items, or as a proportion of maximum stomach capacity (Kimball & Helm 1971). Fullness is expressed on a scale from 0% to 100%. Alternatively, stomach volume and volume of each prey species may be assessed using subjective feeding units (FU). The stomach fullness index (SFI) ranges from 0 to 10, where 10 represents a full stomach. The prey score character (PS) from 0 to 10 represents the contribution of each prey to the bulk. Maximum score is 10, when only one prey species is present. The sum of scores for a stomach containing food must always equal 10. FU is calculated as SFI multiplied by PS for each prey. Maximum units in a stomach are 100. For example: In a half-full stomach (SFI = 5) *Arenicola marina* make up 60% of the bulk (PS = 6), and *Gammarus pulex* make up 40% (PS = 4). FU in this stomach is then calculated as $5 \times 6 + 5 \times 4 = 50$ units. This procedure is continued for all the stomachs, and the percentage prey contribution to the total number of FU can be calculated.

For analysis of prey types the stomach contents are transferred to a container, e.g. a Petri dish, and prey is identified to species, genus or family. A wide range of methods is used to analyse stomach contents.

Occurrence

The simplest way to assess dietary composition of a fish population is to calculate the frequency of occurrence without taking into account either the amounts of each prey or the total quantity of prey. The frequency of occurrence (O_i) is defined as the percentage of stomachs in which a given prey i occurs. The frequency of empty stomachs may change seasonally, so results may differ significantly if the frequency of occurrence is calculated on the basis of the numbers of stomachs with food (O_{fi}), or the total numbers of stomachs (O_{efi}). $O_{fi} = 100 \times N_i/N_f$, where N_i is the number of fish with the prey i in the stomach and N_f is the number of stomachs with food, and $O_{efi} = 100 \times N_i/N_{ef}$, where N_{ef} is the total number of stomachs examined. A modification of the occurrence method is to estimate the occurrence of the dominant food in each stomach. Even in opportunistic fish species, individual fish are often food specialists (Amundsen *et al.* 1995), and the occurrence of dominant food may give a quick and crude qualitative estimation of dietary composition in a population.

Abundance

The relative abundance (A_i) of a prey gives information about the contribution of each prey to the stomach content. Prey categories may be expressed in terms of numbers, volume, or as weight (Windell & Bowen 1978; Bowen 1996): $A_i = 100 \times (\Sigma S_i/\Sigma S_t)$, where S_i is the stomach content (number, A_{Ni}; volume, A_{Vi}; weight, A_{Wi}) of prey i and S_t is total stomach content. Stomach contents may also be squashed to uniform depth and the areas of prey making up the squash preparation measured (Hellawell & Abel 1971).

Importance

The index of importance by numbers (*IN*) is calculated as: $IN = \sqrt{A_{Ni}} \times \sqrt{O_i}$, where A_{Ni} is percentage abundance by numbers and O_i, is percentage occurrence. Similarly, the index of importance by mass (*IM*) is calculated as: $IM = \sqrt{A_{Wi}} \times \sqrt{O_i}$, where A_{Wi} is percentage abundance by weight (Windell 1968; Vesin *et al.* 1981; Castro 1993). The index of relative importance (IRI) is estimated as: $IRI = (A_{Ni} + A_{Vi}) \times O_i$, thereby incorporating percentage abundance by numbers and volume, and percentage occurrence (Pinkas *et al.* 1971; Prince 1975). The relative importance index (RI) is calculated as: $RI = 100 \times AI/\Sigma AI$, for *n* prey categories. AI, termed the absolute importance index, is the sum of percentage frequency of occurrence, percentage by number and percentage by total weight (George & Hadley 1979; Elvira *et al.* 1996; Nicola *et al.* 1996). The abundance index (AI) should not be confused with relative abundance (A_i).

The different occurrence, abundance and importance equations were applied to stomach contents data obtained by sampling sea run brown trout, *Salmo trutta*, in Vejle fjord, south east Denmark during spring 1995 (H.R. Kristiansen, unpublished results). Among the samples, 19 out of 118 stomachs were empty, and 767 prey items with a total weight of 784 g were identified. Abundance by volume was estimated using subjective feeding units, for a total of 4858 units for the 99 stomachs containing food (Table 3.1). Most of the food items were identified to species. Crustaceans (*Gammarus, Idotea, Corophium, Leander*) and fish prey (*Ammodytes, Pomatoschistus, Clupea, Gasterosteus, Pungitus, Limanda*) were usually easy to recognise and discriminate, even when they were located close to the pyloric sphincter, whereas polychaetes (*Nereis, Arenicola*) were often digested and there were few identifiable remains. The data in Table 3.1 illustrate that small crustaceans such as *Gammarus* and *Corophium* were ranked with a high abundance index by number (A_{Ni}), but low abundance index by weight (A_{Wi}). The opposite was the case with large fish prey, such as *Ammodytes* and *Clupea*. This tendency is also reflected in the importance index by number (*IN*) and mass (*IM*), but disappears when the index of relative importance (IRI) or the relative importance index (RI) is used. The IRI and RI ranking systems gave almost the same results and are thus considered suitable ranking indices.

Stomach weight content may be expressed as wet, dry or ash-free dry weight. Dry weight is determined after drying to constant weight at 60°C. Ash-free dry weight involves ashing samples in a muffle oven at 450–500°C. Weight-specific stomach content may be calculated as milligrams dry weight of food per gram fresh weight of fish. As an alternative to bulk weighing, each prey item may be weighed. This may be possible for large prey, but small zooplankton may be difficult to sort and weigh individually (Hyslop 1980). Provided that

Table 3.1 Comparison of methods to analyse stomach contents of 118 sea run brown trout, *Salmo trutta*, sampled from Vejle fjord, Denmark. The explanation of the acronyms used in the table is given in the main body of the text. The first column indicating frequency of occurrence (O_{fi}) is arranged to show the ranking of prey items in descending order: this is denoted by the superscripts. Ranking of the different prey items according to the other methods of analysis is also denoted by superscripts.

Prey species	Occurrence (O_i)		Abundance (A_i)			Importance				
	O_{fi}	O_{efi}	A_{Ni}	A_{Vi}	A_{Wi}	IN	IM	IRI	AI	RI
Nereis diversicolor	69.7[1]	58.5[1]	58.3[1]	60.3[1]	58.3[1]	63.7[1]	63.7[1]	8266.4[1]	186.3[1]	51.4[1]
Gammarus sp.	26.3[2]	22.0[2]	21.8[2]	5.6[4]	2.2[7]	23.9[2]	7.6[4]	720.6[2]	50.3[2]	13.9[2]
Leander adspersus	15.2[3]	12.7[3]	3.7[4]	2.8[7]	3.1[5]	7.5[3]	6.9[6]	98.8[4]	22.0[4]	6.1[4]
Ammodytes lancea	12.1[4]	10.2[4]	2.5[7]	11.2[2]	16.2[2]	5.5[5]	14.0[2]	165.8[3]	30.8[3]	8.5[3]
Arenicola marina	10.1[5]	8.5[5]	3.8[3]	4.8[5]	6.1[4]	6.2[4]	7.8[3]	86.9[5]	20.0[5]	5.5[5]
Pomatoschistus sp.	9.1[6]	7.6[6]	1.7[8]	3.5[6]	2.4[6]	3.9[7]	4.7[7]	47.3[7]	13.2[7]	3.6[7]
Clupea harengus	6.1[7]	5.1[7]	0.9[9]	8.7[3]	8.6[3]	2.3[8]	7.2[5]	58.6[6]	15.6[6]	4.3[6]
Idotea sp.	5.1[8]	4.2[8]	3.7[5]	1.7[8]	2.2[8]	4.3[6]	3.3[8]	27.5[8]	11.0[8]	3.0[8]
unidentified	3.0[9]	2.5[9]	0.4[10]	0.4[10]	0.5[9]	1.1[10]	1.2[9]	2.4[10]	3.9[9]	1.1[9]
Corophium volutator	1.0[10]	0.8[10]	2.7[6]	0.03[13]	0.0[12]	1.6[9]	0.0[12]	2.7[9]	3.7[10]	1.0[10]
Gasterosteus aculeatus	1.0[11]	0.8[11]	0.1[11]	0.08[12]	0.1[11]	0.3[11]	0.3[11]	0.2[13]	1.2[12]	0.3[12]
Pungitus pungitus	1.0[12]	0.8[12]	0.1[12]	0.2[11]	0.0[13]	0.3[12]	0.0[13]	0.3[12]	1.1[13]	0.3[13]
Limanda limanda	1.0[13]	0.8[13]	0.1[13]	0.7[9]	0.3[10]	0.3[13]	0.5[10]	0.8[11]	1.4[11]	0.4[11]
Mysis sp.	1.0[14]	0.8[14]	0.1[14]	0.02[14]	0.0[14]	0.3[14]	0.0[14]	0.1[14]	1.1[14]	0.3[14]
Cardium edule	1.0[15]	0.8[15]	0.1[15]	0.02[15]	0.0[15]	0.3[15]	0.0[15]	0.1[15]	1.1[15]	0.3[15]
Sum			100	100	100	121.8	117.4	9478.5	362.7	100

certain prerequisites are fulfilled (Jobling & Breiby 1986), the initial weight of the prey may be reconstructed using length–weight regression (Langeland *et al.* 1991). A size reconstruction of fish prey in the stomach may be possible using the size of the prey otoliths (L'Abée-Lund *et al.* 1996), or vertebrae (Damsgård & Langeland 1994).

As an alternative to a discrete taxonomic classification of prey, food items may be classified according to the habitat in which they occur (e.g. benthic, pelagic and littoral), and feeding habitat preferences of the predators assessed from the frequencies of food items from different habitats (Damsgård & Ugedal 1997).

3.2.2 *Applications and limitations*

Analysis of stomach contents has been used in examinations of the diets of fish populations (e.g. Hindar & Jonsson 1982; Grønvik & Klemetsen 1987; Damsgård & Langeland 1994), and in studies of daily or seasonal prey preference (e.g. Staples 1975; Frost 1977). When combined with information about rates of evacuation, stomach contents analysis may be used in assessments of the total food consumed by fish populations (e.g. Staples 1975; Eggers 1977; Elliott & Persson 1978; Bromley 1994). Stomach contents analysis may also have application in ecological studies of predation (Damsgård & Langeland 1994; Damsgård & Ugedal 1997), assessments of intra- and interspecific competition (Amundsen & Klemetsen 1988), and in studies of optimal foraging (Pyke *et al.* 1977; Mittelbach 1981; Townsend & Winfield 1985).

Differential digestion of prey in the stomach and their evacuation to the intestine may result in errors in the estimates of the relative importance of prey categories. Zooplankton and other easily digested prey may be underestimated, whereas prey with resistant, and easy recognisable, hard parts may be overestimated. Rates of digestion and evacuation are affected by food deprivation, chemical composition and friability of the prey, meal size, predator size and water temperature (Elliott 1972, 1991; Jobling 1986; dos Santos & Jobling 1992, 1995; Bromley 1994). Several models to calculate food consumption in fish populations are based on the combination of stomach content analysis and information about rates of gastric evacuation (Eggers 1977; Elliott & Persson 1978; Windell & Bowen 1978; Bromley 1994; dos Santos & Jobling 1995). The models estimate consumption over short time periods, usually with intervals of one day. Although several models have been developed for field use few have been tested under controlled conditions (Elliott & Persson 1978; dos Santos & Jobling 1995), so the value of many of these models must remain in doubt.

In ecological studies prey selection may be calculated using the foraging ratio (*FR*) or Ivlev's selection index (E_i). The foraging ratio on a given food type is defined as the ratio of the proportion *r* in the stomach to the proportion *p* of the same food in the environment, i.e. *FR* = *r* / *p* (Jacobs 1974). As an assessment of feeding intensity, Levings and co-workers (1994) expressed *FR* as the ratio of the wet mass of prey in the stomach content to the wet mass of the fish, whereas others have used this as a numerical expression of stomach fullness (Castro 1993). Ivlev's selection index is calculated as: $E_i = (r_i - p_i)/(r_i + p_i)$, where *r* is the number of prey i eaten by the predator, and *p* is the number in the habitat (Ivlev 1961). Prey importance, feeding strategy and the niche width of a fish may be examined graphically on the basis of prey-specific percentage of occurrence and abundance (Mohan & Sankaran 1988; Costello 1990; Amundsen *et al.* 1996).

Stomach contents analysis is suitable for the examination of dietary composition of wild fish populations. Further, samples taken at intervals during a day or season may provide valuable information about daily and seasonal rhythms in feeding. The disadvantage is that the fish are usually killed prior to carrying out stomach analysis, so it is not possible to examine the same fish several times. The analysis of stomach contents is also time-consuming, the length of time required depending on the way the analysis is carried out. Further, direct comparisons between studies may not be possible because differences in methods of analysis and expression of results can make it difficult to distinguish true dietary differences from variation caused by the method used.

The frequency of occurrence method is quick and easy, but it does not provide quantitative information about feeding. The numerical method is suitable when prey are of similar size range, but not when fish have consumed both zooplankton and other fish. For example, although only 11% of the stomachs in a brown trout population contained fish, these prey represented 79% of the total stomach weight content and contributed significantly to the trout diet (Damsgård & Langeland 1994). Similarly, the herring, *Clupea harengus*, was ranked a less important prey for brown trout when using numbers of prey (rank 9) as the criterion than when using volume or weight (rank 3) (Table 3.1). Thus, the numerical method overemphasises the importance of small prey, and it may also be difficult to use if prey items are particularly numerous, or difficult to separate (Hyslop 1980).

Volumetric methods take into account differences in the sizes of the prey, but it may be time consuming to estimate the volume of each prey category. The determination of wet and dry stomach weight is also time-consuming. Weight measurements provide estimates of stomach content, but the dietary contribution of prey with heavy, non-digestible body parts, such as bones and scales, may be overemphasised (Hellawell & Abel 1971).

Stomach pumping is non-destructive, and may enable serial sampling of individual fish to be carried out (e.g. dos Santos & Jobling 1992, 1995; Bromley 1994). However, unless carried out carefully, the method may stress or injure the animals (Foster 1977; Talbot 1985), efficiency may vary for hard-bodied and soft prey (Meehan & Miller 1978; Bromley 1994), and the method may be difficult to use on small fish (Hyslop 1980).

3.3 Dyestuffs and chemical markers

Various dyestuffs and other chemical markers have been added to fish feeds for the study of digestion and rates of gastrointestinal transit, but the addition of marker substances to feeds may have wider application in feeding studies. Techniques involving faecal dilution have been used in studies of livestock nutrition and feed intake for over a century (Edin 1945; Kotb & Luckey 1972), but are most frequently used in digestibility studies in aquaculture (see Chapter 1). These methods have not gained popularity for measurement of feed intake in fish.

3.3.1 *Technical aspects*

The inclusion of dyestuffs in feeds may be used to obtain both qualitative and quantitative information about feeding in fish (Walsh *et al.* 1987; Morris *et al.* 1990; Johnston *et al.* 1994; Unprasert *et al.* 1999). The marked feed is presented to the fish, and the proportion of the population that has fed can be estimated either via examination of faeces or via stomach contents analysis. Stomach contents are usually recovered after the fish have been killed, but for some species non-lethal methods, such as gastric lavage, can provide efficient recovery of the contents of the stomach (Talbot 1985; Bromley 1994). Although the identification of the dyestuff in the stomach contents or faeces can provide information about how many fish have fed on the particular feed, it is far more difficult to obtain reliable data about how much feed has been consumed. There are several limitations with this method that restrict its use, and reduce its value for the quantitative assessment of feed intake in fish.

The faecal dilution techniques are based on the determination of the ratios of indigestible markers in the feed and faeces. Feed intake estimation requires the use of two markers. This is done by having the animal consume a small but known amount of an indigestible external marker each day over a period of time. This marker is usually fed in the morning. The second indigestible marker, which for domestic animals other than fish is usually naturally present in feedstuffs, is fed throughout the day. Faecal samples are collected and the dilution rate of nutrients in the faeces in relation to the indigestible markers determined, giving the apparent digestibility coefficient (ADC) and faecal output (FO). Feed intake (FI) is then calculated as (Harris 1970; Dove & Mayes 1991):

$$FI = FO \times DI^{-1} \tag{3.1}$$

where FO is in weight units, and the so-called dietary indigestibility (DI) is a proportion calculated as $(1 - ADC)$. FO is estimated indirectly with the help of the external marker, using the equation:

$$FO = \frac{\text{External marker consumed (g/day)}}{\text{Concentration of external marker in faeces (g/g)}} \tag{3.2}$$

The DI is determined from the ratio of the internal marker in the feed and the respective faeces, from the following equation:

$$DI\,(g/g) = \frac{\text{Concentration of internal marker in diet (g/g)}}{\text{Concentration of internal marker in faeces (g/g)}} \tag{3.3}$$

This enables simultaneous estimation of intake and digestibility, provided that the daily consumption of external marker is accurately determined and that representative samples of the faeces are obtained (Kotb & Luckey 1972; Gudmundsson & Halldorsdottir 1993). Several compounds have been used as markers in digestibility studies. Markers must be selected carefully, and several compounds – including *n*-alkanes, e.g. straight-chain hydrocarbons – may fulfil the criteria (see Chapter 1; Grace & Body 1981; Cravedi & Tulliez 1986; Dove & Mayes 1991).

To summarise, the chemical markers should be incorporated at the time of feed preparation, and two markers are used; the internal marker is mixed in all of the feed and the external marker is mixed into a small proportion of the feed. The external marker is fed in an exact amount (dose) over the entire measurement period to enable estimation of faecal output of the animals in question. The internal marker is used to estimate apparent digestibility.

The fish need to be acclimatised to tank conditions, the experimental feed, feeding and other procedures prior to the start of any study. During acclimatisation all procedures should be carried out in accordance with the experimental design, and faecal production should be checked every day. Faecal collection can start when the fish are feeding well, faecal production is adequate for analysis of the markers, there is no contamination from previous feed, and the flow of markers through the gut is constant.

The fish are first fed with a small, known amount of feed marked with both the external and internal marker. The exact quantity (e.g. 5–10 g) will depend upon the size and number of fish in the tanks, but the same amount must be fed each day. It is vital that all the feed is consumed. The fish are then fed with feed containing the internal marker in accordance with usual feeding practice. At the end of each daily feeding the tanks and faecal collection devices are cleaned to remove feed waste, and faecal collection is started. During faecal collection the fish are not fed. The number of days over which faeces need to be collected depends on the size and number of fish and how much feed is consumed. The amount of faeces needed depends on the marker used and the digestibility of the feed, e.g. if *n*-alkanes are used, 0.5 g of dry faeces are sufficient for analysis.

It is important to collect the faeces as soon as possible after defecation to avoid losses due to leaching (see Chapter 1; Windell *et al.* 1978; Hajen *et al.* 1993). Two different methods for faeces collection are described by Cho and co-workers (1982) and by Choubert and co-workers (1982). Both the time of collection of samples and the pooling of samples need to be consistent. Faecal material is preferably freeze dried, although oven-drying at temperatures up to 60°C may be adequate. The analyses of feed and faeces are then carried out to determine the relative concentrations of marker(s), dry matter and specific nutrients, and the calculations for the estimation of digestibility and feed intake are then undertaken.

3.3.2 *Applications and limitations*

Feeds marked with dyestuffs can be used to estimate the proportions of fish within a population that have fed, but the use of such feeds for the quantitative assessment of feed intake is more problematic. This is because one prerequisite for quantitative measurement is that the amount consumed in a given meal be easily identifiable. This may not be possible if the dyestuff present in the feed becomes rapidly mixed with any residual stomach content from previous meals. This problem could be overcome by imposing a prolonged period of feed deprivation prior to carrying out the feed intake measurement. This would ensure that the stomach did not contain remains of previous meals, but the imposition of a period of feed deprivation would almost certainly have an influence upon the amount of feed consumed. In other words, the results obtained would not be representative of 'normal' feeding, and, as such, might be considered to have limited value. In addition, the inclusion of dyestuffs in a feed may influence acceptability, due to the change of feed colour (see, however, Unprasert *et al.* 1999). This could lead to an erroneous estimation of 'normal' feed consumption being made when the colour of the feed is changed abruptly immediately prior to the carrying out of a feed intake measurement.

Despite the limited number of studies performed on fish, one might suggest that faecal dilution methods could be used for the estimation of feed intake of many species. There is no requirement for either force-feeding or sacrifice of fish, nor is a feed-deprivation period required prior to feeding the marked feed. Faecal dilution methods need not involve handling or anaesthetisation of the fish, and the methods are environmentally safe provided that the markers chosen are environmentally safe. Feed intake of any size of fish, and any group size, can be measured as long as biomass and faecal collection time are adequate. The methodology can be used to measure consumption of any kind of formulated feed, and has application in the assessment of feed preference. In the latter type of study the fish are presented simultaneously with feeds containing different markers and the ratios of the markers in the faeces provide the information needed for calculation of the relative proportions of each feed consumed. The method can also be used to measure seasonal changes in feeding, and integrates feed intake over long periods of time.

One of the main problems with using the faecal dilution technique to quantify feed intake in fish is related to dosing with the external marker. The daily dose must be known exactly, and the same daily quantities must be provided during both the acclimatisation and faecal collection periods. Fish should not be handled directly for dosing, so a small known amount of feed (which includes markers) must be fed very carefully, by hand. All this feed must be consumed prior to further feeding if reliable results are to be obtained. This can be tedious and time-consuming, especially if many tanks are used.

Moreover, the method requires a reliable estimate of the indigestible component of feeds. Attempts are made to reduce levels of indigestible components in high-quality fish feeds, and this may give high variability in intake estimation. This is because DI is the denominator in Equation (3.1). Indigestibility will obviously become smaller as digestibility increases, leading to increased risks of variability and inaccuracy in the estimation of intake.

The method requires incorporation of two different markers into the feeds: therefore, the chemical analysis can be difficult and tedious if the markers need different analytical methods to be used. This disadvantage can be overcome by using pairs of markers that can be analysed simultaneously, e.g. *n*-alkanes or rare earth elements.

There is always some danger of overestimating digestibility because of leaching of some components from the faeces. This would lead to overestimation of intake because the estimate of indigestibility decreases. There is also the question about what effect there would be if only few fish in the tank consumed most of the external marker, but the remainder of the marked feed was consumed by the rest of the tank population. This has not been studied, and the potential influence of diurnal variation in faecal output of fish is not known. Effects should, however, be limited if the majority of the faeces are collected and care is taken that samples are adequately pooled.

As the method relies on faecal collection it is not possible to measure feed intake of individual fish within a group, nor is the method viable for fish held in cages or ponds. The chemical marker technique permits the quantitative determination of the average feed intake of a group of fish over time. At the same time the technique allows determination of digestibility of feeds or feedstuffs. It permits measurement of intake variations between weeks, months and time of year, but does not enable measurement of feed intake by individual fish in a group.

3.4 Direct observation and video recording

Observation methods used to monitor feed intake and feeding behaviour may involve filming or video recording. Direct observation has been used to study feeding for several decades, both under experimental conditions and in the field. Recent developments of video equipment and within computer programming enable detailed studies of feed intake, feeding behaviour, social interactions and swimming patterns to be made using observational methods. Talbot (1985) and Jobling and co-workers (1995) give summaries of the methods.

3.4.1 Technical aspects

Most commonly the observation method is used in small experimental units holding small numbers of fish, but the method has also been used in aquaculture tanks, stream channels, sea cages, and natural waters. An overview of methodology is given in Fig. 3.2.

If the fish are transferred to specially designed observation tanks from other rearing units, an acclimatisation period will be required prior to the start of an experiment. Acclimatisation is necessary because transportation, anaesthetisation and tagging may cause handling stress, and because the fish require time before they become accustomed to the new environment. The length of time required depends upon several factors, including fish species, size, water temperature and season. A pilot study will usually be required to find the time that must elapse before the fish behave normally.

The observation method may be applied to individuals or at the group level. Inter-individual variation is often large when fish are held in small groups, and observations made on known individuals are desirable. It is often necessary to tag each fish in order to recognise individuals. Usually, external tags are used, e.g. large Peterson disc tags (Tuene & Nordvedt 1995), small coloured tags attached to the dorsal fin, tags of different shape, or strings of

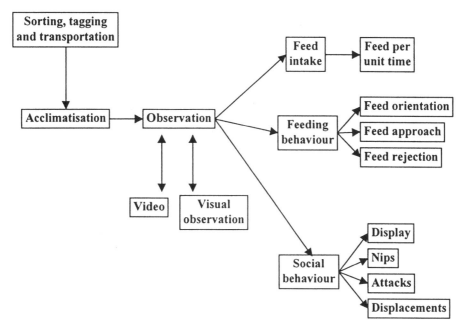

Fig. 3.2   Flow diagram illustrating methods and equipment for direct observation of fish, along with the type of observations made and the results obtained.

plastic beads in unique sequences (Grand 1997). When tagging involves use of colour coding, a preliminary test of the video recordings will be needed to evaluate tag visibility. If observations are made from the side or from beneath the tank, ventral or abdominal colour ink marks (e.g. Alcian Blue) or cold branding may be used for individual identification. In addition to external tags, internal PIT-tags (passive integrated transponder) may be used in combination with direct visual observation (Brännäs & Alanärä 1992).

The length of observation sessions may range from minutes to several hours, but observation should be made at the same time of the day, because feeding responses and several behavioural traits change during the course of the day.

In direct visual observation, the observer takes notes of the feed intake and feeding behaviour. The responses of the fish are usually followed from behind a masking screen (Magnuson 1962; Noakes 1980; Metcalfe *et al.* 1987), or from a darkened adjacent room (Stradmeyer 1989). Observation may be made from the side (Kalleberg 1958; Wankowski 1979; Noakes 1980; Paszkowski & Olla 1985; Stradmeyer 1989), from above (Chapman & Bjornn 1969; Adams *et al.* 1995; Tuene & Nordvedt 1995), from beneath the tank (Winberg *et al.* 1993b), or by using an angled mirror placed above the water (Magnuson 1962). A side view will usually give good information about feeding behaviour, and observations made from beneath the tank will not be affected by disturbances of the water surface (Fig. 3.3).

During the course of the past ten years, video recording has been commonly used in studies of fish behaviour (Abrahams 1989; Kadri *et al.* 1991, 1996; Malmquist 1992; Smith *et al.* 1993, 1995; Juell *et al.* 1994; Hughes & Kelly 1996; Ryer & Olla 1996; Ang & Petrell 1997; Grand 1997; Damsgård & Dill 1998). This technique enables detailed studies of feed intake

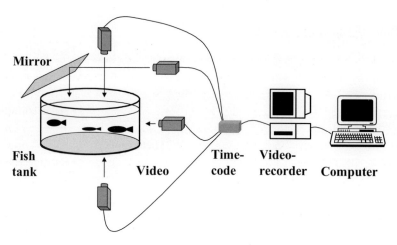

Fig. 3.3   Options for experimental set-ups using video recording.

and feeding behaviour to be undertaken. Video equipment ranges from inexpensive 8-mm camcorders to advanced digital cameras. Most videos have a time limit of approximately 3 h recording, but observation over longer periods is possible using time-lapse techniques.

Underwater videos have been used to study feeding behaviour of Atlantic salmon, *Salmo salar*, in sea cages (Kadri *et al.* 1991, 1996; Smith *et al.* 1993, 1995; Juell *et al.* 1994; Ang & Petrell 1997), with cameras being mounted on a scaffold gantry or moored in or outside the cage. The fish have been viewed either horizontally, from above or from below. The depth of view in sea cages may change, depending on light conditions and water transparency, and this should be taken into account when experiments are being planned.

In field studies, the fish may be observed from directly above the water (Jenkins 1969; Dill *et al.* 1981; Grant 1990), by diving (Wankowski & Thorpe 1979; Stradmeyer & Thorpe 1987; Heggenes *et al.* 1993), or using a video camera placed either in the water column or on the bottom (Hughes & Kelly 1996).

Feed intake of the fish may be estimated by counting the numbers of food items consumed, and intake may be expressed as food eaten per unit time over a short interval, or as total feed intake per day. More detailed studies may include registrations of frequencies of orientations towards food items, approaches towards the food, and frequencies of rejections. Additionally, both patch choice and social interactions may be analysed from videos, including frequencies of aggressive behaviour such as displays, nips, attacks and displacements. The use of computer-aided video systems may ease the task of recording, editing and analysis of the video tapes (Noldus Information Technologies 1997). Behavioural analysis may provide a flexible system for monitoring feed intake or feeding behaviour of individual fish (Brown & Brown 1996). A 'time-code generator' produces a code on the video tapes that the computer uses to record the time and duration of defined events (Fig. 3.3). The observation program analyses frequencies and the mean duration of these events and may, for example, be used to calculate a conflict matrix within a group of fish (Noldus Information Technologies 1997).

Behavioural studies may also include analyses of swimming, using the tracking of the movements of the fish in two or three dimensions (Huse & Skiftesvik 1990; Winberg *et al.* 1993b; Pereira & Oliveira 1994; Skiftesvik *et al.* 1994; Hughes & Kelly 1996).

3.4.2 Applications and limitations

Direct observation has been used to examine the effects of a range of biotic and abiotic factors on feed intake and feeding behaviour: these include the effects of different feed characteristics (Mearns *et al.* 1987; Stradmeyer & Thorpe 1987; Stradmeyer 1989; Smith *et al.* 1995), the distribution of feed (Gotceitas & Godin 1992; Ryer & Olla 1995; Kadri *et al.* 1996; Ang & Petrell 1997), and interactions with conspecifics (Kalleberg 1958; Keenleyside & Yamamoto 1962; Jenkins 1969; Noakes 1980; Abbott & Dill 1985; Grant 1990; Grant & Kramer 1992; Grant & Guha 1993; Adams *et al.* 1995). In addition, several studies of feeding behaviour have included examination of swimming patterns (Skiftesvik *et al.* 1994; Hughes & Kelly 1996).

Direct observation has also been used to obtain estimates of consumption by individual fish used in gastric evacuation studies (Elliott & Persson 1978; dos Santos & Jobling 1995), and in studies of biorhythms (Kadri *et al.* 1991; Smith *et al.* 1993). The method also finds application in studies of habitat selection (Chapman & Bjornn 1969; Jenkins 1969; Wankowski & Thorpe 1979; Dill *et al.* 1981; Heggenes *et al.* 1993; Grand 1997), and in examination of the response of the fish when faced with the dilemma of feeding under threat of predation (Metcalfe *et al.* 1987; Abrahams 1989; Milinski 1993; Ryer & Olla 1996; Damsgård & Dill 1998).

Many studies have involved the examination of responses of fish to feed pellets, but live feed (Chapman & Bjornn 1969; Abrahams 1989; Grand 1997), or natural prey organisms (Keenleyside & Yamamoto 1962; Elliott & Persson 1978; Paszkowski & Olla 1985) have also been used. Juvenile salmonids have been studied frequently, but the observation method has also been used to study feeding behaviour in halibut, *Hippoglossus hippoglossus* (Skiftesvik *et al.* 1994; Tuene & Nordvedt 1995), medaka, *Oryzias latipes* (Magnuson 1962), guppy, *Poecilia reticulata* (Abrahams 1989), zebrafish, *Brachydanio rerio* (Grant & Kramer 1992) and cichlids, *Cichlasoma nigrofasciatum* (Grant & Guha 1993). With the exception of the work conducted on adult Atlantic salmon in sea cages, studies have usually involved either juvenile fish or species that reach small adult body size. Group size is often small, and several studies have been conducted using single fish. It may, however, be possible to study tagged individuals within a larger group (Kadri *et al.* 1996). The time scale of experiments is often short, covering minutes or hours, but the use of video recordings makes longer-term observation possible.

The main advantages of the direct observation method are that it enables continuous records to be made on single, or groups of, fish over time. Compared with other feed intake methods, it is possible to monitor individual feed intake on a minute-to-minute scale. A large day-to-day variation in feed intake is common, and using the observation method intake may be monitored on consecutive days without disturbing the animals. Studies of feed intake may be combined with detailed descriptions of feeding behaviour and social interactions, and this combination of different measurements on individual fish enables elucidation of mechanisms not possible when using other feed intake methods. The method is one of the few that may be adapted to a field situation, and both direct observation and underwater video cameras are well suited for use in field studies (Dill *et al.* 1981; Grant 1990; Hughes & Kelly 1996).

The main disadvantage with the method is that it is time-consuming. This is especially the case with direct visual observation, but also when manual analysis of video tapes is undertaken. Since the method is frequently used on small groups in laboratory-scale systems, it may be difficult to generalise the results to conditions experienced by fish in commercial aquaculture. For example, single fish may behave differently from fish in groups, and feed intake of individuals within small groups may be strongly affected by the social interactions between the fish. Thus, a dominant fish may take a large meal, whereas subordinate fish have reduced access to feed, even though the total feed supply appears to be in excess (Øverli *et al.* 1998). It should be cautioned that both in the laboratory and in the field, the observer may disturb the fish, leading to behavioural changes, an alteration of feeding behaviour and foraging pattern and an underestimation of 'normal' feed intake.

3.5 On-demand feeder with feed waste monitor

Feed intake in fish is influenced by a wide array of factors over both short and long time scales (see Chapters 6, 7 and 8) and on-demand feeding systems have been developed in an attempt to meet the challenge of providing fish with feed without predetermination of either the timing of feed delivery or the quantity delivered. On-demand feeders are of two types, differing in the way by which feed delivery is controlled. In the first type the fish directly 'self-feed' by actuating a triggering mechanism, thereby initiating the release of feed as a reward. In the second type of demand-feeder there is *a posteriori* control of feed delivery, in that feed is dispensed automatically and then a detection system controls future feed delivery via feedback mechanisms.

3.5.1 Technical aspects

In principle, a self-feeder is a very simple feed delivery system: fish actuate a trigger and obtain pellets as a reward. Self-feeders can be classified in two main categories, mechanical and electrical. The mechanical self-feeders deliver feed when a fish actuates a rod (pendulum) attached to the hopper containing the feed pellets. With lateral movement of the rod the position of a plate within the hopper changes and feed is delivered. This kind of feeder is described in detail by Fauré (1983). In electrical self-feeders the trigger is not coupled directly to the feed hopper. Activation of the trigger induces an electrical impulse, and this operates a relay which actuates the feeder (Adron 1972; Landless 1976; Boujard *et al.* 1991, 1992; Sánchez-Vásquez *et al.* 1994). In this kind of self-feeder the activation of the trigger and release of feed from the hopper are separated, and this allows adjustments to be made with regard to the functioning of each. All triggering systems which generate an electrical signal offer the possibility of computerised data acquisition of feeding activity.

The technologies used in interactive feeding systems (feedback feeders) are more diverse than in self-feeders, but the general principle of operation involves data collection which is 'fed back' to a feeding system that adjusts feed delivery accordingly. These feeding systems can be broadly assigned to two main categories: those which detect feed particles; and those which detect changes in the behaviour of the fish. Further, the systems can be either manually operated or automatic.

Feed detection can be made using hydroacoustic transducers which detect the movement of feed particles through the water column (Juell 1991; Juell *et al.* 1993) or along the outflow pipe leaving the tank (Summerfelt *et al.* 1995). Information is relayed to an automated control system which operates by switching the feeder off when excess pellets are detected. Alternatively, infrared sensors may be used to count individual particles of uneaten food falling through the water column (Blyth *et al.* 1993, 1997) or being carried in the water flow along a tank outlet pipe (Chen *et al.* 1999).

Changes in behaviour may be detected using hydroacoustic transducers which collect data on the vertical distribution of the fish. A signal is generated for cessation of feeding once most fish are located below a pre-defined depth in the water column (Bjordal *et al.* 1993). Hydroacoustic transducers are also being developed to detect the sound of fish feeding; the feeding system would be turned off as the level of this sound declined (Mallekh & Lagardère, patent no. 98 09768). Video cameras are also used for data collection in the operation of feedback systems (Foster *et al.* 1993; Ang & Petrell 1997); cameras can be used to observe both uneaten feed and the behaviour of the fish.

Thus, there are several types of feedback system that use different detection criteria, and there are differences between systems in the definition of the point at which fish within the population are satiated. To further complicate matters, many of the automated systems, and all of the manual systems, allow 'satiation point' to be user defined.

3.5.2 *Applications and limitations*

Self-feeders are generally designed to monitor and/or control feeding by populations of fish. If self-feeders can be adjusted to minimise feed wastage, feed intake corresponds to the quantity delivered. In such cases, a good estimation of population feed intake can be obtained by weighing the feed remaining in the hopper at regular intervals. Usually this is done manually but an alternative was presented by Fast *et al.* (1997): a pendulum feeder was suspended from a load cell which measured the weight of feeder and the feed it contained.

Critical to the accurate measurement of feed intake when using on-demand feeding systems is the correct evaluation of feed wastage. In other words, uneaten pellets must be detected or collected in such a way that total losses can be accurately estimated. Under self-feeding conditions, feed wastage is related to the size of the reward (i.e. amount of feed) given at each trigger actuation. If the reward is too high, there may be considerable feed wastage (Brännäs & Alanärä 1994), but when the reward is low the fish may not be able to compensate by increasing trigger activation sufficiently to meet their feed requirement (Alanärä 1994; Alanärä & Kiessling 1996; Gelineau *et al.* 1998). Consequently, the aim is to adjust the reward to match the feeding rate of the fish without imposing any restriction on access. Hoppers exist that offer the possibility to fix reward size to several tens of grams feed (Alanärä 1996; Gelineau *et al.* 1998), down to a few (Hidalgo *et al.* 1988; Gelineau *et al.* 1998) or even a single pellet (Cuenca & de la Higuera 1994; Sánchez-Vásquez *et al.* 1994).

Feed wastage may also result from involuntary contacts of fish with the triggering mechanism (Anthouard *et al.* 1986). Different solutions have been proposed to alleviate this problem, the solutions being based upon the behaviours of the species in question. For example, Boujard *et al.* (1991) located the trigger 2 cm below the water surface to restrict unintentional triggering by atipa, *Hoplosternum littorale*, a bottom-dwelling fish. In studies with rainbow trout, *Oncorhynchus mykiss*, attempts have been made to reduce feed losses by placing the trigger 2 cm above the water surface (Boujard & Leatherland 1992) or by providing the fish with a trigger that has to be bitten to cause the release of feed (Alanärä 1992a, 1994). Covès *et al.* (1998) provided fish with a trigger protected by a cylindrical screen: this proved to be effective in preventing accidental trigger actuations by European sea bass, *Dicentrarchus labrax*.

When uneaten pellets passing from the tank in the outlet water are collected, accurate assessment of feed intake can be made. Some form of waste collection device such as sediment traps or filters may be employed (Boujard & Le Gouvello 1997; Dias *et al.* 1997; Gelineau *et al.* 1998). The automatic collection devices developed by Choubert *et al.* (1982) and Cho *et al.* (1982) have also been used in conjunction with self-feeders to monitor feed waste (Lemarié *et al.* 1996; Twarowska *et al.* 1997; Covès *et al.* 1998). Madrid and co-workers (1997) used a continuous recording device on the outflow pipe leaving the tank to count each uneaten pellet. In all cases special consideration must be given to the rapid drainage of pellets from the tank and rapid counting after collection: this is needed to limit disintegration of feed, so the use of water-stable pellets improves the accuracy of evaluation (Helland *et al.* 1996).

Self-feeders have been used to study feeding of fish of several species, spanning a wide size range (Table 3.2), whereas feedback feeders are most commonly used in cages, particularly during the ongrowing phase of larger fish, e.g. post-smolt salmonids. Feed waste detection is usually the principle used to manage feed delivery by feedback feeders. Detection of feed waste in sea cages is possible using hydroacoustic methods (Juell 1991; Juell *et al.* 1993) and the signals may be used to automatically adjust the feed rations supplied. Blyth and co-workers (1993, 1997) have developed an adaptive feeding system in which the rate of feed input is based on the detection of small amounts of uneaten feed by an underwater infra-red sensor. Underwater video monitoring has also been used to assess feed losses (Foster *et al.* 1993; Ang & Petrell 1997). The limitation of feedback systems to provide an accurate measurement of feed intake is the data used and processing thereof for determination of when to stop the supply of feed. Only a few of these systems show potential as a research tool at this stage.

Self-feeders offer the possibility to record feeding rhythms (see Chapters 8 and 9) of both groups, and isolated fish. Feeding rhythms can also be studied using feedback demand feeders. In both cases attempts should be made to ensure that feed wastage is suppressed, or continuously and accurately recorded. Several authors have used computerised demand feeding systems to study daily (Fig. 3.4) or seasonal feeding patterns (Landless 1976; Tackett *et al.* 1988; Boujard *et al.* 1991; Boujard & Leatherland 1992; Bégout Anras 1995; Sánchez-Vásquez *et al.* 1995a,b; Blyth *et al.* 1997; Fast *et al.* 1997), but feed waste has not been recorded in all cases, thereby casting doubt on the accuracy of the data with regards to feed intake.

For self-feeders to be of any value the fish need to be able to learn how to operate the triggering mechanism. During the learning period neither triggering activity nor feed intake is stable. The learning period may be followed by a 'behavioural overshoot' during which the fish compensate for the previous period of instability. Kentouri *et al.* (1992) reported that European sea bass required 15–20 days to learn how to operate a self-feeder, the duration of the learning period depending upon the social structure of the group. Adron and co-workers (1973) suggested that a training period of at least 10 days was needed by rainbow trout, whereas lake trout, *Salvelinus namaycush*, needed 45 days to stabilise their feed demand (Aloisi 1994). Sometimes, however, it may be difficult to assess when training and adapta-

Table 3.2 Examples illustrating a range of studies in which self-feeders have been used to examine various aspects of fish feeding and growth. Information is given about the type of study, the fish species, weight or age, and rearing conditions.

Type of study or aim of experiment	Fish species	Weight or age	Holding facility and stocking	Reference
Feasibility study	Gilthead sea bream, *Sparus aurata*	0.13–5 g	1.5 m³ indoor square tanks	Divanach *et al.* (1986)
	Lake trout, *Salvelinus namaycush*	6–20 g	85 m³ outdoor raceways, 12 kg/m³	Aloisi (1994)
	Turbot, *Psetta maxima*	20–80 g	1 m² indoor tanks, 21 kg/m³ (8 kg/m²)	Burel *et al.* (1997)
	Chinese catfish, *Clarias fuscus*	Approx. 315 g	1100 diploid or triploid fish stocked in outdoor circular tanks	Fast *et al.* (1997)
Feeding behaviour	*Lithognatus mormyrus*	36 months	Mixed with other species in a 20-m³ tank	Anthouard *et al.* (1986)
	Tilapia, *Oreochromis mossambicus*	7.5–12 months	400-l aquaria	Anthouard & Wolf (1988)
	European catfish, *Silurus glanis*	9 months	400-l aquaria	Anthouard & Wolf (1988)
	Channel catfish, *Ictalurus punctatus*	40–260 g	Stocked at 500 and 1500 kg/hectare	Tackett (1988)
	Rainbow trout, *Oncorhynchus mykiss*	500–3100 g	144 m³ net cages, 20 kg/m³	Alanärä (1992a,b)
	Arctic charr, *Salvelinus alpinus*	48–330 g	1 m³ indoor tanks	Brännäs & Alanärä (1993)
	Cut-throat trout, *Oncorhynchus clarki*	9–98 g	Outdoor raceways (8 m³), 35 kg/m³	Wagner *et al.* (1995)
	Arctic charr, *Salvelinus alpinus*	63–200g	350-l compartments, 2–70 kg/m³	Alanärä & Brännäs (1996)
	Atlantic salmon, *Salmo salar*	10–20 g juveniles	100-l indoor tanks, 10.5 kg/m³	Paspatis & Boujard (1996)
Feeding rhythms	Atipa, *Hoplosternum littorale*	From 49 g	200-l indoor tanks	Boujard *et al.* (1991)
	Atlantic salmon, *Salmo salar*	10–20 g juveniles	100-l indoor tanks, 10.5 kg/m³	Paspatis & Boujard (1996)
	Chinese catfish, *Clarias fuscus*	Approx. 315 g	1100 diploid or triploid fish stocked in outdoor circular tanks	Fast *et al.* (1997)
Feed preference	Rainbow trout, *Oncorhynchus mykiss*	200–1000 g	0.8 m³ indoor tanks	Alanärä & Brännäs (1993)
	Goldfish, *Carassius auratus*	70–163 g	54-l aquarium, single fish or 4 fish per group	Sánchez-Vásquez *et al.* (1998)
Growth performance	Gilthead sea bream, *Sparus aurata*	0.13–5 g	1.5 m³ indoor square tanks, 30 m³ outdoor raceways, 12 kg/m³	Divanach *et al.* (1986)
	Tilapia, *Oreochromis aureus*	30–450 g	200–400 fish per 1-m³ cylindrical cage	Hargreaves *et al.* (1988)
	Annular sea bream, *Diplodus annularis*	few grams–20 g	30 m³ outdoor raceways, 12 kg/m³	Divanach *et al.* (1993)

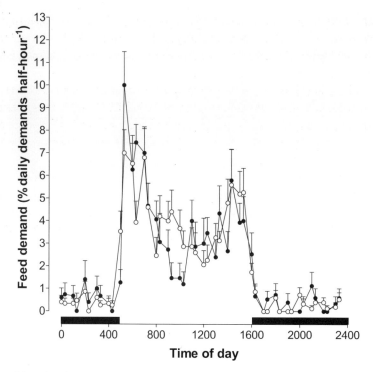

Fig. 3.4   Diel profile of feeding activity (as % of daily demands per half-hour) of Atlantic salmon, *Salmo salar,* fed either high-energy (•) or low-energy (○) feeds by means of self-feeders. Shaded areas indicate the scotophase; vertical bars indicate 1 standard deviation (*n* = 3). (From Paspatis & Boujard 1996.)

tion is complete (Burel *et al.* 1997). Several methods can be used to reduce the length of the training period: Kentouri and co-workers (1986) mentioned the facilitation role played by experienced fish added to a group of naive ones.

Individual feed intake of fish within populations cannot be measured using the self-feeding approach, although individual actuating activity has been monitored using PIT-tagged fish (Alanärä & Brännäs 1993; Brännäs & Alanärä 1993). Feed intake by individual sea bass was studied by Sánchez-Vásquez and co-workers (1995a,b) following three weeks of acclimatisation to isolation. The fish activated a self-feeder which provided one pellet at each food request (Sánchez-Vásquez *et al.* 1994).

Demand-feeders do not permit direct estimation of inter-individual heterogeneity of feed intake. Detailed observation of behaviour and trigger actuation is needed to identify the hierarchical relationships that can modify the feed intake of fish held in groups (Alanärä & Brännäs 1996). For example, trigger activation by a few high-ranked individuals would lead to a situation where a small group of dominant fish 'feed' the rest of the group. The consequences of this can be modified by group size (stocking density) and the way in which the feed is dispersed from the feeder. For example, Alanärä (1996) reported that there was less variation in growth among self-feeding rainbow trout in large cages than among smaller groups held in tanks.

Self-feeders may be used to study feed preferences of fish. Macronutrient or diet selection has been demonstrated in several species (Hidalgo *et al.* 1988; Kentouri *et al.* 1995; Sánchez-

Vásquez *et al.* 1998). Alanärä and Brännäs (1993) attempted to assess the pellet size preferences of rainbow trout using self-feeders, and Boujard and Le Gouvello (1997) demonstrated the capacity of self-fed rainbow trout to discriminate between feeds containing antibiotic incorporated in different forms.

Demand feeders, and especially self-feeding systems combined with feed waste collection, provide flexible systems for the study of feed intake in fish. The different systems allow studies on fish of various species and sizes, and permit investigation of the influence of environmental parameters and feed quality on feeding behaviour. Demand feeders can be adapted to both laboratory and production conditions.

When using self-feeders attention has to be given to the possible influences of hierarchy formation on triggering activity and feed intake variability, especially under laboratory conditions with small groups of fish. Further, consideration must be given to the learning period required before the fish are able to use a self-feeding system effectively, and the feed demand becomes consistent, i.e. feed intake is stable.

In general on-demand feeding systems are particularly suitable for recording feed intake at the population level, and for the study of feeding rhythms and feed preferences.

3.6 X-Radiography

Although X-radiography has been used routinely in medical science for several decades it was not until relatively recently that the technique was adapted for studies of fish feeding and digestion. In the earliest studies the digestive processes of piscivorous fish were examined using X-radiography to follow both the degradation of X-ray-dense skeletal elements of the prey, and their passage through the gut of the predator (Molnar *et al.* 1967). The spiking of soft-bodied prey with X-ray-dense contrast medium (e.g. barium sulphate) permitted qualitative studies of digestion and evacuation to be conducted on a broader range of fish species, and by the late-1970s X-radiography was being widely used to study food transit and gastrointestinal evacuation (Fänge & Grove 1979; Talbot 1985). The replacement of the dispersed contrast medium with X-ray-dense particulate markers opened the way for the quantitative determination of gastrointestinal content (Talbot & Higgins 1983), thereby permitting X-radiography to be used for the monitoring of feed intake in fish.

3.6.1 Technical aspects

The method is based on providing the fish with feed containing particulate X-ray-dense markers. The amount of marker eaten is usually measured by X-raying the fish, and counting the numbers of marker particles present in the gastrointestinal tract (Fig. 3.5). Standard curves for food–marker relationships (mg food/marker) are prepared by X-raying known weights of the labelled feed and counting the numbers of markers present. The amount of feed eaten by the fish is then calculated from the numbers of marker particles in the gastrointestinal tract by reference to the standard curve. The X-radiographic technique may be used in studies requiring accurate information about feed intake by either individual, or groups of, fish (Talbot 1985; Jobling *et al.* 1990, 1993, 1995; McCarthy *et al.* 1993).

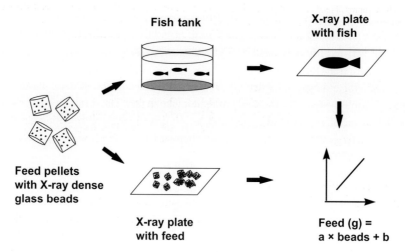

Fig. 3.5   Schematic presentation showing the X-radiographic method for the measurement of feed intake in fish. The upper panel shows the feeding of labelled feed, and the lower panel the method of calculation of the standard curve.

Particulate markers of various types (e.g. electrolytic iron powder, lead glass 'ballotini' spheres of different sizes and lead 'shot' or 'solder') may be used; the types, sizes and concentrations of markers incorporated into the feed are varied depending upon the sizes of the fish used in the experiments and the problem being studied (Talbot 1985; Jobling *et al.* 1990, 1993, 1995; dos Santos & Jobling 1991; McCarthy *et al.* 1993; Amundsen *et al.* 1995; Christiansen & George 1995; Toften *et al.* 1995).

The fact that the X-radiographic technique depends upon there being a radio-opaque marker in the feed means that the technique is better suited to studies carried out using formulated feeds than those in which natural prey are fed. The technique can, however, be employed in studies with natural prey if the prey contain X-ray-dense skeletal elements, or can be 'spiked' with X-ray-dense particulate markers (e.g. Amundsen *et al.* 1995).

The labelled feed should be formulated to be identical to the 'normal' feed in every respect (e.g. nutrient composition, colour, taste, particle size, etc.) with the exception of the inclusion of particulate marker at low concentration (usually 0.1–2.0% by weight). Ideally, the labelled feed should be prepared at the same time as the normal feed. A sample of the feed mix should be taken prior to pelleting, the particulate marker added, and the mixture should then be pelleted in the usual fashion. Alternatively, the labelled feed may be prepared by grinding and homogenising samples of the normal feed and adding known quantities of the X-ray-dense marker. The feed is then re-pelleted and dried. To avoid potential problems relating to changes in feed characteristics, the normal feed to be used in the particular study should also be ground and re-pelleted in the same way as the labelled feed. The fact that the labelled feed differs little, if at all, from the normal feed allows information to be collected about the feed intake of previously unstarved, relatively undisturbed fish fed on a feed identical to that to which they are accustomed.

When the aim is to obtain an accurate assessment of the feed intake of individual fish, the concentration of marker in the feed should be adjusted to a level that results in reasonable numbers (30–300) of marker particles in the gastrointestinal tracts of feeding fish. Too high

Table 3.3   The amounts of lead glass ballotini marker (specific gravity 2.9) of different sizes to be added to the feed (g/kg feed) to give 150 marker particles per fish at designated levels of feed intake.

Feed intake (g/fish)	Diameter of marker (mm)				
	0.2	0.3	0.4	0.5	0.6
0.1	18.2	61.5	–	–	–
0.5	3.6	12.3	29.1	56.9	–
1.0	1.8	6.1	14.6	28.5	49.2
5.0	–	1.2	2.9	5.7	9.8
10.0	–	–	1.5	2.8	4.9

a concentration of marker in the feed makes counting difficult. On the other hand, too low a concentration decreases the accuracy of the method, since one unit change in marker amount has a large influence on the estimate of the amount of feed consumed. In order to avoid these problems the researcher must have some prior knowledge about the approximate level of feed intake by the fish. Once this is known it is possible to calculate inclusion levels of markers of different sizes to be incorporated into the labelled feed (Table 3.3).

The accuracy of the method is dependent upon the marker being mixed homogeneously in the feed. Therefore, careful preparation of the labelled feed is important. The success with which the marker has been mixed into the feed can be checked by taking several subsamples, of about the same weight as the fish are expected to eat, and then calculating the confidence limits for the feed–marker relationship. An alternative method of checking the homogeneity of mixing is to calculate the resolution power of the regression model (standard curve) describing the relationship between marker particle numbers and the weight of the feed (Prairie 1996). Resolution power increases with an increasing coefficient of determination (R^2) of the regression model, so adequate homogeneity of mixing cannot be assumed unless the coefficient of determination of the standard curve is high (ideally, $R^2 > 0.98$).

3.6.2 Applications and limitations

When the X-radiographic technique is used for the quantitative estimation of feed intake by individual fish, a number of prerequisites must be fulfilled. First, the particulate marker must be retained within the gastrointestinal tract for some time, rather than being rapidly passed through the gut and defecated. Any defecation of marker that occurs in the period between the start of feeding and the time at which the fish are X-ray photographed will lead to feed intake being underestimated. This places a limitation on the duration of the feeding period that can be employed. The length of time that elapses between the initial ingestion of the labelled feed and the defecation of the first radio-opaque particles may vary from a few hours to a day or more depending upon the sizes and types of marker used, fish species and the environmental conditions under which the tests are carried out (Talbot 1985; Jobling *et al.* 1990, 1993; Thorpe *et al.* 1990; Sæther & Jobling 1997; Sæther *et al.* 1999).

Thus, prior to using the X-radiographic method for monitoring feed intake on a routine basis some preliminary experiments should be conducted to assess retention and evacuation of the particulate marker from the guts of the test fish (e.g. Jørgensen & Jobling 1988; dos Santos & Jobling 1991; Koskela *et al.* 1997; Sæther & Jobling 1997; Sæther *et al.* 1999). The risk that some of the radio-opaque particles may be lost via defecation means that the duration of the feeding period must be restricted to a maximum of a few hours if an accurate assessment of feed intake is to be obtained. If, however, information is available about the rate at which the particulate marker is passed through the gut, feeding can be carried out over longer periods and corrections made to account for defecation. This enables feed intake to be estimated under conditions where there may have been some loss of marker due to defecation (Thorpe *et al.* 1990).

The prolonged retention of particulate marker within the gut of the fish imposes another limitation on the use of the X-radiographic technique if the aim is to determine feed intake by individuals. Since several days may be required before all the particles of radio-opaque marker have been evacuated from the gut, repeated measures of feed intake are usually conducted at relatively infrequent intervals. In practice, this is not a serious problem since the fish should not be subjected to frequent handling and anaesthetisation if disturbances to feeding are to be avoided. Thus, if repeated measures are to be made on specific groups of fish a period of some days, or weeks, should be allowed to elapse between feed intake determinations. If it is, nevertheless, feared that the time interval is insufficient to allow complete evacuation of particulate marker from the gut, measurements can be carried out by incorporating particles of different sizes into the labelled feed to be provided on each occasion.

Incorporation of markers of different sizes into different batches, or types, of feed allows the amounts of each type of feed consumed to be determined when feeds are presented either simultaneously or as discrete meals offered within a limited time interval (Thorpe *et al.* 1990; Amundsen *et al.* 1995; Christiansen & George 1995; Toften *et al.* 1995; Damsgård & Dill 1998). This enables studies to be carried out when the aim is to assess where, when and how much individual fish have eaten. For example, studies of feed preferences can be carried out using the X-radiographic 'labelled feed' technique, and feed selection by individual fish can be assessed (Amundsen *et al.* 1995; Christiansen & George 1995).

If, on the other hand, the aim is to examine the feed preferences of groups of fish, it is not necessary to handle, anaesthetise and X-ray the fish if the tank system is fitted with a sediment trap to collect feed wastes. In this case the fish may be provided with known quantities of two feeds simultaneously, wastes collected and X-rayed, and the amounts of each feed type consumed calculated from the weights of feed waste recovered (e.g. Koskela *et al.* 1993).

X-Radiography may be used for the quantitative determination of the gastrointestinal content of fish under a range of conditions, and X-radiography has been used to study digestive physiology (Fänge & Grove 1979; Talbot 1985; dos Santos & Jobling 1991), feeding behaviour (Higgins & Talbot 1985; Talbot 1985; Jobling *et al.* 1990, 1993), and feed intake by groups of fish (reviewed by Jobling *et al.* 1990, 1993, 1995). However, the major application of the X-radiographic technique is in studies in which quantitative information about the feed intake of individual fish is required (Talbot 1985; Jobling *et al.* 1990, 1993; McCarthy *et al.* 1992).

When studies are carried out using individually marked, or tagged, fish there is the possibility to perform in-depth examinations of the effects of different rearing environments and feed types on feed intake and growth responses. For example, fish of different species, 'strains' or genetic origins can be reared in a common environment, and their feeding and growth responses compared (Hatlen *et al.* 1997; Jobling *et al.* 1998). Further, physical examination of fish either during the course of an experiment, or *a posteriori*, enables the identification of individuals that may differ in status (e.g. mature versus immature; deformed versus healthy), and the performances of the fish within different categories can be compared (Sæther *et al.* 1996; Toften & Jobling 1996; Tveiten *et al.* 1996). In addition, when both feed intake and growth rates of individual fish have been monitored over time, and scattergrams plotted, it is possible to define criteria for the selection of individuals for further study of a range of physiological parameters (Carter *et al.* 1993; McCarthy *et al.* 1993, 1994).

Further, the repeated measurement of feed intake by individual fish results in the collection of data that permits the examination of both inter-individual variability, and the day-to-day variations in feed intake shown by individuals (Fig. 3.6) (McCarthy *et al.* 1992, 1993; Jobling & Baardvik 1994). Data of this type provide information about differences in feed acquisition, and it has been suggested that the data can be used to provide an indirect assessment of the social environment existing within a group of fish with respect to the establishment of feeding hierarchies (McCarthy *et al.* 1992; Winberg *et al.* 1993a). Thus, it is suggested that the feed intake data collected on individual fish using 'labelled feed' techniques can be of value in the study of social relationships within groups (McCarthy *et al.* 1992; Winberg *et al.* 1993a; Jobling & Baardvik 1994; Jobling & Koskela 1996; Damsgård *et al.* 1997; Damsgård & Dill 1998). This application would seem to be especially useful since it enables the assessment of feeding conditions in much larger groups than is possible using conventional direct observational techniques.

The data relating to individual feeding responses can be treated in a number of ways. One way of evaluating inter-individual (intra-group) variation in feed acquisition is to calculate the coefficient of variation (CV) for the feed intake data collected for the individual fish in the rearing unit (CV = (SD/Mean) × 100). The effects of different experimental treatments on intra-group variability in feed acquisition can then be made by comparison of CVs obtained under the different holding conditions (Fig. 3.7) (Jobling & Koskela 1996; Damsgård *et al.* 1997).

Alternatively, scatterplots may be drawn to show feeding by individual fish on consecutive sampling days, and ranking tests performed to examine whether or not individuals maintain

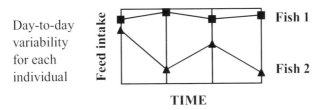

Fig. 3.6 Schematic presentation illustrating inter-individual and intra-individual variations in feed intake in fish.

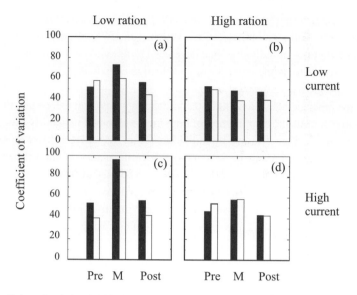

Fig. 3.7 Coefficient of variation (CV) in food intake before manipulation of feed ration and water current speed (Pre), during manipulation (M) and 22 days after reintroduction of pre-manipulation conditions (Post), in Arctic charr, *Salvelinus alpinus* from the Hammerfest strain (filled columns) and the Svalbard strain (open columns). The manipulations were reductions in feed supply (a) and (c) and in water current speed (c) and (d), while plot (b) represents controls without manipulations. (From Damsgård *et al.* 1997.)

their feeding rank over time (Fig. 3.8) (Jobling & Koskela 1996). For example, when groups of rainbow trout were provided with restricted rations, either a single (or a small number of) fish consumed most of the food, and others obtained little or nothing. Those fish which obtained food did so on both sampling days, and there were significant correlations between the individual rankings on the two days (Fig. 3.8a). On the other hand, when the same groups of fish were provided with full rations all the fish within the group fed, and there was no consistency in the ranking of the fish between sampling days (Fig. 3.8b). These differences in feed acquisition between fish fed restricted and full rations were also reflected in the intra-group CV values for feed intake. Thus, when the fish were held on restricted rations, the CV values fell within the range 94 to 123%, but when the same fish were provided with full rations the CV values fell to 38–61% (Jobling & Koskela 1996).

McCarthy *et al.* (1992) presented an alternative method for the analysis of data relating to feed acquisition of individual fish within groups. The analysis relies upon an examination of both inter-individual variability and the day-to-day variations in feed intake shown by individuals. The initial step in the analysis is the calculation of the proportion of the total group feed intake consumed by each individual on each sampling date, i.e. the so-called 'share of meal' eaten by each fish within the group (see Fig. 3.6). When repeated measures of feed intake have been made, and 'shares of meal' calculated for each individual on each sampling date, it is possible to calculate a mean 'share of meal' (MSM) value for each fish within the group. Thus, fish that consistently consume a large proportion of the group intake will have a high MSM, whereas the MSM will be low for those individuals that secure little food on each sampling date. The next step in the analysis is the examination of the variation in feed intake shown by individuals over time (Fig. 3.6). The coefficient of variation in feed intake

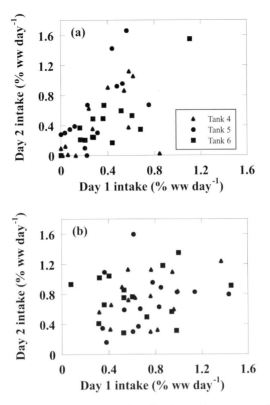

Fig. 3.8 Feed intake by individual rainbow trout, *Oncorhynchus mykiss,* on consecutive sampling dates when fed either (a) restricted rations or (b) full rations. Spearman rank correlations (r_s) for intake by individuals on the different sampling dates were: (a) Tank 4 $r_s = 0.700$; $P < 0.01$; tank 5 $r_s = 0.781$; $P < 0.01$; tank 6 $r_s = 0.689$; $P < 0.01$; and (b) tank 4 $r_s = 0.339$; NS; tank 5 $r_s = 0.443$; NS; tank 6 $r_s = 0.186$; NS. (From Jobling & Koskela 1996.)

over time is then calculated for each individual within the group, the individual CV values are plotted against the MSM values for the same fish; the final step in the analysis is to examine for correlations. The presence of a feeding hierarchy is indicated by a highly significant negative correlation between the intra-individual variations in feed intake (CV values) and the proportions of the feed supply consumed by given individuals (MSM values).

In a series of experiments carried out on rainbow trout fed at various levels of feed restriction there were found to be significant negative correlations between CV values and MSM at all ration levels tested, suggesting that feeding hierarchies were established in all groups of fish. Hierarchy development did, however, appear to be strongest in the groups of fish fed the lowest rations because there was an increase in both intra-individual and inter-individual variability in feed intake as feed availability decreased (McCarthy *et al.* 1992).

The X-radiographic technique, which relies upon the use of labelled feed, permits the quantitative determination of the gastrointestinal content of individual fish, and can also be used for monitoring group feeding responses. It is, however, known that fish may display large day-to-day variations in both feeding activity and feed intake (Smagula & Adelman 1982; Tackett *et al.* 1988). For example, feeding is suppressed in fish recently subjected to disturbance or environmental stressors (Kentouri *et al.* 1994). From this it follows that the information obtained from a single feed intake measurement made on a group of fish should not be used uncritically. If reliable information is to be obtained about the feed intake of groups of fish it is essential that repeated measurements be carried out. However, there may be alternatives to the X-ray method that are more suitable for obtaining information about the total amounts of feed consumed by groups of fish.

The monitoring of feed intake of individual fish held in large groups requires the use of a 'labelled feed' technique. Routine measurement of feed intake using X-radiography provides the data required for the examination of the effects of a variety of biotic and abiotic factors on individual feed acquisition, bioenergetics and energy partitioning (Jobling *et al.* 1993, 1995; McCarthy *et al.* 1993). Although the X-radiographic 'labelled feed' technique is well-suited for the monitoring of individual feed intake, it is essentially a 'one-shot' method that cannot be used for the continuous monitoring of feeding behaviour. The X-radiographic technique is, therefore, not particularly suitable for the collection of information about diel changes in feeding activity, and alternative methods should be sought if this is the aim of a particular study.

3.7 General discussion (see Table 3.4)

The techniques for measuring feed intake in fish presented in this chapter are all environmentally safe. Although some synthetic components have to be added to feeds in the case of the X-radiographic and chemical marker methods, the impact of these on the environment may be considered insignificant in comparison with the potential dangers that arise from methods involving radioisotope-labelled feed. The radioisotope method has not been described here because of its disadvantages: the method cannot be recommended from the health point of view, and considerable care is needed to limit the risk of loss of isotope to the environment.

Handling of fish may be needed when feed intake is measured using the X-radiographic method, but the method is non-invasive and the fish need not be disturbed more than during a standard sampling operation involving anaesthesia, measuring and weighing. The measurement of feed intake using the X-ray and marker methods involves the preparation of special 'labelled' feeds, so these methods are most commonly used in studies in which formulated feeds are used. The X-radiographic method has, however, also been applied to the measurement of consumption of live prey. By contrast, stomach content analysis has its major application in the qualitative and quantitative assessment of the ingestion of live prey. This technique is invasive, and often involves slaughter of the fish sampled for analysis. The techniques involving direct observation and on-demand feeding systems have the advantages that they are non-invasive and do not involve the handling of fish for feed intake measurement.

All the techniques are suitable for use under laboratory test conditions in which relatively small groups of fish are confined in tanks or aquaria, but all have some prerequisites that may be difficult to meet when experiments are scaled-up. Thus, the need to collect faeces or feed waste, and to observe or capture representative samples of fish for measuring feed intake, may create problems when trying to apply the methods to fish held in pens, cages or ponds, or when the aim of a study is to assess the feeding of fish in the wild.

The different techniques enable studies of feed preferences to be undertaken, but two different approaches are adopted. In the first, different feeds are presented to the fish simultaneously, either by hand or from automatic feeders, and records are made of which, and how

Table 3.4 Table indicating the range of applications and limitations of the different techniques used to measure feed intake in fish.

Criteria	Stomach content analysis	Dyestuffs and chemical markers	Direct observation and video recording	On-demand feeder with feed waste monitor	X-Ray
Environmentally safe	Y	Y*	Y	Y	Y*
Invasive	Y	N	N	N	N
Destructive	Y/N	N	N	N	N
Test condition					
Tank/aquarium	Y	Y	Y	Y	Y
Pen cage	Y	N	Y	Y	Y
Pond	Y	N	Y*	Y	Y*
Field	Y	N	Y*	N	N
Live prey	Y	N	Y	N	Y*
Formulated feed	N	Y	Y	Y	Y
Feed preference	Y*	Y	Y*	Y	Y
Groups	N	Y	N	Y	Y
Individuals within a group	Y	N	Y	N	Y
Isolated fish	N	Y*	Y	Y	Y
Total feed intake measurement	Y*	Y	Y	Y	Y
Groups	Y	Y	Y	Y	Y
Individuals within a group	Y	N	Y	N	Y
Isolated fish	Y	Y*	Y	Y	Y
Feeding activity	N	N	Y	Y	Y*
Groups			Y	Y	Y
Individuals within a group			Y	Y*	Y
Isolated fish			Y	Y	Y
Diel feeding rhythms	Y*	N	Y	Y	Y*
Groups	Y		Y	Y	Y
Individuals within a group	N		Y	N	N
Isolated fish	N		Y	Y	Y
Seasonal feeding rhythms	Y*	Y*	Y*	Y	Y*
Groups	Y	Y	Y	Y	Y
Individuals within a group	N	N	Y	N	Y
Isolated fish	N	N	Y	Y	Y

Y* = Yes but with reservations or limitations

much, of the feeds are consumed: this enables an *a posteriori* analysis to be made as to feed choice and preference. The second approach involves the presentation of feeds using several self-feeders, each of which provides the fish with a different feed type; recording of trigger actuations from the different feeders is assumed to reflect feed choice, an assumption that is valid in the absence of feed waste. In cases where waste arises under such a self-feeding regime it must be collected and identified to feed type before an accurate assessment of feed choice can be undertaken.

The marker and on-demand feeder methods may be used to assess the total feed intake by groups of fish, but X-radiography and video recording can also be used for the examination of feed intake by individuals within a group. Unfortunately, the continuous examination of day-to-day variations in feed intake by individuals is not possible using the X-radiographic technique due to the negative consequences of the frequent handling and anaesthesia that would be required for measuring intake. In theory, continuous video recording provides the opportunity to study intra-individual variations in feed intake over time, but the analysis of videotapes is so time-consuming that the method is usually only used for the examination of feed intake in small groups of fish studied over short periods of time.

When the research focus is directed towards the study of feeding activity, diel or seasonal feeding rhythms, which demands that data be collected in temporal series, the measurement method of choice would incorporate a computerised on-demand feeding system. With such systems data may be collected continuously over protracted time periods, and data processing can be carried out quickly and easily. However, the data collected will only provide an assessment of the feeding activity or intake of the entire fish population. On-demand feeding systems may operate on two different principles of feeding: self-feeders rely on the actions of the fish to feed themselves, whereas feedback feeders employ some form of assessment of fish satiation to control rates of feed delivery. Comparative studies involving these feeding systems, both of which are used for feed management in commercial production-scale units, might be useful to reveal some of the factors that have an underlying influence on feed intake.

Each of the methods for measuring feed intake in fish described in this chapter has specific areas of application, and each also has limitations. There are difficulties involved in monitoring the feed intake of individuals within groups on a continuous basis over protracted time periods, and none of the methods currently in use addresses this problem satisfactorily. Nevertheless, the potential for synergy between methods provides opportunities for the carrying out of studies that could lead to increased understanding about how various facets of feeding behaviour influence feed intake. This may shed light on the factors that result in growth differences and size heterogeneity that are observed in groups of fish. Such combinations of methods could include those between on-demand feeding techniques and either video recording or the measurement of feed intake of individual fish using the X-radiographic technique. Studies of this type would seem to hold considerable promise at the laboratory scale, but there remain numerous problems to be solved with respect to the accurate assessment of feed intake by wild fish, and those held in large groups in production-scale facilities.

3.8 References

Abbott, J.C. & Dill, L.M. (1985) Patterns of aggressive attack in juvenile steelhead trout (*Salmo gairdneri*). *Canadian Journal of Fisheries and Aquatic Sciences*, **42**, 1702–1706.

Abrahams, M. (1989) Foraging guppies and the ideal free distribution: the influence of information on patch choice. *Ethology*, **82**, 116–126.

Adams, C.E., Huntingford, F.A., Krpal, J., Jobling, M. & Burnett, S.J. (1995) Exercise, agonistic behaviour and food acquisition in Arctic charr, *Salvelinus alpinus*. *Environmental Biology of Fishes*, **43**, 213–218.

Adron, J.M. (1972) A design for automatic and demand feeders for fish. *Journal du Conseil International pour l'Exploration de la Mer*, **34**, 300–305.

Adron, J.M., Grant, P.T. & Cowey C.B. (1973) A system for the quantitative study of the learning capacity of rainbow trout and its application to the study of food preferences and behaviour. *Journal of Fish Biology*, **5**, 625–636.

Alanärä, A. (1992a) Demand feeding as a self-regulating system for rainbow trout (*Oncorhynchus mykiss*) in net-pens. *Aquaculture*, **108**, 347–356.

Alanärä, A. (1992b) The effect of time restricted demand feeder on feeding activity, growth and feed conversion in rainbow trout (*Oncorhynchus mykiss*). *Aquaculture*, **108**, 357–368.

Alanärä, A. (1994) The effect of temperature, dietary energy content and reward level on the demand feeding activity of rainbow trout (*Oncorhynchus mykiss*). *Aquaculture*, **126**, 349–359.

Alanärä, A. (1996) The use of self-feeders in rainbow trout (*Oncorhynchus mykiss*) production. *Aquaculture*, **145**, 1–20.

Alanärä, A. & Brännäs, E. (1993) A test of the individual feeding activity and food size preference in rainbow trout using demand feeders. *Aquaculture International*, **1**, 47–54.

Alanärä, A. & Brännäs, E. (1996) Dominance in demand-feeding behaviour in Arctic charr and rainbow trout: the effect of stocking density. *Journal of Fish Biology*, **48**, 242–254.

Alanärä, A. & Kiessling, A. (1996) Changes in demand feeding behaviour in Arctic charr, *Salvelinus alpinus* L., caused by differences in dietary energy content and reward level. *Aquaculture Research*, **27**, 479–486.

Aloisi, D.B. (1994) Growth of hatchery-reared lake trout fed by demand feeders. *The Progressive Fish-Culturist*, **56**, 40–43.

Amundsen, P.-A. & Klemetsen, A. (1988) Diet, gastric evacuation rates and food consumption in a stunted population of Arctic charr, *Salvelinus alpinus* L., in Takvatn, northern Norway. *Journal of Fish Biology*, **33**, 697–709.

Amundsen, P.-A., Damsgård, B., Arnesen, A.M., Jobling, M. & Jørgensen, E. (1995) Experimental evidence of cannibalism and prey specialisation in Arctic charr, *Salvelinus alpinus* (L.). *Environmental Biology of Fishes*, **43**, 285–293.

Amundsen, P.-A., Gabler, H.-M. & Staldvik, F.J. (1996) A new approach to graphical analysis of feeding strategy from stomach contents data – modification of the Costello (1990) method. *Journal of Fish Biology*, **48**, 607–614.

Andersen, N.G. (1998) The effect of meal size on gastric evacuation in whiting. *Journal of Fish Biology*, **52**, 743–755.

Ang, K.P. & Petrell, R.J. (1997) Control of feed dispensation in seacages using underwater video monitoring: effects on growth and food conversion. *Aquacultural Engineering*, **16**, 45–62.

Anthouard, M. & Wolf, V. (1988) A computerized surveillance method based on self-feeding measures in fish populations. *Aquaculture*, **71**, 151–158.

Anthouard, M., Desportes, C., Kentouri, M., Divanach, P. & Paris, J. (1986) Etude des modèles comportementaux manifestés au levier par *Dicentrarchus labrax*, *Diplodus sargus*, *Puntazzo puntazzo*, *Sparus aurata*, et *Lithognaus mormyrus* (Poissons Téléostéens) placés dans une situation de nourrissage auto-contrôlé. *Biology of Behaviour*, **11**, 97–110.

Bagge, O., Nielsen, E. & Steffensen, J.F. (1995) Consumption of food and evacuation in dab (*Limanda limanda*) related to saturation and temperature. Preliminary results. *ICES Baltic Fish Committee CM*: J6, 1–8.

Bégout Anras, M.L. (1995) Demand feeding behaviour of sea bass kept in ponds: diel and seasonal patterns, and influences of environmental factors. *Aquaculture International*, **3**, 186–195.

Bjordal, Å., Juell J.E., Lindem, T. & Fernö, A. (1993) Hydroacoustic monitoring and feeding control in cage rearing of Atlantic salmon (*Salmo salar* L.). In: *Fish Farming Technology* (eds H. Reinertsen, L.A. Dahle, L. Jørgensen & K. Tvinnereim), pp. 203–208. Balkema, Rotterdam.

Blyth, P.J., Purser, J.G. & Russell, J.F. (1993) Detection of feeding rhythms in sea caged Atlantic salmon using new feeder technology. In: *Fish Farming Technology* (eds H. Reinertsen, L.A. Dahle, L. Jørgensen & K. Tvinnereim), pp. 209–216. Balkema, Rotterdam.

Blyth, P.J., Purser, J.G. & Russell, J.F. (1997) Progress in fish production technology and strategies: with emphasis on feeding. *Suisanzoshoku*, **45**, 151–161.

Boujard, T. & Leatherland, J.F. (1992) Demand-feeding behaviour and diel pattern of feeding activity in *Oncorhynchus mykiss* held under different photoperiod regimes. *Journal of Fish Biology*, **40**, 535–544.

Boujard, T. & Le Gouvello, R. (1997) Voluntary feed intake and discrimination of diets containing a novel fluoroquinolone in self-fed rainbow trout. *Aquatic Living Resources*, **10**, 343–350.

Boujard, T., Moreau, Y. & Luquet, P. (1991) Entrainment of the circadian rhythm of food demand by infradian cycles of light-dark alternation in *Hoplosternum littorale* (Teleostei). *Aquatic Living Resources*, **4**, 221–225.

Boujard, T., Dugy, X., Genner, D., Gosset, C. & Grig, G. (1992) Description of a modular, low cost, eater meter for the study of feeding behavior and food-preferences in fish. *Physiology and Behavior*, **52**, 1101–1106.

Bowen, S.H. (1996) Quantitative description of diet. In: *Fisheries Techniques* (eds B.R. Murphy & D.W. Willis), pp. 513–532. American Fisheries Society, Bethesda.

Brännäs, E. & Alanärä, A. (1992) Feeding behaviour of the Arctic charr in comparison with the rainbow trout. *Aquaculture*, **105**, 53–59.

Brännäs, E. & Alanärä, A. (1993) Monitoring the feeding activity of individual fish with a demand feeding system. *Journal of Fish Biology*, **42**, 209–215.

Brännäs, E. & Alanärä, A. (1994) Effect of reward level on individual variability in demand feeding activity and growth rate in Arctic charr and rainbow trout. *Journal of Fish Biology*, **45**, 423–434.

Bromley, P.J. (1994) The role of gastric evacuation experiments in quantifying the feeding rates of predatory fish. *Reviews in Fish Biology and Fisheries*, **4**, 36–66.

Brown, G.E. & Brown, J.A. (1996) Does kin-biased territorial behaviour increase kin-biased foraging in juvenile salmonids? *Behavioral Ecology*, **7**, 24–29.

Burel, C., Robin, J. & Boujard, T. (1997) Can turbot, *Psetta maxima*, be fed with self-feeders? *Aquatic Living Resources*, **10**, 381–384.

Carter, C.G., Houlihan, D.F., Brechin, J. & McCarthy, I.D. (1993) The relationships between protein intake and protein accretion, synthesis and retention efficiency for individual grass carp, *Ctenopharyngodon idella* (Val.). *Canadian Journal of Zoology*, **71**, 392–400.

Castro, J.J. (1993) Feeding ecology of chub mackerel *Scomber japonicus* in the Canary Islands area. *South African Journal of Marine Science*, **13**, 323–328.

Chapman, D.W. & Bjornn, T.C. (1969) Distribution of salmonids in streams, with special reference to food and feeding. In: *Symposium on salmon and trout in streams* (ed. T.G. Northcote), pp. 153–177. University of British Columbia, Vancouver.

Chen, W.-M., Purser, J. & Blyth P.J. (1999) Diel feeding rhythms of greenback flounder *Rhombosolea tapirana* (Günther 1862): the role of light-dark cycles and food deprivation. *Aquaculture Research*, **30**, 529–537.

Cho, C.Y., Slinger, S.J. & Bayley, H.S. (1982) Bioenergetics of salmonid fishes: energy intake, expenditure and productivity. *Comparative Biochemistry and Physiology*, **73B**, 25–41.

Choubert, G., de la Noüe, J. & Luquet, P. (1982) Digestibility in fish: improved device for the automatic collection of faeces. *Aquaculture*, **29**, 185–189.

Christiansen, J.S. & George, S.G. (1995) Contamination of food by crude oil affects food selection and growth performance, but not appetite, in an Arctic fish, the polar cod (*Boreogadus saida*). *Polar Biology*, **15**, 277–281.

Costello, M.J. (1990) Predator feeding strategy and prey importance: a new graphical analysis. *Journal of Fish Biology*, **36**, 261–263.

Covès, D., Gasset, E., Lemarié, G. & Dutto, G. (1998) A simple way for avoiding feed wastage in European sea bass, *Dicentrarchus labrax*, under self-feeding conditions. *Aquatic Living Resources*, **11**, 395–401.

Cravedi, J.P. & Tulliez, J. (1986) Metabolism of n-alkanes and their incorporation into lipids in the rainbow trout. *Environmental Research*, **39**, 180–187.

Cuenca, E.M. & de la Higuera, M. (1994) A microcomputer-controlled demand feeder for the study of feeding behavior in fish. *Physiology and Behavior*, **6**, 1135–1136.

Damsgård, B. & Dill, L.M. (1998) Risk-taking behavior in weight-compensating coho salmon, *Oncorhynchus kisutch*. *Behavioral Ecology*, **9**, 26–32.

Damsgård, B. & Langeland, A. (1994) Effects of stocking of piscivorous brown trout, *Salmo trutta*, L., on stunted Arctic charr, *Salvelinus alpinus* (L.). *Ecology of Freshwater Fish*, **3**, 59–66.

Damsgård, B. & Ugedal, O. (1997) The influence of predation risk on habitat selection and food intake in Arctic charr, *Salvelinus alpinus* (L.). *Ecology of Freshwater Fish*, **6**, 95–101.

Damsgård, B., Arnesen, A.M., Baardvik, B.M. & Jobling, M. (1997) State-dependent feed acquisition among two strains of hatchery-reared Arctic charr. *Journal of Fish Biology*, **50**, 859–869.

Dias, J., Gomes, E.F. & Kaushik, S.J. (1997) Improvement of feed intake through supplementation with an attractant mix in European seabass fed plant-protein rich diets. *Aquatic Living Resources*, **10**, 385–389.

Dill, L.M., Ydenberg, R.C. & Fraser, A. (1981) Food abundance and territory size in juvenile coho salmon (*Oncorhynchus kisutch*). *Canadian Journal of Zoology*, **59**, 1801–1809.

Divanach, P., Kentouri, M. & Dewavrin, G. (1986) Sur le sevrage et l'évolution des performances biologiques d'alevins de daurades, *Sparus auratus*, provenant d'élevage extensif, après remplacement des nourrisseurs en continu par des distributeurs libre service. *Aquaculture*, **52**, 21–29.

Divanach, P., Kentouri, M., Charalambakis, G., Pouget, F. & Sterioti, A (1993) Comparison of growth performance of six Mediterranean fish species reared under intensive farming conditions in Crete (Greece), in raceways with the use of self feeders. In: *Production, Environment and Quality* (eds G. Barnabé & P. Kestemont), pp. 285–297. EAS, Oostend.

Dove, H. & Mayes, R.W. (1991) The use of plant wax alkanes as marker substances in studies of the nutrition of herbivores: a review. *Australian Journal of Agricultural Research*, **42**, 913–952.

dos Santos, J. & Jobling, M. (1991) Gastric emptying in cod, *Gadus morhua* L.: emptying and retention of indigestible solids. *Journal of Fish Biology*, **38**, 187–197.

dos Santos, J. & Jobling, M. (1992) A model to describe gastric evacuation in cod (*Gadus morhua* L.) fed natural prey. *ICES Journal of Marine Science*, **49**, 145–154.

dos Santos, J. & Jobling, M. (1995) Test of a food consumption model for the Atlantic cod. *ICES Journal of Marine Science*, **52**, 209–219.

Dubets, H. (1954) Feeding habits of the largemouth bass as revealed by a gastroscope. *The Progressive Fish-Culturist*, **16**, 134–136.

Edin, H. (1945) A summarised description of 'Edin's indicator method' for the determination of digestibility of feeds and feed mixtures. *Annals of the Royal Agricultural College of Sweden*, **12**, 66–71.

Eggers, D.M. (1977) The nature of prey selection by planktivorous fish. *Ecology*, **58**, 46–59.

Elliott, J.M. (1972) Rates of gastric evacuation in brown trout, *Salmo trutta* L. *Freshwater Biology*, **2**, 1–18.

Elliott, J.M. (1991) Rates of gastric evacuation of piscivorous brown trout, *Salmo trutta. Freshwater Biology*, **25**, 297–305.

Elliott, J.M. & Persson, L. (1978) The estimation of daily rates of food consumption for fish. *Journal of Animal Ecology*, **47**, 977–991.

Elvira, B., Nicola, G.G. & Almodovar, A. (1996) Pike and red swamp crayfish: a new case on predator–prey relationship between aliens in central Spain. *Journal of Fish Biology*, **48**, 437–446.

Fänge, R. & Grove, D. (1979) Digestion. In: *Fish Physiology* (eds W.S. Hoar, D.J Randall & J.R. Brett), Vol. VIII, pp.161–260. Academic Press, London.

Fast, A.W., Qin, T. & Szyper, J.P. (1997) A new method for assessing fish feeding rhythms using demand feeders and automated data acquisition. *Aquacultural Engineering*, **16**, 213–220.

Fauré, A. (1983) Interêt et pratique de l'alimentation libre-service en salmoniculture intensive. *La Pisciculture Française*, **74**, 15–26.

Foster, J.R. (1977) Pulsed gastric lavage: an efficient method of removing the stomach contents of live fish. *The Progressive Fish-Culturist*, **39**, 166–169.

Foster, M., Petrell, R., Ito, M.R. & Ward, R. (1993) Detection and counting of uneaten food pellets in a sea cage using image analysis. In: *Techniques for Modern Aquaculture* (ed. J.-K. Wang), pp. 393–402. American Society of Agricultural Engineers, St. Joseph.

Frost, W.E. (1977) The food of charr *Salvelinus willughbii* (Gunther) in Windermere. *Journal of Fish Biology*, **11**, 531–547.

Gélineau, A., Corraze, G. & Boujard, T. (1998) Effects of restricted ration, time restricted access and reward level on voluntary food intake, growth and growth heterogeneity of rainbow trout (*Oncorhynchus mykiss*) fed on demand with self-feeders. *Aquaculture*, **167**, 247–258.

George, E.L. & Hadley, W.F. (1979) Food and habitat partitioning between rock bass (*Amploplites rupestris*) and smallmouth bass (*Micropterus dolomieui*) young of the year. *Transactions of the American Fisheries Society*, **108**, 253–261.

Gotceitas, V. & Godin, J.-G.J. (1992) Effects of location of food delivery and social status on foraging-site selection by juvenile Atlantic salmon. *Environmental Biology of Fishes*, **35**, 291–300.

Grace, N.D. & Body, D.R. (1981) The possible use of long chain (C_{19}–C_{32}) fatty acids in herbage as an indigestible faecal marker. *Journal of Agricultural Science*, **97**, 743–745.

Grand, T. (1997) Foraging site selection by juvenile coho salmon: ideal free distribution of unequal competitors. *Animal Behaviour*, **53**, 185–196.

Grant, J.W.A. (1990) Aggressiveness and the foraging behaviour of the young-of-the-year brook charr (*Salvelinus fontinalis*). *Canadian Journal of Fisheries and Aquatic Sciences*, **47**, 915–920.

Grant, J.W.A. & Guha, R.T. (1993) Spatial clumping of food increases its monopolisation and defense by convict cichlids, *Cichlasoma nigrofasciatum. Behavioral Ecology*, **4**, 293–296.

Grant, J.W.A. & Kramer, D.L. (1992) Temporal clumping of food arrival reduces its monopolization and defence by zebrafish, *Brachydanio rerio. Animal Behaviour*, **44**, 101–110.

Grønvik, S. & Klemetsen, A. (1987) Marine food and diet overlap of co-occurring Arctic charr *Salvelinus alpinus* (L.), brown trout *Salmo trutta* L. and Atlantic salmon *S. salar* L. off Senja, N. Norway. *Polar Biology*, **7**, 173–177.

Gudmundsson, O. & Halldorsdottir, K. (1993) The use of *n*-alkanes as markers for determination of intake and digestibility of fish feed. *Journal of Applied Ichthyology* **11**, 354–358.

Hajen, W.E., Beames, R.M., Higgs, D.A. & Dosanjh, B.S. (1993) Digestibility of various feedstuffs by post-juvenile chinook salmon (*Oncorhynchus tshawytscha*) in sea water. 1. Validation of technique. *Aquaculture*, **112**, 321–332.

Hargreaves, J.A., Rakocy, J.E. & Nair, A. (1988) An evaluation of fixed and demand feeding regimes for cage culture of *Oreochromis aureus*. In: *The Second International Symposium on Tilapia in Aquaculture*. (eds R.S.V. Pullin, T. Bhukaswan, K. Tonguthai & J.L. Maclean), pp. 335–339. ICLARM, Manila.

Harris, L.E. (1970) Measuring intake and digestibility of the range animal's diet. *Nutrition Research Techniques for Domestic and Wild Animals*, **1**, 5501–5507.

Hatlen, B., Arnesen, A.M., Jobling, M., Siikavoupio, S. & Bjerkeng, B. (1997) Carotenoid pigmentation in relation to feed intake, growth and social interactions in Arctic charr (*Salvelinus alpinus* L.) from two anadromous strains. *Aquaculture Nutrition*, **3**, 189–199.

Heggenes, J., Krog, O.M.W., Lindas, O.R., Dokk, J.G. & Bremnes, T. (1993) Homeostatic behavioral responses in a changing environment – brown trout (*Salmo trutta*) become nocturnal during winter. *Journal of Animal Ecology*, **62**, 295–308.

Helland, S.J., Grisdale-Helland, B. & Nerland, S. (1996) A simple method for the measurement of daily feed intake of groups of fish in tanks. *Aquaculture*, **139**, 157–163.

Hellawell, J.M. & Abel, R. (1971) Rapid volumetric method for the analysis of the food of fishes. *Journal of Fish Biology*, **3**, 29–37.

Hidalgo, F., Kentouri, M. & Divanach, P. (1988) Sur l'utilisation du self feeder comme outil d'épreuve nutritionnelle du loup, *Dicentrarchus labrax* – Résultats préliminaires avec la méthionine. *Aquaculture*, **68**, 177–190.

Higgins, P.J. & Talbot, C. (1985) Growth and feeding in juvenile Atlantic salmon (*Salmo salar* L.). In: *Nutrition and Feeding in Fish* (eds C.B. Cowey, A.M. Mackie & J.G. Bell), pp. 243–263. Academic Press, London.

Hindar, K. & Jonsson, B. (1982) Habitat and food segregation of dwarf and normal Arctic charr (*Salvelinus alpinus*) from Vangsvatnet Lake, Western Norway. *Canadian Journal of Fisheries and Aquatic Sciences*, **39**, 1030–1045.

Hughes, N.F. & Kelly, L.H. (1996) New technique for 3-D video tracking of fish swimming movements in still and flowing water. *Canadian Journal of Fisheries and Aquatic Sciences*, **53**, 2473–2483.

Huse, I. & Skiftesvik, A.B. (1990) A PC-aided video based system for behavioural observation of fish larvae and small aquatic invertebrates. *Aquacultural Engineering*, **9**, 131–142.

Hynes, H.B.N. (1950) The food of the freshwater sticklebacks (*Gasterosteus aculeatus* and *Pygosteus pungitius*) with a review of methods used in the studies of the food of fishes. *Journal of Animal Ecology*, **19**, 36–58.

Hyslop, E.J. (1980) Stomach contents analysis – a review of methods and their application. *Journal of Fish Biology*, **17**, 411–429.

Ivlev, V.S. (1961) *Experimental Ecology of the Feeding of Fishes*. Yale University Press, New Haven.

Jacobs, J. (1974) Quantitative measurement of food selection. A modification of the forage ratio and Ivlev's Electivity Index. *Oecologia*, **14**, 413–417.

Jenkins, T.M. (1969) Social structure, position choice and microdistribution of two trout species (*Salmo trutta* and *Salmo gairdneri*) resident in mountain streams. *Animal Behaviour Monographs*, **2**, 56–123.

Jobling, M. (1986) Mythical models of gastric emptying and implications for food consumption. *Environmental Biology of Fishes*, **16**, 35–50.

Jobling, M. & Baardvik, B.M. (1994) The influence of environmental manipulations on inter- and intra-individual variation in food acquisition and growth performance of Arctic charr, *Salvelinus alpinus*. *Journal of Fish Biology*, **44**, 1069–1087.

Jobling, M. & Breiby, A. (1986) The use and abuse of fish otoliths in studies of feeding habits of marine piscivores. *Sarsia*, **71**, 265–274.

Jobling, M. & Koskela, J. (1996) Interindividual variations in feeding and growth in rainbow trout during restricted feeding and in a subsequent period of compensatory growth. *Journal of Fish Biology*, **49**, 658–667.

Jobling, M., Gwyther, D. & Grove, D.J. (1977) Some effects of temperature, meal size and body weight on gastric evacuation time in the dab, *Limanda limanda* (L.). *Journal of Fish Biology*, **10**, 291–298.

Jobling, M., Jørgensen, E.H. & Christiansen, J.S. (1990) Feeding behaviour and food intake of Arctic charr, *Salvelinus alpinus* L., studied by X-radiography. In: *The Current Status of Fish Nutrition in Aquaculture* (eds M. Takeda & T. Watanabe), pp. 461–469. Tokyo University of Fisheries, Tokyo.

Jobling, M., Christiansen, J.S., Jørgensen, E.H. & Arnesen, A.M. (1993) The application of X-radiography in feeding and growth studies with fish: a summary of experiments conducted on Arctic charr. *Reviews in Fisheries Science*, **1**, 223–237.

Jobling, M., Arnesen, A.M., Baardvik, B.M., Christiansen, J.S. & Jørgensen, E.H. (1995) Monitoring feeding behaviour and food intake: methods and applications. *Aquaculture Nutrition*, **1**, 131–143.

Jobling, M., Koskela, J. & Pirhonen, J. (1998) Feeding time, feed intake and growth of Baltic salmon, *Salmo salar* and brown trout, *Salmo trutta* reared in monoculture and duoculture at constant low temperature. *Aquaculture*, **163**, 73–94.

Johnston, W.L., Atkinson, J.L. & Glanville, N.T. (1994) A technique using sequential feedings of different coloured foods to determine food intake by individual rainbow trout, *Oncorhynchus mykiss*: effect of feeding level. *Aquaculture*, **120**, 123–133.

Jørgensen, E. & Jobling, M. (1988) Use of radiographic methods in feeding studies: a cautionary note. *Journal of Fish Biology*, **32**, 487–488.

Juell, J.E. (1991) Hydroacoustic detection of food waste – A method to estimate maximum food intake of fish populations in sea cages. *Aquacultural Engineering*, **10**, 207–217.

Juell, J.E., Furevik, D.M. & Bjordal, Å. (1993) Demand feeding in salmon farming by hydro acoustic food detection. *Aquacultural Engineering*, **12**, 155–167.

Juell, J.E., Fernö, A., Furevik, D. & Huse, I. (1994) Influence of hunger level and food availability on the spatial distribution of Atlantic salmon, *Salmo salar* L., in sea cages. *Aquaculture and Fisheries Management*, **25**, 439–451.

Kadri, S., Metcalfe, N.B., Huntingford, F. & Thorpe, J.E. (1991) Daily feeding rhythms in Atlantic salmon in sea cages. *Aquaculture*, **92**, 219–224.

Kadri, S., Huntingford, F., Metcalfe, N.B. & Thorpe, J.E. (1996) Social interactions and the distribution of food among one sea-winter Atlantic salmon (*Salmo salar*) in a sea-cage. *Aquaculture*, **139**, 1–10.

Kalleberg, H. (1958) Observations in a stream tank of territoriality and competition in juvenile salmon and trout (*Salmo salar* L. and *S. trutta* L.). *Institute of Freshwater Research Drottningholm, Reports*, **39**, 55–98.

Keenleyside, M.H.A. & Yamamoto, F.T. (1962) Territorial behaviour of juvenile Atlantic salmon (*Salmo salar* L.). *Behaviour*, **19**, 139–169.

Kentouri, M., Divanach, P., Batique, O. & Anthouard, M. (1986) Rôle des individus conditionnés dans l'initiation à l'auto-nourrissage et dans l'adaptation à la captivité du loup, *Dicentrarchus labrax*, 0+ sauvage, en période hivernale. *Aquaculture*, **52**, 117–124.

Kentouri, M., Anthouard, M. Divanach, P. & Paspatis M. (1992) Les modalités d'adaptation comportementale de populations de bars (Serranidae: *Dicentrarchus labrax*), soumises à un nourrissage auto-contrôlé. *Ichtyophysiologica Acta*, **15**, 29–42.

Kentouri, M., León, L., Tort, L. & Divanach, P. (1994) Experimental methodology in aquaculture: modification of the feeding rate of the gilthead sea bream *Sparus aurata* at a self-feeder after weighing. *Aquaculture*, **119**, 191–200.

Kentouri, M., Divanach, P., Geurden, I. & Anthouard, M. (1995) Evidence of adaptative behaviour in gilthead sea bream (*Sparus aurata*, L.) in relation to diet composition, in a self-feeding condition. *Ichtyphysiologica Acta*, **18**, 125–143.

Kimball, D.C. & Helm, W.T. (1971) A method of estimating fish stomach capacity. *Transactions of the American Fisheries Society*, **100**, 572–575.

Kotb, A.R. & Luckey, T.D. (1972) Markers in nutrition. *Nutrition Abstracts and Reviews*, **42**, 813–845.

Koskela, J., Pirhonen, J. & Virtanen, E. (1993) Effect of attractants on feed choice of rainbow trout, *Oncorhynchus mykiss*. In: *Fish Nutrition in Practice* (eds S.J. Kaushik & P. Luquet), pp. 419–427. INRA, Paris.

Koskela, J., Jobling, M. & Pirhonen, J. (1997) Influence of the length of the daily feeding period on feed intake and growth of whitefish, *Coregonus lavaretus*. *Aquaculture*, **156**, 35–44.

L'Abée-Lund, J.H., Aass, P. & Sægrov, H. (1996) Prey orientation in piscivorous brown trout. *Journal of Fish Biology*, **48**, 871–877.

Landless, P.J. (1976) Demand feeding behaviour of rainbow trout. *Aquaculture*, **7**, 11–25.

Langeland, A., L'Abée-Lund, J.H., Jonsson, B. & Jonsson, N. (1991) Resource partitioning and niche shift in Arctic charr *Salvelinus alpinus* and brown trout *Salmo trutta*. *Journal of Animal Ecology*, **60**, 895–912.

Lemarié G., Covès, D., Dutto, G. & Gasset, E. (1996) Chronic toxicity of ammonia for European seabass (*Dicentrarchus labrax*) juveniles. In: *Applied Environmental Physiology of Fishes* (eds C. Swanson, P. Young & D. MacKinlay), pp. 65–76. American Fisheries Society, Bethesda.

Levings, C.D., Hvidsten, N.A. & Johnsen, B.Ø. (1994) Feeding of Atlantic salmon (*Salmo salar* L.) postsmolts in a fjord in central Norway. *Canadian Journal of Zoology*, **72**, 834–839.

Madrid, J.A., Azzaydi, M. Zamora, S. & Sánchez-Vásquez, F.J. (1997) Continuous recording of uneaten food pellets and demand-feeding activity: a new approach to studying feeding rhythm in fish. *Physiology and Behavior*, **62**, 689–695.

Magnuson, J.J. (1962) An analysis of aggressive behavior, growth, and competition for food and space in medaka (*Oryzias latipes* (Pisces, Cyprinodontidae)). *Canadian Journal of Zoology*, **40**, 313–363.

Malmquist, H.J. (1992) Phenotype-specific feeding behaviour of two Arctic charr *Salvelinus alpinus* morphs. *Oecologia*, **92**, 354–361.

McCarthy, I.D., Carter, C.G. & Houlihan, D.F. (1992) The effect of feeding hierarchy on individual variability in daily feeding of rainbow trout, *Oncorhynchus mykiss* (Walbaum). *Journal of Fish Biology*, **41**, 257–263.

McCarthy, I.D., Houlihan, D.F., Carter, C.G. & Moutou, K.A. (1993) Variation in individual food consumption rates of fish and its implications for the study of fish nutrition and physiology. *Proceedings of the Nutrition Society*, **52**, 411–420.

McCarthy, I.D., Houlihan, D.F. & Carter, C.G. (1994) Individual variation in protein turnover and growth efficiency in rainbow trout, *Oncorhynchus mykiss* (Walbaum). *Proceedings of the Royal Society of London B*, **257**, 141–147.

Mearns, K.J., Ellingsen, O.F., Døving, K.B. & Helmer, S. (1987) Feeding behaviour in adult rainbow trout and Atlantic salmon parr, elicited by chemical fractions and mixtures of compounds identified in shrimp extract. *Aquaculture*, **64**, 47–63.

Meehan, W.R. & Miller, R.A. (1978) Stomach flushing: effectiveness and influence on survival and condition of juvenile salmonids. *Journal of the Fisheries Research Board of Canada*, **35**, 1359–1363.

Metcalfe, N.B., Huntingford, F.A. & Thorpe, J.E. (1987) The influence of predation risk on the feeding motivation and foraging strategy of juvenile Atlantic salmon. *Animal Behaviour*, **35**, 901–911.

Milinski, M. (1993) Predation risk and feeding behaviour. In: *Behaviour of Teleost Fishes* (ed. T. Pitcher), pp. 285–305. Chapman & Hall, London.

Mittelbach, G.G. (1981) Foraging efficiency and body size: a study of optimal diet and habitat use by bluegills. *Ecology*, **62**, 1370–1386.

Mohan, M.V. & Sankaran, T.M. (1988) Two new indices for stomach content analysis of fishes. *Journal of Fish Biology*, **33**, 289–292.

Molnar, G.Y., Tamassy, E. & Tölg, I. (1967) The gastric digestion of living predatory fish. In: *The Biological Basis of Freshwater Fish Production* (ed. S.D. Gerking) pp. 135–149. Blackwell Scientific, Oxford.

Morris, J.E., D'Abramo, L.R. & Muncy, R.C. (1990) An inexpensive marking technique to assess ingestion of formulated feeds by larval fish. *The Progressive Fish-Culturist*, **52**, 120–121.

Nicola, G.G., Almodovar, A. & Elvira, B. (1996) Diet of introduced largemouth bass, *Micropterus salmoides*, in the natural park of the Ruidera Lakes, central Spain. *Polskie Archiwum Hydrobiologii*, **43**, 179–184.

Noakes, D.L.G. (1980) Social behaviour in young charr. In: *Charrs, Salmonid Fishes of the Genus Salvelinus* (ed. E.K. Balon), pp. 683–701. Dr. W. Junk, The Hague.

Noldus Information Technologies (1997) *The Observer, Support Package for Video Analysis. Version 4.0 for Windows Edition.* Noldus, Wageningen.

Øverli, Ø., Winberg, S., Damsgård, B. & Jobling, M. (1998) Food intake and spontaneous swimming activity in Arctic charr, *Salvelinus alpinus*: Role of brain serotonergic activity and social interactions. *Canadian Journal of Zoology*, **76**, 1366–1370.

Parker, R.R. (1963) Effects of formalin on length and weight of fishes. *Journal of the Fisheries Research Board of Canada*, **20**, 144–155.

Paspatis, M. & Boujard, T. (1996) A comparative study of automatic feeding and self-feeding in juvenile Atlantic salmon (*Salmo salar*) fed diets of different energy. *Aquaculture*, **145**, 245–257.

Paszkowski, C.A. & Olla, B.L. (1985) Foraging behaviour of hatchery-produced coho salmon (*Oncorhynchus kisutch*) smolts on live prey. *Canadian Journal of Fisheries and Aquatic Sciences*, **42**, 1915–1921.

Pereira, P. & Oliveira, R. F. (1994) A simple method using a single video camera to determine the 3-dimensional position of a fish. *Behavior Research Methods Instruments & Computers*, **26**, 443–446.

Pinkas, L., Oliphant, M.S. & Iverson, I.L.K. (1971) Food habits of albacore, bluefin tuna and bonito in Californian Waters. *Californian Fish and Game Bulletin*, **152**, 1–105.

Prairie, Y.T. (1996) Evaluating the predictive power of regression models. *Canadian Journal of Fisheries and Aquatic Sciences*, **53**, 490–492.

Prince, E.D. (1975) Pinnixid crabs in the diet of young-of-the-year Copper Rockfish (*Sebastes caurinus*). *Transactions of the American Fisheries Society*, **104**, 539–540.

Pyke, G.H., Pulliam, H.R. & Charnov, E.L. (1977) Optimal foraging: a selective review of theory and tests. *The Quarterly Review of Biology*, **52**, 137–154.

Ruttner-Kolisko, A. (1977) Suggestions for biomass calculation of plankton rotifers. *Archiv für Hydrobiologie*, **8**, 71–76.

Ryer, C.H. & Olla, B.L. (1995) The influence of food distribution upon the development of aggressive and competitive behaviour in juvenile chum salmon, *Oncorhynchus keta*. *Journal of Fish Biology*, **46**, 264–272.

Ryer, C.H. & Olla, B.L. (1996) Social behaviour of juvenile chum salmon, *Oncorhynchus keta*, under risk of predation: the influence of food distribution. *Environmental Biology of Fishes*, **45**, 75–83.

Sæther, B.-S. & Jobling, M. (1997) Gastrointestinal evacuation of inert particles by turbot, *Psetta maxima*: evaluation of the X-radiographic method for use in feed intake studies. *Aquatic Living Resources*, **10**, 359–364.

Sæther, B.-S., Johnsen, H.K. & Jobling, M. (1996) Seasonal changes in food consumption and growth of Arctic charr exposed to either simulated natural or a 12:12 LD photoperiod at constant water temperature. *Journal of Fish Biology*, **48**, 1113–1122.

Sæther, B.-S., Christiansen, J.S. & Jobling, M. (1999) Gastrointestinal evacuation of particulate matter in polar cod, *Boreogadus saida*. *Marine Ecology Progress Series*, **188**, 201–205.

Sánchez-Vásquez, F.J., Martínez, M., Zamora, S. & Madrid, J.A. (1994) Design and performance of an accurate demand feeder for the study of feeding behaviour in sea bass, *Dicentrarchus labrax* L. *Physiology and Behavior*, **56**, 789–794.

Sánchez-Vásquez, F.J., Martínez, M., Zamora, S. & Madrid, J.A. (1995a) Light-dark and food restriction cycles in seabass: effect of conflicting zeitgebers on demand-feeding rhythms. *Physiology and Behavior*, **58**, 705–714.

Sánchez-Vásquez, F.J., Madrid, J.A. & Zamora, S. (1995b) Circadian rhythms of feeding activity in sea bass, *Dicentrarchus labrax* L.: dual phasing capacity of diel demand-feeding pattern. *Journal of Biological Rhythms*, **10**, 256–266.

Sánchez-Vásquez, F.J., Yamamoto, T., Akiyama, T., Madrid, J.A. & Tabata, M. (1998) Selection of macronutrients by goldfish operating self-feeders. *Physiology and Behavior*, **65**, 211–218.

Skiftesvik, A.B., Bergh, Ø. & Opstad, I. (1994) Activity and swimming speed at time of first feeding of halibut (*Hippoglossus hippoglossus*) larvae. *Journal of Fish Biology*, **45**, 349–351.

Smagula, C.M. & Adelman, I.R. (1982) Day-to-day variation in food consumption by largemouth bass. *Transactions of the American Fisheries Society*, **111**, 543–548.

Smith, I.P., Metcalfe, N.B., Huntingford, F.A. & Kadri, S. (1993) Daily and seasonal patterns in the feeding behaviour of Atlantic salmon (*Salmo salar* L.) in a sea cage. *Aquaculture*, **117**, 165–178.

Smith, I.P., Metcalfe, N.B. & Huntingford, F.A. (1995) The effects of food pellet dimensions on feeding response by Atlantic salmon (*Salmo salar* L.) in a marine net pen. *Aquaculture*, **130**, 167–175.

Staples, D.J. (1975) Production biology of an upland bully *Philypnodon breviceps* Stokell in a small New Zealand lake. 1. Life history, food, feeding and activity rhythms. *Journal of Fish Biology*, **7**, 1–24.

Stradmeyer, L. (1989) A behavioural method to test feeding responses of fish to pelleted diets. *Aquaculture*, **79**, 303–310.

Stradmeyer, L. & Thorpe, J.E. (1987) Feeding behaviour of wild Atlantic salmon, *Salmo salar* L., parr in mid- to late summer in a Scottish river. *Aquaculture and Fisheries Management*, **18**, 33–49.

Summerfelt, S.T., Holland, K.H., Hankins, J.A. & Durant, M.D. (1995) A hydroacoustic waste feed controller for tank systems. *Water Science and Technology*, **31**, 123–129.

Tackett, D.L., Carter, R.R. & Allen, K.O. (1988) Daily variation in feed consumption by channel catfish. *The Progressive Fish-Culturist*, **50**, 107–110.

Talbot, C. (1985) Laboratory methods in fish feeding and nutritional studies. In: *Fish Energetics – New Perspectives* (eds P. Tytler & P. Calow), pp.125–154. Croom-Helm, London.

Talbot, C. & Higgins, P.J. (1983) A radiographic method for feeding studies on fish using metallic iron powder as a marker. *Journal of Fish Biology*, **23**, 211–220.

Thorpe, J.E., Talbot, C., Miles, M.S., Rawlings, C. & Keay, D.S. (1990) Food consumption in 24 hours by Atlantic salmon (*Salmo salar* L.) in a sea cage. *Aquaculture*, **90**, 41–47.

Toften, H. & Jobling, M. (1996) Development of spinal deformities in Atlantic salmon and Arctic charr fed diets supplemented with oxytetracycline. *Journal of Fish Biology*, **49**, 668–677.

Toften, H., Jørgensen, E.H. & Jobling, M. (1995) The study of feeding preferences using radiography: Oxytetracycline as a feeding deterrent and squid extract as a stimulant in diets for Atlantic salmon. *Aquaculture Nutrition*, **1**, 145–149.

Townsend, C.R. & Winfield, I.J. (1985) The application of optimal foraging theory to feeding behaviour in fish. In: *Fish Energetics – New Perspectives* (eds P. Tytler & P. Calow), pp. 67–98. Croom-Helm, London.

Tuene, S. & Nortvedt, R. (1995) Feed intake, growth and feed conversion efficiency of Atlantic halibut, *Hippoglossus hippoglossus* (L.). *Aquaculture Nutrition*, **1**, 27–35.

Tveiten, H., Johnsen, H.K. & Jobling, M. (1996) Influence of maturity status on the annual cycles of feeding and growth in Arctic charr reared at constant temperature. *Journal of Fish Biology*, **48**, 910–924.

Twarowska, J.G., Westerman, P.W. & Losordo, T.M. (1997) Water treatment and waste characterization evaluation of an intensive recirculating fish production system. *Aquacultural Engineering*, **16**, 133–147.

Twomey, H. & Giller, P.S. (1990) Stomach flushing and individual Panjet tattooing of salmonids: an evaluation of the long-term effects on two wild populations. *Aquaculture and Fisheries Management*, **21**, 137–142.

Unprasert, P., Taylor, J.B. & Robinette, H.R. (1999) Competitive feeding interactions between small and large channel catfish cultured in mixed-size populations. *North American Journal of Aquaculture*, **61**, 336–339.

Vesin, J.-P., Leggett, W.C. & Able, K.W. (1981) Feeding ecology of capelin (*Mallotus villosus*) in the estuary and western Gulf of St. Lawrence and its multispecies implications. *Canadian Journal of Fisheries and Aquatic Sciences*, **38**, 257–267.

Wagner, E.J., Ross, D.A., Routledge, D., Scheer, B. & Bosakowski, T. (1995) Performance and behaviour of cutthroat trout (*Oncorhynchus clarki*) reared in covered raceways or demand fed. *Aquaculture*, **136**, 131–140.

Walsh, M., Huguenin, J.E. & Ayers, K.T. (1987) A fish feed consumption monitor. *The Progressive Fish-Culturist*, **49**, 133–136.

Wankowski, J.W.J. (1979) Morphological limitations, prey size selection, and growth response of juvenile Atlantic salmon, *Salmo salar*. *Journal of Fish Biology*, **14**, 89–100.

Wankowski, J.W.J. & Thorpe, J.E. (1979) Spatial distribution and feeding in Atlantic salmon, *Salmo salar* L. juveniles. *Journal of Fish Biology*, **14**, 239–247.

Winberg, S., Carter, C.G., McCarthy, I.D., He, Z.-Y., Nilsson, G.E. & Houlihan, D.F. (1993a) Feeding rank and brain serotonergic activity in rainbow trout, *Oncorhynchus mykiss* (Walbaum). *Journal of Experimental Biology*, **179**, 197–211.

Winberg, S., Nilsson, G.E., Spruijt, B.M. & Höglund, U. (1993b) Spontaneous locomotor activity in Arctic charr measured by a computerized imaging technique: role of brain serotonergic activity. *Journal of Experimental Biology*, **179**, 213–232.

Windell, J.T. (1968) Food analysis and rate of digestion. In: *Methods for Assessment of Fish Production in Fresh Waters* (ed. W.E. Ricker), pp. 197–203. Blackwell Scientific, Oxford.

Windell, J.T. & Bowen, S.H. (1978) Methods for study of fish diets based on analysis of stomach contents. In: *Methods for Assessment of Fish Production in Fresh Waters* (ed. T. Bagenal), pp. 219–226. Blackwell Scientific, Oxford.

Windell, J.T., Foltz, J.W. & Sarokon, J.A. (1978) Methods of faecal collection and nutrient leaching in digestibility studies. *The Progressive Fish-Culturist*, **40**, 51–55.

Chapter 4
Experimental Design in Feeding Experiments

Kari Ruohonen, Juhani Kettunen and Jonathan King

4.1 Introduction

Experiments are frequently used to study feed intake in fish, but whether these experiments are beneficial is open to question. The pragmatic answer is that experiments have value because they reduce the amount of uncertainty. To an academic researcher, lowered uncertainty means increased reliability of scientific results, whereas to the aquaculture industry it means lower risks of errors in estimation of economic returns. Consequently, the fundamental target of experimental research is to lower uncertainty: this calls for high-quality data, which can only be obtained from correctly designed experiments. Therefore, careful design must precede the implementation of experimental studies.

There are several ways to approach a design problem. The main differences between them from a statistical point of view relate to the methods that will be used to process the data. Thus, it is important to decide on the statistical methods in advance. Time series analysis, for example, requires a different form of data acquisition from regression analysis or analysis of variance (ANOVA). Generally, the question posed will define the most powerful statistical approach, and thereafter, the chosen statistical tool will determine the best experimental design. The sequence of events, from defining the biological problem through to data analysis, is shown in Fig. 4.1. One should be aware of short cuts that endanger effective experimental design. The first of these is to base the experiment on intuition; this rarely results in effective data collection. The second, and probably most common, short cut occurs when experimental design is based solely on the biology of the system. This often results in the correct choice of variables for study, but rarely leads to an optimal allocation of resources or powerful statistical data. To overcome these shortcomings, it is important that *the biological problem is transformed into a statistical one.* Having done this, the design aims are to achieve an experimental set-up that will maximise statistical information with least cost.

In this chapter the theory of ANOVA will be used to illustrate the design process. We have chosen this family of statistical tests because most fish biologists use some sort of ANOVA to analyse their data. It therefore provides a familiar environment within which to introduce the concepts of experimental design. We will use an imaginary problem as an example throughout the chapter to examine various aspects of statistical methods and to illustrate how the effects of practical limitations, such as limited resources, can be minimised by careful choice of experimental conditions.

Fig. 4.1 Experimental research process as a sequence of critical tasks. Processing contains both biological and statistical phases and they should be differentiated for successful outcome.

The imaginary problem is as follows: trials on plaice, *Pleuronectes platessa*, have shown that feeds in which soybean meal replaces fish meal result in poor food consumption and growth. Therefore, a study of feed additives is initiated to find out whether the consumption of the feeds containing soybean can be increased. We assume that extracts from three natural prey organisms are available; these are added to the soybean feed at realistic concentrations, and there is also a control – to which nothing has been added. We will go through the various steps in the design of this simple experiment to familiarise the reader with the concepts necessary for the more complicated design problems.

Using this experiment as a basis, various constraints and expansions will be introduced. First, we assume that there are too few tanks to allow testing of all feeds simultaneously. The carrying out of series of similar experiments sequentially may solve this problem. Second, the effects of additives are studied, but we are now also interested in studying their effects on different stocks of plaice. Next, we modify the data by making available individual measurements of feed intake, and finally we examine ways of dealing with some more complicated situations. It is not possible to present these problems in the same degree of detail as the simplest case, and the reader may need to consult statistical textbooks to refresh the basics of the various forms of ANOVA.

4.2 How does one design an experiment?

In the first case we will discuss, there are no restrictions of resources, leading to a type of experiment that is commonly called a *completely randomised design*.

Before we proceed to details, we should look at the general procedure of designing an experiment; what steps are involved? We can identify at least five important steps:

(1)   The *biological* hypothesis for the problem should be clearly identified, so that *statistical* hypotheses – the null and the alternative hypotheses – can be clearly stated.

(2)   All prior information regarding the problem and the experiment, and especially the variability in our experimental population, should be evaluated.

(3)   It helps if the experiment can be outlined either graphically or in the form of a data table.

(4)   We should build a *statistical model* for the experiment and summarise the information by formulating an analysis of variance table, or the essential parts of it, corresponding to the model.

(5)   Because of the stochastic nature of our environment we try to evaluate the *probabilities* and *uncertainties* connected to the outcome of our experiment, especially to any statistical hypothesis.

These five steps suggest that the statistical design of experiments is a *decision-making* problem, as summarised in Fig. 4.2.

How does one set about determining which statistical test is appropriate? The first thing to look at is the number of factors (i.e. independent variables) and levels (the number of different values for each factor) in the study. It is usually worth trying to study several factors simultaneously, because possible interactions between factors can then be detected. The OVAT (One Variable At a Time) (Box *et al.* 1978) approach can easily lead to false interpretations and conclusions if interactions between factors exist. It is important to consider care-

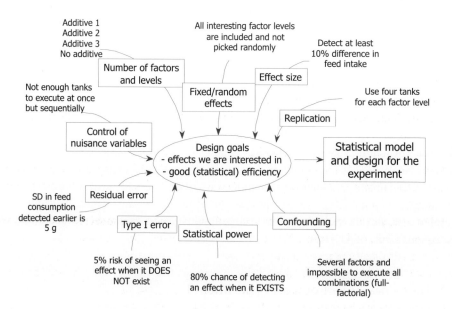

Fig. 4.2   Decisions required for designing an experiment. Major statistical questions for successful experimental design are given. These relate to identification of statistical hypotheses, evaluation of probabilities and uncertainties and acknowledging various constraints affecting the design. See text for details.

fully how many factor levels to include, because increasing the number of levels decreases the statistical efficiency of the design. The additive experiment has one factor with four levels: the levels are the feed types (three different additives and a control).

Having established the number of factors and levels, one can further delimit the selection of statistical design by considering how the levels of each factor are chosen. Commonly the levels are fixed by deliberate choice. Accordingly, if the experiment were to be repeated, the *same* factor levels would be chosen. This approach leads to a *fixed effect design*. However, sometimes factor levels may be taken randomly from a large set of possible levels. This is known as a *random effect design*. In this case, if the experiment were repeated the factor levels would not be the same as at first, but a new random set would be selected from all possible factor levels (see Winer *et al.* 1991; Lindman 1992 for further discussion). The most important difference between the two designs is that the outcome from a fixed effect experiment cannot be generalised outside the factors and levels that were included. On the other hand, the random effect design allows generalisation to the range of factors and levels from which the selection was made. A further category of experiments involving multiple factors contains both fixed and random effects: these result in statistical models called *mixed models*. The choice of the type of design depends on the biological problem and on the need for generalisations. However, researchers often attempt to choose fixed effect models because they are statistically more efficient and they lead to lower uncertainties than other model types.

In the imaginary feed additive example, we have a *fixed effect model*. The type of model we have is of central importance because the expected values for mean squares of the factors are different in fixed, random and mixed models. This, in turn, determines which of the mean squares will act as the denominator in the F-test for the statistical hypotheses (see later).

Having decided which model of ANOVA we must use, there is a series of further decisions to be made (Fig. 4.2). The first is to decide upon the smallest difference to be detected reliably: this is known as the *effect size* (d). There are various ways of expressing effect size, e.g. relative to the mean or relative to the standard deviation (e.g. Cohen 1988; Kirk 1995), but we define it here simply as the absolute difference between the largest and smallest treatment mean. The choice of an effect size for a particular experiment depends on the objectives. From a statistical point of view infinitesimal differences in treatment means could be detected, but whether these are biologically meaningful may be in doubt. Let us assume in our example that earlier findings suggest that growth of plaice fed soybean meal would match that of fish fed fish meal if feed intake could be increased by 10%. Assuming an intake of 100 g over a given time, we would wish to detect a difference between treatments of at least 10 g. Smaller differences in intake do not fit the objectives of the study. The differences to be detected are usually taken as being between *any two* of the treatments, not specific pairs. We will return to this later.

Having determined the effect size that we require the next step is to set the number of replicate tanks for each factor level. Replication is *always required* because otherwise we cannot estimate the residual error for our statistical model. Usually, the efficiency of the experiment can be manipulated by changing the number of replicates. In our example, we start by using four replicate tanks for each treatment. Whether the number is sufficient is something we will need to evaluate at a later stage of the design process.

For completely randomised designs, there is no need to control nuisance variables. Later, we will study a situation where nuisance variables need to be accounted for. Further, there

Table 4.1   Decision table and possible errors for hypothesis testing. H_0 refers to null hypothesis, and H_a to alternative hypothesis.

	State of nature	
Conclusion	H_0	H_a
H_0	Correct	Type II error
H_a	Type I error	Correct

is no need to consider the problem of confounding effects, as this is only a problem if we want to undertake an experiment with several factors and we have insufficient resources to consider all simultaneously. Confounding effects are discussed further in a fractional factorial example.

The three remaining components shown in Fig. 4.2 deal with various types of errors inherent in our experiment. The first of these, *residual error*, is the inherent variability we expect to see in the response variable, e.g. feed intake; the unexplained variability that is not due to a treatment effect. This corresponds to the population variance of the particular response variable, and may be expressed as the standard deviation (SD) or coefficient of variation (CV). Let us assume that previous experiments with plaice have shown that the SD for feed intake among tanks is 5 g.

The second error, *type I error* (usually abbreviated as α), refers to uncertainty in decision-making, i.e. which of the statistical hypotheses – the null or the alternative hypothesis – is more likely to be true in the light of the experimental data. Specifically, it refers to the probability of rejecting the null hypothesis when it is true (Table 4.1). Finally, there is *type II error* (β) that refers to the probability of accepting the null hypothesis when it is false. In experimental design it is more common to refer to $1 - \beta$, the *power of the test*, because this gives the probability of rejecting the null hypothesis when it is false. A value of 0.05 is usually chosen for the type I error, and 0.80 for the power. The choice of values is subject to change from experiment to experiment, and values should be chosen according to the objectives of the experiment.

4.3 The structural model equation

The statistical model is formalised in the structural model equation. The model equation is both valuable and necessary because it shows the design of the experiment and how the data are to be analysed. It is also used for deriving the expected values for mean squares of the model components.

Let μ represent the true population mean of the observations denoted by Y_{an} (a = number of factor levels and n = number of observations within each factor level). Each observation could be represented by the grand mean, μ, and the deviation from the grand mean, $\varepsilon_{n(a)}$. Further, if we let A_a represent the deviation of each treatment mean from the grand mean, we could write the model equation as follows:

$$Y_{an} = \mu + A_a + \varepsilon_{n(a)} \tag{4.1}$$

The true values of the parameters μ, A_a and $\varepsilon_{n(a)}$ are unknown, but they can be estimated from the data gained in the experiment. The structural equation of such sampled data has the form (4.2):

$$Y_{an} = \bar{Y} + (\bar{Y}_{a.} - \bar{Y}) + (Y_{an} - \bar{Y}_{a.})$$ (4.2)

where Y is the mean of all observations, and Y_a is the mean of a certain factor level.

4.4 Sums of squares

By looking at Equation 4.2, it is easy to see that if $Y_{an} = \bar{Y}$ with all values of a and n, the experiment gives us very little new information. In such case, treatments do not affect the response at all. In addition, we cannot get any estimate of the experimental error. Therefore, it is likely that the most interesting information gathered in the experiment is in terms $(\bar{Y}_{a.} - \bar{Y})$ and $(Y_{an} - \bar{Y}_{a.})$ in Equation 4.2. The first of these measures the effect of treatment on feed intake, whereas the latter estimates the experimental error. A key topic of experimental design is to maximise the ratio of total treatment effects and experimental errors. For several mathematical and statistical reasons – these are discussed in more detail by Kirk (1995) – we choose the design that with the necessary constraints will maximise the ratio between mean squares of treatment effects and experimental errors. To obtain this design, we must calculate the total sums of squares of the experiment SS_T, treatment effects SS_A and experimental error SS_e:

$$SS_T = \sum_{i=1}^{a}\sum_{j=1}^{n}(Y_{ij} - \bar{Y})^2$$ (4.3)

$$SS_A = n\sum_{i=1}^{a}(\bar{Y}_{i.} - \bar{Y})^2$$ (4.4)

$$SS_e = \sum_{i=1}^{a}\sum_{j=1}^{n}(Y_{ij} - \bar{Y}_{i.})^2$$ (4.5)

The mean squares used in the experimental design can be obtained by dividing Equation 4.4 and Equation 4.5 by respective degrees of freedom:

$$MS_A = SS_A/(a-1)$$ (4.6)

$$MS_e = SS_e/(a(n-1))$$ (4.7)

4.5 Evaluation of the experimental design

The evaluation of the design consists of checking whether the decisions made earlier will give the desired statistical efficiency. In practice, the design is often fine-tuned by adjusting the number of replicates. However, if there are economic or technical constraints, and it is not possible to have a sufficient number of replicates, a compromise between power, type I error and effect size must be made.

The *F*-distribution will be used as the benchmark to test whether there are differences in feed intake caused by feed additives. The test statistics is given by the ratio:

$$F = MS_A / MS_e \qquad\qquad (4.8)$$

The power of the *F*-test depends on the probability distribution of *F* when the null hypothesis is false. This is the *non-central F distribution*. The core parameter for this distribution is the *non-centrality parameter* that depends on the effect size, residual error and number of replicates. Its calculation is illustrated in Fig. 4.3. The probability density of the non-central *F* distribution corresponds to the power of the *F* test. It cannot be calculated directly, but may be either estimated with the help of the central *F* distribution (Winer *et al*. 1991; Lindman 1992) or obtained from published tables (Kirk 1995).

In our example, we wish to have an effect size of 10 g, a type I error of 0.05 and four replicate tanks for each factor level. We also expected that the SD of feed intake would be 5 g, i.e. the expected residual error variance would be 25 g². In our example, the non-centrality parameter has a value of 8.0, and the power estimate is 0.50, i.e. we have a 50% chance of detecting if the null hypothesis is false (Fig. 4.4a). This is not what is usually regarded as sufficient, i.e. a power ~0.80 (means that four out of five experiments will detect the treatment effect if it exists). It is possible to calculate how many replicates we would need to achieve the desired power of 0.80. The design could then be corrected accordingly. The calculations

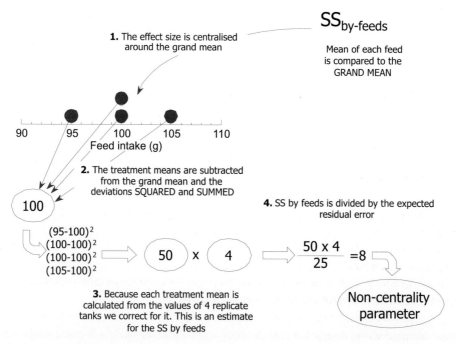

Fig. 4.3   Non-centrality parameter for estimation of power of the *F* test is calculated from the anticipated sums-of-squares caused by treatments (SS$_{by\text{-}feeds}$). Anticipated SS$_{by\text{-}feed}$ is computed by centralising the adopted effect size around the expected grand mean to get expected treatment means. Squared deviations of treatment means are then summed and corrected for the number of replicates. The non-centrality parameter is computed by dividing this sum with the expected residual error.

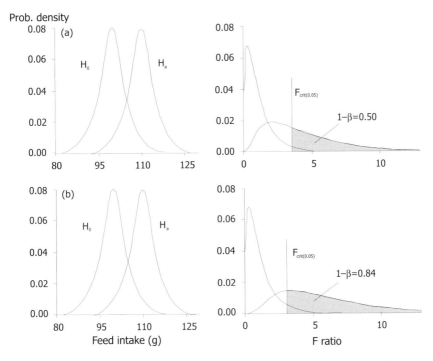

Fig. 4.4   The sampling distributions for null and alternative hypotheses and corresponding central and non-central *F* distributions with the critical *F* value. The normal distributions on the left represent the sampling distributions when the treatment means differ by the effect size. The distributions on the right are central *F*-distribution corresponding to the case that null hypothesis is true and the non-central F-distribution for the case that alternative hypothesis is true. F_{crit} marks the critical value for *F* ratio. The shaded area denotes the probability density of the power of the test. (a) Four replicates and (b) seven replicates. The difference between H_0 and H_a is the effect size.

show that if we used seven replicate tanks for each factor level we achieve a power of 0.84, which is satisfactory for our requirements (Fig. 4.4b).

4.6 The compromise

Above, the experimental requirements – type I error (α), power of the test ($1 - \beta$), effect size (*d*) and number of replicates (*n*) – were decided upon with no consideration of the dependencies between them. The question of design could be thought of as a multi-attribute decision problem in which all the elements can be fixed simultaneously in a compromise that offers the best solution in relation to requirements. First, it is necessary to examine the factors that should be taken into consideration in the decision process.

Let us suppose that we have chosen too strict an α-level. In such a case we might encounter situations in which we accept the null hypothesis even when the additives have an influence on feeding. Here, we would reject a correct result and conclude that our efforts have been in vain. We could avoid this by compromising between α and power of the test.

On the other hand, if we opt for too high a power ($1 - \beta$) and underestimate the α-risk, we might end up with a conclusion that the additives have positive effects on fish feeding and growth when they do not. In this case, we would be induced to allocate resources to further studies, all to no avail. Therefore, even though it might be desirable to detect small differences in feed intake (because, hypothetically even a small increase may make it economically viable to use the additives tested) it may not be possible to do so without an unacceptable α-risk.

As discussed earlier, the number of replicates is also open to choice, and by increasing replication we can strengthen our design. There is, however, a limit to the resources that can be invested in an experiment.

Let us now consider the design as a compromise between all the decision variables discussed above. To limit the number of possible combinations, let us fix a set of constraints. Let us assume that we can accept all type I errors in the range of 0.025 to 0.150. Similarly, let the power of the test be between the values 0.70 and 0.95, and let the minimum effect size that we consider biologically meaningful be 5% and the maximum 30%. If we assume that all decision variables in such intervals are equally important, we can work out the number of replicates required as a function of the rest of the decision process. Figure 4.5 summarises this analysis in the form of a ternary plot that allows us to draw all four dimensions in a two-dimensional graph (Gorman & Hinman 1962).

From the presentation given in Fig. 4.5 it can be concluded that if the effect size is fixed at 15%, α is 0.05 and power is 0.80, more than three (=4) replicates are needed to reach the goal. Alternatively, with an effect size of 10%, an α-risk of 0.05 and a power of 0.85, we end up with a requirement of 10 replicates.

With these conclusions as a starting point it is now possible to finalise the compromise. Assume that the research budget makes it possible to perform an experiment in 16 tanks, giving four replicates for each of the four levels. This provides a set of different alternatives (shaded area in Fig. 4.5) that can be evaluated in relation to the different types of errors, and a final choice can then be made.

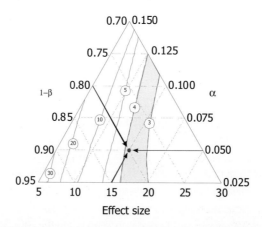

Fig. 4.5   Number of replicates as a function of effect size (% of the grand mean), type I error (α) and power of the test (1–β), when SD is assumed to be 5 g. Arrows point to the required number of replicates when effect size is 15%, α is 0.05 and power is 0.80. Shaded area corresponds to the 'decision' area if a total 16 tanks are available.

4.7 Sensitivity analysis

How the compromise between the decision variables can be made has been discussed, but there is a further point that should be considered before we can start our experiments: how do assumptions about the expected standard deviation (SD) influence decisions? We have assumed a SD of 5 g, based on previous knowledge, but let us now consider what happens to the design if the SD is 10 g or 15 g. Obviously, the increase in residual error will set new requirements for the design as summarised in Fig. 4.6. If the same constraints as used previously are retained, we should be prepared to increase the number of replicates as SD increases. In fact, the numbers of replicates required become prohibitively high if we want to detect either a small effect, keep α down, or require a high power of the test.

Above, we have used ternary plots to illustrate the compromising nature of the design process. We have assumed that we are playing a zero-sum game, where stricter requirement, e.g. in α-risk, would force us to diminish the effect size or test power. Very often, however, a zero-sum game is not completely true, as experimenters can change several decisions simultaneously. Because of this, the ternary plots shown are not generally applicable but, nevertheless, they remind us that statistical design is always a compromise.

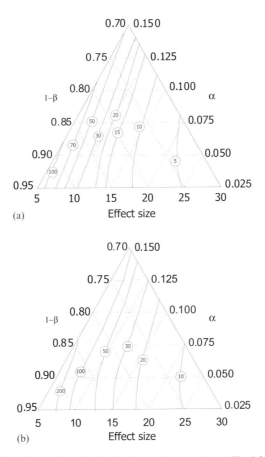

Fig. 4.6 The effect of SD on the number of required replicates in comparison to Fig. 4.5. SD is assumed to be 10 g (a) or 15 g (b). Effect size (% of the grand mean), type I error (α) and power of the test ($1-\beta$) as in Fig. 4.5.

4.8 Nuisance variables and ways of controlling them

Sometimes there are insufficient resources, e.g. tanks, to undertake an experiment in a completely randomised design. Alternatively, the arrangement, or type, of tanks could introduce increased random variation, or systematic errors, into the study. Slight differences in environmental conditions, e.g. temperature, oxygen concentrations, and measurements undertaken at different times or seasons also create problems. All of this could threaten the validity of our inference making. Variables of this kind are commonly called *nuisance variables*. Usually, attempts to tackle the problem of nuisance variables rely on distributing unsuspected sources of variation over the entire experiment, i.e. randomisation, or on holding the nuisance variable constant for all subjects. As an alternative, nuisance variables can be introduced as factors in the experimental design. The simplest design takes the form of *randomised blocks* in which the experimental units are divided into blocks in such a way that the magnitude of the nuisance variable can be estimated. Balanced blocking requires that all treatments are present in all blocks in at least one experimental unit, and leads to the following model equation

$$Y_{abn} = \mu + A_a + B_b + \varepsilon_{n(ab)} \tag{4.9}$$

where B_b represents the effect of blocks. All other symbols are as in Equation 4.1, with an addition that we now have n replicates within a factor levels and b blocks.

Typical cases where blocking might help are:

(1)   Repeated measures; where measurement time could be handled as a block.
(2)   Heterogeneous experimental units, e.g. an experiment run at several laboratories or farms, in different types of tanks, or under different conditions; a series of homogeneous blocks are established and treatments randomised within the blocks.
(3)   Heterogeneous experimental material, e.g. different fish stocks, different acclimatisation history; by distributing the material amongst homogeneous blocks, the effects of heterogeneity could be controlled.

The completely randomised design for the feed additive experiment gave seven replicate tanks for each feed to ensure sufficient statistical power; so $4 \times 7 = 28$ tanks are needed to execute this experiment. What may be done if only 16 tanks are available? How may we proceed: shall we compromise on the objectives and use only four tanks per treatment? What if a satisfactory compromise cannot be found? Four replicate tanks per treatment would give only about a 50% chance of detecting if the null hypothesis is false (if an effect size of 10 g and a type I error of 0.05 are retained), which may mean that the experiment is not worth undertaking. However, we could undertake sequential experiments and introduce time as a blocking variable into the model equation. By having four replicate tanks for each feed in two experiments, we could increase the number of replicates from seven to eight, and control for the possible increase in variation caused by the spread in time.

Compared to the completely randomised design, a blocking design may 'explain' variation that was previously 'unexplained' (Fig. 4.7): this reduces the residual error in the model. One word of caution: blocking does not always make the experiment more effective, and it could *decrease* the power of the test in some cases. This decrease in power may arise because

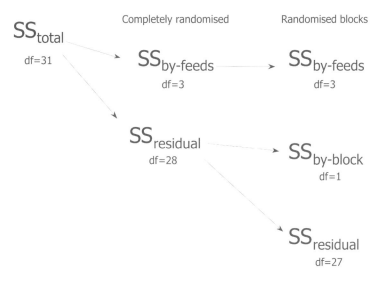

Fig. 4.7　Breaking the total sums-of-squares into sum-of-square components of the completely randomised design and further into components of the randomised block design. Corresponding degrees of freedom (df) are also shown.

the blocking factor reduces the degrees of freedom of the residual error in the analysis of variance; if no block effect existed, the residual error mean square would be larger than it would be without blocking (Fig. 4.7).

It is also possible to control several nuisance variables simultaneously by blocking. Blocking in two and three dimensions are called Latin squares and Graeco-Latin squares, respectively. Kirk (1995), for example, discusses them in more detail.

If the nuisance variable is a continuous one, e.g. fish weight, blocking is not suitable. In such cases *analysis of covariance* (a mixture of ANOVA and regression) is the appropriate method for the control of nuisance variables (Winer *et al.* 1991; Lindman 1992; Kirk 1995).

4.9　Adding extra factors: why do it and what considerations are necessary?

Suppose that a feed company desires to incorporate soybean meal into feeds that could be marketed throughout Europe. The plaice used in the initial trials may have been from the Irish Sea stocks. Plaice of this stock eat chiefly *Abra abra* and *Pectinaria koreni* (Basimi & Grove 1985), but the dominant prey species of plaice vary with geographic distribution (Kuipers 1977; Palsson 1983; Poxton *et al.* 1983; Clerck & Buseyne 1989). In order to check that fish from different stocks respond in the same way to the feed additives used, feeding trials were carried out using fish from three well-defined stocks. How does this affect the experimental design?

In this case the experimental design needed is that for *factorial experiments*, involving two or more factors with two or more levels in each. All levels of all factors are studied in combination, and experimental units are assigned randomly to factor–factor combinations. Thus, factorial experiments make efficient use of available resources, and also enable the evaluation of interaction effects between factors. However, full factorials become cumbersome and 'resource-hungry' as the number of factors increases, but some of the problems can be overcome by using *fractional factorials*. Fractional factorials are flexible and optimise the use of resources for large numbers of factors. However, fractioning confounds certain effects, and this makes the intepretation more difficult (see later).

Our example of a factorial design includes four feeds and three plaice stocks. The structural model equation can be written

$$Y_{asn} = \mu + A_a + S_s + AS_{as} + \varepsilon_{n(as)} \tag{4.10}$$

where S_s represents the effect of fish stock and AS_{as} represents the interaction between feed and stock. All other symbols are as in Equation 4.1, with an addition that we now have n replicates within a feed levels and s plaice stocks.

The evaluation of the experimental design must be performed separately for the statistical hypothesis $H_0(a) = 0$, $H_0(s) = 0$ and $H_0(as) = 0$. The power for each of these hypotheses can be calculated in the same way as outlined for a completely randomised one-factor experiment. The non-centrality parameter λ for *any fixed effect hypothesis* $[H_0(g)]$ with an F-test variable MS_g/MS_e and MS_g with v degrees of freedom can be calculated as follows (Lindman 1992):

$$\lambda = \frac{vMS_g}{MS_e} - 1 \tag{4.11}$$

Using the same assumptions as previously, i.e. an expected variance of 25 g², an effect size of 10 g for both feed and fish stock effect, and a type I error of 0.05, gives a power of 0.90 for the feed effect and 0.99 for the stock effect (as a fixed effect) with three replicate tanks for each factor–factor combination. Having only two replicate tanks decreases the power for the feed effect to 0.64. Thus, we would need 36 tanks (4 feeds × 3 stocks × 3 replicates) to run this experiment as a completely randomised two-way design. We might also evaluate the experiment according to the interaction hypothesis $H_0(as) = 0$, but there is seldom enough prior information to enable precise formulation of the alternative hypothesis. Consequently, factorials are usually evaluated by the main effects only. Further, the test for the interaction hypothesis is usually more powerful than for the main effect hypotheses, so it is sufficient to ensure power for the latter. If however, the interaction hypothesis needs to be evaluated, the calculation procedure for the interaction is, in principle, similar to that for the main effects.

4.10 Measuring individual feed intake – what are the benefits?

Thus far, we have restricted ourselves to measurements of feed intake of groups of fish, but what if we could measure feed intake of individuals, e.g. with the X-ray technique described in Chapter 3? The observation unit would change from a tank to an individual, but the experimental unit would remain the same, i.e. a group of fish in a tank. This is so because we cannot randomise our factors and levels at an individual level, but only at the level of tanks. There are now two types of residual error to be dealt with: the variability among the individuals within a tank, and the variability among the tanks. If we now return to the original design with feed as the factor, the addition of observations from individual fish results in something that resembles a completely randomised design with two factors, feeds and tanks. However, a closer look at the data table reveals that the design is incomplete and we do not have observations for all factor–factor combinations (Table 4.2). Each tank appears under only one feed, so we cannot evaluate a particular tank for all feeds but only for one. The tank effect appears nested within the feed effect and thus these designs are commonly called *nested designs*.

If we suppose that we have 20 fish in each tank and seven replicate tanks for each of our four feeds, we can write the structural model as follows:

$$Y = \mu + A_a + T_{t(a)} + \varepsilon_{n(at)} \tag{4.12}$$

where $T_{t(a)}$ represents the tank effect. All other symbols are as in Equation 4.1, with an addition that we now have n replicates within a feed levels and t tanks. In this particular case $n = 20$, $a = 4$ and $t = 7$.

Compare this structural model to Equations 4.1 and 4.9. Our one-factor completely randomised design has grown by one component, $T_{t(a)}$. The bracketed a in the subscript indicates that the levels of t are nested within levels of a. There are also some similarities with Equation 4.9 for randomised blocks. The difference, however, is that in the case of a nested design the 'tank blocks' vary from feed to feed; there is a different set of tanks for each feed.

To evaluate our new design we need to check the points indicated in Fig. 4.2. We still have a feed factor with four levels, but within each feed we also have another factor: tanks with seven levels within each feed. The feeds were deliberately chosen, but what about the tanks? We assign the tanks randomly to each of the feeds and the fish randomly to each of the tanks. If the experiment were to be repeated, it is likely that different tanks would be allocated to each of the feeds. Thus, it appears that our tank factor is a *random effect*. Since the feed factor is a *fixed effect*, our structural model is really a *mixed model*. As stated earlier, the type of model used affects how the F-test variable is calculated for hypothesis testing. This involves

Table 4.2 Data table for observations from an experiment with measurements from individual fish. We assume four feeds, four replicate tanks for each and twenty fish in each tank. Numbers denote observations made on individual fish and dashes denote no observations. Tanks are assigned randomly to each feed and fish randomly to each tank.

	Feed 1	Feed 2	Feed 3	Feed 4
Tank 1	—	1–20	—	—
Tank 2	1–20	—	—	—
Tank 3	—	1–20	—	—
Tank 4	—	—	1–20	—
Tank 5	—	—	1–20	—
Tank 6	—	—	—	1–20
Tank 7	—	—	—	1–20
Tank 8	1–20	—	—	—
Tank 9	—	—	—	1–20
Tank 10	1–20	—	—	—
Tank 11	—	—	1–20	—
Tank 12	1–20	—	—	—
Tank 13	—	1–20	—	—
Tank 14	—	—	1–20	—
Tank 15	—	—	—	1–20
Tank 16	—	1–20	—	—

deriving the expected mean squares for the terms in the model. Rules for deriving expected mean squares can be found in Hicks (1982), Winer *et al.* (1991), Lindman (1992) and Kirk (1995). The expected mean squares for the model are given in Table 4.3 (see Ruohonen 1998 for additional detail). For a random effect, τ_K^2 in Table 4.3 equals σ_K^2, the variance of K, but for a fixed effect the relation is:

$$\tau_K^2 = \frac{I}{I-1}\sigma_K^2 \tag{4.13}$$

where I is the number of factor levels.

If we retain the same effect size (10 g), type I error (0.05) and power of the test (0.80) criteria as previously, we can evaluate the replication of our nested design. For our main effect, the feeds, this means evaluating the F ratio MS_{Feed}/MS_{Tanks} (see Table 4.3). To calculate MS_{Tanks} we need two variance estimates: one for the between tanks variability and one for the residual error variance at the individual level. These may be estimated if data from earlier experiments are available. If, for example, we have previously carried out individual feed intake measurements using four tanks with 100 fish in each, and run an ANOVA, we might have obtained the results shown in Table 4.4. The residual error variance at the individual level is the MS_{Error} in Table 4.4, i.e. 300 g². The variance estimate for tanks may be calculated with the help of the expected mean squares. The expected mean square for the tank effect is the sum of the residual error variance σ_e^2 and $n\tau_{T(A)}^2$ (Table 4.3), and this transforms to:

$$\tau_{T(A)}^2 = \frac{E(MS_{T(A)}) - \sigma_e^2}{n} \tag{4.14}$$

Using the information given in Table 4.4 it is calculated that $\tau_{T(A)}^2 = 10$ g², with feed intake being measured for 100 fish measured in each tank ($= n$). Following calculation of the expected mean squares for the F ratio, the power of the test for nested designs may be calculated according to the outlines described previously, with Equation 4.11 being used for estimating the non-centrality parameter.

Table 4.3   Expected mean squares $E(MS)$ and proper F-test variables for the effects in Equation 4.12. n = number of fish in each tank, j = number of tanks within each diet, σ_e^2 = residual error variance, τ_K^2 = variance component of effect K.

Term	Symbol	Expected mean squares	F-test variable
Feed	A_a	$\sigma_e^2 + n\tau_{T(A)}^2 + nt\tau_A^2$	MS_{Feed}/MS_{Tanks}
Tanks	$T_{t(a)}$	$\sigma_e^2 + n\tau_{T(A)}^2$	MS_{Tanks}/MS_{error}
Residual error	$e_{n(at)}$	σ_e^2	

Table 4.4   Example of analysis of variance table for measurements of individual feed intake from four tanks, 100 fish in each, for estimating variance components for experimental design (see text for details).

Source	Sum-of-squares	df	Mean-square
Tanks	3 900	3	1300
Error	118 800	396	300

Table 4.5 Tanks/individuals: combinations for achieving 0.80 power of the test in the example experiment. Number of tanks refers to the number needed for each feed and number of individuals is the fish within each tank (see text for details).

No. of tanks	No. of individuals
4	338
5	64
6	36
7	26
8	20
9	16
10	14
11	12
12	8
13	7
14	7
15	6

Examination of the expected mean squares in Table 4.3 shows that the variance component τ_A^2 in the expected MS_A is multiplied by both n, number of fish in each tank, and j, number of tanks with each feed. This means that a change in replication at *either level* affects the power of the test for the main effect. Consequently, there are several pairs of combinations of numbers of fish in each tank and numbers of tanks with each feed that give the same statistical power. This could be exploited to find an optimal economic solution to the design for running the experiment when the relative costs of adding an additional tank or additional individuals vary (Ruohonen 1998). Biological as well as other constraints may cause problems, such as the minimum number of fish in tank or the maximum cost of the experiment, and they can be introduced into the optimisation. Examples of tank and fish number combinations that result in the same objectives with the previous variance estimates for tanks and individuals are given in Table 4.5.

4.11 What can be done when life becomes more complicated?

The examples discussed hitherto have been relatively simple, and experiments of fish feeding are often more complicated. For example, one seldom relies on a single measure of feed intake, usually several measurements are made over time. This results in an experimental design called *repeated measures*. There are several approaches to repeated measurements; one is to consider such measurements as a special case of blocking (see Hicks 1982). Here, the experimental units act as blocks, and this allows separation of the group-to-group or individual-to-individual variability from the residual error. In practice, repeated measures are often used in growth trials, where fish are weighed at regular intervals to monitor weight change over time. These changes, and consequently growth rate, will have quite large variability from individual to individual. Let us imagine an experiment in which we measure the growth of fish held in individual tanks over a period of time. If we could measure the effects of different feeds in a time series so that each individual receives each of the feeds in turn, we could include the effect of an individual fish as a block in the structural model. Either randomised blocks or Latin square designs can be used to control for nuisance variables. The simplest case could be written as follows:

$$Y_{afn} = \mu + A_a + F_f + \varepsilon_{n(af)} \tag{4.15}$$

where F_f represents the effect of individual fish. All other symbols are as in Equation 4.9.

Further discussion of repeated measures can be found in Winer *et al.* (1991) and Kirk (1995).

However, if we perform repeated measurements of feed intake and growth, the situation is somewhat different to that discussed above. We would not be changing the treatments between the measurements, but would be measuring the same treatment–experimental unit combination all the time. This type of repeated measures design is the one that is most common in feeding and growth trials (e.g. McCarthy *et al.* 1992; Jørgensen *et al.* 1996; Sæther *et al.* 1996; Koskela *et al.* 1997). An example showing two treatments and three repeated measurements for each of ten experimental units is illustrated in Fig. 4.8. The structural model for this could be written:

$$Y_{atmn} = \mu + A_a + T_{t(a)} + M_m + AM_{am} + MT_{mt(a)} + \varepsilon_{n(atm)} \tag{4.16}$$

where A_a represents treatment, $T_{t(a)}$ experimental unit (tank) and M_m time of measurement. All other symbols are as in Equation 4.9.

Note that experimental units are nested within treatments A_a but not within the time of measurement M_m because although the response (feed intake) of all experimental units is measured at all times the response of any given unit is measured under one particular treatment (treatment 1 or 2).

This example shows an important point about the experimental designs we have discussed so far. The basic designs can be seen as building blocks for more complicated designs, and in the last example we have mixed randomised blocks and nested designs to provide an adequate description of the experiment.

A complication that is often encountered involves the choice of variables from the large group of possible alternatives that may affect the response we are interested in studying. Experiments that are useful in this regard are *screening experiments*, which include all factors

Treatment 1			Treatment 2		
1st meas-urement	2nd meas-urement	3rd meas-urement	1st meas-urement	2nd meas-urement	3rd meas-urement
Tanks 1 to 10			Tanks 11 to 20		

Fig. 4.8   Illustrative problem of the repeated measures design in feed intake studies. The responses of two treatments are measured on three occasions. The effect of each treatment as well as the response of each tank is measured at all measurement times on the same experimental units. Each tank is measured only under one particular treatment, which makes any tank effect nested within a treatment.

at two levels – 'high' and 'low' level (they are denoted 2^k-experiments, where k indicates the number of factors studied). However, when the number of factors exceeds three or four the full factorial design becomes impractical because of the large size of the experiment required, and fractional factorials can be used to reduce the number of treatment combinations. As the name implies, fractional factorial designs include a fraction of the full design, e.g. 1/8, 1/4 or 1/2 of the possible treatment combinations in the complete factorial design. The fractionation should be undertaken according to standard rules rather than intuitively because the latter seldom leads to a design with good statistical properties. Fractionation should be carried out in such a way that the experimental space remains as high and representative as possible (Fig. 4.9). Kirk (1995) gives the following notes and guidelines regarding the use of fractional factorials:

(1)   Fractional factorials are appropriate when the experiment contains many treatments, giving a prohibitively large number of treatment combinations. Fractional factorial designs are rarely used for less than four or five treatments.

(2)   The number of treatment levels should, if possible, be equal for all treatments. With the exception of a fractional factorial using a Latin square building block design, analysis procedures for designs with mixed numbers of treatment levels are relatively complex.

(3)   There should be some *a priori* reason for believing that higher-order interactions are zero, or small, relative to main effects. In practice, fractional factorial designs, with the exception of those based on a Latin square, are most often used with treatments having either two or three levels. The use of a restricted number of levels increases the likelihood that interactions will be insignificant.

(4)   Fractional factorial designs are most useful for exploratory research when follow-up experiments may be performed. Thus, a large number of treatments can be investigated in an initial experiment, and subsequent experiments can then be designed to focus on those that seem most promising.

Hitherto, fractional factorials have been used infrequently in fish biology research but they would seem to have the potential for quite wide application.

$$2^k \qquad\qquad\qquad 2^{k-1}$$

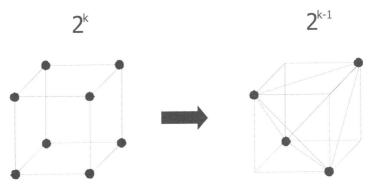

Fig. 4.9   Comparison of 2^k full factorial and 2^{k-1} fractional factorial designs. In full factorial design the response is measured at all combinations of two levels of all factors whereas fractional factorial design aims at reducing the number of combinations while spanning effectively the experimental space.

4.12 Conclusions

The statistical design is an essential part of the experimental design. Systematic statistical design leads to efficient use of resources and also facilitates – and speeds up – the analysis of data. This is due to the fact that when statistical design is completed, the structural model and relevant estimation methods have also been set up prior to any field work. Also, because the effect size, power of the test, type I error, residual error and the limitations of randomisation are acknowledged and predetermined during the design stage, precise interpretation of the results is possible and the conclusions that are drawn are likely to be valid.

Above, we have emphasised two main features of the statistical design. First, statistical design is a subtask of the experimental design separate from the biological problem. Biological problems are reduced to statistical hypotheses that can be solved with statistical tools and then interpreted biologically. Second, statistical design process is a compromise that involves several factors simultaneously. One should not consider only α-risk but all other elements of uncertainty, nuisance variables, restrictions of experimental capacity and structure of the statistical problem, i.e. the model.

Principally, all ANOVA designs are combinations of few basic building blocks. Completely randomised design suits for most situations when there are no restrictions of randomisation or capacity. It forms the basis for all other designs that are applied in cases with some limitations in the problem. Block designs are applicable for controlling nuisance variables. For example, an experiment may be split to several laboratories or field sites or performed with several fish stocks and response variables can be measured repeatedly. These all are applications of blocking. Factorial designs are useful to study the effects of several factors simultaneously, and especially when factors interact with each other. The limitation of factorial designs is that they tend to be very 'hungry' in resources when number of variables is increased. In such cases fractional factorials can be used.

The characteristics of a well-conceived and properly designed experiment are that it is as simple as possible, it has high probability of achieving its objectives and it avoids systematic and biased errors.

4.13 References

Basimi, R.A. & Grove, D.J. (1985) Estimates of daily food intake by an inshore population of *Pleuronectes platessa* L. off eastern Anglesey, North Wales. *Journal of Fish Biology*, **27**, 505–520.

Box, G.E.P., Hunter, W.G. & Hunter, J.S. (1978) *Statistics for Experimenters*. John Wiley & Sons, New York.

Clerck, R. De & Buseyne, D. (1989) On the feeding of plaice (*Pleuronectes platessa* L.) in the southern North Sea. ICES, CM 1989/G:23.

Cohen, J. (1988) *Statistical Power Analysis for the Behavioral Sciences*. Lawrence Elbaum Associates, Hillsdale.

Gorman, J.W. & Hinman, J.E. (1962) Simple lattice designs for multicomponent systems. *Technometrics*, **4**, 463–487.

Hicks, C.R. (1982) *Fundamental Concepts in the Design of Experiments*. 3rd edition. CBS College Publishing, New York.

Jørgensen, E.H., Baardvik, B.M., Eliassen, R. & Jobling, M. (1996) Food acquisition and growth of juvenile Atlantic salmon (*Salmo salar*) in relation to spatial distribution of food. *Aquaculture*, **143**, 277–289.

Kirk, R.E. (1995) *Experimental Design: Procedures for the Behavioral Sciences*. Brooks/Cole Publishing, Pacific Grove.

Koskela, J., Pirhonen, J. & Jobling, M. (1997) Feed intake, growth rate and body composition of juvenile Baltic salmon exposed to different constant temperatures. *Aquaculture International*, **5**, 351–360.

Kuipers, B.R. (1977) On the ecology of juvenile plaice on a tidal flat in the Wadden Sea. *Netherlands Journal of Sea Research*, **11**, 56–91.

Lindman, H.R. (1992) *Analysis of Variance in Experimental Design*. Springer-Verlag, New York.

McCarthy, I.D., Carter, C.G. & Houlihan, D.F. (1992) The effect of feeding hierarchy on individual variability in daily feeding of rainbow trout, *Oncorhynchus mykiss* (Walbaum). *Journal of Fish Biology*, **41**, 257–263.

Palsson, O.K. (1983) The feeding habits of demersal fish species in Icelandic waters. *Rit-Fiskideild*, **7**, 1–60.

Poxton, M.G., Eleftheriou, A. & McIntyre, A.D. (1983) The food and growth of 0-group flat fish on nursery grounds in the Clyde Sea area. *Estuarine and Coastal Shelf Science*, **17**, 319–337.

Ruohonen, K. (1998) Individual measurements and nested designs in aquaculture experiments: a simulation study. *Aquaculture*, **165**, 149–157.

Sæther, B.-S., Johnsen, H.K. & Jobling, M. (1996) Seasonal changes in food consumption and growth of Arctic charr exposed to either simulated natural or a 12:12 LD photoperiod at constant water temperature. *Journal of Fish Biology*, **48**, 1113–1122.

Winer, B.J., Brown, D.R. & Michels, K.M. (1991) *Statistical Principles in Experimental Design*. 3rd edition. McGraw-Hill, New York.

Chapter 5
Gustation and Feeding Behaviour

Charles F. Lamb

5.1 Introduction

Chemoreception in fish, and the effect of chemical stimulation on feeding behaviour, has been a topic of research for more than one hundred years. This is due partly to commercial and recreational attempts to stimulate feeding in fish, but also because of the diversity and sensitivity of chemoreceptor systems in fishes. Teleost fishes possess several chemosensory systems, including gustation, olfaction, common chemical sense, and solitary chemoreceptor cells. Distinguishing between the roles played by each of these sensory systems in fish is often difficult because each system responds to aqueous chemical stimuli, and different systems in a given species might respond to common stimuli. Most of the evidence gathered to date suggests that olfaction and solitary chemoreceptor cells are important for what might be called social behaviours – intraspecific and interspecific communication, homing, etc. Little is known about the function of the common chemical sense – generally defined as the detection of chemicals through stimulation of free-nerve endings of the trigeminal or spinal nerves – but it likely plays a role in the avoidance of noxious stimuli. The gustatory system might also be involved in these behaviours, but taste is most important in the acquisition and ingestion of food and the rejection of potentially harmful or toxic substances.

Gustation is a sense common to all vertebrates – with shared features including modified epithelial receptor cells; input to the central nervous system through sensory fibres of the facial, glossopharyngeal and vagal nerves; and similar pathways through the brainstem. In many teleosts, the gustatory system is highly developed, with a proliferation of taste buds throughout the epithelium and a substantial enlargement of brain regions responsible for processing taste information. The most studied of the teleost fishes are members of the siluriform (catfishes) and cypriniform (minnows and carps) groups of fishes. These have proven to be fruitful models for anatomical, biochemical, physiological and behavioural studies of taste and its roles in feeding. This chapter will summarise the information provided by multidisciplinary research involving a few of these well-studied species, as well as what is known about the diversity of taste systems possessed by other teleosts.

5.2 Peripheral gustatory sensation

Gustatory stimuli are detected by receptor cells located in epithelial taste buds. In teleosts, these taste buds are not only found in the oropharyngeal cavity, but can also be quite numerous in the epithelium of the head and whole body. Taste bud densities in ictalurids and cyprinids have been estimated at up to several hundred per square millimetre of epithelium surface (Atema 1971; Kiyohara *et al.* 1980). Teleost taste buds, regardless of their location on the body, are flask-shaped collections of 30–100 epithelial cells. Taste buds communicate with the external environment at the apical surface (Fig. 5.1). There are several different cell types within the taste bud, and these are distinguished by cytological features (Reutter 1978; Jakubowski & Whitear 1990). The majority of cells in a taste bud are elongated cells which are oriented perpendicular to the epithelial surface and have apical microvilli. These cells are generally categorised into two classes – light cells and dark cells, with reference to their relative electron opacity (Reutter 1978). Light cells contain a single, larger apical microvillus (0.5 μm thick, 2 μm long) and have presynaptic specialisations associated with dendritic processes of the gustatory sensory nerves along their lateral and basal surfaces. Dark cells possess numerous smaller apical microvilli (0.2 μm thick, 1 μm long) and seem to have synaptic

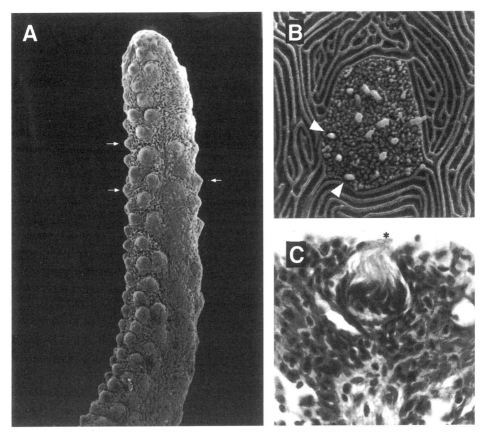

Fig. 5.1   Taste buds from teleost epithelium. A: Scanning electron micrograph (SEM) of channel catfish (*Ictalurus punctatus*) maxillary barbel showing papillae (arrows) containing taste buds (the barbel is approximately 250 μm across). B: SEM of an apical pore of a taste bud from the facial epithelium of an African cyprinid, *Phreathichthys andruzzi,* with long (arrowheads) and short microvillar processes at the surface (the pore is 5 μm × 7 μm). C: Light micrograph of a pharyngeal taste bud from the goldfish (*Carassius auratus*). Note the characteristic flask shape and the intrinsic cells of the bud with apical processes (asterisk) extending out through the pore.

connections with both taste nerves and other cells in the taste bud. These anatomic features suggest that light cells are taste receptor cells, while dark cells function as supporting cells or, possibly, as receptor cells (Ezeasor 1982; Jakubowski & Whitear 1990; Reutter & Witt 1993). A third class of cells consists of smaller, horizontally oriented cells found in the basal portions of the taste bud. These basal cells contain serotonin (5-HT) and seem to possess synaptic connections with light and dark cells (reciprocally) and with nerve fibres (Toyoshima *et al.* 1984; Reutter & Witt 1993). Their features have led them to be called Merkel-like cells by some researchers, and suggest that they might act as modifiers of taste activity as it passes from taste bud to taste nerve (Reutter 1978; Toyoshima *et al.* 1984).

Taste stimulation is thought to result from dissolved chemical stimuli interacting with re-ceptors on the apical membrane of taste receptor cells, leading to a transduction event within the receptor cell and modification of taste nerve activity at the basal portion of the taste bud. Biochemical and physiological studies, mostly with the facial taste system of channel catfish (*Ictalurus punctatus*), have helped to identify multiple transduction mechanisms potentially associated with stimulation of gustatory receptor cells in teleosts. Unfortunately, it has been difficult to correlate structural and functional data from single taste cells in fish, so whether different morphologically defined cell types possess unique functions is still an open ques-tion. Information has been collected using receptor binding and inhibition studies, electro-physiology of reconstituted membrane patches, and electrophysiological recording of peri-pheral taste nerve fibres.

Receptor binding studies with membrane extracts from channel catfish epithelium suggest that there are multiple, independent high-affinity receptors for amino acids associated with taste buds. L-Alanine and L-arginine both bind with dissociation constants in the micromolar range ($K_D = 1.5\,\mu M$ and $1.3\,\mu M$, respectively), while L-arginine also has a higher-affinity state ($K_D = 18\,nM$) (Brand *et al.* 1987; Kalinoski *et al.* 1989). Competitive inhibition studies with these amino acids, their D-enantiomers, and amino acid analogues indicate the presence of two major classes of taste receptors for amino acids. L-Alanine, D-alanine, and several neutral amino acids compete for L-[^{3}H]alanine binding, while L-arginine and related compounds do not (Cagan 1986; Brand *et al.* 1987). Similarly, L-arginine, D-arginine, and several arginine analogues compete for L-[^{3}H]arginine binding, and neutral amino acids are relatively inef-fective (Kalinoski *et al.* 1992). Studies of plant lectin binding to taste epithelium also sup-port the presence of distinct alanine and arginine receptor types. Binding of L-[^{3}H]alanine is selectively inhibited by lectin from *Dolichos biflorus* (DBA), while L-[^{3}H]arginine binding is selectively inhibited by *Phaseolus vulgaris* lectin (PHA-E) (Kalinoski *et al.* 1989). Ad-ditional binding studies using a monoclonal antibody (G-10) to a fraction derived from catfish taste epithelium have shown L-[^{3}H]alanine, but not L-[^{3}H]arginine, binding can be selectively inhibited by the presence of this antibody (Goldstein & Cagan 1982). Double-labelling stud-ies of catfish taste epithelium with G-10 and PHA-E indicate that the receptors affected by these two inhibitors are found on the apical surfaces of separate taste cells, and that individual taste buds have cells with each type of receptor present (Finger *et al.* 1996). In other words, individual taste cells generally express either L-arginine or L-alanine receptors, while a given taste bud will contain cells of both types.

In addition to there being distinct receptors for L-alanine and L-arginine, different trans-duction pathways are also activated by these taste stimulants. Micromolar concentrations of L-alanine, but not L-arginine, activate G protein-dependent production of two second messenger molecules: inositol 1,4,5-trisphosphate and cyclic AMP (Caprio *et al.* 1993). Electrophysiological recordings from reconstituted membrane from taste epithelium of channel catfish revealed that L-arginine activated a non-specific cation channel that led to depolarisation, while L-alanine did not (Teeter *et al.* 1990). The threshold for L-arginine stimulation was 0.5 µM and saturation occurred at 200 µM. The membrane conductance produced by L-arginine stimulation was competitively suppressed by D-arginine at concentrations from 1 to 100 µM, suggesting that the two enantiomers bind to the same receptor. A similar selectivity has been found for L-proline, an effective taste stimulant that produces an increase in conductance at millimolar concentrations (Kumazawa *et al.* 1998). The L-proline-activated channels were unaffected by L- or D-arginine, but were blocked by millimolar concentrations of D-proline. It is not known whether there are other amino acid receptors or transduction pathways in catfish, or whether different species possess different mechanisms. This type of information is necessary for gaining an increased understanding of the initial events involved in teleost taste.

Much of what is known about taste sensitivity in teleosts has been gathered through extracellular recording from fibres of the facial, glossopharyngeal or vagal nerves that lead from the taste epithelium to the hindbrain. The first electrophysiological recordings of taste activity were performed on the facial nerve fibres of catfish barbels nearly seventy years ago (Hoagland 1933), and a variety of teleost preparations have been tested since then. Until about thirty years ago, stimuli used for fish taste electrophysiology were patterned after human taste sensitivities, i.e. compounds producing sweet, sour, salty, or bitter tastes. Subsequent studies with biologically relevant stimuli have shown that amino acids are much more effective taste stimuli for many teleost species (Table 5.1), with most species studied possessing electrophysiological thresholds to amino acid stimulation from 10^{-6} M to 10^{-9} M (Sorensen & Caprio 1997). Given the caveat that most of the data collected on amino acid gustation in fish come from studies considering only the facial nerve, some interesting generalisations can be drawn.

First of all, L-amino acids are typically more stimulatory than their corresponding D-isomers (Caprio 1978), supporting the enantiomeric specificity identified in receptor-binding studies (see above). A notable exception to this is the responsiveness of facial taste fibres of the sea catfish (*Arius felis*) to D-alanine, which is the most effective stimulus for a large portion of the fibres (Michel & Caprio 1991). This apparently reflects a specialisation to marine food items commonly encountered by the sea catfish (Sorensen & Caprio 1997). More information is necessary to determine whether this is a species-specific adaptation, or if marine teleosts are generally more responsive to D-amino acids than are freshwater species.

Secondly, short-chained neutral amino acids (e.g. alanine, glycine, proline) and basic amino acids (arginine) are more stimulatory than are acidic amino acids (aspartate). While specificity for taste stimuli can vary greatly across species, most species examined are highly sensitive to neutral amino acids (Marui & Caprio 1992; Hara 1993, 1994). A derivative of glycine, betaine (*N*-trimethylglycine), is also very stimulatory in a number of teleost species (see Marui & Caprio 1992). Those species that respond only to a few (usually neutral) amino acids, such as the puffer (*Fugu pardalis*) (Kiyohara *et al.* 1975) and several salmonids (Marui *et al.* 1983a; Hara *et al.* 1993), have been classified as possessing a 'limited response range', while those such as the channel catfish, that respond to a wider variety of amino acids (Caprio 1975), are considered to have a 'wide response range' (Marui & Caprio 1992; Hara 1993,

Table 5.1   Effective amino acid stimuli from electrophysiological recording of gustatory nerves in various fish species.

Species	Effective stimuli[1]	Reference(s)
Japanese eel *Anguilla japonica*	Arg, Gly, Ala, Pro, Lys, Ser, Abu	Yoshii *et al.* (1979)
Sea catfish *Arius felis*	Ala, Gly, D-Ala, Pro, Arg, His	Michel & Caprio (1991)
Sea bream *Chrysophrys major*	Ala, Gly, Arg, Ser, Lys, Thr, Bet, Leu, Pro	Goh & Tamura (1980a)
Common carp *Cyprinus carpio*	Pro, Ala, Cys, Glu, Bet, Gly	Marui *et al.* (1983b)
Puffer *Fugu pardalis*	Pro, Ala, Gly, Bet	Kiyohara *et al.* (1975)
Yellow and brown bullhead *Ictalurus natalis* and *I. nebulosus*	Cys, Tau, Ala, Phe, Leu, Glu	Bardach *et al.* (1967)
Channel catfish *Ictalurus punctatus*	Ala, Arg, Ser, Gly, Pro, Abu, Gln, D-Ala	Caprio (1975, 1978); Davenport & Caprio (1982); Kanwal & Caprio (1983); Kohbara *et al.* (1992); Wegert & Caprio (1991)
Tomcod *Microgadus* sp.	Cys, Tau, Ala, Phe, Asp, Glu	Bardach *et al.* (1967)
Mullet *Mugil cephalus*	Arg, Lys, Ala, Ser, Bet, Leu, Pro, Gly	Goh & Tamura (1980a)
Rainbow trout *Oncorhynchus mykiss*	AGPA, Pro, Pro-OH, Bet, Leu, Ala, Phe	Marui *et al.* (1983a)
Isaki grunt *Parapristipoma trilineatum*	Neutral and basic amino acids[2]	Ishida & Hidaka (1987)
Japanese sea catfish *Plotosus anguillaris*	Pro, Bet, Ala	Sorensen & Caprio (1997)
Searobin *Prionotus carolinus*	Cys, Tau, Ala, Phe	Bardach *et al.* (1967)
Topmouth minnow *Pseudorasbora parva*	Pro, Ala, Ser	Kiyohara *et al.* (1981)
Arctic charr *Salvelinus alpinus*	Pro, Pro-OH, Ala	Hara *et al.* (1993)
Brook charr *Salvelinus fontinalis*	Pro, Pro-OH, Ala	Hara *et al.* (1993)
Lake trout *Salvelinus namaycush*	Pro, Pro-OH, Ala	Hara *et al.* (1993)
Chub mackerel *Scomber japonica*	Neutral and basic amino acids[2]	Ishida & Hidaka (1987)
Amberjack *Seriola dumerili*	Pro, Try, Bet, Ala	Ishida & Hidaka (1987)
Yellowtail *Seriola quinqueradiata*	Pro, Bet, Try, Val, Ala	Hidaka *et al.* (1985)
Aigo rabbitfish *Siganus fuscescens*	Neutral, basic, acidic amino acids[2]	Ishida & Hidaka (1987)
Tigerfish *Therapon oxyrhynchus*	Neutral, basic, acidic amino acids[2]	Hidaka & Ishida (1985)
African cichlid *Tilapia nilotica*	Neutral, basic, acidic amino acids	Marui & Caprio (1992)
Maaji jack mackerel *Trachurus japonicus*	Neutral and basic amino acids[2]	Ishida & Hidaka (1987)

[1] Abbreviations: Abu, L-α-aminobutyric acid; AGPA, L-α-amino-β-guanidino propionic acid; Ala, L-alanine; Arg, L-arginine; Asp, L-aspartic acid; Bet, betaine; Cys, L-cysteine; D-Ala, D-alanine; Gln, L-glutamine; Glu, L-glutamic acid; Gly, glycine; Leu, L-leucine; Phe, L-phenylalanine; Pro, L-proline; Pro-OH, hydroxy-L-proline; Ser, serine; Tau, taurine; Thr, L-threonine; Try, L-tryptophan; Val, L-valine.
[2] From Marui & Caprio (1992).

1994). The functional significance of this has yet to be tested, but the differences in amino acid spectra likely represent differences in ecological niches of the species studied (Marui & Caprio 1992).

Finally, there is evidence from studies of peripheral nerve electrophysiology indicating the presence of receptor types that distinguish neutral from basic amino acids, and there may even be receptors that distinguish between different neutral amino acids (Marui & Caprio 1992). The identity of receptor types in the epithelium can be deduced from studies utilising single-fibre analyses to characterise the response patterns of individual taste nerve fibres to various stimuli (Kiyohara *et al.* 1985a; Michel & Caprio 1991; Kohbara *et al.* 1992), or from cross-adaptation experiments which involve the simultaneous (or nearly so) application of two different stimuli to see whether they are recognised as separate stimuli by the taste system

(Yoshii *et al.* 1979; Kiyohara & Hidaka 1991; Wegert & Caprio 1991). Results from these studies indicate the presence of taste fibres that are tuned to particular amino acids or classes of amino acids. Most of the species studied possess fibres that respond best to stimulation by a number of neutral amino acids, and fibres that respond best to a much narrower group of basic or acidic amino acids (Marui & Caprio 1992). Numbers of receptor types identified for a given species range from one in Arctic charr (*Salvelinus alpinus*) (Hara *et al.* 1993), to three in rainbow trout (*Oncorhynchus mykiss*) (Marui *et al.* 1983a) and puffer (Kiyohara & Hidaka 1991), and eight in channel catfish (Wegert & Caprio 1991). However, direct investigations of receptors in the taste epithelium, such as those for L-alanine, L-arginine and L-proline in the channel catfish (see above), are necessary to identify and characterise the receptors involved in these systems.

In addition to amino acids, many other biologically relevant compounds have been tested on teleost taste nerve preparations, and there appear to be species-specific sensitivities to different classes of taste stimuli. An example of such selectivity is the responsiveness to bile salts, which are very effective taste stimuli in several salmonid species but not for common carp (*Cyprinus carpio*), Japanese sea catfish (*Plotosus lineatus*) or Nile tilapia (*Tilapia nilotica*) (Marui & Caprio 1992). Electrophysiological recording from the rainbow trout (Marui *et al.* 1983a; Hara *et al.* 1984) and from brook (*Salvelinus fontinalis*), lake (*Salvelinus namaycush*) and Arctic charrs (Hara 1993, 1994) has shown that these salmonids detect bile salts down to 10^{-12} M. Interestingly, bile salts are also very effective olfactory stimulants in salmonid species, and it is possible that gustatory and olfactory detection of these compounds might act together in the control of the homing behaviours exhibited by these species (Doving & Selset 1980; Doving *et al.* 1980). Carboxylic acids are another class of taste stimulants that are effective at very low concentrations, but these have been tested in a limited number of teleosts. Atlantic salmon (*Salmo salar*) (Sutterlin & Sutterlin 1970), Japanese eel (*Anguilla japonica*) (Yoshii *et al.* 1979) and carp (Marui & Caprio 1992) can detect aliphatic carboxylic acids at concentrations down to 10^{-7} M, and respond better to these compounds at 1 mM than they do to amino acids at that concentration. The responses in the eel are not cross-adapted by amino acid stimulation, suggesting that this species possesses unique receptors for carboxylic acids (Yoshii *et al.* 1979). A final class of compounds that have been identified as effective taste stimulants in teleosts are the nucleotides. Nucleotide-stimulated activity has been recorded from facial nerve fibres in the puffer (Kiyohara *et al.* 1975), tigerfish (*Therapon oxyrhynchus*) (Hidaka & Ishida 1985), yellowtail (*Seriola quinqueradiata*) (Hidaka *et al.* 1985; Harada 1986), oriental weatherfish (*Misgurnus anguillicaudatus*) (Harada 1986), greater amberjack (*Seriola dumerili*), club mackerel (*Scomber japonicus*), grunt (*Parapristipoma trilineatum*), jack mackerel (*Trachurus japonicus*), rabbitfish (*Siganus fuscescens*) (Ishida & Hidaka 1987), carp (Marui & Caprio 1992) and channel catfish (Littleton *et al.* 1989). While thresholds for nucleotide and carboxylic acid stimulation of teleost taste fibres are generally two to three orders of magnitude higher than thresholds for amino acid stimulation (Sorensen & Caprio 1997), these compounds appear to be important feeding stimulants for many fishes.

Mechanical stimulation – touching, stroking, etc. – also is an effective stimulus for gustatory nerves of fishes. While studies with mammals have linked mechanosensation to trigeminal nerve innervation of the face and oral cavity, several electrophysiological studies of teleost gustatory nerves showed a bimodal (taste/touch) responsiveness of facial (Davenport &

Caprio 1982; Kiyohara *et al.* 1985a), glossopharyngeal and vagal (Kanwal & Caprio 1983) nerves in fishes. The bimodal sensitivity of the incoming gustatory afferents in teleosts is augmented by overlapping input from trigeminal fibres in the primary gustatory centres (Marui & Funakoshi 1979; Kiyohara *et al.* 1986, 1999; Puzdrowski 1988), and taste and tactile information are functionally linked throughout the gustatory pathways of the teleost brain (Marui & Caprio 1982; Hayama & Caprio 1989; Lamb & Caprio 1992, 1993b; see below).

5.3 Gustatory pathways in the central nervous system

Gustatory pathways in the brain of teleosts were described almost 100 years ago (Herrick 1905). Teleost gustatory systems have been studied most extensively in catfishes and carps (see reviews by Finger 1983; Kanwal & Finger 1992; Wulliman 1997). All teleost species whose gustatory pathways have been examined possess an ascending gustatory lemniscus similar to that found in other vertebrates: a primary gustatory centre in the medulla oblongata with input from the facial, glossopharyngeal and vagal nerves, a secondary gustatory centre in the isthmic region, a complex tertiary gustatory centre in the diencephalon, and various forebrain gustatory centres (Finger 1984; Wulliman 1997) (Fig. 5.2). Taste-specialist species have enlarged gustatory centres and tracts of fibres connecting them. There are also elaborate reflexive connections between those centres and other brain regions, that reflect the differences in feeding ecology between species (Herrick 1905; Finger & Morita 1985). Some fishes (e.g. catfishes) possess enlarged facial portions of the gustatory lemniscus associated with extensive facial nerve afferents from extraoral taste buds (Fig. 5.3B), while other species (e.g. carps and suckers) possess enlarged vagal portions of the taste pathways associated with numerous taste buds in the oropharyngeal cavity (Wullimann 1997) (Fig. 5.3C).

The most derived features of the gustatory pathways of taste-specialist teleosts are the primary gustatory centres of the medulla. These structures consist of enlargements of the dorsal plate, and can comprise up to 40% of the total mass of the brain (Miller & Evans 1965). In species that lack enhanced taste capabilities, the primary gustatory centre is a relatively

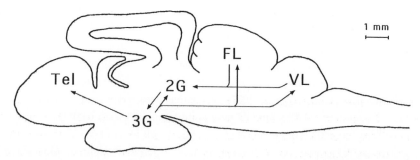

Fig. 5.2 Schematic diagram of gustatory pathways in the brain of teleosts (anterior/rostral is to the left). Taste information enters the hindbrain through termination of the facial, glossopharyngeal and vagal nerves in the facial (FL) and vagal (VL) lobes. The secondary gustatory tract projects from the FL and VL to the superior secondary gustatory nucleus (2G) in the rostral hindbrain. Cells from 2G project to the tertiary gustatory centre (3G) in the diencephalon, where there are ascending connections to the telencephalon (Tel) and descending pathways back to the FL and VL.

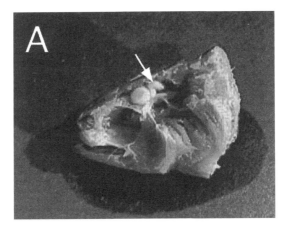

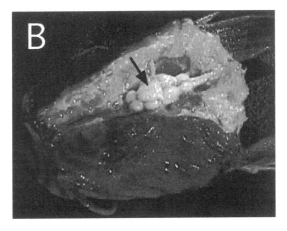

Fig. 5.3 External morphology of the brains of three fishes showing different specialisations for gustation (arrows indicate the cerebellum for reference). (A) The brain of a silver salmon (*Oncorhynchus kisutch*) which is not specialised for taste. Note the enlarged optic tecta (rostral to the cerebellum), for vision, and the lack of any enlargements on the medulla (caudal to the cerebellum). (B) Fishes specialised for extraoral taste, such as the Japanese sea catfish (*Plotosus lineatus*), possess enlargements of the dorsal medulla (caudal to the cerebellum). These represent the facial lobe and the relatively smaller vagal lobe, for extraoral and oropharyngeal taste input, respectively. (C) Many cyprinids and catostomids, such as the river carpsucker (*Carpiodes carpio*), possess extremely large vagal lobes corresponding to their highly developed oropharyngeal taste systems. These species also possess an enlarged palatal organ (asterisk) on the dorsal wall of the pharynx, a muscular pad used for sorting food from other materials by taste and texture. (*Continued.*)

undifferentiated rostral portion of the visceral sensory and motor columns of the brainstem (Fig. 5.3A) (Wullimann 1997). In the catfishes (Herrick 1905; Atema 1971; Finger 1978; Marui *et al.* 1988), goatfish (*Parupenus multifaciatus*) (Barry & Norton 1989) and rocklings (*Gaidropsarus mediterraneus* and *Ciliata mustela*) (Kotrschal & Whitear 1988), and other species which possess increased facial nerve input from extraoral taste buds, the primary gustatory centre is dominated by an enlarged facial lobe (Fig. 5.3B). The facial lobe of some of these species is organised in columns, or discernible collections of cell bodies, each receiving input from a discrete region of the taste epithelium (Fig. 5.4). For example, the facial lobe of channel catfish contains six longitudinal columns of cells that respond to taste or tactile

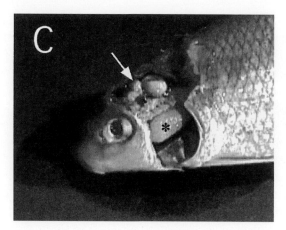

Fig. 5.3   (*Continued.*)

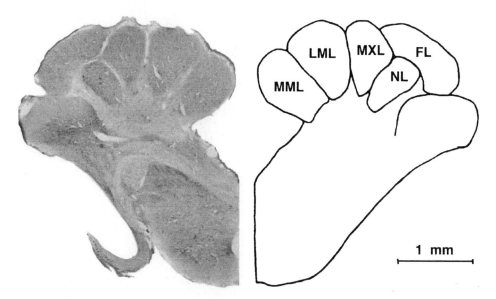

Fig. 5.4   A cross-sectional view of the facial lobe of the Japanese sea catfish (*Plotosus lineatus*), showing the distinct lobules associated with taste input from facial nerve branches from different regions of the body surface. Abbreviations: FL, flank lobule; LML, lateral mandibular barbel lobule; MML, medial mandibular barbel lobule; MXL, maxillary barbel lobule; NL, nasal barbel lobule.

stimulation of either the flank and face, the pectoral fin, or one of the four pairs of barbels surrounding the mouth (Marui & Caprio 1982; Hayama & Caprio 1989). Individual cells within each column respond to stimulation of even more restricted receptive fields, usually limited to only a few mm² of the receptive field for that column (Biedenbach 1973; Hayama & Caprio 1989). Another characteristic of the facial lobe in these species is that facial afferents terminate in a topographical representation of the extraoral body surface (Finger 1978; Marui & Caprio 1982; Kiyohara & Caprio 1996; Kiyohara *et al.* 1996), so that the receptive fields of neurones within a column form a map of the surface onto that column (Marui & Caprio

1982; Hayama & Caprio 1989). The restricted receptive fields within the facial lobe, and their topographical organisation, facilitate taste-generated food search behaviours by these fishes through connections between the facial lobe and descending motor pathways (Kanwal & Finger 1997). This is evidenced by behavioural studies of catfish, in which ablation of the facial lobe prevents the fish from locating food items (Atema 1971). Many cyprinid species also possess enlarged facial lobes in which the extraoral surface is represented topographically (Marui 1977; Marui & Funakoshi 1979; Kiyohara *et al.* 1985b; Puzdrowski 1987), although they lack the lobular organisation common in catfishes.

The primary gustatory centre of catostomid fishes, and many cyprinid species, is dominated by an enlarged vagal lobe (Fig. 5.3C), corresponding to increased vagal input from taste epithelium on specialised oropharyngeal structures (Evans, H.M. 1931; Evans, H.E. 1952; Miller & Evans 1965). The vagal lobe of these species is one of the most highly organised structures in their brain, with multiple layers of cells and fibres forming a complex lamination of sensory and motor regions within the lobe (Fig. 5.5) (Morita *et al.* 1983; Morita & Finger 1985). Essentially, the layers of the vagal lobe form three concentric zones: a superficial sensory zone covering the lateral and dorsal surfaces of the lobe, a deeper fibre zone containing both afferent and efferent connections to the lobe, and a medial motor zone containing motor neurones which innervate the musculature of the pharynx. Sensory fibres of the vagal nerve enter the brainstem ventral to the base of the vagal lobe and travel dorsally into the lobe through the central fibre zone, where they turn outward to terminate in the sensory zone.

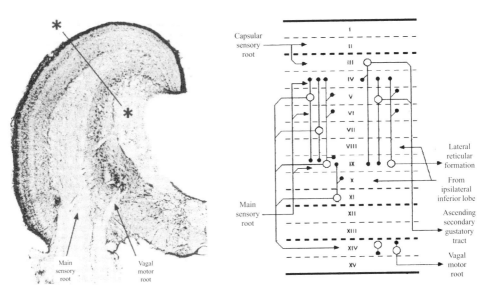

Fig. 5.5 A transverse section through the vagal lobe of the goldfish (*Carassius auratus*), showing the sensory and motor roots of the vagus nerve and the laminar structure of the vagal lobe. The schematic diagram of the vagal lobe circuitry represents a radial section (between the asterisks on the micrograph) with more superficial layers on top and deeper layers below. Cells within each layer (open circles) have distinct dendritic fields (filled circles) and projection patterns (arrows). Termination of afferent sensory fibres is indicated on the left, while connections of different layers of the lobe with other portions of the brainstem are indicated on the right. A prominent feature of the vagal lobe organisation is the localised reflex circuitry between sensory input and motor output through interneurones in layers V, VII, IX and XI (see text).

Input to the sensory zone is topographically organised, so that the pharyngeal epithelium is mapped onto the vagal lobe (Morita & Finger 1985). The sensory zone contains several layers of interneurones which relay vagal information to the motor zone and to the reticular formation and the secondary gustatory centre (Morita *et al.* 1983). Most of the interneurones in the sensory zone project radially inward to innervate motor neurones that are topographically organised as to the location of the muscles they innervate (Morita & Finger 1985). The vagal lobe maps for sensory input and motor output are in register, so that the pathway just described forms localised reflex circuits whereby taste or tactile stimulation of a particular portion of the pharyngeal epithelium produces muscle activity in that region of the pharynx (Finger 1988). This derived organisation of the cyprinid and catostomid vagal lobe allows the pharyngeal processing of potential food items characteristic of feeding behaviours in these species (Finger 1997). Support for this has been obtained following electrical stimulation of the vagal lobe of goldfish (*Carassius auratus*), which produces the display of typical feeding patterns (Fig. 5.6) (C.F. Lamb & L.S. Demski, unpublished results), and from ablation of the catfish vagal

Fig. 5.6   Appetitive and consummatory feeding behaviours in goldfish elicited by electrical stimulation of the vagal lobe. These behaviours include a postural orientation for bottom-searching and random intake and processing behaviours in the absence of any food items. They are produced only during the electrical stimulation.

lobe, which results in the inability of a feeding fish to ingest food (Atema 1971). Electrical stimulation of the vagal lobe also produces tachycardia in goldfish, indicating a role in non-gustatory visceral functions (Hornby & Demski 1988).

Neurones in the facial and vagal lobes project to the secondary gustatory centre in the isthmic region of the hindbrain, sometimes called the superior secondary gustatory nucleus (see Fig. 5.2). This nucleus relays taste and tactile information from the primary centre to the diencephalic tertiary gustatory centre (Herrick 1905; Finger 1978; Morita *et al.* 1980). Cells in the secondary gustatory nucleus respond to similar chemical and mechanical stimuli that are effective on peripheral fibres and on cells in the facial and vagal lobes, but receptive fields are much larger for secondary gustatory neurones (Marui 1981; Lamb & Caprio 1992). This suggests that discrete localisation of a stimulus is important in processing taste information within the medulla, but this feature is not maintained in the ascending lemniscal pathway (Lamb & Caprio 1992).

The tertiary gustatory centre consists of a collection of nuclei that has been a source of considerable confusion because of the apparent species-dependent orientation of cell groups and connections (Wulliman 1997). Ascending fibres from the secondary gustatory nucleus project to several nuclei in the diencephalon, often through collateral branches of single fibres (Lamb & Caprio 1993a; Lamb & Finger 1996). These nuclei have been identified in non taste-specialists green sunfish (*Lepomis cyanellus*) (Wullimann 1988) and Nile tilapia (*Oreochromis [Tilapia] niloticus*) (Yoshimoto *et al.* 1998), the extraoral-taste specialised channel catfish (Kanwal *et al.* 1988; Lamb & Caprio 1993a; Lamb & Finger 1996), and the oropharyngeal-taste specialised crucian carp (*Carassius carassius*) (Morita *et al.* 1980, 1983) and goldfish (Rink & Wullimann 1998). Some of the nuclei are responsible for lemniscal projections to the telencephalon – these include the nucleus centralis of the inferior lobe and the nucleus lobobulbaris (= nucleus of the posterior thalamus) of the channel catfish (Kanwal *et al.* 1988; Striedter 1990; Lamb & Caprio 1993a), nucleus centralis and nucleus diffusus of the inferior lobe (Airhart 1987) or the nucleus of the posterior thalamus in the goldfish (Rink & Wullimann 1998), the nucleus of the posterior thalamus in the common carp (Murakami *et al.* 1986), and the preglomerular tertiary gustatory nucleus in tilapia (Yoshimoto *et al.* 1998). Other nuclei in the tertiary gustatory centre project back to the primary and secondary centres to provide feedback to the hindbrain. The facial and vagal lobes receive descending input from the nucleus lobobulbaris in catfish (Finger 1978; Kanwal *et al.* 1988; Lamb & Caprio 1993a) and the nucleus of the posterior thalamus in goldfish (Rink & Wullimann 1998) and crucian carp (along with nucleus diffusus of the inferior lobe; Morita *et al.* 1983), while the secondary gustatory nucleus of catfish receives descending input from the nucleus of the lateral thalamus (= preglomerular tertiary gustatory nucleus) (Lamb *et al.* 1987; Kanwal *et al.* 1988; Lamb & Caprio 1993a). These descending projections appear to be derived features for the taste-specialist species, as none of the other teleosts examined possesses these connections (Wullimann 1997). While functional data for this region of the teleost brain are scarce, electrophysiological recording from channel catfish identified gustatory responses to amino acids and nucleotides in the nucleus lobobulbaris, nucleus of the lateral thalamus, and the nucleus centralis of the inferior lobe (Lamb & Caprio 1993b). Gustatory responses from these nuclei were generally more complex than those from the primary or secondary gustatory centres, suggesting a higher level of processing of taste information in the tertiary centre. Electrical stimulation of this region produced feeding responses in bluegill (*Lepomis macrochirus*) (Demski & Knigge 1971), blackjaw mouthbrooder (*Tilapia macrocephala*) (Demski 1973), and goldfish (Savage & Roberts 1975). Further, lesioning of this region in goldfish produced animals with feeding deficiencies (aphagia or hypophagia) (Roberts & Savage 1978). These functional studies of the tertiary gustatory centre suggest its importance in taste-related decision making in teleosts.

Telencephalic gustatory areas have been identified in channel catfish (Kanwal *et al.* 1988; Lamb & Caprio 1993a), goldfish (Airhart 1987; Rink & Wulliman 1998), Nile tilapia (Yoshimoto *et al.* 1998) and common carp (Murakami *et al.* 1986). Generally, gustatory nuclei in this region include medial and central portions of the area dorsalis, which receive projections from the tertiary gustatory centre in catfish, goldfish and tilapia (Kanwal *et al.* 1988; Lamb & Caprio 1993a; Rink & Wulliman 1998; Yoshimoto *et al.* 1998), and the intermediate portion of area ventralis, which receives direct projections from the secondary gustatory nucleus in tilapia (Yoshimoto *et al.* 1998). While it is difficult to compare telencephalic organisation between teleosts and other vertebrates, these connections might represent the equivalent of gustatory pathways to the amygdala of mammals (Northcutt & Davis 1983; Kanwal *et al.* 1988; Yoshimoto *et al.* 1998). However, additional functional studies of both the tertiary and telencephalic gustatory centres in teleosts are necessary before possible homologies can be proposed.

5.4 Taste and feeding behaviours

Feeding behaviours of fishes are generally categorised into three sequential phases – beginning with an alerting or arousal phase, followed by an appetitive phase where potential food items are identified and located, and concluding with a consummatory phase consisting of food intake and either ingestion or rejection (Atema 1971; Jones 1992). Corresponding to the electrophysiological data discussed above, amino acids are potent stimulants for each of these phases in many teleost species (Table 5.2). While multiple chemosensory systems, and other sensations as well, can play important roles in any of these phases of feeding, this discussion will emphasise the roles gustation plays in each phase.

The arousal phase serves as a 'primer' for the chemosensory systems that are used in the subsequent appetitive responses to the presence of food. This phase can be initiated through any of the sensory systems available to a particular species. Arousal consists of increased muscular activity of the mouth, pharynx, appendages (fins, barbels, etc.) and flank, and results in twitching or exaggerated movements (Jones 1992). These responses are especially notable in fishes with specialised extraoral gustatory structures. Catfishes stiffen their bodies, twitch their barbels, and increase respiratory movements following chemical detection of food items or amino acids (Herrick 1904; Parker & Sheldon 1912; Olmsted 1918; Bardach *et al.* 1967; Atema 1971; Valentincic & Caprio 1994a). A similar pattern of arousal is displayed by Hawaiian goatfish (*Parupeneus porphyreus*) upon presentation of food extracts (Holland 1978). Gustatory stimulation is responsible for arousal behaviours in both channel catfish (Valentincic & Caprio 1994b) and goatfish (Holland 1978), as the responses are identical in both intact and anosmic fish. Cod (*Gadus morhua*) respond to food extracts and amino acids with exaggerated oropharyngeal activity and postural movements (Ellingsen & Doving 1986). Arousal responses, in some species, involve changes in activity patterns that resemble behaviours

Table 5.2 Effective amino acid stimuli from behavioural experiments with various fish species. (Adapted from Jones, 1992.)

Species	Effective stimuli[1]	Reference
Zebrafish *Brachydanio rerio*	Ala	Scarfe *et al.* (1985)
Goldfish *Carassius auratus*	Ala, Arg, Pro	Lamb & Finger (1995)
Sea bream *Chrysophrys major*	Ala, Val, Gly, Ser, Pro, Arg, Gln	Goh & Tamura (1980b)
Herring *Clupea harengus*	Glu, Asp, Gly, Met, Ala, Pro	Dempsey (1978)
Puffer *Fugu pardalis*	Ala, Gly, Pro, Ser, Bet	Hidaka *et al.* (1978)
Killifish *Fundulus heteroclitus*	Cys, Ser, Ala, His, Gly, Thr, Leu, Ile, Val, Glu, Tau, Arg, Met, Asn, Lys, Trp, Asp, GABA, Pro-OH	Sutterlin (1975)
Cod *Gadus morhua*	Gly, Ala, Arg, Pro	Ellingsen & Doving (1986)
Yellow and brown bullhead *Ictalurus natalis* and *I. nebulosus*	CysH	Bardach *et al.* (1967)
Channel catfish *Ictalurus punctatus*	Ala, Arg, Pro	Valentincic & Caprio (1992, 1994b)
Silverside *Menidia menidia*	Ala, Thr, Met	Sutterlin (1975)
Oriental weatherfish *Misgurnus anguillicaudatus*	His, Arg, Gly, Lys, Phe, Tyr, Ile, Thr, Cys, Asn	Harada *et al.* (1987)
Rainbow trout *Oncorhynchus mykiss*	Leu, Ile, Pro, Met, Glu, Try, Arg, Phe, Tau	Jones (1989, 1990)
Pigfish *Orthopristis chrysopterus*	Bet	Carr *et al.* (1977)
Searobin *Prionotus carolinus*	Phe, Pro, Asp, Try, Glu, Gly	Bardach & Case (1965)
Flounder *Pseudooleuronectes americanus*	Gly, Met, Ala, Asn, Cys, Glu, Leu, Asp, Ser, Bet, GABA	Sutterlin (1975)
Arctic charr *Salvelinus alpinus*	Ala, Ser	Jones & Hara (1985)
Lake trout *Salmo salar*	Ala, Arg, Glu, His, Leu, Ile, Lys, Met, Pro, Gly, Val	Mearns (1989)
Brown trout *Salmo trutta*	Ala, Arg, Pro, Gly, His	Mearns (1989)
Yellowtail *Seriola quinqueradiata*	His, Arg, Lys, Gly, Val, Thr, Met, Gln, Asn, Orn, Cyt	Harada *et al.* (1987)
Sole *Solea solea*	Bet	Mackie & Mitchell (1982)
African cichlid *Tilapia zillii*	Glu, Asp, Lys, Ala, Ser	Johnsen & Adams (1986)

[1] Abbreviations: Abu, L-α-aminobutyric acid; AGPA, L-α-amino-β-guanidino propionic acid; Ala, L-alanine; Arg, L-arginine; Asp, L-aspartic acid; Bet, betaine; Cys, L-cysteine; D-Ala, D-alanine; Gln, L-glutamine; Glu, L-glutamic acid; Gly, glycine; Leu, L-leucine; Phe, L-phenylalanine; Pro, L-proline; Pro-OH, hydroxy-L-proline; Ser, serine; Tau, taurine; Thr, L-threonine; Try, L-tryptophan; Val, L-valine.

displayed during the appetitive phase, with the exception that arousal behaviours are non-directed (Jones 1992).

Continued chemical stimulation results in searching behaviours that signal the appetitive phase of feeding. These typically include swimming patterns or other movements that are directed toward locating a chemical stimulus. Gustatory input has a profound effect on appetitive behaviours, especially in fishes with highly developed extraoral (facial) taste systems. Species with gustatory barbels or fin rays respond to the presence of taste stimuli with an increase in swimming activity so that those structures are used to search for the source of the stimulation. Catfish drag their barbels along the substrate and turn abruptly once the stimulus is located (Herrick 1904). Goatfish actively probe the substrate with their barbels as they swim rapidly across the bottom (Holland 1978). Cod initiate a 'bottom search behaviour' including a backward swimming motion with the body oriented so the chin barbel drags the bottom (Ellingsen & Doving 1986). Other gadid fishes, such as hake (*Urophycis tenuis*), tomcod (*Microgadus tomcod*) and rocklings, use the gustatory sensitivity of their pelvic fins (and barbels) to locate food sources (Herrick 1904), as does the gourami (*Trichogaster trichopterus*) (Scharrer *et al.* 1947). It is important to note that, in all of these studies, tactile information provided by touching the source of the taste stimuli was more effective in producing consummatory behaviours than was gustatory information alone. This has led to the suggestion that gustation is more important as a close-range chemosensory system, while olfaction is probably more effective at greater distances. However, catfish respond with searching behaviours to low concentrations of chemical stimuli, even in the absence of olfactory input (Bardach *et al.* 1967; Valentincic & Caprio 1992, 1994a; Valentincic *et al.* 1994). Other species, such as rainbow trout (Valentincic & Caprio 1997) and Arctic charr (Jones & Hara 1985; Olsen *et al.* 1986), perform searching behaviours when stimulated with amino acids (L-alanine, L-proline, L-arginine, L-serine, glycine and betaine), but these species appear to detect the stimuli through olfaction and not gustation (Valentincic & Caprio 1997).

Once a potential food source has been identified and located, it is taken into the oropharynx where its palatability is assessed through intraoral gustation. Fishes with highly developed vagal gustatory systems may display complex consummatory feeding behaviours (Fig. 5.7), but even species that do not rely on taste information for the arousal and appetitive phases of feeding utilise taste information in deciding which items to ingest and which to reject. A notable example from the former category is the goldfish, whose vagal taste system has been described above. The consummatory phase of feeding in goldfish consists of taking food into the mouth, followed by behaviours allowing the fish to select palatable from unpalatable particles, and culminating in the ingestion of food items and the expulsion of those that are unpalatable (Fig. 5.7). The selection (sorting) behaviours, 'rinsing' and 'backwashing', allow goldfish to identify food through gustatory stimulation of vagal taste buds in the pharyngeal epithelium. For example, particles containing the amino acids L-alanine, L-arginine or L-proline are selectively ingested, while particles containing quinine or caffeine are rejected

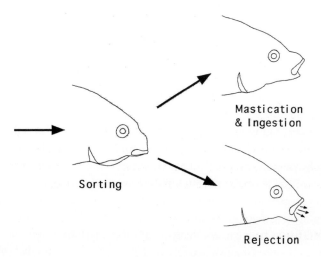

Fig. 5.7   Feeding behaviour pattern of cyprinid and catostomid fishes with specialised oropharyngeal taste systems. Food is sucked into the oral cavity and is held in the pharynx where it is sorted from non-food items. Food is moved back to the oesophagus for ingestion, while non-food items, or those deemed unpalatable, are rejected through the mouth and opercular slits.

(Lamb & Finger 1995). When goldfish take particles containing a mixture of these stimuli, the selection behaviours are performed repeatedly as the fish attempts to separate the palatable from the unpalatable components of the particle. This consummatory pattern is common to other cyprinid (Sibbing *et al.* 1986) and catostomid (C.F. Lamb, unpublished data) species that possess elaborate vagal lobes. Channel catfish also exhibit biting, snapping, gulping and other consummatory behaviours when stimulated by solutions of L-alanine, L-arginine or L-proline (Valentincic & Caprio 1992, 1994b). Examples of the non-taste-specialised fishes that respond to amino acids with consummatory behaviours include the rainbow trout (Adron & Mackie 1978; Jones 1989, 1990), the Arctic charr (Jones & Hara 1985), and the herbivorous redbelly tilapia (*Tilapia zillii*) (Johnsen & Adams 1986). These species produce similar consummatory responses to several amino acids, including L-alanine, L-aspartate, L-glutamate, L-phenylalanine, L-leucine, L-lysine and L-proline. Differences in relative effectiveness that exist between the species reflect differences in the amino acid composition of their preferred food sources, as is generally the case with many teleost species (Carr *et al.* 1996).

5.5 Conclusions

Many fishes possess highly developed gustatory, along with other chemosensory, systems, providing evidence for the importance of chemoreception in an aquatic environment. Much of what is known about the sense of taste in vertebrates comes from the use of fish species as models. While it is tempting to look at the body of research on taste in fish and draw generalisations and simplifications, it is important to remember that even the ray-finned fishes comprise a very large and diverse collection of species. It should not be surprising that a wide variety of species respond to chemicals in the environment in order to find food and to communicate. The result has been the evolution of taste systems with a varying degree of species-specificity in terms of the spectra of effective taste stimuli, peripheral structures and mechanisms, central pathways involved in gustatory processing, and feeding behaviours. However, some taste features appear to be common to most teleosts studied. Teleosts are most sensitive and responsive to small organic molecules such as L-amino acids and certain derivatives, nucleotides, and other molecules such as bile salts and carboxylic acids. Additionally, teleosts share a common vertebrate neural organisation for processing gustatory information in the brain, along with fundamental similarities in behavioural responses to taste stimulation. Further studies of taste in fish should expand on the investigations made in catfishes and carps to divergent species. Anatomical, physiological and behavioural studies are needed in combination to provide a systematic analysis of gustation within a species. As comprehensive studies are completed on a variety of fish species, the roles that taste plays in the survival and success of this diverse group of organisms will be better understood.

5.6 Acknowledgements

The author thanks Dr Anne Hansen for supplying images for Fig. 5.1, and Bradley Young at South Dakota State University and Dr Jay Hatch at the Bell Museum of Natural History for supplying several catostomid specimens. Some of the catostomid work was conducted by Nathan Steinle, Michelle Sulzbach, Brett Theeler, and Michelle Vieyra in the author's laboratory at Black Hills State University. These studies were supported, in part, by grants from the Faculty Research Committee and the Joseph F. and Martha P. Nelson Scholarship Fund at Black Hills State University.

5.7 References

Adron, J.W. & Mackie, A.M. (1978) Studies on the chemical nature of feeding stimulants for rainbow trout, *Salmo gairdneri* Richardson. *Journal of Fish Biology*, **12**, 303–310.

Airhart, M.J. (1987) Axonal connections between telencephalon and hypothalamus in goldfish. *Anatomical Record*, **218**, 5A.

Atema, J. (1971) Structures and functions of the sense of taste in the catfish, *Ictalurus natalis. Brain Behaviour and Evolution*, **4**, 273–294.

Bardach, J.E. & Case, J. (1965) Sensory capabilities of the modified fins of squirrel hake (*Urophysis chuss*) and searobins (*Prionotus carolinus* and *P. evolans*). *Copeia*, **1965**(2), 194–206.

Bardach, J.E., Todd, J.H. & Crickmer, R. (1967) Orientation by taste in fish of the genus *Ictalurus. Science*, **155**, 1276–1278.

Barry, M.A. & Norton, L.E. (1989) Organization of primary gustatory nuclei in a goatfish, *Parupenus multifaciatus. American Zoologist*, **29**, 13A.

Biedenbach, M.A. (1973) Functional properties and projection areas of cutaneous receptors in catfish. *Journal of Comparative Physiology*, **84**, 227–250.

Brand, J.G., Bryant, B.P., Cagan, R.H. & Kalinoski, D.L. (1987) Biochemical studies of taste sensation. XIII. Enantiomeric specificity of alanine taste receptor sites in catfish, *Ictalurus punctatus. Brain Research*, **416**, 119–128.

Cagan, R.H. (1986) Biochemical studies of taste sensation. XII. Specificity of binding of taste ligands to a sedimentable fraction from catfish taste tissue. *Comparative Biochemistry and Physiology*, **85A**, 355–358.

Caprio, J. (1975) High sensitivity of catfish taste receptors to amino acids. *Comparative Biochemistry and Physiology*, **52A**, 247–251.

Caprio, J. (1978) Olfaction and taste in the channel catfish: an electrophysiological study of the responses to amino acids and derivatives. *Journal of Comparative Physiology*, **123A**, 357–371.

Caprio, J., Brand, J.G., Teeter, J.H., *et al.* (1993) The taste system of channel catfish: from biophysics to behavior. *Trends in Neuroscience*, **16**, 192–197.

Carr, W.E.S., Blumenthal, K.M. & Netherton, J.C., III (1977) Chemoreception in the pigfish, *Orthopristis chrysopterus*: the contribution of amino acids and betaine to stimulation of feeding behavior by various extracts. *Comparative Biochemistry and Physiology*, **58A**, 69–73.

Carr, W.E.S., Netherton, J.C., III, Gleeson, R.A. & Derby, C.D. (1996) Stimulants of feeding behavior in fish: analyses of tissues of diverse marine organisms. *Biological Bulletin*, **190**, 149–160.

Davenport, C.J. & Caprio, J. (1982) Taste and tactile recordings from the ramus recurrens facialis innervating flank taste buds in the catfish. *Journal of Comparative Physiology*, **147A**, 217–229.

Dempsey, C.H. (1978) Chemical stimuli as a factor in feeding and intraspecific behaviour of herring larvae. *Journal of the Marine Biological Association of the UK*, **58**, 739–747.

Demski, L.S. (1973) Feeding and aggressive behavior evoked by hypothalamic stimulation in a cichlid fish. *Comparative Biochemistry and Physiology*, **44A**, 685–692.

Demski, L.S. & Knigge, K.M. (1971) The telencephalon and hypothalamus of the bluegill (*Lepomis macrochirus*): Evoked feeding, aggressive and reproductive behavior with representative frontal sections. *Journal of Comparative Neurology*, **143**, 1–16.

Doving, K.B. & Selset, R. (1980) Behavior patterns in cod released by electrical stimulation of olfactory tract bundlets. *Science*, **207**, 559–560.

Doving, K.B., Selset, R. & Thommesen, G. (1980) Olfactory sensitivity to bile acids in salmonid fishes. *Acta Physiologica Scandinavica*, **108**, 123–131.

Ellingsen, O.F. & Doving, K.B. (1986) Chemical fractionation of shrimp extracts inducing bottom food search behavior in cod (*Gadus morhua* L.). *Journal of Chemical Ecology*, **12**, 155–168.

Evans, H.E. (1952) The correlation of brain pattern and feeding habits in four species of cyprinid fishes. *Journal of Comparative Neurology*, **97**, 133–142.

Evans, H.M. (1931) A comparative study of the brains in British cyprinoids in relation to their habits of feeding, with special reference to the anatomy of the medulla oblongata. *Proceedings of the Royal Society, London (B)*, **108**, 233–257.

Ezeasor, D.N. (1982) Distribution and ultrastructure of taste buds in the oropharyngeal cavity of the rainbow trout, *Salmo gairdneri* Richardson. *Journal of Fish Biology*, **20**, 53–68.

Finger, T.E. (1978) Gustatory pathways in the bullhead catfish. II. Facial lobe connections. *Journal of Comparative Neurology*, **180**, 691–706.

Finger, T.E. (1983) The gustatory system in teleost fish. In: *Fish Neurobiology, Vol. 1* (eds R.G. North-cutt & R.E. Davis), pp. 285–310. University of Michigan Press, Ann Arbor.

Finger, T.E. (1984) Gustatory nuclei and pathways in the central nervous system. In: *Neurobiology of Taste and Smell* (eds T.E. Finger & W.L. Silver), pp. 331–353. John Wiley & Sons, New York.

Finger, T.E. (1988) Sensorimotor mapping and oropharyngeal reflexes in goldfish, *Carassius auratus*. *Brain Behaviour and Evolution*, **31**, 17–24.

Finger, T.E. (1997) Feeding patterns and brain evolution in ostariophysean fishes. *Acta Physiologica Scandinavica*, **161** (Suppl. **638**), 59–66.

Finger, T.E. & Morita, Y. (1985) Two gustatory systems: facial and vagal gustatory nuclei have different brainstem connections. *Science*, **227**, 776–778.

Finger, T.E., Bryant, B.P., Kalinoski, D.L., *et al.* (1996) Differential localization of putative amino acid receptors in taste buds of the channel catfish, *Ictalurus punctatus*. *Journal of Comparative Neurology*, **373**, 129–138.

Goh, Y. & Tamura, T. (1980a) Olfactory and gustatory responses to amino acids in two marine teleosts: red sea bream and mullet. *Comparative Biochemistry and Physiology*, **66C**, 217–224.

Goh, Y. & Tamura, T. (1980b) Effect of amino acids on the feeding behaviour in red sea bream. *Comparative Biochemistry and Physiology*, **66C**, 225–229.

Goldstein, N. & Cagan, R.H. (1982) Biochemical studies of taste sensation. X. Monoclonal antibody against L-alanine binding activity of catfish taste epithelium. *Proceedings of the National Academy of Science USA*, **79**, 7595–7597.

Hara, T.J. (1993) Chemoreception. In: *The Physiology of Fishes* (ed. D.H. Evans), pp. 191–218. CRC Press, Boca Raton.

Hara, T.J. (1994) Olfaction and gustation in fish: an overview. *Acta Physiologica Scandinavica*, **152**, 207–217.

Hara, T.J., Macdonald, S., Evans, R.E., Marui, T. & Arai, S. (1984) Morpholine, bile acids, and skin mucus as possible chemical cues in salmonid homing: electrophysiological re-evaluation. In: *Mechanisms of Migration in Fishes* (eds J.D. McCleave, G.P. Arnold, J.J. Dodson & W.H. Neill), pp. 363–378. Plenum, New York.

Hara, T.J., Sveinsson, T., Evans, R.E. & Klaprat, D.A. (1993) Morphological and functional characteristics of the olfactory and gustatory organs of three *Salvelinus* species. *Canadian Journal of Zoology*, **71**, 414–423.

Harada, K. (1986) Feeding attraction activities of nucleic acid-related compounds for abalone, oriental weatherfish, and yellowtail. *Bulletin of the Japanese Society of Scientific Fisheries*, **52**, 1961–1968.

Harada, K., Eguchi, A. & Kurosaki, Y. (1987) Feeding attraction activities in the combinations of amino acids and other compounds for abalone, oriental weatherfish, and yellowtail. *Nippon Suisan Gakkaishi*, **53**, 1483–1489.

Hayama, T. & Caprio, J. (1989) Lobule structure and somatotopic organization of the medullary facial lobe in the channel catfish, *Ictalurus punctatus*. *Journal of Comparative Neurology*, **285**, 9–17.

Herrick, C.J. (1904) The organ and sense of taste in fishes. *Bulletin of the U.S. Fisheries Communication* (for 1902), **22**, 237–272.

Herrick, C.J. (1905) Central gustatory paths in brains of bony fishes. *Journal of Comparative Neurology*, **15**, 375–456.

Hidaka, I. & Ishida, Y. (1985) Gustatory response in the shimaisaki (tigerfish) *Therapon oxyrhynchus*. *Bulletin of the Japanese Society of Scientific Fisheries*, **51**, 387–391.

Hidaka, I., Ohsugi, T. & Kubomatsu, T. (1978) Taste receptor stimulation and feeding behaviour in the puffer, *Fugu pardalis* I. Effect of single chemicals. *Chemical Sensitivity and Flavour*, **3**, 341–354.

Hidaka, I., Ohsugi, T. & Yamamoto, Y. (1985) Gustatory response in the young yellowtail *Seriola quinqueradiata*. *Bulletin of the Japanese Society of Scientific Fisheries*, **51**, 21–24.

Hoagland, H. (1933) Specific nerve impulses from gustatory and tactile receptors in catfish. *Journal of General Physiology*, **16**, 385–393.

Holland, K. (1978) Chemosensory orientation to food by a Hawaiian goatfish (*Parupeneus porphyreus*, Mullidae). *Journal of Chemical Ecology*, **4**, 173–186.

Hornby, P.J. & Demski, L.S. (1988) Functional-anatomical studies of neural control of heart rate in goldfish. *Brain Behaviour and Evolution*, **31**, 181–192.

Ishida, Y. & Hidaka, I. (1987) Gustatory response profiles for amino acids, glycinebetaine, and nucleotides in several marine teleosts. *Nippon Suisan Gakkaishi*, **53**, 1391–1398.

Jakubowski, M. & Whitear, M. (1990) Comparative morphology and cytology of taste buds in teleosts. *Zietschrift Mikroskopie Anatomica Forscheng (Leipzig)*, **104**, 529–560.

Johnsen, P.B. & Adams, M.A. (1986) Chemical feeding stimulants for the herbivorous fish, *Tilapia zillii*. *Comparative Biochemistry and Physiology*, **83A**, 109–112.

Jones, K.A. (1989) The palatability of amino acids and related compounds to rainbow trout, *Salmo gairdneri* Richardson. *Journal of Fish Biology*, **34**, 149–160.

Jones, K.A. (1990) Chemical requirements of feeding in rainbow trout, *Oncorhynchus mykiss* (Walbum); palatability studies on amino acids, amides, amines, alcohols, aldehydes, saccharides, and other compounds. *Journal of Fish Biology*, **37**, 413–423.

Jones, K.A. (1992) Food search behaviour in fish and the use of chemical lures in commercial and sports fishing. In: *Fish Chemoreception* (ed. T.J. Hara), pp. 288–320. Chapman & Hall, London.

Jones, K.A. & Hara, T.J. (1985) Behavioural responses of fish to chemical cues: results from a new bioassay. *Journal of Fish Biology*, **27**, 495–504.

Kalinoski, D.L., Bryant, B.P., Shaulsky, G., Brand, J.G. & Harpaz, S. (1989) Specific L-arginine taste receptor sites in the catfish, *Ictalurus punctatus*: biochemical and neurophysiological characterization. *Brain Research*, **488**, 163–173.

Kalinoski, D.L., Johnson, L.C., Bryant, B.P. & Brand, J.G. (1992) Selective interactions of lectins with amino acid taste receptor sites of the channel catfish. *Chemical Senses*, **17**, 381–390.

Kanwal, J.S. & Caprio, J. (1983) An electrophysiological investigation of the oro-pharyngeal (IX-X) taste system in the channel catfish, *Ictalurus punctatus*. *Journal of Comparative Physiology*, **150A**, 345–357.

Kanwal, J.S. & Finger, T.E. (1992) Central representation and projections of gustatory systems. In: *Fish Chemoreception* (ed. T.J. Hara), pp. 79–102. Chapman & Hall, London.

Kanwal, J.S. & Finger, T.E. (1997) Parallel medullary gustatospinal pathways in a catfish: possible neural substrates for taste-mediated food search. *Journal of Neuroscience*, **17**, 4873–4885.

Kanwal, J.S., Finger, T.E. & Caprio, J. (1988) Forebrain connections of the gustatory system in ictalurid catfishes. *Journal of Comparative Neurology*, **278**, 353–376.

Kiyohara, S. & Caprio, J. (1996) Somatotopic organization of the facial lobe of the sea catfish *Arius felis* studied by transganglionic transport of horseradish peroxidase. *Journal of Comparative Neurology*, **368**, 121–135.

Kiyohara, S. & Hidaka, I. (1991) Receptor sites for alanine, proline, and betaine in the palatal taste system of the puffer, *Fugu pardalis*. *Journal of Comparative Physiology*, **69A**, 523–530.

Kiyohara, S., Hidaka, I. & Tamura, T. (1975) Gustatory response in the puffer - II. Single fiber analyses. *Bulletin of the Japanese Society of Scientific Fisheries*, **41**, 383–391.

Kiyohara, S., Yamashita, S. & Kitoh, J. (1980) Distribution of taste buds on the lips and inside the mouth in the minnow, *Pseudorasbora parva*. *Physiological Behaviour*, **24**, 1143–1147.

Kiyohara, S., Yamashita, S. & Harada, S. (1981) High sensitivity of minnow gustatory receptors to amino acids. *Physiological Behaviour*, **26**, 1103–1108.

Kiyohara, S., Hidaka, I., Kitoh, J. & Yamashita, S. (1985a) Mechanical sensitivity of the facial nerve fibers innervating the anterior palate of the puffer, *Fugu pardalis*, and their central projection to the primary taste center. *Journal of Comparative Physiology*, **157A**, 705–716.

Kiyohara, S., Shiratani, T. & Yamashita, S. (1985b) Peripheral and central distribution of major branches of the facial taste nerve in the carp. *Brain Research*, **325**, 57–69.

Kiyohara, S., Houman, H., Yamashita, S., Caprio, J. & Marui, T. (1986) Morphological evidence for a direct projection of trigeminal nerve fibers to the primary gustatory center in the sea catfish *Plotosus anguillaris*. *Brain Research*, **379**, 353–357.

Kiyohara, S., Kitoh, J., Shito, A. & Yamashita, S. (1996) Anatomical studies of the medullary facial lobe in the sea catfish *Plotosus lineatus*. *Fisheries Science*, **62**, 511–519.

Kiyohara, S., Yamashita, S., Lamb, C.F. & Finger, T. (1999) Distribution of trigeminal fibers in the primary facial gustatory center of channel catfish, *Ictalurus punctatus*. *Brain Research*, **841**, 93–100.

Kohbara, J., Michel, W. & Caprio, J. (1992) Responses of single facial taste fibers in the channel catfish, *Ictalurus punctatus*, to amino acids. *Journal of Neurophysiology*, **68**, 1012–1026.

Kotrschal, K. & Whitear, M. (1988) Chemosensory anterior dorsal fin in rocklings (*Gaidropsarus* and *Ciliata*, teleostei, gadidae): somatotopic representation of the ramus recurrens facialis as revealed by transganglionic transport of HRP. *Journal of Comparative Neurology*, **268**, 109–120.

Kumazawa, T., Brand, J.G. & Teeter, J.H. (1998) Amino acid-activated channels in the catfish taste system. *Biophysical Journal*, **75**, 2757–2766.

Lamb, C.F. & Caprio, J. (1992) Convergence of oral and extraoral information in the superior secondary gustatory nucleus of the channel catfish. *Brain Research*, **588**, 201–211.

Lamb, C.F. & Caprio, J. (1993a) Diencephalic gustatory connections in the channel catfish. *Journal of Comparative Neurology*, **337**, 400–418.

Lamb, C.F. & Caprio, J. (1993b) Taste and tactile responsiveness of neurons in the posterior diencephalon of the channel catfish. *Journal of Comparative Neurology*, **337**, 419–430.

Lamb, C.F. & Finger, T.E. (1995) Gustatory control of feeding behavior in goldfish. *Physiological Behaviour*, **57**, 483–488.

Lamb, C.F. & Finger, T.E. (1996) Axonal projection patterns of neurons in the secondary gustatory nucleus of channel catfish. *Journal of Comparative Neurology*, **365**, 585–593.

Lamb, C.F., Marui, T. & Kasahara, Y. (1987) Anatomical and electrophysiological investigation of the superior secondary gustatory nucleus of the Japanese sea catfish. *Society of Neuroscience Abstracts*, **13**, 406.

Littleton, J.T., Kohbara, J., Michel, W. & Caprio, J. (1989) Gustatory responses of the channel catfish, *Ictalurus punctatus*, to nucleotides and related substances. *Chemical Sensitivity*, **14**, 732.

Mackie, A.M. & Mitchell, A. I. (1982) Further studies on the chemical control of feeding behaviour in the Dover sole, *Solea solea*. *Comparative Biochemistry and Physiology*, **73A**, 89–93.

Marui, T. (1977) Taste responses in the facial lobe of the carp, *Cyprinus carpio* L. *Brain Research*, **130**, 287–298.

Marui, T. (1981) Taste responses in the superior secondary gustatory nucleus of the carp, *Cyprinus carpio* L. *Brain Research*, **217**, 59–68.

Marui, T. & Caprio, J. (1982) Electrophysiological evidence for the topographic arrangement of taste and tactile neurons in the facial lobe of the channel catfish. *Brain Research*, **231**, 185–190.

Marui, T. & Caprio, J. (1992) Teleost Gustation. In: *Fish Chemoreception* (ed. T.J. Hara), pp. 171–198. Chapman & Hall, London.

Marui, T. & Funakoshi, M. (1979) Tactile input to the facial lobe of the carp, *Cyprinus carpio* L. *Brain Research*, **177**, 479–488.

Marui, T., Evans, R.E., Zielinski, B. & Hara, T.J. (1983a) Gustatory responses of the rainbow trout (*Salmo gairdneri*) palate to amino acids and derivatives. *Journal of Comparative Physiology*, **153A**, 423–433.

Marui, T., Harada, S. & Kasahara, Y. (1983b) Gustatory specificity for amino acids in the facial taste system in the carp, *Cyprinus carpio* L. *Journal of Comparative Physiology*, **153A**, 299–308.

Marui, T., Kiyohara, S., Caprio, J. & Kasahara, Y. (1988) Topographical organization of taste and tactile neurons in the facial lobe of the sea catfish, *Plotosus lineatus*. *Brain Research*, **446**, 178–182.

Mearns, K.J. (1989) Behavioural responses of salmonid fry to low amino acid concentrations. *Journal of Fish Biology*, **34**, 223–232.

Michel, W. & Caprio, J. (1991) Responses of single facial taste fibers in the sea catfish, *Arius felis*, to amino acids. *Journal of Neurophysiology*, **66**, 247–260.

Miller, R.J. & Evans, H.E. (1965) External morphology of the brain and lips in catostomid fishes. *Copeia*, **1965**(4), 467–487.

Morita, Y. & Finger, T.E. (1985) Reflex connections of the facial and vagal gustatory systems in the brainstem of the bullhead catfish, *Ictalurus nebulosus*. *Journal of Comparative Neurology*, **231**, 547–558.

Morita, Y., Ito, H. & Masai, H. (1980) Central gustatory paths in the crucian carp, *Carassius carassius*. *Journal of Comparative Neurology*, **191**, 119–132.

Morita, Y., Murakami, T. & Ito, H. (1983) Cytoarchitecture and topographic projections of the gustatory centers in a teleost, *Carassius carassius*. *Journal of Comparative Neurology*, **218**, 378–394.

Murakami, T., Ito, H. & Morita, Y. (1986) Telencephalic afferent nuclei in the carp diencephalon, with special reference to fiber connections of the nucleus preglomerulosus pars lateralis. *Brain Research*, **382**, 97–103.

Northcutt, R.G. & Davis, R.E. (1983) Telencephalic organization in ray-finned fishes. In: *Fish Neurobiology, Vol. 2* (eds R.G. Northcutt & R.E. Davis), pp. 203–236. University of Michigan Press, Ann Arbor.

Olmsted, J.M.D. (1918) Experiments on the nature of the sense of smell in the common catfish, *Amiurus nebulosus* (LeSeur). *American Journal of Physiology*, **46**, 443–458.

Olsen, K.H., Karlsson, L. & Helander, A. (1986) Food search behavior in arctic charr, *Salvelinus alpinus* (L.), induced by food extracts and amino acids. *Journal of Chemical Ecology*, **12**, 1987–1998.

Parker, G.H. & Sheldon, R.E. (1912) The sense of smell in fishes. U.S. Bur. Fish. Document Number. 775. *Bulletin of the Bureau of Fisheries*, **32**, 33–46.

Puzdrowski, R.L. (1987) The peripheral distribution and central projections of the sensory rami of the facial nerve in goldfish, *Carassius auratus*. *Journal of Comparative Neurology*, **259**, 382–392.

Puzdrowski, R.L. (1988) Afferent projections of the trigeminal nerve in the goldfish, *Carassius auratus*. *Journal of Morphology*, **193**, 131–147.

Reutter, K. (1978) Taste organ in the bullhead (Teleostei). *Advances in Anatomy, Embryology and Cell Biology*, **55**, 1–98.

Reutter, K. & Witt, M. (1993) Morphology of vertebrate taste organs and their nerve supply. In: *Mechanisms of Taste Transduction* (eds S.A. Simon & S.D. Roper), pp. 29–82. CRC Press, Boca Raton.

Rink, E. & Wullimann, M.F. (1998) Some forebrain connections of the gustatory system in the goldfish *Carassius auratus* visualized by separate DiI application to the hypothalamic inferior lobe and the torus lateralis. *Journal of Comparative Neurology*, **394**, 152–170.

Roberts, M.G. & Savage, G.E. (1978) Effects of hypothalamic lesions on the food intake of the goldfish (*Carassius auratus*). *Brain Behaviour and Evolution*, **15**, 150–164.

Savage, G.E. & Roberts, M.G. (1975) Behavioural effects of electrical stimulation of the hypothalamus of the goldfish (*Carassius auratus*). *Brain Behaviour and Evolution*, **12**, 42–56.

Scarfe, A.D., Steele, C.W. & Rieke, G.K. (1985) Quantitative chemobehavior of fish: an improved methodology. *Environmental Biology of Fishes*, **13**, 183–194.

Scharrer, E., Smith, S.W. & Palay, S.L. (1947) Chemical sense and taste in the fishes, *Prionotus* and *Trichogaster*. *Journal of Comparative Neurology*, **68**, 183–198.

Sibbing, F.A., Osse, J.W.M. & Terlouw, A. (1986) Food handling in the carp (*Cyprinus carpio* L.), its movement patterns, mechanisms and limitations. *Journal of Zoology (London), Series A*, **210**, 161–203.

Sorensen, P.W. & Caprio, J. (1997) Chemoreception. In: *The Physiology of Fishes*, 2nd edn. (ed. D.H. Evans), pp. 375–405. CRC Press, Boca Raton.

Striedter, G.F. (1990) The diencephalon of the channel catfish, *Ictalurus punctatus*: I. Nuclear organization. *Brain Behaviour and Evolution*, **36**, 329–354.

Sutterlin, A.M. (1975) Chemical attraction of some marine fish in their natural habitat. *Journal of the Fisheries Research Board of Canada*, **32**, 729–738.

Sutterlin, A.M. & Sutterlin, N. (1970) Taste responses in Atlantic salmon (*Salmo salar*). *Journal of the Fisheries Research Board of Canada*, **28**, 565–572.

Teeter, J.H., Brand, J.G., & Kumazawa, T. (1990) A stimulus-activated conductance in isolated taste epithelial membranes. *Biophysical Journal*, **58**, 253–259.

Toyoshima, K., Nada, O. & Shimamura, A. (1984) Fine structure of monoamine-containing basal cells in the taste buds on the barbels of three species of teleosts. *Cell and Tissue Research*, **235**, 479–484.

Valentincic, T. & Caprio, J. (1992) Gustatory behavior of channel catfish to amino acids. In: *Chemical Signals in Vertebrates VI* (eds R.L. Doty & D. Muller-Schwarze), pp. 365–369. Plenum Press, New York.

Valentincic, T. & Caprio, J. (1994a) Chemical and visual control of feeding and escape behaviors in the channel catfish *Ictalurus punctatus*. *Physiological Behaviour*, **55**, 845–855.

Valentincic, T. & Caprio, J. (1994b) Consummatory feeding behavior in intact and anosmic channel catfish *Ictalurus punctatus* to amino acids. *Physiological Behaviour*, **55**, 857–863.

Valentincic, T. & Caprio, J. (1997) Visual and chemical release of feeding behavior in adult rainbow trout. *Chemical Senses*, **22**, 375–382.

Valentincic, T., Wegert, S. & Caprio, J. (1994) Learned olfactory discrimination versus innate taste responses to amino acids in channel catfish (*Ictalurus punctatus*). *Physiological Behaviour*, **55**, 865–873.

Wegert, S. & Caprio, J. (1991) Receptor sites for amino acids in the facial taste system of the channel catfish. *Journal of Comparative Physiology*, **168A**, 201–211.

Wullimann, M.F. (1988) The tertiary gustatory nucleus in sunfishes is not nucleus glomerulosus. *Neuroscience Letters*, **86**, 6–10.

Wullimann, M.F. (1997) The central nervous system. In: *The Physiology of Fishes*, 2nd edn. (ed. D.H. Evans), pp. 245–282. CRC Press, Boca Raton.

Yoshii, K., Kamo, N., Kurihara, K. & Kobatake, Y. (1979) Gustatory responses of eel palatine receptors to amino acids and carboxylic acids. *Journal of General Physiology*, **74**, 301–317.

Yoshimoto, M., Albert, J.S., Sawai, N., Shimizu, M., Yamamoto, N. & Ito, H. (1998) Telencephalic ascending gustatory system in a cichlid fish, *Oreochromis (Tilapia) niloticus*. *Journal of Comparative Neurology*, **392**, 209–226.

Chapter 6
Environmental Factors and Feed Intake: Mechanisms and Interactions

Patrick Kestemont and Etienne Baras

6.1 Introduction

Feeding by fish is dependent upon their sensory capacities to locate food, their ability to capture, handle and ingest food items, and their physiological and biochemical capacities to digest and transform the ingested nutrients. All of these may depend on environmental factors, which are either physical, chemical or biological in nature (Fig. 6.1). Some factors may impact on a single step in the feeding process, whereas others, such as temperature, are known to influence food availability, probability of food capture and processes involved in digestion, absorption and nutrient transformations.

Alone or in combination, environmental factors may be classified as lethal, controlling, limiting, masking or directive (Fry 1971), and their influence depends on fish species, life stage, acclimatisation and/or experience. Some factors may impact on feed intake to give a maximum at some intermediate value and thresholds above or below which intake declines,

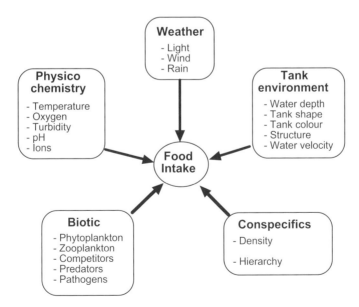

Fig. 6.1 Overview of the complexity of environmental influences on feed intake of fish. Factors are grouped into functional units, but for clarity, possible interactions between factors are not illustrated.

whereas others are truly limiting with distinct thresholds resulting in cessation of feeding. Not infrequently, the influence of a given factor is masked, buffered or amplified by another, making analysis of influences extremely difficult.

Here, we deal with the influence of abiotic and biotic factors considered in isolation, relying on results of experimental research of interest to aquaculture.

6.2 Abiotic factors

6.2.1 Light

Among the abiotic factors likely to influence feeding behaviour of fish, light has received much attention, but generalisations regarding effects remain difficult. This results from a multiplicity of properties of light (quality or light spectrum, light intensity, daily and seasonal variations of day length) and interactions with other environmental (e.g. temperature) or physiological factors. According to Fry's (1971) classification, light is a directive factor, i.e. an environmental entity which exerts its effect on the organism by stimulating some transductive response (Fry 1971). It is assumed that directive factors give a basis for behavioural regulation, and for anticipatory adjustments in physiological regulation.

The perception of light depends on stimulation of several organs, including both the retina and the pineal gland. This, in turn, leads to stimulation of the brain-pituitary axis and, in turn, the endocrine and nervous systems. As far as influences on feeding are concerned, it has been demonstrated that light (photoperiod) has effects on the production and secretion of growth hormone, steroid and thyroid hormones (Fry 1971; Spieler 1979; Noeske & Spieler 1984; Spieler & Noeske 1984).

Photoperiod and feeding

Day–night alternation is one of the most important zeitgebers entraining physiological variables: several reviews have covered this topic (Rusak & Zucker 1975; Aschoff 1979, 1981; Boujard & Leatherland 1992a; see Chapters 8 to 10). Results of early research led to the conclusion that photoperiod was the most pervasive of the stimuli that could entrain daily rhythms, but meal timing can also be a potent entraining stimulus in both mammals (Moore-Ede *et al.* 1982) and fish (Spieler & Noeske 1984; Alanärä 1992; Boujard & Leatherland 1992b,c; see Chapters 8 to 10). Several authors (Komourdjian *et al.* 1976; Higgins & Talbot 1985; Saunders *et al.* 1989; Stefanson *et al.* 1990; Jørgensen & Jobling 1992) report that a long (or increasing) photoperiod stimulates feeding activity, whereas a short (or decreasing) photoperiod leads to a reduction in feeding. Such studies have usually been carried out on salmonids, or other temperate zone species, so universal generalisation would be premature.

Light intensity and light spectrum

In comparison with photoperiod, the effects of light intensity and spectrum on feeding have received less attention, although both have been reported to influence behaviour (Holanov & Tash 1978; Mills *et al.* 1984; Tandler & Helps 1985; Barlow *et al.* 1993; Gehrke 1994; Fermin *et al*. 1996; Fermin & Seronay 1997). Many planktivorous fish are visual feeders that seem to be restricted to feeding when illumination is sufficient for perception of the prey (e.g. Asian sea bass *Lates calcarifer*; Davis 1985; Russell & Garrett 1985; Barlow *et al*. 1993). Threshold intensities vary between species, being as high as 1500 lux in Eurasian perch *Perca fluviatilis* (Dabrowski 1982) or 860 lux in turbot *Psetta maxima* (Huse 1994), whereas Atlantic cod *Gadus morhua* are able to feed at low light intensities (0.1–1.0 lux). Changes in light intensity may result in a shift to feeding on food items having different characteristics (e.g. size, motion, transparency, etc.). For example, Mills *et al*. (1984) observed that young yellow perch *Perca flavescens* fed on large daphnids at low light intensity and shifted to smaller prey as light intensity increased. As an additional example, the capture of small walleye *Stizostedion vitreum* by larger siblings was reduced by 40% when light intensity was reduced from 680 to 140 lux (Colesante 1989).

In several cases in which light has been reported to influence feeding indirect mechanisms relating to the attraction of prey to a light source have probably been involved. For example, Fermin and Seronay (1997) reported that larval Asian sea bass reared in illuminated cages (20–300 lux) displayed higher feeding activity than those reared in dark cages. Zooplanktonic organisms were attracted in greater numbers to the surface of illuminated cages, and increased light levels probably also enhanced prey detection and capture by the fish larvae. Similar effects of light on the attraction and subsequent capture of zooplankton by young fish have been reported in freshwater species such as coregonids (Mamcarz & Szczerbowski 1984; Rösch & Eckman 1986; Champigneulle & Rojas-Beltran 1990; Mamcarz 1995).

Illumination from outside a rearing unit may lead to generation of shadows during human activity or bird passage, and this disturbance to the fish may result in cessation of feeding for several hours (walleye, Nagel 1978, 1996; yellow perch, Malison & Held 1992). Malison and Held (1996) took account of this in developing a weaning procedure for yellow perch using a combination of illumination conditions (including internal or floating lights) and sinking pellets that mimicked the movements of live prey. Different combinations of light intensity and tank colour modify the feeding behaviour of, and efficiency of food detection by, fish larvae. Inappropriate contrast may result in reduced feeding. For example, use of dark-walled tanks can facilitate prey detection by improving the contrast between the food and the background (Blaxter 1980; Hinshaw 1986; Ronzani-Cerqueira 1986; Ounaïs-Guschemann 1989). Larvae or juveniles that show a phototactic response may be attracted by bright light reflecting from tank walls, and this may influence feeding (sea bass *Dicentrachus labax*, Barnabé 1976; walleye, Corazza & Nickum 1981; yellow perch, Hinshaw 1985). However, not all species respond in the same way; in several species, no effect of tank colour on feeding behaviour has been noticed (Ronzani-Cerqueira 1986; Ounaïs-Guschemann 1989), and some authors recommend the use of tanks with white or light-coloured walls (Houde 1973; Tamazouzt 1995).

Turbidity interferes with light penetration and the ability of fish to detect food, so is generally deemed to have a negative effect on feeding (Ang & Petrell 1997; Mallekh *et al*. 1998). However, in species such as walleye and pikeperch *Stizostedion lucioperca*, the reduced light intensity resulting from increased turbidity has been shown to promote feeding under some circumstances. This appeared to be the result of a combination of reduced reflection of light from tank walls, increased dispersal of larvae, improved contrast of food particles and better illumination of each food particle due to light scattering (Bristow & Summerfelt 1994; Hilge & Steffens 1996; Summerfelt 1996) (Fig. 6.2).

The effects of light intensity, tank wall colour and water turbidity on visual feeders are interdependent. Visual feeding also depends on the food characteristics (colour, movement,

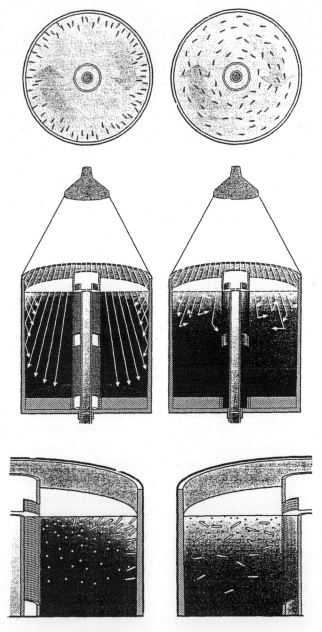

Fig. 6.2 Contrasting behaviour of walleye larvae (*Stizostedion vitreum*) in clear and turbid water. In tanks with clear water (left), fish are attracted to light reflected from side walls and the meniscus. In turbid water (right), the light-scattering effect of suspended clay particles diffuses the light, reducing the reflected light from the side walls (middle). Fish distribute across the tanks (bottom, right) and their orientation is into the current (top right). Arrows in the top figures indicate the direction of water current. (Reproduced courtesy of Bristow & Summerfelt 1994.)

shape, etc.) and on the developmental stage of the fish. Brightly coloured items are detected more easily by larvae than are dull items (e.g. whitefish *Coregonus lavaretus*, Rösch 1992), probably because the retina of most larvae at hatching consists predominantly of cones while rods develop later (O'Connell 1981; Blaxter 1986). Colour does not concern only tank wall or food items, but also feeding devices. For example, Paspatis *et al.* (1994) provided evidence that sea bass used self-feeding devices more frequently when the trigger was coloured with achromatic colours (mainly black) than with chromatic colours. In the so-called visual species, vision and vision-related factors (light, contrast, colour) may be of importance at short range only, and chemical cues may be used for longer-range detection of food (Kolkovski *et al.* 1997). This was demonstrated by Rottiers and Lemm (1985) who investigated the feeding behaviour of underyearling walleye in a Y-shaped maze. When fish were given both olfactory and visual cues, the chemical stimulus was important for orientation towards the prey, whereas vision became important for ingestion when fish were in the immediate vicinity of food.

6.2.2 Temperature

According to the classification of Fry (1971), temperature is the most important of the controlling factors (termed 'tonic effects' by Blackman (1905)), governing metabolic rate by its influence on molecular activation of the components of the metabolic chain. Thus, temperature influences several processes which are directly or indirectly related to food demand and feeding activity (Brett 1979; Jobling 1994, 1997).

Except for some representatives of the tuna family (Carey *et al.* 1971), fish are heterothermic (ectothermic) animals that do not regulate their body temperature, and body temperature fluctuates with that of the environment (with a difference of about 0.5°C). Fish are often classified as being coldwater or warmwater stenotherms, or eurytherms. Stenotherms have a limited range of tolerance to temperature, whereas eurytherms tolerate greater ranges of temperature. According to Elliott (1981), the optimum temperature range is defined as 'the range over which feeding occurs and where there are no signs of abnormal behaviour linked to thermal stress'. However, as far as growth is concerned, optimum temperature often coincides with the final preferendum zone for temperature, i.e. the temperature zone in which individuals of a given species will ultimately congregate (Jobling 1981). The correspondence between the thermal preferendum and the optimum temperature for growth is generally closer than the correspondence between preferred temperature and the temperature at which food consumption or food conversion are maximised (Jobling 1997).

Food intake by fish increases with increasing temperature, reaches a peak and then falls more or less dramatically at supra-optimal temperatures (Brett 1979), whereas metabolic rate may show a continuous increase up to the upper thermal limit for growth (Fig. 6.3). It is possible that the suppression of appetite observed at high temperatures reflects limitations in the capacity of the respiratory and circulatory systems to deliver oxygen to the respiring tissues under conditions of very high oxygen demand (Jobling 1997). When comparing the optimum temperature for growth with the temperatures at which feed intake and conversion efficiency are maximised (Table 6.1), it appears that the optimum temperature for growth is usually slightly lower than the temperature at which feed intake is highest but slightly higher than that corresponding to the best food conversion efficiency (Brett 1971, 1979; Elliott 1976,

Table 6.1   Comparison between temperature of growth optimum and temperature at which feed intake is maximum in different species (°C).

Species	Growth optimum	Feed intake maximum	References
Oncorhynchus nerka	15–16	19	Brett (1971)
Atlantic *Salmo salar*	16–19	16–17	Farmer *et al.* (1983)
Baltic *Salmo salar*	15.6	17.8	Koskela *et al.* (1997a)
Salmo trutta	16–17	18	Elliott (1976)
Oncorhynchus mykiss	16.5	19.5	Wurtsbaugh & Davis (1977)
Anarhicas lupus	11	>14?	McCarthy *et al.* (1998)
Psetta maxima	16–19	17–20	Burel *et al.* (1996); Mallekh *et al.* (1998)
Carassius auratus	28	>28?	Kestemont (1995)
Esox lucius	18–26	>20	Craig (1996); Salam & Davies (1994)
Perca fluviatilis	23	>23	Kestemont *et al.* (1996); Mélard *et al.* (1996)
Stizostedion vitreum	23.5	25.9	Summerfelt & Summerfelt (1996)
Oreochromis niloticus	28–30	>30?	Mélard (1986)

1981; Cox & Coutant 1981; Wurtsbaugh & Cech 1983; Jobling 1994, 1997; Mélard *et al.* 1996; McCarthy *et al.* 1998) (Fig. 6.3). It is worth noting however, that fish continue to feed at temperatures that exceed those at which intake is highest.

For example, Grande and Andersen (1991) and Koskela *et al.* (1997a) indicated that juvenile Atlantic salmon *Salmo salar* might not stop feeding until temperature reached 28–29°C, whereas the temperature at which feeding reached its maximum was about 16–18°C.

Fish lose appetite in the lower part of the temperature tolerance range (Brett 1979), but feed intake may have been underestimated, and the minimum temperatures for feeding have probably been overestimated in some cases due to masking effects of other factors such as prey availability, light intensity, acclimation processes or life-history pattern. For example, feed intake of Baltic salmon *Salmo salar* and brown trout *Salmo trutta* held at 2°C for two months increased progressively during the course of experiment, suggesting that acclimatisation to the rearing conditions required several weeks at low temperature (Koskela *et al.* 1997b). The increase of feed intake was a result of an increase in both the proportions of fish that fed and in feed intake amongst feeding fish. Higgins and Talbot (1985) provided evidence that feeding activity and growth of Atlantic salmon parr at low temperature (0.9–4.5°C) is dependent on fish size: large fish, which will undergo the parr-smolt transformation in spring, continue to feed throughout the winter months, whereas the smaller individuals reduce feed intake and consume no more than maintenance rations during the winter (Metcalfe *et al.* 1987). Koskela *et al.* (1997c) estimated that juvenile Baltic salmon would feed at a temperature as low as about 0.35°C, whereas Elliott (1991) provided an estimate of 3.8°C for another salmon strain, and observations of foraging behaviours in the wild have led to suggestions that salmon may cease feeding when temperature falls to 6–7°C (Gardiner & Geddes 1980; Jensen & Johnsen 1986). However, there is now evidence that Atlantic salmon tend to exhibit nocturnal foraging when exposed to low temperatures (Fraser *et al.* 1993, 1995): at temperatures below 6°C over 90% of the feeding attempts occurred at night, and few fish attempted to feed by day, suggesting that juvenile salmon switch from diurnal to nocturnal foraging as the temperature drops. Similar observations of feeding at very low temperatures (approaching 0°C) were also reported in wild (Cunjack & Power 1987; Heggenes *et al.* 1993) and in captive (Koskela *et al.* 1997d) brown trout.

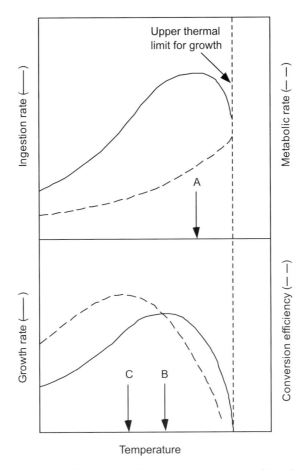

Fig. 6.3 Rate–temperature curves illustrating the effects of temperature on rates of ingestion, metabolism, growth and conversion efficiency (defined as growth per unit food ingested). Temperature at which ingestion rate reaches its maximum (A) is a few degrees higher than the optimum temperature for growth (B) which, in turn, is a few degrees higher than the one at which the most efficient conversion (C) is achieved. (Modified from Jobling 1997.)

Acute variations of temperature can induce marked changes in the feeding activity of fish. A steep increase of water temperature consecutively to a cold period no longer than a few hours induced hyperphagia in European catfish *Silurus glanis* and tilapia (Anthouard *et al.* 1994). Sensitivity in the feeding activity of sea bass to very small thermal changes – both diurnal or nocturnal – was also observed by using a computer-assisted demand feeder (Anthouard *et al.* 1993).

6.2.3 *Other physical factors*

Waves and water currents can impact on fish feeding (Bégout-Anras 1995; Juell 1995; Mallekh *et al.* 1998; see Chapter 7). Both wave height and frequency can have an impact on feeding of fish held in non-rigid cages, which dissipate very little of the wave energy. For example, rainbow trout *Oncorhynchus mykiss* held in cages stop feeding at wave frequencies greater than 0.28 Hz (Srivastava *et al.* 1991). Gusting winds and rainfall may lead fish to move away from the water surface, and this may affect feeding by restricting fish to certain parts of the water column (Juell 1995). Mallekh *et al.* (1998) reported that wind direction could influence feeding: turbot seemed to feed more when there was an onshore wind than when the wind was offshore, but the reason for this was not known. Whether wind and rain influence feeding primarily through a modification of water currents, or via changes in noise levels, is open to question given the findings of Lagardère *et al.* (1994) relating to the influence of noise on swimming and foraging of sole *Solea solea*.

Excessive water velocity may reduce the ability of fish to capture passing food items (Flore & Keckeis 1998), whereas the generation of moderate water currents in fish tanks may result in improved feed distribution and give better feeding conditions (Christiansen & Jobling 1990; Jørgensen *et al.* 1996). Fish exercising in moderate water currents frequently consume more food (Leon 1986; Totland *et al.* 1987; Jørgensen & Jobling 1993), and show faster growth. Additional reasons for faster growth, and improved food conversion efficiency (FCE: weight gain per unit weight of food consumed), may include the hypertrophy of the swimming muscle mass (Totland *et al.* 1987), changes in rates of protein synthesis and deposition (Houlihan & Laurent 1987), lower levels of aggression (Christiansen & Jobling 1990), an absence of dominance hierarchies and the more even distribution of food within a group of exercised fish (Jobling & Baardvik 1994; see also Biotic factors).

6.2.4 Chemical factors

While the lethal limits of factors such as oxygen, pH and nitrogenous compounds have been documented for several cultured fish species, their sublethal effects on feeding have rarely been investigated (see also Chapter 7).

Oxygen

Aerobic metabolism predominates in fish, so dissolved oxygen is potentially a limiting environmental factor (Fry 1971), especially at high temperature (Jobling 1997). Most studies on the effects of oxygen on feeding and growth have concentrated on hypoxic conditions, and information about possible effects of hyperoxia is scarce. Neither Edsall and Smith (1990) nor Caldwell and Hinshaw (1994) found growth differences between rainbow trout reared under normoxic and hyperoxic (180% saturation) conditions. By contrast, Tsadik and Kutty (1987) reported reduced feeding and growth in juvenile Nile tilapia *Oreochromis niloticus* held under hyperoxia (200% saturation) compared to fish under normoxic conditions (90% saturation).

Temporary, local oxygen depletion can occur in eutrophic or stratified lakes, ponds, oxbows or backwaters, and this generally induces an avoidance reaction (Coutant 1985; Kramer 1987). If this results in the fish becoming separated from their primary forage, feeding and growth may be substantially decreased (e.g. brown and rainbow trout, Hampton & Ney 1993). When avoidance is not possible, the effect of low oxygen concentration on fish metabolism, feeding and growth may vary substantially between species and life stages (reviews in Doudoroff & Shumway 1970; Davis 1975). Early life stages may be less sensitive to changes in oxygen than older fishes, as oxygen can diffuse passively through the body wall, but their feeding activity can be affected under conditions of reduced oxygen. For example, the initiation of exogenous feeding in several freshwater fish species has been found to be delayed under low oxygen concentrations (Siefert & Spoor 1974; Carlson *et al.* 1974, in Kramer 1987).

Fish have adapted to low aquatic oxygen concentrations in various ways including greater affinity of haemoglobin for oxygen, increased gill surface, or by resorting to air-breathing using the gas bladder, mouth, stomach or intestine as air-breathing organs (Lewis 1970; Braum & Junk 1982). Some species having little or no air-breathing capacity may move to the surface and resort to aquatic surface respiration (ASR), which enables them to exploit the oxygen-rich air–water interface (Lewis 1970; Kramer & Mehegan 1981; Kramer & McClure 1982; Saint-Paul & Soares 1987). ASR may allow feeding and growth to be maintained even if water is hypoxic (Weber & Kramer 1983). Frequent swimming at the surface during ASR (so-called 'ram-assisted ASR') permits more oxygenated water to contact the gills, and improves ASR efficiency (Chapman *et al.* 1994). It has been demonstrated that mudminnows *Umbra limi* engulf air bubbles at the surface, and stock these in their lung-like gas bladder, giving them the opportunity of foraging in hypoxic or anoxic environments (Rahel & Nutzman 1994). The frequent occurrence of fish in anoxic or strongly hypoxic waters (yellow perch, Hasler 1945; Hubert & Sandheinrich 1983; rainbow trout, Luecke & Teuscher 1994, in Rahel & Nutzman 1994) suggests that compensation behaviours may be more frequent than previously thought. ASR and air-gulping are short-term behavioural and physiological responses to a deficit in oxygen. In fishes exposed to chronic hypoxic conditions, there is an induction of physiological and haematological changes, that, at least partially, offset the negative effects of hypoxia. These include changes in haematocrit, red blood cell size and haemoglobin content (e.g. Saint-Paul 1984).

Except for extreme hypoxia, the transition from normoxia to hypoxia rarely results in a complete cessation of feeding by all fish in a population, but some individuals may do so (e.g. channel catfish *Ictalurus punctatus*, Randolph & Clemens 1976). Effects of hypoxia on feeding can vanish soon after return to normoxic conditions, or persist over longer periods of time, depending on species, life stage, acclimation and sometimes on dietary quality. For example, only rainbow trout fed with an animal diet returned to normal feeding after exposure to hypoxia, whereas those fed with proteins of vegetal origin did not (Pouliot & de la Noüe 1988, 1989).

Nitrogenous compounds, pH and salinity

Depending on temperature, oxygen and pH, non-lethal concentrations of nitrogenous compounds (unionised ammonia, nitrite) may have effects on gill structure and epidermal mucus (Kamstra *et al.* 1996), but their effect on feeding has rarely been quantified. Beamish and Tandler (1990) found that the feeding of juvenile lake trout *Salvelinus namaycush* was unaffected by ammonia concentrations less than 0.1 ppm NH_3, but was depressed at 0.3 ppm NH_3, and the decrease in feed intake was greater when fish were given a high-protein diet. The exposure of lake trout to 0.1 ppm NH_3 only resulted in a temporary decrease of feeding. Similarly, juvenile rainbow trout resumed normal feeding a few days after they were first exposed to 0.25 ppm NH_3 (Lang *et al.* 1987).

The effects of non-lethal pH environments have mainly been investigated in salmonids, within the context of acid rain threats in northern latitudes. Acid-stressed brook trout *Salveli-*

nus fontinalis (pH 4.15; Tam & Zhang 1996) and Arctic charr *Salvelinus alpinus* (pH 4.5, Jones *et al.* 1987) showed reduced feeding and depressed growth, partly as a consequence of suppression of somatotrope secretory activity (MacKett *et al.* 1992). As for most chemical factors, the effects of pH may depend on acclimation. For example, a progressive increase to pH 9.5 in 6 h was lethal to rainbow trout, whereas they showed only a temporary loss of appetite when pH was not increased beyond 9.3 during the same period (Murray & Ziebell 1984). When the acclimation period was extended to five days, rainbow trout could acclimate to pH as high as 9.8.

The exposure of fish to an extreme salinity or to a steep salinity gradient imposes energetic constraints, the severity of which vary among species (euryhaline versus stenohaline), life stages (e.g. parr versus smolt versus adults for salmonids), season and with acclimation. Generally, food intake and growth reach a maximum at the mid-point of the salinity tolerance range (e.g. tolerance of 5–35 ppm, maximum growth, FCE and food intake at 15–25 ppm in juvenile *Liza parsia*; Paulraj & Kiron 1988), but this is not universal. For example, both the feed intake and growth of juvenile of red Florida tilapia (hybrid tilapia) increase with increasing salinity (Watanabe *et al.* 1988), and in some cyprinids (e.g. grass carp *Ctenopharyngodon idella* and common carp *Cyprinus carpio*), feed intake continues to increase at salinities above the optimum for growth (Wang *et al.* 1997). It has also been proposed that rearing freshwater species at the upper end of the salinity range for growth may be profitable since these conditions apparently permit some fish (i.e. common and Chinese carps) to endure lower oxygen concentrations (Wang *et al.* 1997).

In typical diadromous, amphihaline species such as migratory salmonids, contrasted situations have been reported, with either no effect of salinity on growth (Atlantic salmon, Saunders & Henderson 1969, 1970), maximum growth at intermediate salinities (coho salmon *Oncorhynchus kisutch*, Otto 1971), enhanced (rainbow trout; Smith & Thorpe 1976) or depressed growth in sea water (rainbow trout, McKay & Gjerde 1985; Atlantic salmon, Arnesen *et al.* 1998). As exemplified by studies on Arctic charr, this apparent disparity may originate from the complex dynamics of acclimation of salmonids to seawater, which is dependent on the mode of transfer (gradual versus abrupt) and on season (Arnesen *et al.* 1993a,b). During winter, the abrupt transfer of charr to sea water results in a marked short-term reduction in feed intake and weight loss, while gradual transfer mitigates the effect of salinity change. By contrast, an abrupt transfer in summer may only induce small changes in feeding and growth. This indicates that charr (and possibly other salmonids) display seasonal changes in hypo-osmoregulatory capacity which intimately influence the effect of salinity changes on feeding.

Other toxicants and pollutants

Numerous chemicals have been identified as feed intake modifiers, via actions on food palatability, digestibility, appetite, metabolism and/or sensory systems (Christiansen & George 1995; Boujard & Le Gouvello 1997; see also Chapters 2 and 5). In open systems, such as cages or ponds, water quality is not monitored continuously, and it may be difficult to assess possible effects of pollutants. In recirculating systems, substances originating from prophylactic and therapeutic treatments (e.g. chloramine, malachite green, formalin) may influence feeding, but effects are variable (see Chapter 7).

For example, no change of appetite was observed when juvenile rainbow trout were exposed to therapeutic (200 ppm) levels of formalin (Speare & MacNair 1996) or to prophylactic (10 ppm over 1 h, twice weekly) chloramine-T treatment (Sanchez *et al.* 1996). Similarly, long-term exposure to oxytetracycline (OTC) did not affect feeding and growth of Arctic charr, whereas Atlantic salmon undergoing the same treatment had reduced feeding and growth, and the antibiotic induced spinal deformities (Toften & Jobling 1996). OTC is used at high concentration to mark otoliths in some growth studies, so it must be borne in mind that repeated exposure to this antibiotic may introduce a bias.

Attention has been drawn by ecologists and aquaculturists to possible effects of crude oil pollution on fish stocks. Christiansen and George (1995) provided evidence that food contamination by crude oil affected the growth, but not feeding, of polar cod *Boreogadus saida*. The responses differed between males and females; females consumed more contaminated food than males, and their growth was proportionally more compromised, possibly because the enzymes responsible for detoxification of xenobiotics are strongly inhibited in female fish undergoing sexual maturation (Jimenez & Stegeman 1990).

Many pesticides and herbicides are applied directly to aquatic habitats, or may reach water bodies in run-off from the land. These chemicals may potentially affect feeding and growth of fish and other aquatic organisms. For example, the feed intake of *Lepidocephalichthys thermalis* was reduced by 49–63% during exposure to 2.0 ppm Nuvacron or 0.8 ppm of Rogor (Muniandy & Sheela 1993). Other compounds such as pentachlorophenol (PCP) may affect feeding and growth in different ways. PCP is a biocide, used as a wood preservative, and many water bodies are contaminated with this compound. Samis *et al.* (1993) reported that feed intake of bluegill *Lepomis macrochirus* exposed to 48 µg/L PCP for 22 days was reduced by 10% relative to controls, whereas growth was reduced by 26%. At 173 µg/L PCP, feed intake was 29% lower and growth was 75% lower. The compound impacted on feed intake by effects on the ability of the fish to capture passing prey (Samis *et al.* 1993).

6.3 Biotic factors

The abundance and the temporal and spatial distributions of food have major impacts on feed intake in fish, and these factors also interact with other biotic factors such as predator avoidance, social interactions and responses to human disturbances.

6.3.1 Stocking density

High stocking density is often considered to be a stressor with detrimental effects on feeding, growth and a range of physiological processes (review in Wedemeyer 1997). However, in some studies, it has been shown that stocking fish at high density may lead to increased survival, better growth and reduced size heterogeneity within the population (Nile tilapia, Mélard 1986; Arctic charr, Jørgensen *et al.* 1993; sharptooth catfish *Clarias gariepinus*, Kaiser *et al.* 1995; Hecht & Uys 1997; vundu catfish *Heterobranchus longifilis*, Baras *et al.* 1998). The discrepancies may in part arise due to a lack of clarity relating to the so-called density-dependent effects. Some detrimental effects may arise as a consequence of a deterioration in water quality, or reduced adequacy of food distribution, at higher stocking density rather than being a direct consequence of an increase in stocking density. For example, in environments where feeding places are limited in number, surface or volume, or when the food reward is limited, high stocking density may depress growth, increase inter-individual variations, and eventually promote cannibalism, since not all fish gain access to food (Atlantic salmon, Thorpe *et al.* 1990). Conversely, increasing the reward as a response to increased stocking density can reduce feeding and growth in species that are more efficient at striking isolated pellets (e.g. Arctic charr, Brännäs & Alanärä 1992).

Direct effects of stocking density on territorial fish species include changes in behaviour because the energetic investments required for territorial defence are influenced by the numbers of potential competitors, i.e. density (Grant 1997). Increasing stocking density beyond a threshold may result in a reduction of agonistic interactions (e.g. Kaiser *et al.* 1995). Similarly, there may be a reduction in frequency of agonistic interactions under conditions of reduced light (Jørgensen & Jobling 1993; Baras *et al.* 1998) or when fish are exposed to moderate water currents and commence schooling (Jobling *et al.* 1993). As a corollary, the improved growth resulting from exposure to water currents or reduced light levels may be of lesser proportions in fish reared at high stocking density than in those reared at low density (Jørgensen & Jobling 1993; Baras *et al.* 1998). To date, most information relates to the effects of stocking density on survival and growth, and feed intake has been monitored systematically in only a few experiments (e.g. Jørgensen *et al.* 1993; Jobling & Baardvik 1994). However, the higher growth rates of fish held at higher stocking density may not only be the result of a higher feed intake, especially when there is a reduction of agonistic interactions in high-density groups (Jørgensen *et al.* 1993).

6.3.2 Social structure

Social environment may be influenced not only by population density, but also by factors such as size heterogeneity and sex ratio. Toguyéni *et al.* (1996) provided evidence that monosex groups of male Nile tilapia fed using self-feeders grew faster and showed better feed conversion efficiency than sibling females, although food demand was similar. When fish were reared in mixed-sex groups, food demand was higher than in monosex groups. Females in mixed-sex groups grew as well as females in monosex groups, but males in mixed-sex groups grew less well, presumably because of competition for access to females in the mixed groups. The initial establishment of a social hierarchy does not require there to be size differences among the fish, but may be promoted by size heterogeneity. Social hierarchy generally results in a variable access of individual fish to food resources (Alanärä & Brännäs 1993, 1996; Greenberg *et al.* 1997; Hakoyama & Iguchi 1997), and promotes growth heterogeneity. In some species, or contexts, size-based hierarchies result in feed intake and specific growth rates that are proportional to fish size and weight, resulting in increasing size heterogeneity (e.g. Alanärä & Brännäs 1996). In other contexts, agonism is not aimed towards all fish, but chiefly towards those that are just slightly smaller, and the end result may be a slight decrease of size heterogeneity within the population (Baras *et al.* 2000).

Competition for food is usually promoted under conditions of restricted supply in terms of quantity, time or space (McCarthy *et al.* 1992, 1993; Carter *et al.* 1996; Grant 1997). A number of workers have investigated the competition for food that occurs at different rates of input: the monopolisation of food by dominants is usually greatest at low input rates (e.g.

Brännäs and Alanärä 1994; Hakoyama & Iguchi 1997). When access to feed is restricted to a certain period of the day, better competitors that gain early access to food may have digested their first meal and feed again before the end of the feeding period, whereas this opportunity is denied to fish that ingest their first meal later. For example, Kadri *et al.* (1997a,b) provided evidence that juvenile Atlantic salmon that fed early in a feeding session were more successful and showed less variation in feed intake on different days than did conspecifics that started to feed later in the day.

6.3.3 Predators

Fish employ a variety of behaviours or tactics to avoid, deter or evade predators (Pitcher & Parrish 1992; Godin 1997). The risk of predation may be higher when a fish is foraging (Metcalfe *et al.* 1987; Milinski 1992; Hughes 1997), and costs of sheltering or avoiding predators must be weighed against lost feeding opportunities. Risk-taking while foraging varies with the physiological state of the fish. For example, when food-deprived fish are exposed to a predator they have shorter response distances, and resume feeding earlier than do well-fed conspecifics (sand gobies *Pomatoschistus minutus*, Magnhagen 1988; bluntnose minnow *Pimephales notatus*, Morgan 1988; coho salmon, Damsgård & Dill 1998) (Fig. 6.4). Threespine stickleback *Gasterosteus aculeatus* parasitised by the cestode *Schistocephalus solidus* have a greater energy need than uninfected fish, and they show changes in foraging behaviour that lead to increased exposure to the final host (a piscivorous bird) (Godin & Sproul 1988).

Risk-taking behaviour is also influenced by social tendencies. In shoaling or schooling species, the reduction of foraging activity and feed intake in the presence of a predator may be inversely proportional to shoal size (bluntnose minnow, Morgan & Colgan 1987; Morgan 1988), probably because increased shoal size reduces individual risk of predation. Foraging

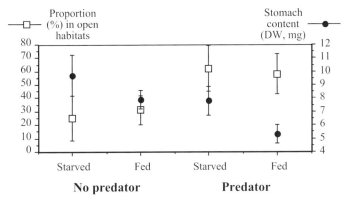

Fig. 6.4 Effect of hunger level and predator (Atlantic cod *Gadus morhua*) presence on the foraging habits (proportion in open habitats, open squares) and food intake (dry weight, DW, closed circles) by sand goby *Pomatoschistus minutus*. (Redrawn from Magnhagen 1988.) Gobies exposed to predators decreased their feed intake, but the decrease was less steep in starved fish than in fed fish. Predators were placed in an inner chamber, and gobies in an outer chamber, with a combination of transparent and opaque walls that gave gobies the possibility of hiding from the predator. Open habitats refer to places where the predator was within sight. Starved fish had been deprived of food for one week prior to the experiment, whereas the other fish had been fed continuously. Data points for habitat use are mean (±1 SD) of four experiments. Stomach contents are means ± SE of samples of at least forty fish.

under the threat of predation may also be dependent on the energetic investments (somatic or reproductive) already made by the fish, with fish that have made a greater investment taking fewer risks than others. For example, mature Eurasian perch spend less time foraging in the presence of a predator than do immature conspecifics (Utne *et al.* 1997). Large juvenile coho salmon are more averse to foraging under predation risk than are smaller conspecifics (Reinhardt & Healey 1997), and large threespine sticklebacks emerged later from refuges following exposure to a predator than did smaller individuals (Krause *et al.* 1998). The response to predation may be influenced by the social status of the fish, and predation risk may also impact on social structure. For example, the number of aggressive interactions between large and small brown trout was found to be reduced in the presence of a pike *Esox lucius* predator (Greenberg *et al.* 1997), and Reinhardt and Healey (1997) suggested that predation pressure might be partly responsible for the reduction of growth-rate differences among size classes, due to differential effects on foraging behaviour.

However, investment-dependent risk-taking behaviour is not universal. For example, Utne *et al.* (1997) demonstrated that perch of a slow-growing phenotype spent a lower proportion of their time in a foraging patch when a predator was present than did fast-growers, and this enhanced size heterogeneity within the population. These types of differences (see Huntingford *et al.* (1988) for a parallel with life history strategies in Atlantic salmon parr), make it particularly difficult to predict the effect of predation risk on inter-individual differences in food intake and growth.

6.3.4 *Human disturbance*

Man is a major biotic factor in aquaculture environments, and fish behaviour may be influenced in a number of ways by routine farming procedures such as handling, cleaning of tanks, prophylaxis and disease treatments or provision of food. It is generally believed that acute disturbances, such as weighing or the transfer of fish from one rearing unit to another, result in reduced feed intake for periods of several hours to days (e.g. Boujard *et al.* 1992). Temporary perturbations such as movements or noise near the tanks may have little effect on feeding: fish may move away from feeders, but then resume feeding within minutes of the end of the perturbation (Baras 1997). Prolonged human activity near tanks may impact on fish in various way. Eurasian perch that are disturbed by prolonged human activity during the early morning, prior to taking their first meal of the day, reduce their frequency of feeding over the entire day, and the daily ration demanded is only about 40% of that under 'normal' conditions (E. Baras & M. Anthouard, unpublished results). When perch are disturbed later in the day, after having fed, they resume feeding soon after the perturbation, and the daily ration is little affected. By contrast, the timing of a disturbance seems to have little effect on other species, such as tilapia *Oreochromis aureus*, which adjust their feeding in accordance to food availability, and are willing to feed at any time of the day (Baras *et al.* 1996).

Whereas acute disturbance usually has negative effects on feeding, some forms of disturbance may have a positive effect. For example, juvenile rainbow trout reared in tanks closer to a fish farm walkway were found to demand more food than those in tanks further from the walkway (Speare *et al.* 1995) (Fig. 6.5). Reasons for this greater food demand may involve the association of the presence of humans with food distribution.

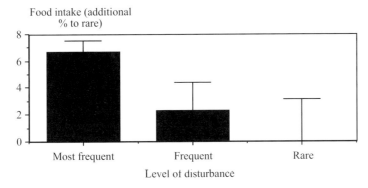

Fig. 6.5   Illustration of tank effect on the feed intake of juvenile rainbow trout *Oncorhynchus mykiss*, fed by hand to satiation, with respect to the level of disturbance. (Redrawn after Speare *et al.* 1995.) Fish that were frequently disturbed consumed ca. 7% more food than those subjected to infrequent disturbance. Levels of disturbance: 'most frequent' refer to tanks sited at the intersection of two common-use walkways, 'frequent' to tanks sited along a single common-use walkway, and 'rare' to tanks at a distance from both walkways.

6.4 Conclusions

This brief review illustrates the diversity of environmental influences that may affect feed intake in fish, as well as the dearth of knowledge on the effects of many of them despite their probable importance in outdoor systems (e.g. wind, rain and noise; see Chapter 7). By contrast, the rearing of fish in indoor tank systems may closely resemble the conditions in many experimental studies, allowing more straightforward predictions of the influences of environmental factors on feed intake.

Nevertheless, the unpredictability relating to environmental influences on feeding makes accurate prediction of daily rations most difficult, and this has consequences for feed management (see Chapter 14). This provides further arguments in favour of the use of on-demand feeding devices in commercial aquaculture to minimise waste of feed and reduce pollution of the environment.

6.5 References

Alanärä, A. (1992) The effects of time-restricted demand feeding on feeding activity, growth and feed conversion in rainbow trout (*Oncorhynchus mykiss*). *Aquaculture*, **108**, 357–368.

Alanärä, A. & Brännäs, E. (1993) A test of the individual feeding activity and food size preference in rainbow trout using demand feeders. *Aquaculture International*, **1**, 47–54.

Alanärä, A. & Brännäs, E. (1996) Dominance in demand-feeding behaviour in Arctic charr and rainbow trout: the effect of stocking density. *Journal of Fish Biology*, **48**, 242–254.

Ang, K.P. & Petrell, R.J. (1997) Control of feed dispensation in seacages using underwater video monitoring: effects on growth and food conversion. *Aquaculture Engineering*, **16**, 45–62.

Anthouard, M., Kentouri, M. & Divanach, P. (1993) An analysis of feeding activities of sea-bass (*Dicentrarchus labrax*, Moronidae) raised under different lighting conditions. *Ichthyophysiologica Acta*, **16**, 59–73.

Anthouard, M., Divanach, P. & Kentouri, M. (1994) L''auto-nourrissage': une méthode moderne d'alimentation des poissons en élevage. In: *Measures for Success – Metrology and Instrumentation in Aquaculture Management* (eds P. Kestemont, J. Muir, F. Sevilla & P. Williot), pp. 285–289. CEMAGREF Editions, Anthony, France.

Arnesen, A.M., Jørgensen, E.H. & Jobling, M. (1993a) Feed intake, growth and osmoregulation in Arctic charr, *Salvelinus alpinus* (L.), following abrupt transfer from freshwater to more saline water. *Aquaculture*, **114**, 327–338.

Arnesen, A.M., Jørgensen, E.H. & Jobling, M. (1993b) Feed intake, growth and osmoregulation in Arctic charr, *Salvelinus alpinus* (L.) transferred from freshwater to saltwater at 8°C during summer and winter. *Fish Physiology and Biochemistry*, **12**, 281–292.

Arnesen, A.M., Johnsen, H.K., Mortense, A. & Jobling, M. (1998) Acclimation of Atlantic salmon (*Salmo salar* L.) smolts to 'cold' sea water following direct transfer from fresh water. *Aquaculture*, **168**, 351–367.

Aschoff, J. (1979) Circadian rhythms: general features and endocrinological aspects. In: *Endocrine Rhythms* (ed. D.T. Krieger), pp. 1–61. Raven Press, New York.

Aschoff, J. (1981) Free-running and entrained circadian rhythms. In: *Handbook of Behavioral Neurobiology, Vol. 4, Biological Rhythms* (ed. J. Aschoff), pp. 81–93. Plenum Press, New York.

Baras, E. (1997) Application of telemetry techniques to remotely measure the behaviour of unrestrained cultured tilapias. *Northeast Regional Agricultural Series*, **106**, 701–712.

Baras, E., Thoreau, X. & Mélard, C. (1996) How do cultured tilapias *Oreochromis aureus* adapt their activity budget to variations in meal timing and abundance? In: *Underwater Biotelemetry – Proceedings of the First Conference and Workshop on Fish Telemetry in Europe* (eds E. Baras & J.C. Philippart), pp. 195–202. Liège, Belgium.

Baras, E., Tissier, F., Westerloppe, L., Mélard, C. & Philippart, J.-C. (1998) Feeding in darkness alleviates density-dependent growth of juvenile vundu catfish *Heterobranchus longifilis* (Clariidae). *Aquatic Living Resources*, **11**, 335–340.

Baras E., Malbrouck, C., Colignon, Y., *et al.* (2000) Application of PIT-tagging to the study of interindividual competition among juvenile perch *Perca fluviatilis:* Methodological steps and first results. In: *Advances in Fish Telemetry* (eds A. Moore & I.C. Russell), pp. 79–88. CEFAS, Lowestoft, UK.

Barlow, C.G., Rodgers, L.J., Palmer, P.J. & Longhurst, C.J. (1993) Feeding habits of hatchery-reared barramundi *Lates calcarifer* (Bloch) fry. *Aquaculture*, **109**, 131–144.

Barnabé, G. (1976) *Contribution à la connaissance de la biologie du loup* Dicentrarchus labrax *L. (Serranidae) de la région de Sète*. PhD thesis, Université des Sciences et Techniques du Languedoc, Montpellier, France.

Beamish, F.W.H. & Tandler, A. (1990) Ambient ammonia, diet and growth in lake trout. *Aquatic Toxicology*, **17**, 155–156.

Bégout Anras, M.L. (1995) Demand-feeding behaviour of sea bass kept in ponds: diel and seasonal patterns, and influences of environmental factors. *Aquaculture International*, **3**, 186–195.

Blackman, F.F. (1905) Optima and limiting factors. *Annals of Botany* (London), **19**, 282–295.

Blaxter, J.H.S. (1980) Vision and feeding of fishes. In: *Fish behavior and its use in the capture and culture of fishes* (eds J.E. Bardach, J.J. Magnuson, R.C. May & J.M. Reinhart), pp. 32–56. ICLARM, Manila, Philippines.

Blaxter, J.H.S. (1986) Development of sense organs and behavior of teleost larvae with special references to feeding and predator avoidance. *Transactions of the American Fisheries Society*, **115**, 98–114.

Boujard, T. & Leatherland, J.F. (1992a) Circadian rhythms and feeding time in fishes. *Environmental Biology of Fishes*, **35**, 109–131.

Boujard, T. & Leatherland, J.F. (1992b) Demand-feeding behaviour and diel pattern of feeding activity in *Oncorhynchus mykiss* held under different photoperiod regimes. *Journal of Fish Biology*, **40**, 535–544.

Boujard, T. & Leatherland, J.F. (1992c) Circadian pattern of hepatosomatic index, liver glycogen and lipid content, plasma non-esterified fatty acid, glucose, T_3, T_4, growth hormone and cortisol concentrations in *Oncorhynchus mykiss* held under different photoperiod regimes and fed using demand-feeders. *Fish Physiology and Biochemistry*, **10**, 111–122.

Boujard, T. & Le Gouvello, R. (1997) Voluntary feed intake and discrimination of diets containing a novel fluoroquinone in self-fed rainbow trout. *Aquatic Living Resources*, **10**, 343–350.

Boujard, T., Dugy, X., Genner, D., Gosset, C. & Grig, G. (1992) Description of a modular, low cost, eater meter for the study of feeding behavior and food preferences in fish. *Physiology and Behavior*, **52**, 1101–1106.

Brännäs, E. & Alanärä, A. (1992) Feeding behaviour of the Arctic charr in comparison with the rainbow trout. *Aquaculture*, **105**, 53–59.

Brännäs, E. & Alanärä, A. (1994) Effect of reward level on individual variability in demand feeding activity and growth rate in Arctic charr and rainbow trout. *Journal of Fish Biology*, **45**, 423–434.

Braum, E. & Junk, W.J. (1982) Morphological adaptation of two Amazonian Characoids (Pisces) for surviving in oxygen deficient waters. *Internationale Revue Gesamte Hydrobiologie*, **67**, 869–886.

Brett, J.R. (1971) Energetic responses of salmon to temperature. A study of some thermal relations in the physiology and freshwater ecology of sockeye salmon (*Oncorhynchus nerka*). *American Zoologist*, **11**, 99–113.

Brett, J.R. (1979) Environmental factors and growth. In: *Fish Physiology*, vol. 8 (eds W.S. Hoar, D.J. Randall & J.R. Brett), pp. 599–675. Academic Press, New York.

Bristow, B.T. & Summerfelt, R.C. (1994) Performance of larval walleye cultured intensively in clear and turbid water. *Journal of the World Aquaculture Society*, **25**, 3, 454–464.

Burel, C., Person-Le Ruyet, J., Gaumet, F., Le Roux, A., Sévère, A. & Bœuf, G. (1996) Effects of temperature on growth and metabolism in juvenile turbot. *Journal of Fish Biology*, **49**, 678–692.

Caldwell, C.A. & Hindshaw, J. (1994) Physiological and haematological responses in rainbow trout subjected to supplemental dissolved oxygen in fish culture. *Aquaculture*, **126**, 183–193.

Carey, F.G., Teal, J.M., Kanwisher, J.W., Lawson, K.D. & Becket, J.S. (1971) Warm-bodied fish. *American Zoologist*, **11**, 137–145.

Carlson, A.R., Siefert, R.E. & Herman, L.J. (1974) Effects of lowered dissolved oxygen concentrations on channel catfish (*Ictalurus punctatus*) embryos and larvae. *Transactions of the American Fisheries Society*, **103**, 623–626.

Carter, C.G., Purser, G.J., Houlihan, D.F. & Thomas, P. (1996) The effect of decreased ration on feeding hierarchies in groups of greenback flounder (*Rhombosolea tapirina*: Teleostei). *Journal of the Marine Biology Association UK*, **76**, 505–516.

Champigneulle, A. & Rojas-Beltran, R. (1990) First attempts to optimize the mass rearing of whitefish (*Coregonus lavaretus* L.) larvae from Leman and Bourget Lakes (France) in tanks and cages. *Aquatic Living Resources*, **3**, 217–228.

Chapman, L.J., Kaufman, L.S. & Chapman, C.A. (1994) Why swim upside down? A comparative study of two mochokid catfishes. *Copeia*, **1994**, 130–135.

Christiansen, J.S. & George, S.G. (1995) Contamination of food by crude oil affects food selection and growth performance, but not appetite, in an Arctic fish, the polar cod (*Boreogadus saida*). *Polar Biology*, **15**, 277–281.

Christiansen, J.S. & Jobling, M. (1990) The behaviour and the relationship between food intake and growth in juvenile Arctic charr, *Salvelinus alpinus* L., subjected to sustained exercise. *Canadian Journal of Zoology*, **68**, 2185–2191.

Colesante, R.T. (1989) Improved survival of walleye fry during the first 30 days of intensive rearing on brine shrimp and zooplankton. *The Progressive Fish-Culturist*, **51**, 109–111.

Corazza, T. & Nickum, J. (1981) Positive phototaxis during initial feeding stages of walleye larvae. In: *The Early Life History of Fish: Recent Studies* (eds R. Lasker & K. Sherman), pp. 492–494.

Coutant, C.C (1985) Striped bass: environmental risks in fresh and salt waters. *Transactions of the American Fisheries Society*, **114**, 1–31.

Cox, D.K. & Coutant, C.C. (1981) Growth dynamics of juvenile striped bass as functions of temperature and ration. *Transactions of the American Fisheries Society*, **110**, 226–238.

Cunjak, R.A. & Power, G. (1987) The feeding and energetics of stream resident trout in winter. *Journal of Fish Biology*, **31**, 493–511.

Craig, J. F. (ed.) (1996) *Pike biology and exploitation*. Chapman & Hall, London.

Dabrowski, K. (1982) The influence of light intensity on feeding fish larvae and fry. II. *Rutilus rutilus* (L.), *Perca fluviatilis* (L.). *Zoological Journal of Physiology*, **86**, 353–360.

Damsgård, B. & Dill, M.W. (1998) Risk-taking behavior in weight-compensating coho salmon, *Oncorhynchus kisutch*. *Behavioral Ecology*, **9**, 26–32.

Davis, J.C. (1975) Minimal dissolved oxygen requirements of aquatic life, with emphasis on Canadian species: a review. *Journal of the Fisheries Research Board of Canada*, **32**, 2295–2332.

Davis, T.L.O. (1985) The food of barramundi, *Lates calcarifer* (Bloch), in coastal and inland waters of Van Diemen Gulf and the Gulf of Carpentaria, Australia. *Journal of Fish Biology*, **26**, 669–682.

Doudoroff, P. & Shumway, D.L. (1970) *Dissolved oxygen requirements of freshwater fishes*. FAO Fishery Technical Paper 86, Rome.

Edsall, D.A. & Smith, C.E. (1990) Performance of rainbow trout and Snake River cutthroat trout reared in oxygen-supersaturated water. *Aquaculture*, **90**, 251–259.

Elliott, J.M. (1976) The energetics of feeding metabolism and growth of brown trout (*Salmo trutta* L.) in relation to body weight, water temperature and ration size. *Freshwater Biological Association*, **45**, 923–948.

Elliott, J.M. (1981) Thermal stress on freshwater teleosts. In: *Stress and Fish* (ed. A.D. Pickering), pp. 209–245. Academic Press, London, New York.

Elliott, J.M. (1991) Tolerance and resistance to thermal stress in juvenile Atlantic salmon *Salmo salar*. *Freshwater Biology*, **25**, 61–70.

Farmer, G.J., Ashfield, D. & Goff, T.R. (1983) A feeding guide for juvenile Atlantic salmon. *Canadian Report of Fisheries and Aquatic Sciences*, **1718**, 1–13.

Fermin, A.C. & Seronay, G.A. (1997) Effects of different illumination levels on zooplankton abundance, feeding periodicity, growth and survival of the Asian sea bass, *Lates calcarifer* (Bloch), fry in illuminated floating nursery cages. *Aquaculture*, **157**, 227–237.

Fermin, A.C., Bolivar, M.E. & Gaitan, A. (1996) Nursery rearing of the Asian sea bass, *Lates calcarifer*, fry in illuminated floating net cages with different feeding regimes and stocking densities. *Aquatic Living Resources*, **9**, 43–49.

Flore, L. & Keckeis, H. (1998) The effect of water current on foraging behaviour of a rheophilic cyprinid, *Chondrostoma nasus* (L.), during ontogeny: evidence of a trade-off between energetic gain and swimming costs. *Regulated Rivers: Research & Management*, **14**, 141–154.

Fraser, N.H.C., Metcalfe, N.B. & Thorpe, J.E. (1993) Temperature-dependent switch between diurnal and nocturnal foraging in salmon. *Proceedings of the Royal Society of London*, **B 252**, 135–139.

Fraser, N.H.C., Heggenes, J., Metcalfe, N.B. & Thorpe, J.E. (1995) Low summer temperatures cause juvenile Atlantic salmon to become nocturnal. *Canadian Journal of Zoology*, **73**, 446–451.

Fry, F.E.J. (1971) The effects of environmental factors on the physiology of fish. In: *Fish Physiology*, Vol. 6 (eds W.S. Hoar & D.J. Randall), pp. 1–98. Academic Press, New York.

Gardiner, W.R. & Geddes, P. (1980) The influence of body composition on the survival of juvenile salmon. *Hydrobiologia*, **69**, 67–72.

Gehrke, P.C. (1994) Influence of light intensity and wavelength on phototactic behavior of larval silver perch, *Bidyanus bidyanus*, and golden perch, *Macquaria ambigua*, and the effectiveness of light traps. *Journal of Fish Biology*, **44**, 741–751.

Godin, J.-G.J. (1997) Evading predators. In: *Behavioural Ecology of Teleost Fishes* (ed. J.-G.J. Godin), pp. 191–236. Oxford University Press, Oxford.

Godin, J.-G.J. & Sproul, C.D. (1988) Risk taking in parasitized sticklebacks under threat of predation: effects of energetic need and food availability. *Canadian Journal of Zoology*, **66**, 2360–2367.

Grande, M. & Andersen, S. (1991) Critical thermal maxima for young salmonids. *Journal of Freshwater Ecology*, **6**, 275–280.

Grant, J.W.A. (1997) Territoriality. In: *Behavioural Ecology of Teleost Fishes* (ed. J.-G.J. Godin), pp. 81–103. Oxford University Press, Oxford.

Greenberg, L.A., Bergman, E. & Eklöv, A.G. (1997) Effects of predation and intraspecific interactions on habitat use and foraging by brown trout in artificial streams. *Ecology of Freshwater Fish*, **6**, 16–26.

Hakoyama, H. & Iguchi, K. (1997) Why is competition more intense if food is supplied more slowly? *Behavioral Ecology and Sociobiology*, **40**, 159–168.

Hampton, T.M. & Ney, J.J. (1993) Effect of summer habitat limitation on trout in Lake Moomaw, Virginia. *Virginia Journal of Sciences*, **44**, 147.

Hasler, A.D. (1945) Observation on the winter perch population of Lake Mendota. *Ecology*, **26**, 90–94.

Hecht, T. & Uys, W. (1997) Effect of density on the feeding and aggressive behaviour in juvenile African catfish, *Clarias gariepinus*. *South African Journal of Sciences*, **93**, 537–541.

Heggenes, J., Krog, O.M.W., Lindas, O.R., Dokk, J.G. & Bremnes, T. (1993) Homeostatic behavioural responses in a changing environment: brown trout (*Salmo trutta*) become nocturnal during winter. *Journal of Animal Ecology*, **62**, 295–308.

Higgins, P.J. & Talbot, C. (1985) Growth and feeding in juvenile Atlantic salmon (*Salmo salar* L.). In: *Nutrition and feeding in Fish* (eds C.B. Cowey, A.M. Mackie & J.G. Bell), pp. 243–263. Academic Press, London.

Hilge, V. & Steffens, W. (1996) Aquaculture of fry and fingerling of pike-perch (*Stizostedion lucioperca* L.) – a short review. *Journal of Applied Ichthyology*, **12**, 3–4, 167–170.

Hinshaw, M. (1985) Effects of illumination and prey contrast on survival and growth of larval yellow perch (*Perca flavescens*). *Transactions of the American Fisheries Society*, **114**, 540–545.

Hinshaw, M. (1986) *Factors affecting survival and growth of larval and early juvenile perch* (Perca flavescens, *Mitchill*). PhD Thesis, University North Carolina State, USA.

Holanov, S.H. & Tash, J.C. (1978) Particulate and filter feeding in threadfin shad, *Dorosoma petenense*, at different light intensities. *Journal of Fish Biology*, **13**, 619–625.

Houde, E.D. (1973) Some recent advances and unsolved problems in the culture of marine fish larvae. *Proceedings of the World Mariculture Society*, **3**, 83–103.

Houlihan, D.F. & Laurent, P. (1987) Effects of exercise training on the performance, growth and protein turnover of rainbow trout (*Salmo gairdneri*). *Canadian Journal of Fisheries and Aquatic Sciences*, **44**, 1614–1621.

Hubert, W.A. & Sandheinrich, M.B. (1983) Patterns of variation in gill-net catch and diet of yellow perch in a stratified Iowa lake. *North American Journal of Fisheries Management*, **3**, 156–162.

Hughes, R.N. (1997) Diet selection. In: *Behavioural Ecology of Teleost Fishes* (ed. J.-G.J. Godin), pp. 134–162. Oxford University Press, Oxford.

Huntingford, F.A., Metcalfe, N.B. & Thorpe, J.E. (1988) Choice of feeding station in Atlantic salmon, *Salmo salar*, parr: effects of predation risk, season and life history strategy. *Journal of Fish Biology*, **33**, 917–924.

Huse, I. (1994) Feeding at different illumination levels in the larvae of three marine teleost species: cod *Gadus morhua* L., plaice, *Pleuronectes platessa* L., and turbot, *Scophthalmus maximus* L. *Aquaculture and Fisheries Management*, **25**, 687–695.

Jensen, A.J. & Johnsen, B.O. (1986) Different adaptation strategies of Atlantic salmon (*Salmo salar*) populations to extreme climates with special reference to some cold Norwegian rivers. *Canadian Journal of Fisheries and Aquatic Sciences*, **43**, 980–984.

Jimenez, B.D. & Stegeman, J.J. (1990) Detoxification enzymes as indicators of environmental stress on fish. *American Fisheries Society Symposium*, **8**, 67–79.

Jobling, M. (1981) Temperature tolerance and the final preferendum – a rapid method for the assessment of optimum growth temperatures. *Journal of Fish Biology*, **19**, 439–455.

Jobling, M. (1994) *Fish Bioenergetics*. Chapman & Hall, London.

Jobling, M. (1997) Temperature and growth: modulation of growth rate via temperature change. In: *Global Warming: Implications for freshwater and marine fish*. (eds C.M. Wood & D.G. McDonald), pp. 225–253. Cambridge University Press, Cambridge.

Jobling, M. & Baardvik, B.M. (1994) The influence of environmental manipulations on inter- and intra-individual variation in food acquisition and growth performance of Arctic charr, *Salvelinus alpinus*. *Journal of Fish Biology*, **44**, 1069–1087.

Jobling, M., Baardvik, B.M., Christiansen, J.S. & Jørgensen, E.H. (1993) The effects of prolonged exercise training on growth performance and production parameters in fish. *Aquaculture International*, **1**, 95–111.

Jones, K.A., Brown, S.B. & Hara, T.J. (1987) Behavioral and biochemical studies of onset and recovery of acid stress in Arctic charr (*Salvelinus alpinus*). *Canadian Journal of Fisheries and Aquatic Sciences*, **44**, 373–381.

Jørgensen, E.H., Baardvik, B.M., Eliassen, R. & Jobling, M. (1996) Food acquisition and growth of juvenile Atlantic salmon (*Salmo salar*) in relation to spatial distribution of food. *Aquaculture*, **143**, 277–289.

Jørgensen, E.H. & Jobling, M. (1990) Feeding modes in Arctic charr, *Salvelinus alpinus* L.: the importance of bottom feeding for the maintenance of growth. *Aquaculture*, **86**, 379–385.

Jørgensen, E.H. & Jobling, M. (1992) Feeding behaviour and effect of feeding regime on growth of Atlantic salmon, *Salmo salar*. *Aquaculture*, **101**, 135–146.

Jørgensen, E.H. & Jobling, M. (1993) Feeding in darkness eliminates density-dependent growth suppression in Arctic charr. *Aquaculture International*, **1**, 90–93.

Jørgensen, E.H., Christiansen J.S. & Jobling, M. (1993) Effects of stocking density on food intake, growth performance and oxygen consumption in Arctic charr (*Salvelinus alpinus*). *Aquaculture*, **110**, 191–204.

Juell, J.E. (1995) The behaviour of Atlantic salmon in relation to efficient cage rearing. *Reviews in Fish Biology and Fisheries*, **12**, 1–18.

Kadri, S., Metcalfe, N.B., Huntingford, F.A. & Thorpe, J.E. (1997a) Daily feeding rhythms in Atlantic salmon. I: feeding and aggression in parr under ambient environmental conditions. *Journal of Fish Biology*, **50**, 267–272.

Kadri, S., Metcalfe, N.B., Huntingford, F.A. & Thorpe, J.E. (1997b) Daily feeding rhythms in Atlantic salmon. II: size-related variation in feeding patterns of patterns of post-smolts under constant environmental conditions. *Journal of Fish Biology*, **50**, 273–279.

Kaiser, H., Weyl, O. & Hecht, T. (1995) The effect of stocking density on growth, survival and agonistic behaviour of African catfish. *Aquaculture International*, **3**, 217–225.

Kamstra, A., Span, J.A. & van Weerd, J.H. (1996) The acute toxicity and sublethal effects of nitrite on growth and feed utilization of European eel, *Anguilla anguilla* (L.). *Aquaculture Research*, **27**, 903–911.

Kestemont, P. (1995) Influence of feed supply, temperature and body size on the growth of goldfish *Carassius auratus* larvae. *Aquaculture*, **136**, 341–349.

Kestemont, P., Mélard, C., Fiogbé, E., Masson, G. & Vlavonou, R. (1996) Nutritional and animal husbandry aspects of rearing early life stages of Eurasian perch *Perca fluviatilis*. *Journal of Applied Ichthyology*, **12**, 157–166.

Kolkovski, S., Arieli, A. & Tandler, A. (1997) Visual and chemical cues stimulate microdiet ingestion in sea bream larvae. *Aquaculture International*, **5**, 527–536.

Komourdjian, M.P., Saunders, R.L. & Fenwick, J.C. (1976) Evidence for the role of growth hormone as a part of a 'light-pituitary axis' in growth and smoltification of Atlantic salmon (*Salmo salar*). *Canadian Journal of Zoology*, **54**, 544–570.

Koskela, J., Pirhonen, J. & Jobling, M. (1997a) Feed intake, growth rate and body composition of juvenile Baltic salmon exposed to different constant temperatures. *Aquaculture International*, **5**, 351–360.

Koskela, J., Pirhonen, J. & Jobling, M. (1997b) Variations in feed intake and growth of Baltic salmon and brown trout exposed to continuous light at constant low temperature. *Journal of Fish Biology*, **50**, 837–845.

Koskela, J., Pirhonen, J. & Jobling, M. (1997c) Effect of low temperature on feed intake, growth rate and body composition of juvenile Baltic salmon. *Aquaculture International*, **5**, 479–488.

Koskela, J., Pirhonen, J. & Jobling, M. (1997d) Growth and feeding responses of a hatchery population of brown trout (*Salmo trutta* L.) at low temperatures. *Ecology of Freshwater Fish*, **6**, 116–121.

Kramer, D.L. (1987) Dissolved oxygen and fish behaviour. *Environmental Biology of Fishes*, **18**, 81–92.

Kramer, D.L. (1988) The behavioral ecology of air-breathing by aquatic animals. *Canadian Journal of Zoology*, **66**, 89–94.

Kramer, D.L. & McClure, M. (1982) Aquatic surface respiration, a widespread adaptation to hypoxia in tropical freshwater fishes. *Environmental Biology of Fishes*, **7**, 47–55.

Kramer, D.L. & Mehegan, J.P. (1981) Aquatic surface respiration, an adaptive response to hypoxia in the guppy, *Poecilia reticulata* (Pisces, Poecilidae). *Environmental Biology of Fishes*, **6**, 299–313.

Krause, J., Loader, S.P., McDermott, J. & Ruxton, G.D. (1998) Refuge use by fish as a function of body length-related metabolic expenditure and predation risks. *Proceedings of the Royal Society of London*, **B 265**, 2373–2379.

Lagardère, J.-P., Bégout, M.L., Lafaye, J.Y. & Villotte, J.-P. (1994) Influence of wind-produced noise on orientation in the sole *Solea solea*. *Canadian Journal of Fisheries & Aquatic Sciences*, **51**, 1258–1264.

Lang, T., Peters, G., Hoffmann R. & Meyer, E. (1987) Experimental investigations on the toxicity of ammonia on ventilation frequency, growth, epidermal mucous cells, and gill structure of rainbow trout *Salmo gairdneri*. *Diseases of Aquatic Organisms*, **3**, 159–165.

Leon, K.A. (1986) Effects of exercise on feed consumption, growth food conversion and stamina of brook trout. *The Progressive Fish-Culturist*, **48**, 43–46.

Lewis, W.M. Jr. (1970) Morphological adaptations of cyprinodontoids for inhabiting oxygen deficient waters. *Copeia*, **1970**, 319–326.

Luecke, C. & Teuscher, D. (1994) Habitat selection by lacustrine rainbow trout within gradients of temperature, oxygen and food availability. In: *Theory and Application in Feeding Ecology* (eds D.J. Stouder, K.L. Fresh & R.J. Feller), University of South Carolina Press, Columbia, South Carolina, USA.

MacKett, D.B., Tam, W.H. & Fryer, J.N. (1992) Histological changes in insulin-immunoreactive pancreatic beta cells, and suppression of insulin secretion and somatotrope activity in brook trout (*Salvelinus fontinalis*) maintained on reduced food intake or exposed to acidic environment. *Fish Physiology and Biochemistry*, **10**, 229–243.

Magnhagen, C. (1988) Changes in foraging as a response to predation risk in two gobiid fish species, *Pomatoschistus minutus* and *Gobius niger*. *Marine Ecology Progress Series*, **49**, 21–26.

Malison, J.A. & Held, J.A. (1992) Effects of fish size at harvest, initial stocking density and tank lighting conditions on the habituation of pond-reared yellow perch (*Perca flavescens*) to intensive culture conditions. *Aquaculture*, **104**, 67–78.

Malison, J.A. & Held, J.A. (1996) Habituating pond-reared fingerlings to formulated feed. In: *Walleye Culture Manual* (ed. R.C. Summerfelt), pp. 199–204. Iowa State University, Ames.

Mallekh, R., Lagardère, J.P., Bégout-Anras, M.L. & Lafaye, J.Y. (1998) Variability in appetite of turbot *Scophthalmus maximus* under intensive rearing conditions: the role of environmental factors. *Aquaculture*, **165**, 123–138.

Mamcarz, A. (1995) Changes in zooplankton structure around illuminated cage culture. *Aquaculture Research*, **26**, 515–525.

Mamcarz, A. & Szczerbowski, J.A. (1984) Rearing of coregonid fishes (Coregonidae) in illuminated cages: 1. Growth and survival of *Coregonus lavaretus* L. and *Coregonus peled* Gmel. *Aquaculture*, **40**, 135–145.

McCarthy, I.D., Carter, C.G. & Houlihan D.F. (1992) The effect of feeding hierarchy on individual variability in daily feeding of rainbow trout, *Oncorhynchus mykiss* (Walbaum). *Journal of Fish Biology*, **41**, 257–263.

McCarthy, I.D., Carter, C.G. & Houlihan, D.F. (1993) Individual variation in consumption in rainbow trout measured using radiography. In: *Fish Nutrition in Practice* (eds S.J. Kaushik & P. Luquet), pp. 85–88. INRA Editions, Paris.

McCarthy, I.D., Moksness, E. & Pavlov, D.A. (1998) The effects of temperature on growth rate and growth efficiency of juvenile common wolffish. *Aquaculture International*, **6**, 207–218.

McKay, L.R. & Gjerde, B. (1985) The effect of salinity on growth of rainbow trout. *Aquaculture*, **49**, 325–331.

Mélard, C. (1986) Les bases biologiques de l'élevage du tilapia du Nil. *Cahiers d'Ethologie Appliquée*, **6**, 1–224.

Mélard, C., Kestemont, P. & Grignard, J.C. (1996) Intensive culture of juvenile and adult Eurasian perch (*P. fluviatilis*): effects of major biotic and abiotic factors on growth. *Journal of Applied Ichthyology*, **12**, 175–180.

Metcalfe, N.B., Huntingford, F.A. & Thorpe, J.E. (1987) Predation risk impair diet selection in juvenile salmon. *Animal Behaviour*, **35**, 931–933.

Milinski, M. (1992) Predation risk and feeding behaviour. In: *Behaviour of Teleost Fishes* (ed. T.J. Pitcher), pp. 285–305. Chapman & Hall, London.

Mills, E.L., Confer, J.L. & Ready, R.C. (1984) Prey selection by young yellow perch: the influence of capture success, visual acuity, and prey choice. *Transactions of the American Fisheries Society*, **113**, 579–587.

Moore-Ede, M.C., Syulzman F.M. & Fuller, C.A. (1982) *The Clocks that Time Us*. Harvard University Press, Cambridge, Massachusetts, USA.

Morgan M.J. (1988) The influence of hunger, shoal size and predator presence on foraging in bluntnose minnows. *Animal Behaviour*, **36**, 1317–1322.

Morgan, M.J. & Colgan, P.W. (1987) The effects of predator presence and shoal size on foraging in bluntnose minnows, *Pimephales notatus*. *Environmental Biology of Fishes*, **20**, 105–111.

Muniandy, S. & Sheela, S. (1993) Studies on the effects of pesticides on food intake, growth and food conversion efficiencies of a fish *Lepidocephalichthyes thermalis*. *Comparative Physiology and Ecology*, **18**, 92–95.

Murray, C.A. & Ziebell, C.D. (1984) Acclimation of rainbow trout to high pH to prevent stocking mortality in summer. *The Progressive Fish-Culturist*, **46**, 176–179.

Nagel, T.O. (1978) *Walleye Production in Controlled Environments*. Ohio Department of Natural Resources Division of Wildlife, Ohio, USA.

Nagel, T.O. (1996) Intensive culture of fingerling walleye on formulated feeds. In *Walleye Culture Manual* (ed. R.C. Summerfelt), pp. 205–207. Iowa State University, Ames.

Noeske, T.A. & Spieler, R.E. (1984) Circadian feeding time affects growth of fish. *Transactions of the American Fisheries Society*, **113**, 540–544.

O'Connell, C.P. (1981) Development of sense organ systems in the northern anchovy *Engraulis mordax* and other teleosts. *American Zoologist*, **21**, 429–446.

Otto, R.G. (1971) Effects of salinity on the survival and growth of pre-smolt coho salmon (*Oncorhynchus kisutch*). *Journal of the Fisheries Research Board of Canada*, **28**, 343–349.

Ounaïs-Guschemann, N. (1989) *Définition d'un modèle d'élevage larvaire intensif pour la dorade Sparus auratus*. PhD thesis, Université Aix-Marseille II, France.

Paspatis, M., Kentouri, M., Anthouard, M. & Divanach, P. (1994) 2-choice discrimination and preference of sea bass *Dicentrarchus labrax* among 9 colours. In: *Measures for success – Metrology and
instrumentation in aquaculture management* (eds P. Kestemont, J. Muir, F. Sevilla & P. Williot), pp.
291–295. CEMAGREF Editions, Anthony, France.

Paulraj, R. & Kiron, V. (1988) Influence of salinity on the growth and feed utilization in *Liza parsia*
fry. In: *Proceedings of the First Indian Fisheries Forum* (ed. M.M. Joseph), pp. 61–63. Mangalore,
India.

Pitcher, T.J. & Parrish, J.K. (1992) Functions of shoaling behaviour in teleosts. In: *Behaviour of Teleost
Fishes* (ed. T.J. Pitcher), pp. 363–439. Chapman & Hall, London.

Pouliot, T. & de la Noüe, J. (1988) Apparent digestibility in rainbow trout (*Salmo gairdneri*): influence
of hypoxia. *Canadian Journal of Fisheries and Aquatic Sciences*, **45**, 2003–2009.

Pouliot, T. & de la Noüe, J. (1989) Feed intake, digestibility and brain neurotransmitters of rainbow trout
under hypoxia. *Aquaculture*, **89**, 317–327.

Rahel, F.J. & Nutzman, J.W. (1994) Foraging in a lethal environment: fish predation in hypoxic waters
of a stratified lake. *Ecology*, **75**, 1246–1253.

Randolph, K.N. & Clemens, H.P. (1976) Some factors influencing the feeding behavior of channel
catfish in culture ponds. *Transactions of the American Fisheries Society*, **105**, 718–724.

Reinhardt, U.G. & Healey, M.C. (1997) Size-dependent foraging behaviour and use of cover in juvenile
coho salmon under predation risk. *Canadian Journal of Zoology*, **75**, 1642–1651.

Ronzani-Cerqueira, V. (1986) *L'élevage larvaire intensif du loup* Dicentrarchus labrax. *Influence de la
lumière, de la densité en proies et de la température sur l'alimentation, sur le transit digestif et sur
les performances zootechniques*. PhD thesis, Université Aix-Marseille II, France.

Rösch, R. (1992) Food intake and growth of larvae of *Coregonus lavaretus*: effect of diet and light.
Polskie Archiv für Hydrobiologie, **39**, 671–676.

Rösch, R. & Eckman, R. (1986) Survival and growth of prefed *Coregonus lavaretus* L. held in illuminated cages. *Aquaculture*, **52**, 245–252.

Rottiers, D.V. & Lemm, C.A. (1985) Movement of underyearling walleyes in response to odor and
visual cues. *The Progressive Fish-Culturist*, **47**, 34–41.

Rusak, B. & Zucker, I. (1975) Biological rhythms and animal behavior. *Annual Review of Psychology*,
26, 137–171.

Russell, D.J. & Garrett, R.N. (1985) Early life history of barramundi, *Lates calcarifer* (Bloch) in north-
eastern Queensland. *Australian Journal of Marine and Freshwater Research*, **36**, 191–201.

Saint-Paul, U. (1984) Physiological adaptation to hypoxia of a neotropical characoid fish *Colossoma
macropomum*, Serrasalmidae. *Environmental Biology of Fishes*, **11**, 53–62.

Saint-Paul, U. & Soares, B.M. (1987) Diurnal distribution and behavioral responses of fish to extreme
hypoxia in an Amazon floodplain lake. *Environmental Biology of Fishes*, **20**, 91–104.

Salam, A. & Davies, P.M.C. (1994) Effects of body weight and temperature on the maximum daily food
consumption of *Esox lucius*. *Journal of Fish Biology*, **44**, 165–167.

Samis, A.J.W., Colgan, P.W. & Johansen, P.H. (1993) Pentachlorophenol and reduced food intake of
bluegill. *Transactions of the American Fisheries Society*, **122**, 1156–1160.

Sanchez, J.G., Speare, D.J., MacNair, N. & Johnson, G. (1996) Effects of prophylactic chloramine-T
treatment on growth performance and condition indices of rainbow trout. *Journal of Aquatic Animal
Health*, **8**, 278–284.

Saunders, R.L. & Henderson, E.B. (1969) Growth of Atlantic salmon smolts and post-smolts in relation
to salinity, temperature and diet. *Fisheries Research Board of Canada Technical Report*, **149**, 31 p.

Saunders, R.L. & Henderson, E.B. (1970) Influence of photoperiod on smolt development and growth in Atlantic salmon (*Salmo salar*). *Journal of the Fisheries Research Board of Canada*, **27**, 1295–1311.

Saunders, R.L., Specker, J.L. & Komourdjan, M.P. (1989) Effects of photoperiod on growth and smolting in juvenile Atlantic salmon (*Salmo salar*). *Aquaculture*, **82**, 103–117.

Siefert, R.E. & Spoor, W.A. (1974) Effects of reduced oxygen on embryos and larvae of the white sucker, coho salmon, brook trout, and walleye. In: *The Early Life History of Fish* (ed. J.H.S. Blaxter), pp. 487–495. Springer-Verlag, New York,.

Smith, M.A.K. & Thorpe, J.E. (1976) Nitrogen metabolism and trophic input in relation to growth in freshwater and saltwater of *Salmo gairdneri*. *Biological Bulletin*, **150**, 139–151.

Speare, D.J. & MacNair, N. (1996) Effects of intermittent exposure to therapeutic levels of formalin on growth characteristics and body condition of juvenile rainbow trout. *Journal of Aquatic Animal Health*, **8**, 58–63.

Speare, D.J., MacNair, N. & Hammell, K.L. (1995) Demonstration of tank effect on growth indices of juvenile rainbow trout (*Oncorhynchus mykiss*) during an ad libitum feeding trial. *American Journal of Veterinary Research*, **56**, 1372–1379.

Spieler, R.E. (1979) Diel rhythms of circulating prolactin, cortisol, thyroxine, and triiodothyronine levels in fishes: a review. *Revue Canadienne de Biologie*, **38**, 301–315.

Spieler, R.E. & Noeske, T.A. (1984) Effects of photoperiod and feeding schedule on diel variations of locomotor activity, cortisol, and thyroxine in goldfish. *Transactions of the American Fisheries Society*, **113**, 528–539.

Srivastava, R., Brown, J.A. & Allen, J. (1991) The influence of wave frequency and wave height on the behaviour of rainbow trout (*Oncorhynchus mykiss*) in cages. *Aquaculture*, **97**, 143–153.

Stefanson, S.O., Nortvedt, R., Hansen, T.J. & Taranger, G.L. (1990) First-feeding of Atlantic salmon, *Salmo salar* L., under different photoperiods and light intensities. *Aquaculture and Fisheries Management*, **21**, 435–441.

Summerfelt, R.C. (1996) Intensive culture of walleye fry. In: *Walleye Culture Manual* (ed. R.C. Summerfelt), pp. 161–185. Iowa State University, Ames.

Summerfelt, S.T. & Summerfelt, R.C. (1996) Aquaculture of walleye as food fish. In: *Walleye Culture Manual* (ed. R.C. Summerfelt), pp. 215–230. Iowa State University, Ames.

Tam, W.H. & Zhang, X. (1996) The development and maturation of vitellogenic oocytes, plasma steroid hormone levels, and gonadotrope activities in acid-stressed brook trout (*Salvelinus fontinalis*). *Canadian Journal of Zoology*, **74**, 587–594.

Tamazouzt, L. (1995) *L'alimentation artificielle de la perche* Perca fluviatilis *en milieux confinés (eau recyclée, cage flottante): incidence sur la survie, la croissance et la composition corporelle.* PhD thesis, Université de Nancy 1, France.

Tandler, A. & Helps, S. (1985) The effects of photoperiod and water exchange rate on growth and survival of gilthead sea bream (*Sparus aurata*, Linnaeus; Sparidae) from hatching to metamorphosis in mass rearing systems. *Aquaculture*, **48**, 71–82.

Thorpe, J.E., Talbot, C., Miles, M.S., Rawlings, C. & Keay, D.S. (1990) Food consumption in 24 hours by Atlantic salmon (*Salmo salar* L.) in a sea cage. *Aquaculture*, **90**, 41–47.

Toften, H. & Jobling, M. (1996) Development of spinal deformities in Atlantic salmon and Arctic charr fed diets supplemented with oxytetracycline. *Journal of Fish Biology*, **49**, 668–677.

Toguyéni, A., Baroiller, J.F., Fostier, A., *et al.* (1996) Consequences of food restriction on short-term growth variation and on plasma circulating hormones in *Oreochromis niloticus* in relation to sex. *General and Comparative Endocrinology*, **103**, 167–175.

Totland, G.K., Kryvi, H., Jødestøl, K.A., Christiansen, E.N., Tangerås, A. & Slinde, E. (1987) Growth and composition of the swimming muscle of adult Atlantic salmon (*Salmo salar* L.) during long-term swimming. *Aquaculture*, **66**, 299–313.

Tsadik, G.G & Kutty, M.N. (1987) Influence of ambient oxygen on feeding and growth of the tilapia, *Oreochromis niloticus* (Linnaeus). UNDP-FAO-NIOMR, 13 p.

Utne, A.C.W., Brännäs, E. & Magnhagen, C. (1997) Individual responses to predation risk and food density in perch (*Perca fluviatilis* L.). *Canadian Journal of Zoology*, **75**, 2027–2035.

Wang, J.Q., Lui, H., Po, H. & Fan, L. (1997) Influence of salinity on food consumption, growth and energy conversion efficiency of common carp (*Cyprinus carpio*) fingerlings. *Aquaculture*, **148**, 115–124.

Watanabe, W.O, Ellingson, L.J., Wicklund, R.I. & Olla, B.L. (1988) The effects of salinity on growth, food consumption and conversion in juvenile, monosex male Florida red tilapia. In: *Proceedings of the Second International Symposium on Tilapia in Aquaculture* (eds R.S.V. Pullin, T. Bhukaswan, K. Tonguthai & J.L. Maclean), pp. 515–523. ICLARM, Manila, Philippines.

Weber, J.-M. & Kramer, D.L. (1983) Effects of hypoxia and surface access on growth, mortality, and behavior of juvenile guppies *Poecilia reticulata*. *Canadian Journal of Fisheries and Aquatic Sciences*, **40**, 1583–1588.

Wedemeyer, G.A. (1997) Effects of rearing conditions on the health and physiological quality of fish in intensive culture. In: *Fish Stress and Health in Aquaculture* (eds K. Iwama, A.D. Pickering, J.P. Sumpter & C.B. Schreck), pp. 35–71. Cambridge University Press, Cambridge.

Wurtsbaugh, W.A. & Cech, J.J., Jr. (1983) Growth and activity of juvenile mosquitofish: temperature and ration effects. *Transactions of the American Fisheries Society*, **112**, 653–660.

Wurtsbaugh, W.A. & Davis, G.E. (1977) Effects of temperature and ration level on the growth and food conversion efficiency of *Salmo gairdneri* Richardson. *Journal of Fish Biology*, **11**, 87–98.

Chapter 7

Environmental Factors and Feed Intake: Rearing Systems

Marie-Laure Bégout Anras, Marilyn Beauchaud, Jon-Erik Juell, Denis Covès and Jean-Paul Lagardère

7.1 Introduction

Feed represents a major cost in intensive aquaculture, and efficient feed use is therefore of prime economic importance for the industry (Knights 1985; Goddard 1996). To achieve efficient feed utilisation it is important to understand the mechanisms controlling food intake. However, food intake of fish can be influenced by a large number of extraneous factors (see Chapter 6). Commercial production of fish involves rearing in a heterogeneous environment with high fish numbers and large water volumes. These conditions are different from those of a controlled laboratory environment, and these differences may be expected to influence feeding and swimming and other aspects of behaviour, including social interactions among fish within the rearing units. There is therefore a question about the extent to which results of laboratory studies can be generalised to commercial conditions.

Exogenous factors that may influence food intake include photoperiod and light intensity, temperature, water current and weather-related factors (wind, rain and pressure), salinity, oxygen, CO_2, NH_3, pH, food abundance, availability and composition, and social interactions. How might these factors influence fish in culture systems?

There is no general answer to this question because culture systems differ greatly in environmental complexity (Fig. 7.1). There is a wide diversity of aquaculture production systems world-wide, and the systems can be considered to represent a continuum from culture in

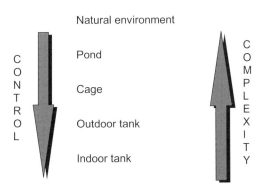

Fig. 7.1 Relative scale of complexity and possibility of controlling environmental factors in fish culture systems.

open waters to rearing in closed systems that are almost entirely independent of the natural environment.

This chapter divides the continuum into three sections each presenting a rearing system: ponds, cages, and tanks. Each section of the chapter describes the environment in the culture systems, summarises current knowledge about the relative importance of different environmental factors, and provides guidelines on the assessment of the impact of environmental factors on feed intake within each system.

7.2 Feed intake of fish in pond systems

The focus of this section is on the environmental factors in ponds and the particular consequences of these for feed intake of fish.

The world finfish production is currently dominated by warm-water omnivorous/ herbivorous species (85%, with cyprinids constituting 70%); carnivorous species represent only approximately 15% of total production. Semi-intensive or extensive pond-based culture accounts for approximately 80% of the world finfish production (Tacon & De Silva 1997).

The following classification is often adopted for purposes of convenience: warmwater, low-intensity or high-intensity fish production ponds, coolwater ponds and brackish-water ponds, including mangrove swamps. Although the principles of culture are similar irrespective of the pond type, there are physicochemical and operational differences, including origin (natural versus man-made), stocking density, type and amount of fertilisation, fallow periods, supplemental feeding, and the fish species used either in monoculture or polyculture (for background literature see Tucker 1985; Lannan *et al.* 1986; Little & Muir 1987; Horvath *et al.* 1992; Egna & Boyd 1997). For example, in Asia, pond culture is usually characterised by polyculture carried out in fertilised ponds integrated with other agrosystems (Joseph 1990; Pillay 1990; Billard 1995a). By contrast, pond culture in Western Europe is mainly based on the monoculture of salmonids, cyprinids (Billard 1995b), sea bass (*Dicentrarchus labrax*) and sea bream (*Sparus aurata*).

7.2.1 Pond characteristics

The types of ponds used for fish culture differ in origin, structure and degree of management, and as a consequence exhibit marked differences in production. Production is largely governed by the interactions between physical, chemical and biological conditions of the water, sediment and benthos, and the species present (Chang 1986).

Adequate design of ponds for aquaculture requires information about meteorology, hydrology, hydraulics, soil mechanics and biology. Pond design includes features relating to the pond itself, the water collection system, water conveyance system and water control structure (Boyd 1995). The most common sources of water are surface streams, possibly diverted towards a dam and flowing by gravity into the pond, or tidal streams that are dependent on the tidal cycle. Coastal ponds usually communicate with the sea through sluice gates, and they have an intermittent marine water supply. The latter differs according to pond elevation, distance to the sea and any freshwater or agricultural run-off.

Unlined ditches are most commonly used to distribute the water, and the water level within ponds is usually controlled by an overflow system. The design of levees for ponds depends on the desired depth of the water and the probability of overtopping due to flooding. Pond depths typically range from 30 to 200 cm (Colt 1986). Selection of depth is often based on modelling of pond stratification, pond production and energy balance (Losordo & Piedrahita 1991) or dissolved oxygen characteristics (Boyd *et al.* 1978; Romaire & Boyd 1979).

Pond areas typically range from 0.1 to 2.0 ha, but extend up to several hundred hectares in Central and Eastern Europe. Choice of pond size usually depends upon harvesting and operational flexibility. Nursery ponds for milkfish (*Chanos chanos*) or carp (*Cyprinus carpio*) may be only 10–50 m^2, whereas production ponds for milkfish may range from 4 to 6 ha.

7.2.2 Variability of environmental factors

In pond systems, most abiotic factors are – at best – only partially controlled, although biotic factors may be managed. The natural fluctuations in many environmental factors are considerable. These fluctuations may introduce marked day-to-day variations in feed intake of the fish and a production that varies seasonally depending upon latitude.

Ponds in the tropics differ considerably from those in temperate regions. Lowland ponds near the equator may show only slight seasonal variations compared with those in the temperate zone which exhibit large seasonal variations in temperature, photoperiod and light intensity (Fig. 7.2). However, many tropical ponds also experience strong seasonal variations (wet versus dry season). Equatorial high-altitude coolwater ponds, mainly Andean, may show large diel changes in temperature, but with limited seasonal change in mean temperatures. As for salinity, brackish or saltwater ponds in the temperate zone exhibit high fluctuations (2 to 40 ppt) between winter and summer.

Physical and chemical factors

The hydrology of ponds is often characterised by short-term spatial and temporal fluctuations in physicochemical parameters due to the fact that ponds are shallow isolated water bodies, and are therefore very sensitive to environmental changes. These changes may relate to rates of water renewal, which modifies water quality, or to meteorological factors, which affect the air–water interface and modify the hydrological characteristics.

The water renewal rate in ponds is directly determined by refill frequency and the volume exchanged, and the difference in density between the incoming and resident water. Warm or less saline renewal water will remain as a layer on top of resident water, and may exit the next time the sluice gate opens, whereas colder or more saline renewal water will sink to the bottom of the pond.

Atmospheric conditions and meteorological factors such as rain, air pressure and wind influence water temperature. Due to the low area/volume ratio in most ponds there is little delay between an atmospheric change and a water temperature change (Losordo & Piedrahita 1991). Wind, by increasing mixing at the air–water interface, modifies oxygen content (Boyd & Teichert-Coddington 1992), increases turbidity and creates water currents (Howerton & Boyd 1992). Salinity may be increased due to evaporation caused by high air temperatures, or decreased due to dilution by precipitation or freshwater run-off. Stratification may be ob-

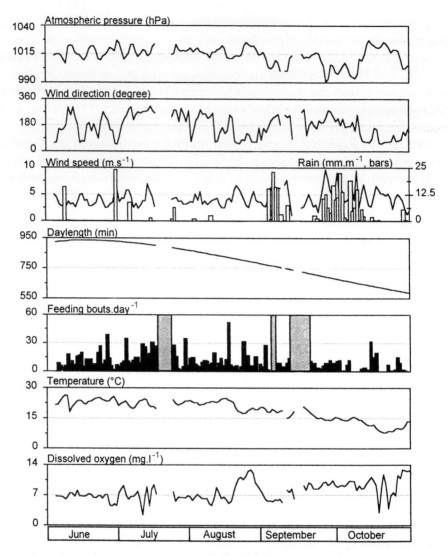

Fig. 7.2 Time series of mean daily values of environmental variables and demand feeding by sea bass recorded in an earth pond on the Atlantic coast (46°9'N, 1°9'W) from June 1 to October 31 1993. The hatched areas in July and September correspond to failure in the recording system. (From Bégout Anras 1995.)

served after heavy rain, and precipitation also directly modifies water temperature through advection. Light energy warms water directly but the degree of warming depends on wave action, and cloud cover limits heat losses. Light energy is important because of influences on the dynamics of dissolved oxygen (Romaire & Boyd 1979), photosynthesis and primary production. The latter, in turn, determines water column and benthos production capacity (Ali *et al.* 1987; Teichert-Coddington & Green 1993; Boyd 1995).

These interrelationships have been modelled along with management practices (liming, fertilisation, manuring, draining and drying), feeding and stocking rates, water exchange, and plankton production (Nath *et al.* 1995; Piedrahita & Balchen 1987; Losordo & Piedrahita 1991), as have their impact on fish growth (Cuenco *et al.* 1985; Prein 1985).

Both temperature and salinity have an influence on oxygen solubility. Biological activity, both in the water column and at the sediment–water interface, has an influence on oxygen and CO_2 concentrations and pH via the processes of photosynthesis and metabolism. There may be considerable diurnal variations in these parameters, a morning minimum and a late afternoon maximum usually being observed for oxygen (Losordo & Piedrahita 1991). Besides these commonly measured variables, others like nutrients also greatly influence the primary production capacities of ponds (Boyd 1995).

Pesticides, heavy metals, blue-green algae, pathogenic bacteria or industrial pollution may impose a major constraint in some areas; for example marine wetlands are often bordered by freshwater marshes devoted to agriculture (Bamber *et al.* 1993) and as such would be exposed to agricultural run-off.

Biotic factors and culture practices

Pond aquaculture is often carried out as polyculture by combining fish species that have complementary feeding habits, or feed in different zones. Thus, both space and food are utilised effectively, thereby increasing fish production (Spataru 1977; Hogendoorn & Koops 1983; Shang 1986; Milstein 1992; Teichert-Coddington 1996; Engle & Brown 1998). Rational stocking procedures depend on knowledge about the food preferences of the fish (Colt 1986) because utilisation of natural food is central to increasing pond productivity, and the fish exert an influence on plankton populations (Torrans 1986), nutrient cycling, benthos populations (Castel & Lasserre 1982; Flos *et al.* 1998), turbidity and macrophytes (King & Garling 1986).

Supplemental feed may be added to the pond, but the majority of the food consumed by fish reared in ponds is usually from natural sources. For example, during the seasonal dispersal of eggs and larvae of benthic invertebrate species in coastal zones, regular water renewal within coastal ponds ensures entrapment and settlement of these species. In the northern temperate zone, June is the month of highest settlement of annelids and molluscs (Reymond 1991). In summer, benthic macro-faunal biomass may reach 50 g ash-free dry weight per m^2 (Hussenot & Reymond 1990), although values of ca. 30 g/m^2 may be more usual (Labourg 1979). However, natural productivity of benthos in ponds is rarely sufficient to sustain semi-intensive rearing of fish, so fertilisation or supplemental feeding is usually employed.

Fertilisation of ponds with inorganic fertiliser, animal manures or plant material is commonly used to increase primary and secondary production (Pillay 1990). Good fertilisation practices depend on a knowledge of water chemistry, nutrient levels and oxygen balance (Galemoni & Ngokaka 1989), and the food web in the pond (Colt 1986).

For example, Schroeder (1974, in Hepher 1988) studied the effect of organic manures on the production of natural food organisms and on growth of common carp in ponds. The biomasses of organisms in manured and non-manured ponds, with and without fish, were compared. When stocked at a density of 5000 per hectare, the carp grew 25–100% faster in the manured ponds than in the non-manured ponds. The effect of the manure on the biomass of food organisms was apparent mainly in ponds without fish, while there was only a slight difference in the biomass in ponds stocked with fish due to the consumption of these organisms by the fish.

Supplemental feed may be both exploited by the fish, and act as an organic fertiliser to increase prey production. One technique involves the addition of feed pellets (ca. 35% protein) at $1–1.5$ g/m^2 per day (Hussenot *et al.* 1991, 1993). This enrichment induces an increase in benthic biomass, mainly macroinvertebrates, such as polychaetes. Insects may also play an important trophic role in fish production in ponds: for example, chironomids and diptera (*Ephydra* sp.) larvae may represent an important food source for juvenile fish of several species (Ceretti *et al.* 1987). Studies conducted in freshwater lakes demonstrate that primary and secondary production are boosted by organomineral inputs (Perschbacher & Strawn 1984; Welch *et al.* 1988).

7.2.3 Variations in feed intake in relation to pond environment

The measurement of feed intake of fish in ponds has rarely been achieved. Feeding activity of the fish is usually not monitored, although yield, growth in weight and length, and changes in body composition or condition factor, are often recorded (Tucker *et al.* 1979; Jana & Das 1980; Prein 1985; Sarkar & Konar 1989; Pfeiffer & Lovell 1990; Li & Lovell 1992a; Munsiri & Lovell 1993; Robinson *et al.* 1995).

In studies of fish feeding in ponds, the methods used have either been direct and invasive, such as stomach contents analysis (Labourg & Stequert 1973; Spataru 1977; Udrea 1978; Cremer & Smitherman 1980; De Lemos Vasconcelos Filho *et al.* 1980; Spataru *et al.* 1980; Ferrari & Chieregato 1981; Castel 1985; Chiu *et al.* 1986; Andrade *et al.* 1996), or indirect and non-invasive, such as monitoring the use of demand feeders by the fish (Randolph & Clemens 1976a, 1978; Spataru *et al.* 1980, Tackett *et al.* 1988, Bégout Anras 1995; Robinson *et al.* 1995).

The importance of natural food items

Even though increased prey production induces an increase in the fish standing crop, the effect may be limited. With an increase in standing crop, there is an associated increase in food requirement, and grazing/predation may become more efficient. At a certain point, predation will impair production of food organisms and hinder any further increase in standing crop (Hepher 1988).

Nevertheless, natural prey are often of the utmost importance for the production of fish in pond systems. The culture of sea bass in Atlantic coastal saltmarshes will be used as an example. A study was conducted to follow feeding in juvenile sea bass raised in saltmarshes to determine the importance of natural food items vs. supplemental pellets over the first two years of rearing (Reymond 1989a,b).

0-group bass (weight ca. 1 g) were introduced into a 1000-m^2 pond in early July at a stocking density of 22 fish per m^2. The fish were provided with between 1.1 and 2.5 kg feed each day, divided into three meals (09:00, 12:00 and 17:00). Fish had reached a mean weight of 12.3 g by early November, and production was 2213 kg/ha.

When fish were sampled after 62 days of culture, stomach content analysis revealed 62% pellets versus 38% natural prey (Fig. 7.3A). However, at this time, the size frequency distribution of the fish was bimodal, with small fish weighing ca. 2.0 g and large fish ca. 7.5 g. The large fish were feeding mainly on pellets (75%), whereas the small fish fed mainly on natural prey (62%). Furthermore, prey type differed depending on the size of the fish, the small fish taking mostly polychaete larvae whereas the large fish consumed a greater proportion of insects (Fig. 7.3a).

A 500-m² pond was stocked with 1-group bass (mean 2 g, range 0.5–20 g) in early June at a density of 21 fish per m². Fish were provided with 1.5–4.8 kg feed daily, and had reached a mean weight of 47.1 g by the end of October, with production being 5831 kg/ha.

After 78 days of culture, fish were sampled for stomach content analysis. The contents of the stomachs consisted of 57% pellets and 43% prey (Fig. 7.3b). Once again, the large fish (24–37 g) fed on pellets, whereas small fish (8–15 g) had consumed natural food items, mainly crustaceans (amphipods, isopods and copepods, 82%, Fig. 7.3b).

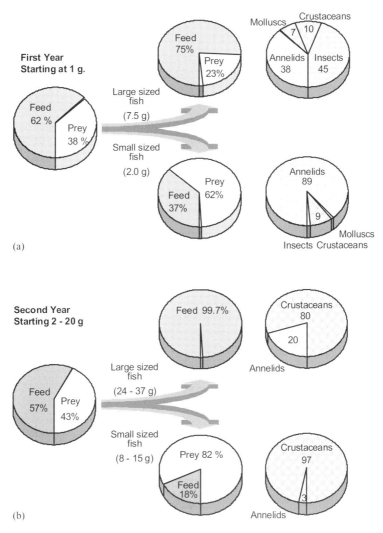

Fig. 7.3 Stomach contents of juvenile sea bass, *Dicentrarchus labrax*, raised in Atlantic saltmarshes. (a) 0-Group sea bass feeding regime from a sampling after 62 days of culture for all fish together or depending on the size of the fish. (b) 1-Group sea bass feeding regime from a sampling after 78 days of culture for all fish together or depending on the size of the fish. Stomach content is expressed as volumetric frequency of each feed type (%). Feed/prey composition is expressed as numerical frequency Cn (%). (After Reymond 1989a,b.)

Thus, two food sources were clearly demonstrated in the ponds, with the larger fish mainly consuming pellets. The role of natural food items was important because the natural prey represented a food source for the small fish, which might otherwise have died. Dual food sources have also been described for milkfish, which fed on natural food during the day and on artificial food during the night (Chiu *et al.* 1986). When carp were provided with supplementary food by means of a demand feeder, they changed their feeding pattern. They tended to consume prey from the surface layer, and to take the most accessible natural food items (Spataru *et al.* 1980). Andrade *et al.* (1996) demonstrated dual use of food sources by sea bream, but concluded that despite abundant natural prey in the ponds, the dependence of sea bream on pelleted food was high.

Consequences of environmental factors: spatiotemporal variability

Given the marked variations imposed on ponds by meteorological conditions, greater effort should be directed towards the study of spatiotemporal variability and seasonal changes on food access and fish distribution.

Timing of feeding and feeding regime (restricted versus satiation) has been investigated in pond-cultured channel catfish, and most producers feed their fish once daily to satiation (Robinson & Li 1995). In most studies, growth was not affected by time of feeding or feeding regime, but strong interactions appeared between these factors and feed quality (Li & Lovell 1992b; Phelps *et al.* 1992; Munsiri & Lovell 1993; Robinson *et al.* 1995).

The variability in feed demand has usually been explained on the basis of changes in water temperature, oxygen levels and social interactions (Randolph & Clemens 1976a; Tucker *et al.* 1979; Tackett *et al.* 1987, 1988). In a study of the use of demand-feeders by pond-reared channel catfish, Randolph & Clemens (1976a) found that the fish did not operate the feeders until the temperature rose above 12°C, and they stopped feeding during the autumn when water temperature was ca. 22°C. During the summer, low oxygen concentrations (<5 mg/l) often caused the fish to adjust or miss their daily feeding period. They also showed that large fish were dominant and that small fish usually had to wait in order to feed. Fish within each category fed at approximately the same time day after day, and so became accustomed to feeding under different temperature and oxygen conditions (Fig. 7.4). Feeding activity may also be linked to other environmental factors such as day length, wind direction and rain (see Fig. 7.2). These factors are more important than temperature or oxygen contents in determining feed demand by sea bass in ponds (Bégout Anras 1995).

When fish feed on natural food items they must search for patchily or randomly distributed prey, but they do not use space uniformly. They may maintain position in restricted areas characterised by landmarks, such as inlet or outlet pipes, a self-feeder, vegetation or stay at a particular depth (Randolph & Clemens 1976b; Cunjak & Power 1987; Bégout Anras & Lagardère 1995), they may select areas sheltered from wind (Lagardère *et al.* 1994) or avoid areas affected by wind-induced turbidity (Chapman & Mackay 1984), or may choose areas characterised by certain water temperatures, high oxygen contents or shade (Johnsen & Hasler 1977; Mosneron Dupin & Lagardère 1990; Lagardère *et al.* 1996).

Not only do environmental factors affect space use, they also influence swimming activity. Low atmospheric pressure, rain and high wind speed influence the movements of sea bass, sea bream and turbot (*Scophthalmus maximus*) (Bégout & Lagardère 1993; Lagardère *et al.*

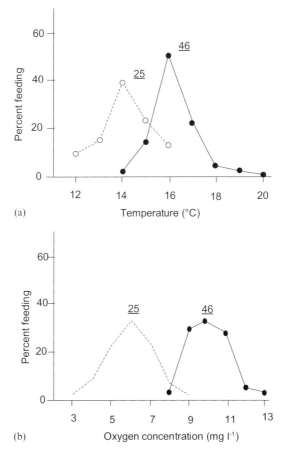

Fig. 7.4 (a) Percentage of twenty marked (25 and 46 cm in length) channel catfish feeding at various temperatures in 1.6-ha ponds during March, regardless of time of the day and showing that small and large fish were acclimatised to different ranges (*n* = 632). (b) Percentage of twenty marked (25 and 46 cm in length) channel catfish feeding at various oxygen concentrations in 1.6-ha ponds during July and August, regardless of time of the day and showing that small and large fish were acclimated to different ranges (*n* = 840). (Figures from Randolph & Clemens 1976a.)

1995; Bégout Anras & Lagardère 1998) and water renewal may lead to increased activity in sole (*Solea solea*) (Lagardère *et al.* 1992). Social factors also affect fish swimming behaviour (Bégout Anras *et al.* 1997).

Various automated water-quality data acquisition systems have been developed for aquaculture purposes (e.g. Losordo *et al.* 1988; Poxton 1991), but there is a need for information about the correspondence between sampling sites and places actually occupied by fish, especially in heterogeneous environments. This necessitates the simultaneous recording of environmental (meteorological and hydrological) factors, and fish feeding and swimming activity in culture systems. In semi-natural culture environments such as ponds, lagoons and saltmarshes, fish usually have few possibilities to buffer the impact of environmental variations through vertical or horizontal migrations. Changes in spontaneous activity or swimming in response to environmental stressors are often observed but are difficult to quantify (Scherer 1992). Various techniques have been used to assess the behaviour of cultured fish (for review see Poncin & Ruwet 1994). Among these techniques, underwater biotelemetry may be used in ponds to study space use and swimming activity, and they may also provide opportunities for monitoring the physiology of free-swimming fish (for review see Baras & Lagardère 1995). Since biotelemetry is based on the examination of individual behaviour, it is complementary to other techniques such as the use of demand-feeders to estimate feeding and to monitor the well-being of fish (Lagardère *et al.* 1996).

7.3 Feed intake in fish cages

At present, ca. 1.5 million tons of fish are raised in cages, of which ca. 65% are in fresh water (Beveridge 1996). Cage culture is, however, also of prime importance for intensive ongrowing of high-value marine species such as salmonids, sea bass and yellowtail (*Seriola quinqueradiata*). Net cages or pens used in fish farming vary in size from freshwater nursery cages less than 1 m³ to 30 000 m³ offshore cages. Beveridge (1996) provides an overview of cage culture operations, including cage design, site selection, environmental impacts and management.

Compared to a land-based system, cage-rearing may provide a cost-effective water exchange – something that is a prerequisite when rearing fish at high densities. In marine farming this has led to a move away from small cages sited in sheltered areas to the use of larger cages located at more exposed (offshore) sites. The intensive forms of cage culture rely on inputs of high-quality fish feed, and feed costs may present from 40 to 60% of production costs. Efficient feed management is therefore essential for cage farming to be profitable.

A good understanding of how feed intake is influenced by single environmental factors has been developed (see Chapter 6), but the relative importance of the different factors within the context of a complex cage environment has yet to be established. The main motivation for studying feed intake of fish in production-scale cage units is to develop practical methods of feed management, to use feeding as an indicator of performance that can be used to develop alternative production strategies, and to identify the relative importance of interactions between environmental factors in governing the feeding and growth responses of the test species.

7.3.1 Feed intake in a complex, semi-controlled environment

The disadvantage of using the low-cost natural environment for intensive fish farming is that it can be highly unpredictable. The environment within a fish cage is determined by its latitude and location, e.g. whether the cage is sited in fresh, brackish or sea water, and whether it is at an inshore or offshore location. Further, the range of seasonal variations in temperature will differ considerably between the tropics and temperate areas. At any given site the type of cage (floating, submersible or closed) and its geometry (depth, circumference and form) will also have an influence upon the suite of environmental factors to which the fish is exposed. The cage environment is not homogeneous, but varies spatially, especially with depth, and with time. For example, depending upon meteorological and hydrographic conditions, the temperature close to the surface may be higher or lower than in deeper parts of the cage. Surface temperatures will also show greater, and more rapid, variation than water deeper in the cage.

The extent to which a farmer can control these factors to produce a stable rearing environment that promotes efficient production is usually a question of available technology and the economic viability of using this technology. For example, efforts to reduce environmental variation have been made via the development of submersible cages, where the fish are subjected to deep water, and more stable, environments (see Beveridge 1996) and by the construction of floating closed cages supplied with water pumped from deeper water layers (Skaar & Bodvin 1993).

Relatively few studies have been conducted to measure feed intake of fish in cages, although there are many studies on growth of fish – mainly salmonids – in cages. There has been a call for an increased effort in this field to complement experimental studies (Juell 1995; Tacon 1995). Several methodological approaches have been adopted to examine feeding by fish in net pens, including direct observation and the use of behavioural indicators. For example, in a short-term experiment Thorpe *et al.* (1990) monitored food consumption of some of the fish in a group of 2068 Atlantic salmon (*Salmo salar*) using the X-radiographic method (see Chapter 3) and assessed how feeding method influenced the ability of individuals to obtain a meal. At the other end of the time-scale, Fernö *et al.* (1988) attempted to map seasonal variations in feed intake of large groups of salmon (3200–5000 fish) using direct observations made from the surface and underwater video cameras. Juell and Westerberg (1993) used telemetry to track individual fish in a production cage to obtain indications of feeding. In smaller experimental cages, Kadri *et al.* (1996a) used video recordings of 19 individually recognisable salmon to observe the effect of social interactions on food-sharing. Bjordal *et al.* (1993) and Juell *et al.* (1993) used acoustic measurements of changes in fish density within the feeding area as an indicator of appetite within groups of 30 000 salmon in 1800-m³ cages. These examples serve to illustrate the range of methods that have been used to study feeding behaviour of fish in cages: the choice of method is often determined by the purpose and the duration of the experiment, as well as by the fish density and size of the cage.

7.3.2 Environmental variation and feeding activity

Some work has been carried out to examine feeding of fish in cages in relation to long-term variations in environmental factors.

Physical and chemical factors

The water temperature in a cage fluctuates with season, and there is a general relationship between mean temperature, and feeding and growth (e.g. Austreng *et al.* 1987; Borghetti & Canzi 1993; Juell *et al.* 1998). For example, the demand feeding activity of rainbow trout (*Oncorhynchus mykiss*) was strongly correlated with water temperature in summer (Alanärä 1992a), but the possibility of greater feed waste at high temperature precluded the drawing of firm conclusions about the relationship between feed intake and temperature (Alanärä 1992b). Using behavioural indices of feeding (number of fish responding, and time used to consume pellets), Smith *et al.* (1993) found that feeding activity of Atlantic salmon increased in late winter/early spring although the temperature was stable: they concluded that the change in photoperiod had a greater influence on feeding than did temperature during this period. In another study, variations in daily feed intake of salmon were not closely correlated with day-to-day variations in temperature (Fig. 7.5), but feeding appeared to be more dependent upon the time since the previous meal and the amount eaten in that meal (Juell *et al.* 1994a). Temperature also seems to have little effect on the actual rate of consumption in a meal (Talbot *et al.* 1999). In all of the above studies water temperature was measured at a single depth, and there might be considerable thermal stratification in cages (Fig. 7.6). The fish may distribute themselves within particular thermal strata (Sutterlin & Stevens 1992; Fernö *et al.* 1995). Studies relating to the influence of temperature on feeding should include observations of fish distribution in relation to thermal stratification, and problems arising from environmental heterogeneity should be kept in mind when conducting experiments in cages.

Light experienced by fish in cages is easier to control than either temperature or salinity. The photoperiod can be extended by the use of artificial lights, but a reduction in day length

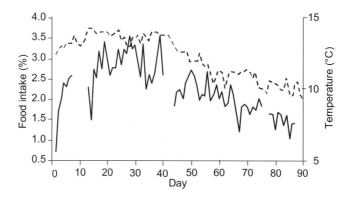

Fig. 7.5   Plots of average daily food intake (% of biomass) of two groups of caged Atlantic salmon fed to satiation (solid line), and temperature at 2 m depth (dotted line) recorded over a three-month period. Although there was a general long-term relationship between temperature and food intake there was considerable day-to-day variation. Breaks in the feeding line denote days of feed deprivation. (From Juell *et al.* 1994b.)

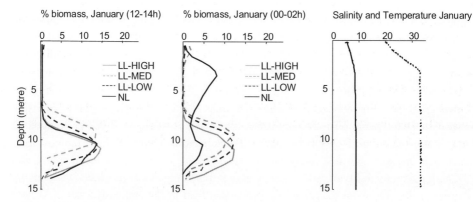

Fig. 7.6   The vertical distributions of groups of 2300 Atlantic salmon (1.8 kg) in four 15 m-deep sea cages in January (61°N). One group (NL) was held on a natural photoperiod while three groups (LL-LOW, LL-MED, LL-HIGH) were subjected to 24L:00D of different intensities. Salinity (dashed line) and temperature (continuous line) profiles show environmental stratification. LL-groups swam in deep warmer water during both day and night, while the NL-group ascended to the thermocline at night (F. Oppedal *et al.* unpublished results).

is more difficult to achieve. Light intensity can be reduced during daytime by shading nets or covers (Huse *et al.* 1990) and increased at night by use of artificial light. Regardless of any artificial manipulations of light conditions, changes of season and cloud cover will give large variations in light intensity during the daytime.

Oppedal *et al.* (1997) reported that post-smolt Atlantic salmon subjected to a 24-h photoperiod from January to June grew faster in spring than fish held on natural photoperiod, and suggested that the light intensity applied must be above a certain threshold to trigger this effect. Manipulation of photoperiod is now widely used to influence sexual maturation in salmonids (Hansen *et al.* 1992), and trials are under way on other species such as cod (*Gadus morhua*), sea bass and mackerel (*Scomber scombrus*). Exposure to continuous light may lead to alterations in the behaviour of the fish (Fig. 7.6). Salmon exposed to a natural photoperiod usually swim in circular schools during daytime, but schools disperse at dusk, and the fish rise towards the surface (Juell 1995). Thus, when held on a 24L:00D photoperiod in winter, the period of active swimming may be increased, and the effect of this could be increased energy expenditure and reduced growth in comparison with salmon held on natural photoperiod at this time (Oppedal *et al.* 1997, 1999).

In the rearing of planktivorous fish the use of artificial lights at night may serve to concentrate their food source. This was suggested to have a significant effect on feeding, growth and mortality of juvenile Asian sea bass, *Lates calcifer*, in nursery cages (Fermin & Seronay 1997). Plankton and other particles in the water column will interact with light to determine the visual environment in the cage (see Chapter 6). High turbidity may result in problems related to the detection of food and thus, indirectly, reduce feeding (Ang & Petrell 1997). Absorption of light passing through the water column leads to changes in both intensity and spectral qualities. Light preferences for most cultured species are not known, but their existence should be considered when developing management strategies (Huse & Holm 1993). For example, it has been suggested that the depth at which salmon swim in cages is based on a trade-off between surface avoidance (high light intensity and predators) and attraction to food (Juell *et al.* 1994b; Fernö *et al.* 1995).

Fish reduce their feeding activity when exposed to hypoxic conditions. Oxygen concentrations in fish cages are, however, difficult to monitor and record. Oxygen concentrations in a water mass will vary with temperature, and a high biomass of fish inside a cage may create horizontal gradients in oxygen levels due to metabolism. However, tidal or wind-driven currents may provide an effective water exchange between the cage and the surrounding medium. However, neither Kadri *et al.* (1991) nor Smith *et al.* (1993) found any clear relationship between feeding behaviour and stage of the tidal cycle. Schooling fish can, by the force of their tail-beats, set up strong vertical currents (Kils 1989; Juell *et al.* 1998) and these may enhance the mixing of water. Hypoxic conditions are most likely to be encountered in freshwater cages during periods of algal blooms and low water exchange rates.

Biotic factors

Fish reared in cages interact with other biota: nets may be used to protect the fish against avian and aquatic predators and scavengers competing for fish-food. Although protective nets may reduce predator-related mortality to a minimum, predators may still be present around the cage. For example, birds flying over a cage, attacks by predatory fish and mammals, or the close presence of schools of 'scavenging' fish may lead to short-term reductions in feeding activity by the fish inside the cage. Planktonic organisms may also interact with the caged fish either in the form of toxic algal blooms, ectoparasites (Heuch *et al.* 1995) or as alternative food sources (Juell *et al.* 1998). Further, the fouling of nets with macroalgae, tunicates, molluscs and other organisms that use the net as a substrate may alter the environment within the cage by reducing water exchange rates.

7.3.3 Daily and seasonal feeding patterns

There has been some effort in identifying daily and seasonal feeding patterns in fish held in cages. Using an on-demand feeding system incorporating a waste feed detector, Blyth *et al.* (1993) found that during the winter months Atlantic salmon consumed 60% of their daily ration during the first hours of daylight, ceased feeding during the middle of the day, and fed again around dusk (Fig. 7.7). A similar pattern was observed regardless of season by Blyth *et al.* (1999). The pattern of daily feeding could however, be disrupted by external disturbances such as human activities or the presence of predators. Periods of feeding around dawn and dusk were also reported by Kadri *et al.* (1991) in a summer trial carried out on Atlantic salmon, and by Alanärä (1992a) who worked with rainbow trout. A morning peak in feeding was also observed in Atlantic salmon in wintertime, with feed intake decreasing gradually as the day progressed (Juell *et al.* 1994a). By contrast Smith *et al.* (1993) did not find a consistent diurnal pattern in feeding behaviour between months. Feeding patterns of caged fish species other than salmonids have not been investigated in any detail, but some work is in progress. Whether or not the observed patterns reflect natural feeding rhythms or are artefacts of the feeding regimes employed on the farms, for example relating to the cessation of feed distribution at night, is a question that deserves closer examination.

7.3.4 Cage management practices

Management practices include making decisions about factors such as food availability, fish density and handling procedures, all of which may influence feed intake of fish in cages. Whether or not feed is available to fish may be determined by factors such as pellet size, the amount provided, and the spatial and temporal distribution of the feed. Hitherto, most feeding trials have been carried out in small cages with a low number of fish, and without accurate measurement of feed intake. Observations are usually restricted to growth measurements and aspects of carcass composition (e.g. Tacon *et al.* 1991; Borghetti & Canzi 1993; Webster *et al.* 1995; Hemre *et al.* 1997; Johansen & Jobling 1998).

Smith *et al.* (1995) used variations in a series of behavioural indices to examine effects of pellet dimensions on feeding responses of Atlantic salmon: they found that long pellets were rejected more often than shorter ones, although the long pellets were initially more attractive. In some studies the focus has been on the effects of temporal and spatial distribution of feed on intake, growth and size variation. Thorpe *et al.* (1990) concluded that highly localised feeding may influence the feed-sharing within groups of caged salmon leading to increased heterogeneity. This conclusion was supported by Johansen & Jobling (1998) who compared the growth of salmon fed either by automatic feeders or by hand. By contrast, Juell *et al.* (1994b) did not find that feeding intensity had a marked influence on size variation among salmon in production

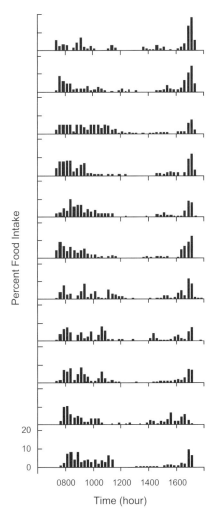

Fig. 7.7   Daily feeding patterns of Atlantic salmon in Tasmania shown as quarter-hour food distribution (% of daily feed intake) over eleven consecutive days. (From Blyth *et al.* 1993.)

cages, irrespective of whether the fish were fed to satiation or on restricted rations which were predicted to provoke intensive competition for feed. This indicates that the effects of social interactions on feed-sharing observed in experimental studies with small groups of fish (e.g. Kadri *et al.* 1996a) may not be present when high fish density and moderate food dispersal makes feed supplies indefensible (Grant 1993; Jørgensen *et al.* 1993). Individuals within a group may, however, feed and grow at different rates due to inter-individual variation in other factors such as maturation status (Kadri *et al.* 1996b; Johansen & Jobling 1998).

7.4 Feed intake in tanks

Intensive fish culture in tanks, often referred to as land-based fish farming, is carried out world-wide. Tank culture is used to produce freshwater and marine species, and benthic and open-water species. In European countries freshwater species are more commonly reared in tanks than are marine finfish. Rainbow trout and freshwater stages of other salmonids represent the bulk of production in Europe, but marine species such as sea bass, sea bream and turbot are also raised in tanks.

Tanks differ widely in design, e.g. shape, size and depth, and tank production systems differ in the types of water management employed, i.e. flow-through single or multiple pass systems or recirculating systems. In contrast with pond and cage culture, tank culture offers opportunities for the control and manipulation of a range of environmental parameters. Water exchange rates can be adjusted in relation to tank biomass in order to meet the water quality requirements of each species or developmental stage. Tank culture will also almost inevitably involve feeding the fish exclusively on formulated feeds. This sets tanks apart from ponds and open net cages.

The focus of this section is the management of environmental factors in tanks, and the particular consequences of this for feed intake of fish during the ongrowing phase.

7.4.1 Tank culture systems

The classification of tanks according to their shape – rectangular (raceway) or circular – is oversimplistic, because it is the combined effects of shape, size, depth and water inlet and outlet arrangements that influence water flow characteristics and quality, fish behaviour and biological performance.

Within the raceway family, plug-flow tanks differ from cross-flow tanks in section profile and in the way by which the water is introduced (Watten 1990; Ross *et al.* 1995; Ross & Watten 1998). Within the circular tank family, differences are often related to the design of inlet and outlet flow structures (reviewed by Timmons *et al.* 1998).

Studies dealing with the improvement of tank design usually focus on decreasing water residence time and water quality gradients, and at increasing the efficacy of solid waste (faeces and uneaten pellets) removal. Special attention is often directed towards water current management, because currents may influence feed distribution, fish swimming orientation and activity, schooling behaviour and agonistic responses (Jobling *et al.* 1993). In closed recirculating systems some by-products resulting from fish metabolism and/or water treatment (Fig. 7.8) may influence the quality of the inlet water entering the fish-rearing tank. However, the low volumes of make-up water added to such systems provide opportunities for increased manipulation of water parameters via mechanical filtration, gas stripping or thermal adjustments.

The interactions between these multiple factors define the environment experienced by the fish, which will influence feed intake, and contribute to carrying capacity limitation (Losordo & Westers 1994). Feeding regimes used in commercial production represent compromises: automatic feeding systems often deliver predetermined amounts of feed in accordance with predictive models based on fish size and temperature, but such models have major weaknesses. On the other hand, on-demand feeding systems have been developed to meet feed requirements without predetermination of either the timing of feed delivery or the quantity delivered (see Chapter 3). In each case the objective of feed management is to promote fish growth and reduce feed waste (see Chapter 15). When feed waste detection and measurement

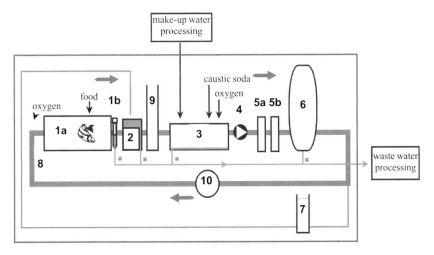

Fig. 7.8   Water processing units in a recirculating system for fish production. 1a, rearing tank; 1b, sediment trap; 2, mechanical filter; 3, pumping tank; 4, pump; 5a, UV; 5b, optional ozonation system; 6, nitrifying biofilter; 7, denitrifying biofilter; 8, oxygenation device; 9, CO_2 degassing column; 10, heat exchanger; *, collection of suspended matters and material that settles.

systems are operational and efficient, feed intake can be recorded. Under these conditions the relation between feed intake and environmental parameter fluctuations can be studied.

7.4.2 Environmental variation and feeding activity

Physical and chemical factors

Light is characterised by its quality (different wavelengths), quantity (different intensities) and periodicity (daily and seasonal cycles).

There are several reports covering photoperiod effects on the growth of juvenile and adult fish. Some workers have failed to find any effect of different photoperiods on growth, for example in halibut *Hippoglossus hippoglossus* (Hallaraker *et al.* 1995) and turbot (Pichavant *et al.* 1998), but marked influences have been recorded in some studies, e.g. on Atlantic salmon (Villareal *et al.* 1988; Rottiers 1992), perch (*Perca fluviatilis*) (Wieser & Medgyesy 1991) and on pike (*Esox lucius*) (Beauchaud *et al.* 1999). The photoperiodic effects on growth may be a result of a combination of increased feed intake and better food conversion rather than just greater feed intake.

Light is known to be the most important zeitgeber influencing diurnal and seasonal rhythms of locomotion, reproduction and feeding (Boujard & Luquet 1990; Boujard & Leatherland 1992; see also Chapter 8). There are clear relationships between day length and feeding activity in several species (Eriksson & Van Veen 1980; Boujard *et al.* 1991), but the influence of light may be modified by changes in other factors such as temperature or availability of oxygen. For example, some salmonids, cyprinids and sea bass may display either diurnal or nocturnal feeding behaviour depending on light and temperature conditions (Fraser *et al.* 1993, 1995; Sanchez Vasquez *et al.* 1997).

Fish of most species are not able to maintain body temperature by physiological means, and their body temperature fluctuates with the temperature of the surrounding water. Within the range of temperatures under which growth is possible, metabolic rate tends to increase with increasing temperature, but at high temperature, the metabolic rates of fish feeding maximally may stabilise (level off) or even decline (Jobling 1997). Increases in temperature initially lead to increased rates of ingestion, feed intake peaks at some intermediate temperature, and then declines precipitously as the temperature continues to rise. Thus, manipulation of temperature within tank systems can be used to manipulate rates of feeding and growth (see Jobling 1997 for review).

The term total ammonia nitrogen (TAN) refers to the sum of the unionised fraction, NH_3-N (UIA-N) and the ionised fraction, NH_4^+. The NH_3 molecule, which is non-polar and soluble in lipids, diffuses more readily through lipoprotein membranes, such as those in fish gills, than does NH_4^+. This is one of the reasons that NH_3 is the more toxic form. The UIA-N:TAN ratio depends on pH, temperature and, to a lesser extent, on salinity (Person Le Ruyet & Bœuf 1998). Many factors may affect the nitrogen excretion of fish, including body weight (Kikuchi *et al.* 1990), physiological and nutritional status (Wiggs *et al.* 1989) and rearing conditions. Both the quantity (Jobling 1981; Poxton & Lloyd 1989) and quality of ingested protein (Forsberg & Summerfelt 1992; Li & Lovell 1992a) influence rates of nitrogen excretion. Excretion may increase severalfold after feeding (Dosdat *et al.* 1998). When water is reused there is a danger that ambient ammonia concentration may increase and become a limiting factor for feeding and growth (Lemarie *et al.* 1996; Rasmussen & Korsgaard 1996).

The pH of the water will influence the form in which TAN occurs, and under some conditions concentrations of unionised ammonia may become high enough to be toxic (Bergerhouse 1992). The ionic imbalances that result under unfavourable pH conditions, leading to circulatory failure, are believed to be the primary cause of the death in acid-exposed fish (Wood 1989). Acidification of water may also affect the ability of fish to detect and/or to respond to a chemical feeding stimulus (Lemly & Smith 1985). Some acclimatisation to both high pH (Jordan & Lloyd 1964) and low pH (Balm & Pottinger 1993; Van Ginneken *et al.* 1997) may be possible.

Nitrite is toxic to fish, although toxicity varies widely among species and test conditions (Russo & Thurston 1991). Salmonids appear to be among the most susceptible. Some workers have attempted to study the effects of long-term exposure to sublethal levels of nitrite on feeding and growth (Colt & Tchobanoglous 1978; Saroglia *et al.* 1981; Kamstra *et al.* 1996), but data have not proved consistent.

Nitrate is not very toxic for fish (Pierce *et al.* 1993), but nitrate can become a limiting factor in recirculating systems because of effects on the nitrifying efficiency of the bacteria in the biofilter.

Metabolically derived carbon dioxide may accumulate within intensive rearing systems (Sanni *et al.* 1993), and under some circumstances may limit feeding and growth (Fivelstad *et al.* 1998).

Feed intake and growth each decrease in fish exposed to hypoxia, as demonstrated for both sea bass (Thetmeyer *et al.* 1999) and turbot (Pichavant *et al.* 1998). In silver bream (*Sparus sarba*), feeding decreased markedly when oxygen saturation was reduced from 95 to 40% (Chiba 1983). Some authors (Pouliot & De la Noue 1989; Neji *et al.* 1993) have reported that dietary composition may have an influence on feeding in rainbow trout exposed to hypoxia.

Some work has been carried out to examine the physiological effects of hyperoxia on fish, but information about the feeding of fish exposed to hyperoxia is sparse. Edsall and Smith (1990) observed that there was no difference in growth or feed conversion among groups of juvenile rainbow trout reared at 180 and 95% dissolved oxygen saturation for 125 days. Similar conclusions with regard to the effects of hyperoxia were reached by Caldwell and Hinshaw (1994), while Mallekh *et al.* (1998) showed that oxygen supersaturation was of no advantage for feed intake in turbot under intensive farming in concrete tanks.

Discoloration of the water has often been described in recirculating systems, but may also occur in flow-through systems (Hirayama 1974; Meade 1976; Muir & Roberts 1982). Muir and Roberts (1982) suggested that the transformation process occurring in biofilters might be responsible for the discoloration. Hirayama *et al.* (1988) found a correlation between food input and accumulation of slowly decomposed colouring organic substances in closed recirculating systems, and Schuster (1994) suggested that browning products produced during the manufacture of fish meal might be the source of discoloration. Takeda and Kiyono (1990) suggested that extreme accumulation of these substances might have detrimental effects on fish and other organisms within closed recycling systems.

Biotic factors and culture practices

Stocking density can influence feed intake and growth of fish, both directly and indirectly via effects on water quality and on the social behaviour of the animals. In coho salmon *Oncorhynchus kitsutch* (Schreck *et al.* 1985) and brook charr *Salvelinus fontinalis* (Vijayan & Leatherland 1988) high density conditions seem to lead to reduced feeding possibly as a result of fish stress. On the other hand, growth of channel catfish, vundu catfish (*Heterobranchus longifilis*) or Arctic charr (*Salvelinus alpinus*) was slower among fish stocked at low density (Jobling & Wandsvik 1983; Davis *et al.* 1984; Kerdchuen & Legendre 1992; Baras *et al.* 1998), than at high density, possibly because of lower levels of agonistic interactions under high rearing density (Wallace *et al.* 1988; Baker & Ayles 1990; Christiansen & Jobling 1990; Brown *et al.* 1992; Jorgensen *et al.* 1993). Ewing *et al.* (1998) found interactions between rearing density and raceway conformation on growth, and survival of juvenile Chinook salmon (*Oncorhynchus tshawytscha*).

Size-sorting, grading and other routine handling procedures may also have an influence on feeding and growth rates (Jobling & Reinsnes 1987; Baardvik & Jobling 1990). Further, the presence of pathogens, such as bacteria, viruses and parasites, may lead to a decrease in feed intake and growth, and some disease treatments may have negative influences upon feeding both during the treatment period and for some time after medication has been terminated (Bloch & Larsen 1993; Neji *et al.* 1993).

7.4.3 Tank management practices and possibility of control

The possibility to control the environmental factors differs depending on whether flow-through or recirculation systems are used (Table 7.1), but greater manipulation is possible in tanks than in ponds or cages. Lighting can be controlled and temperature can be regulated, especially within recirculation systems. Inlet water can be treated by mechanical filtration and with ultraviolet (UV) irradiation to remove particulate matter, fouling organisms and

Table 7.1 Flow-through system versus recirculating system: control possibilities of environmental factors ($+$ or $-$) and procedure indication.

Limiting factors	Flow-through system	Recirculating system
Light	$+$	$+$
Temperature	$-$	$+$
	(+ if access to power plant or ground water)	(depending on the degree of recycling)
Salinity	$-$	$+$
	(+ if access to ground water)	(depending on the degree of recycling)
Pathogens		
Bacteria	$+$	$+$
	(prevention by vaccines and curative by oral antibiotics)	(prevention in make-up water by UV and/or ozone, prevention by vaccines and curative by oral antibiotics)
Virus	$-$	$+$
		(prevention in make-up water by UV and/or ozone)
Parasites	$+$	$+$
	(curative only)	(only prevention by mechanical filtration of the make-up water)
Limiting factors		
O_2	$+$	$+$
	(before or inside tank only)	(before tank and biofilter)
CO_2	$+$	$+$
	(biomass: water residence time)	(degassing column)
Ammonia	$+$	$+$
	(biomass: water residence time)	+ (nitrifying biofilter)
Suspended matters and material that settles	$+$	$+$
	(biomass: water residence time and tank shape)	+ (mechanical filter and ozone)
pH	$+$	$+$
	(biomass: water residence time)	(caustic soda supply and CO_2 degassing column, denitrifying biofilter)
Nitrite	$-$	$+$
		(nitrifying biofilter)
Nitrate	$-$	$+$
		(denitrifying biofilter)
Coloured water (yellow substances)	$-$	$+$
		(ozone and UV)

pathogens. Changes in water quality will be linked to fish metabolism, so water flow must be adjusted in relation to the standing biomass to prevent waste accumulation. The inlet water may be oxygenated to reduce the need for water turnover, because oxygen is often the first limiting factor in tank culture.

Despite the imposition of some degree of environmental control in tank systems, several factors may show fluctuations. Some of these fluctuations are directly related to the feeding and activity cycles of the fish, which lead to rhythmic patterns of O_2 demand, CO_2 production and nitrogenous excretion. Some other fluctuations remain uncontrolled and are due to weather-related factors, such as wind conditions, which may have an influence on fish even under intensive rearing conditions in land-based tanks (Mallekh *et al.* 1998).

7.5 Conclusions

It is evident from this review that there is considerable scope for increasing knowledge about food intake of cultivated fish in each of the rearing systems considered.

Short-term variations in feed intake are unpredictable in commercial rearing systems where the fish are subjected to fluctuating abiotic and biotic factors, as well as handling procedures. This has led to an increased focus on the development of feeding systems that are capable of monitoring these short-term changes in feed demand and adjusting feed provisioning accordingly (see Chapters 3 and 15). Calculation of food conversion ratios requires the cumulative food intake of the fish to be recorded throughout the production cycle. Environmental factors in ponds and cages are highly variable in space and time, as is fish behaviour (space use and activity rhythms), and mapping these requires the use of an array of direct and indirect methods – some of which have been described in this chapter. Case studies should be carried out under realistic culture conditions with respect to fish density and rearing systems in order to strengthen the link between academic and industrial approaches.

In semi-natural environments such as ponds and cages, several basic factors (light, temperature and oxygen) will have major influences, but recordings of natural prey abundance, meteorological conditions and human disturbances should be of primary concern and should not be neglected. Combining these types of information with effects on growth and quality of the fish at the end of the production cycle may help in the development of production strategies that promote economic efficiency through good feed management, and also improve animal welfare and ecosystem health.

7.6 References

Alanärä, A. (1992a) Demand feeding as a self-regulating feeding system for rainbow trout (*Oncorhynchus mykiss*) in net pens. *Aquaculture*, **108**, 347–356.

Alanärä, A. (1992b) The effect of time-restricted demand feeding on feeding activity, growth and feed conversion in rainbow trout (*Oncorhynchus mykiss*). *Aquaculture*, **108**, 357–368.

Ali, M.M., Rahmatullah, S.M. & Habib, M.A.B. (1987) Abundance of benthic fauna in relation to meteorological and physicochemical factors of water. *Bangladesh Journal of Agriculture*, **12**, 239–247.

Andrade, J.P., Erzini, K. & Palma, J. (1996) Gastric evacuation rate and feeding in the gilthead sea bream reared under semi-intensive conditions. *Aquaculture International*, **4**, 129–141.

Ang, K.P. & Petrell, R.J. (1997) Control of feed dispensation in sea cages using underwater video monitoring: effects on growth and food conversion. *Aquacultural Engineering*, **16**, 45–62.

Austreng, E., Storebakken, T. & Åsgård, T. (1987) Growth rate estimates for cultured Atlantic salmon and rainbow trout. *Aquaculture*, **60**, 157–160.

Baardvik, B.M. & Jobling, M. (1990) Effect of size-sorting on biomass gain and individual growth rates in Arctic charr, *Salvelinus alpinus* L. *Aquaculture*, **90**, 11–16.

Baker, R.F. & Ayles, G.B. (1990) The effect of varying density and loading level on the growth of Arctic charr (*Salvelinus alpinus* L.) and rainbow trout (*Oncorhynchus mykiss*). *World Aquaculture*, **21**, 58–61.

Balm, P.H.M. & Pottinger, T.G. (1993) Acclimation of rainbow trout (*Oncorhynchus mykiss*) to low environmental pH does not involve an activation of the pituitary-interrenal axis, but evokes adjustments in branchial ultrastructure. *Canadian Journal of Fisheries and Aquatic Sciences*, **50**, 2532–2540.

Bamber, R.N., Batten, S.D. & Bridgwaters, N.D. (1993) Design criteria for the creation of brackish lagoons. *Biodiversity and Conservation*, **2**, 127–137.

Baras, E. & Lagardère, J.P. (1995) Fish telemetry in aquaculture: review and perspectives. *Aquaculture International*, **3**, 77–102.

Baras, E., Tissier, F., Westerloppe, L., Melard, C. & Philippart, J.C. (1998) Feeding in darkness alleviates density-dependent growth of juvenile vundu catfish *Heterobranchus longifilis* (Clariidae). *Aquatic Living Resources*, **11**, 335–340.

Beauchaud, M., Rehailia, M., Petit, G., Bouchut, C. & Buisson, B. (1999) La croissance chez les brochetons (*Esox lucius*) en fonction de l'éclairement. In: *Quelques aspects de la Chronobiologie*. (ed. B. Buisson). Presse Universitaire, Saint-Etienne, France.

Bégout Anras, M.L. (1995) Demand-feeding behaviour of sea bass kept in ponds: diel and seasonal patterns, and influences of environmental factors. *Aquaculture International*, **3**, 186–195.

Bégout, M.L. & Lagardère, J.P. (1993) Acoustic telemetry: a new technology to control fish behaviour in culture conditions. In: *Production, environment and quality* (eds G. Barnabé & P. Kestemont), pp. 167–175. European Aquaculture Society, Ghent, Belgium.

Bégout, M.L. & Lagardère, J.P. (1995) An acoustic telemetry study of seabream (*Sparus aurata* L.): first results on activity rhythm, effects of environmental variables and space utilization. *Hydrobiologia*, **300–301**, 417–423.

Bégout Anras, M.L. & Lagardère, J.P. (1998) Variabilité météorologique: conséquences sur l'activité natatoire d'un poisson marin. *Comptes Rendus de l'Académie des Sciences, Série III, Sciences de la Vie*, **321**, 641–648.

Bégout Anras, M.L., Lagardère, J.P. & Lafaye, J.Y. (1997) Diel activity rhythm of seabass tracked in a natural environment: group effects on swimming patterns and amplitudes. *Canadian Journal of Fisheries and Aquatic Sciences*, **54**, 162–168.

Bergerhouse, D.L. (1992) Lethal effects of elevated pH and ammonia on early life stages of walleye. *North American Journal of Fisheries Management*, **12**, 356–366.

Beveridge, M. (1996) *Cage Aquaculture*. Blackwell Science Ltd, Oxford.

Billard, R. (1995a) Les systèmes de production aquacole et leurs relations avec l'environnement. *Cahier d'Agriculture*, **4**, 9–28.

Billard, R. (ed.) (1995b) *Les carpes, biologie et élevage*. INRA Editions, Versailles, France.

Bjordal, Å., Lindem, T., Juell, J.E. & Fernö, A. (1993) Hydroacoustic monitoring and feeding control in cage rearing of Atlantic salmon (*Salmo salar* L.). In: *Fish Farming Technology* (eds H. Reinertsen, L.A. Dahle, L. Jørgensen & K. Tvinnereim), pp. 203–208. Balkema, Rotterdam.

Bloch, B. & Larsen, J.L. (1993) An iridovirus-like agent associated with systemic infection in cultured turbot *Scophthalmus maximus* fry in Denmark. *Diseases of Aquatic Organisms*, **15**, 235–240.

Blyth, P.J, Purser, J.G. & Rusell, J.F. (1993) Detection of feeding rhythms in seacaged Atlantic salmon using new feeder technology. In: *Fish Farming Technology*. (eds H. Reinertsen, L.A. Dahle, L. Jørgensen & K. Tvinnereim), pp. 209–216. Balkema, Rotterdam.

Blyth, P.J., Kadri, S., Valdimarsson, S.K., Mitchell, D.F. & Purser, J.G. (1999) Diurnal and seasonal variation in feeding patterns of Atlantic salmon in sea cages. *Aquaculture Research*, **30**, 539–544.

Borghetti, J.R. & Canzi, C. (1993) The effect of water temperature and feeding rate on the growth rate of pacu (*Piaractus mesopotamicus*) raised in cages. *Aquaculture*, **114**, 93–101.

Boujard, T. & Leatherland, F. (1992) Circadian rhythms and feeding time in fishes. *Environmental Biology of Fishes*, **35**, 109–131.

Boujard, T. & Luquet, P. (1990) Diel cycle in *Hoplosternum littorale* (Teleostei). Evidence for synchronisation of locomotor, breathing and feeding activity by circadian alternation of light and dark. *Journal of Fish Biology*, **36**, 133–140.

Boujard, T., Moreau, Y. & Luquet, P (1991) Entrainment of the circadian rhythm of food demand by infradian cycles of light-dark alternation in *Hoplosternum littorale* (Teleostei). *Aquatic Living Resources*, **4**, 221–225.

Boyd, C.E. (ed.) (1995) *Bottom Soils, Sediment, and Pond Aquaculture*. Chapman & Hall, New York.

Boyd, C.E. & Teichert-Coddington, D. (1992) Relationship between wind speed and reaeration in small aquaculture ponds. *Aquacultural Engineering*, **11**, 121–131.

Boyd, C.E., Romaire, R.P. & Johnston, E. (1978) Predicting early morning dissolved oxygen concentration in channel catfish production ponds. *Journal of Environment Quality*, **107**, 484–492.

Brown, G.E., Brown, J.A., & Srivastava, R.K. (1992) The effect of stocking density on the behaviour of Arctic charr (*Salvelinus alpinus* L.). *Journal of Fish Biology*, **41**, 955–963.

Caldwell, C.A. & Hinshaw, J. (1994) Physiological and haematological responses in rainbow trout subjected to supplemental dissolved oxygen in fish culture. *Aquaculture*, **126**, 183–193.

Castel, J. (1985) Importance des copépodes meiobenthiques lagunaires dans le régime alimentaire des formes juvéniles de poissons euryhalins. *Bulletin d'Ecologie*, **16**, 169–176.

Castel, J. & Lasserre, P. (1982) Régulation biologique du méiobenthos d'un écosystème lagunaire par un alevinage expérimental en soles (*Solea vulgaris*). *Oceanologica Acta*, **4**, 243–251.

Cerreti, G., Ferrarese, U., Francescon, A. & Barbaro, A. (1987) Chironomids (Diptera: Chironomidae) in the natural diet of gilthead seabream (*Sparus aurata* L.) farmed in the Venice lagoon. *Entomologica Scandinavica*, **29**, 289–292.

Chang, W.Y.B. (1986) Biological principles of pond culture: an overview. In: *Principles and Practices of Pond Aquaculture* (eds J.E. Lannan, R.O. Smitherman & G. Tchobanoglous), pp. 1–5. Oregon State University Press, Corvallis, Oregon.

Chapman, C.A. & Mackay, W.C. (1984) Versatility in habitat use by a top aquatic predator, *Esox lucius* L. *Journal of Fish Biology*, **25**, 109–115.

Chiba, K. (1983) The effect of dissolved oxygen on the growth of young silver bream. *Bulletin of the Japanese Society of Scientific Fisheries*, **49**, 601–610.

Chiu, Y.N., Macahilig, M.P. & Sastrillo, M.A.S. (1986) Preliminary studies of factors affecting the feeding rhythm of milkfish (*Chanos chanos* Forskal). In: *The First Asian fisheries forum, Manila (Philippines), 26 May 1986*. (eds J.L. Maclean, L.B. Dizon &. L.V. Hosillos), pp. 547–550. Asian Fisheries Society, Manila, Philippines.

Christiansen, J.S. & Jobling, M. (1990) The behaviour and the relationship between food intake and growth rate of juvenile Arctic charr, *Salvelinus alpinus,* subjected to sustained exercise. *Canadian Journal of Zoology*, **68**, 2185–2191.

Colt, J. (1986) Pond culture practices. In: *Principles and Practices of Pond Aquaculture* (eds J.E Lannan, R.O. Smitherman & G. Tchobanoglous), pp. 191–203. Oregon State University Press, Corvallis, Oregon.

Colt, J. & Tchobanoglous, G. (1978) Chronic exposure of channel catfish, *Ictalurus punctatus*, to ammonia: effects on growth and survival. *Aquaculture*, **15**, 353–372.

Cremer, M.C. & Smitherman, R.O. (1980) Food habits and growth of silver and bighead carp in cages and ponds. *Aquaculture*, **20**, 57–64.

Cuenco, M.L., Stickney, R.R. & Grant, W.E. (1985) Fish bioenergetics and growth in aquaculture ponds: 3. Effects of intraspecific competition, stocking rate, stocking size and feeding rate on fish productivity. *Ecological Modelling*, **28**, 73–96.

Cunjak, R.A & Power, G. (1987) Cover use by stream-resident trout in winter: a field experiment. *North American Journal of Fisheries Management*, **7**, 539–544.

Davis, K.B., Suttle, M.A. & Parker, N.C. (1984) Biotic and abiotic influences on corticosteroid hormone rhythms in channel catfish. *Transactions of the American Fisheries Society*, **113**, 414–421.

De Lemos Vasconcelos Filho, A., Eskinazi-Leca, E. & de Souza, A.E. (1980) Feeding behaviour of Mugilidae cultured in ponds of the region of Itamaraca (Pernambuca, Brazil). In: *Simposio Brasileiro Aquicultura* (ed. Academia Brasileira de Ciencas), pp. 121–130. Rio de Janeiro, Brazil.

Dosdat, A., Metailler, R., Tetu, N., Servais, F., Chartois, H., Huelvan, C. & Desbruyeres, E. (1998) Nitrogenous excretion in juvenile turbot, *Scophtalmus maximus* L., under controlled conditions. *Aquaculture Research*, **26**, 639–650.

Edsall, D.A. & Smith, C.E. (1990) Performance of rainbow trout and Snake river cutthroat trout reared in oxygen-supersaturated water. *Aquaculture*, **90**, 251–259.

Egna, H.S. & Boyd, C.E. (eds) (1997) *Dynamics of Pond Aquaculture*. CRC Press, Boca Raton, New York.

Engle, C.R. & Brown, D. (1998) Growth, yield, dressout, and net returns of bighead carp *Hypophthalmichthys nobilis* stocked at three densities in catfish *Ictalurus punctatus* ponds. *Journal of the World Aquaculture Society*, **29**, 414–421.

Eriksson, L.O. & Van Veen, T. (1980) Circadian rhythms in the brown bullhead, *Ictalurus nebulosus* (Teleostei). Evidence for an endogenous rhythm in feeding, locomotor, and reaction time behaviour. *Canadian Journal of Zoology*, **58**, 1899–1907.

Ewing, R.D., Sheahan, J.E., Lewis, M.A. & Palmisano, A.N. (1998) Effects of rearing density and raceway conformation on growth, food conversion, and survival of juvenile Spring Chinook salmon. *The Progressive Fish-Culturist*, **60**, 167–178.

Fermin, A.C. & Seronay, G.A. (1997) Effects of different illumination levels on zooplankton abundance, feeding periodicity, growth and survival of the Asian sea bass, *Lates calcarifer* (Bloch), fry in illuminated floating nursery cages. *Aquaculture*, **157**, 227–237.

Fernö, A., Furevik, D., Huse, I. & Bjordal, Å. (1988) A multiple approach to behaviour studies of salmon reared in marine net pens. *ICES C.M.* F:15, 15 pp.

Fernö, A., Huse, I., Juell, J.E. & Bjordal, Å. (1995) The vertical distribution of Atlantic salmon in net pens: trade-off between surface light avoidance and food attraction. *Aquaculture*, **132**, 285–296.

Ferrari, I. & Chieregato, A.R. (1981) Feeding habits of juvenile stages of *Sparus auratus* L., *Dicentrarchus labrax* L. and Mugilidae in a brackish embayment of the Po river delta. *Aquaculture*, **25**, 243–257.

Fivelstad, S., Haavik, H., Lovik, G. & Olsen, A. B. (1998) Sublethal effects and safe levels of carbon dioxide in seawater for Atlantic salmon postsmolts *Salmo salar* L.: ion regulation and growth. *Aquaculture*, **160**, 305–316.

Flos, R., Hussenot, J. & Lagardère, F. (1998) Les soles, quelles recherches, pour quels types d'élevage et de production ? In: *Marais maritimes et aquaculture, Actes de colloques 19* (eds J. Hussenot & V. Buchet), pp. 155–164. Editions IFREMER, Plouzané, France.

Forsberg, J. A. & Summerfelt, R. C. (1992) Ammonia excretion by fingerling walleyes fed two formulated diets. *The Progressive Fish-Culturist*, **54**, 45–48.

Fraser, N.H.C., Heggenes, J., Metcalfe, N.B. & Thorpe, J.E. (1995) Low summer temperatures cause juvenile Atlantic salmon to become nocturnal. *Canadian Journal of Zoology*, **73**, 446–451.

Fraser, S.A., Wisenden, B.D. & Keenleyside, M.H.A. (1993) Aggressive behaviour among cichlid (*Cichlasoma nigrofasciatum*) fry of different sizes and its importance to brood adoption. *Canadian Journal of Zoology*, **71**, 2358–2362.

Galemoni, F. & Ngokaka, C. (1989) Evolution of three environmental factors (oxygen, temperature, pH) in an intensive culture pond at the Djoumouna fish farm. *Pisciculture Française*, **98**, 27–32.

Goddard, S. (ed.) (1996) *Feed Management in Intensive Aquaculture*. Chapman & Hall, London.

Grant, J.W.A. (1993) Whether or not to defend – the influence of resource distribution. *Marine Behaviour and Physiology*, **23**, 137–153.

Hallaraker, H., Folkword, A. & Stefansson, S.O. (1995) Growth of *Hippoglossus hippoglossus* L. related to temperature, light period, and feeding regime. *ICES Marine Sciences Symposium*, **201**, 196.

Hansen, T., Stefansson, S.O. & Taranger, G.L. (1992) Growth rate and sexual maturation in Atlantic salmon, *Salmo salar*, reared in sea cages at two different light regimes. *Aquaculture and Fisheries Management*, **23**, 275–280.

Hemre, G.I., Juell, J.E., Hamre, K., Lie, Ø., Strand, B., Arnesen, P. & Holm, J. C. (1997) Cage feeding of Atlantic mackerel (*Scomber scombrus*): effect on muscle lipid content, fatty acid composition, oxidation status and vitamin E concentration. *Aquatic Living Resources*, **10**, 365–370.

Hepher, B. (ed.) (1988) *Nutrition of Pond Fishes*. Cambridge University Press, Cambridge.

Heuch, P.A., Parsons, A. & Boxaspen, K. (1995) Diel vertical migration: A possible host finding mechanism in salmon louse (*Lepeophtheirus salmonis*) copepodids? *Canadian Journal of Fisheries and Aquatic Sciences*, **52**, 681–689.

Hirayama, K. (1974) Water control by filtration in closed aquaculture systems. *Aquaculture*, **4**, 369–385.

Hirayama, K., Mizuma, H. & Mizue, Y. (1988) The accumulation of dissolved organic substances in closed recirculation systems. *Aquacultural Engineering*, **7**, 73–87.

Hogendoorn, H. & Koops, W.J. (1983) Growth and production of the African catfish, *Clarias lazera* (C& V) I. Effects of stocking density, pond size and mixed culture with tilapia (*Sarotherodon niloticus* L.) under extensive field conditions. *Aquaculture*, **34**, 253–263.

Horvath, L, Tamas, G. & Seagrave, C. (eds) (1992) *Carp and pond fish culture. Including Chinese herbivorous species, pike, tench, zander, wels catfish and goldfish*. Fishing News Books, Oxford.

Howerton, R.D. & Boyd, C.E. (1992) Measurement of water circulation in ponds with gypsum blocks. *Aquacultural Engineering*, **11**, 141–155.

Huse, I. & Holm, J.C. (1993) Vertical distribution of Atlantic salmon (*Salmo salar*) as a function of illumination. *Journal of Fish Biology*, **43 A**, 147–156.

Huse, I., Bjordal, Å. Fernö, A. & Furevik, D. (1990) The effect of shading in pen rearing of Atlantic salmon (*Salmo salar* L.). *Aquacultural Engineering*, **9**, 235–244.

Hussenot, J. & Reymond, H. (1990) Accroissement de la production des élevages semi-extensifs de crevette impériale en marais salés atlantiques (France). *Proceedings of FAO-EIFAC Symposium on Production Enhancement in Still-Water Pond Culture, Prague 15 May 1990* (ed. The Research Institute of Fish Culture and Hydrobiology), pp. 194–198. Vodnany, Czechoslovakia.

Hussenot, J., Martin, J.L., Gouleau, D., Ravail, B. & Eveno, A. (1991) Effects of a complement diet managed like an organic stimulant on ponds sediment. *European Aquaculture Society Special Publication*, **14**, 159–160.

Hussenot, J., Lagardère, J.P., Blachier, P., Halley, R. & Gautier, D. (1993) Use of a carp diet (35% protein) in semi-extensive culture of *Peneaus japonicus* both fertilizer and alternative feed. *European Aquaculture Society Special Publication*, **19**, 137.

Jana, B.B. & Das, R.N. (1980) Relationship between environmental factors and fish yield in a pond with airbreathing fish *Clarias batrachus*. *Aquacultura Hungarica*, **2**, 139–146.

Jobling, M. (1981) Some effects of temperature, feeding and body weight on nitrogenous excretion in young plaice *Pleuronectes platessa* L. *Journal of Fish Biology*, **18**, 87–96.

Jobling, M. (1997) Temperature and growth: modulation of growth via temperature change. *Society of Experimental Biology Seminar Series*, **61**, 225–253.

Jobling, M. & Reinsnes, T.G. (1987) Effect of sorting on size frequency distribution and growth of Arctic charr, *Salvelinus alpinus* L. *Aquaculture*, **60**, 27–31.

Jobling, M. & Wandsvik, A. (1983) Effect of social interactions on growth rates and conversion efficiency of Arctic charr, *Salvelinus alpinus* L. *Journal of Fish Biology*, **22**, 379–386.

Jobling, M., Baardvik, B.M., Christiansen, J.S. & Jørgensen, E.H. (1993) The effects of prolonged exercise training on growth performance and production parameters in fish. *Aquaculture International*, **1**, 95–111.

Johansen, S.J.S. & Jobling, M. (1998) The influence of feeding regime on growth and slaughter traits of cage-reared Atlantic salmon. *Aquaculture International*, **6**, 1–17.

Johnsen P.B. & Hasler, A.D. (1977) Winter aggregations of carp (*Cyprinus carpio*) as revealed by ultrasonic tracking. *Transactions of the American Fisheries Society*, **106**, 556–559.

Jordan, D.H.M. & Lloyd, R. (1964) The resistance of rainbow trout (*Salmo gairdneri* Richardson) and roach (*Rutilus rutilus* L.) to alkaline solutions. *International Journal of Air and Water Pollution*, **8**, 405–409.

Jørgensen, E.H., Christiansen, J.S. & Jobling, M. (1993) Effects of stocking density on food intake, growth performance and oxygen consumption in Arctic charr (*Salvelinus alpinus*). *Aquaculture*, **110**, 191–204.

Joseph, M.M. (ed.) (1990) *Aquaculture in Asia*. Asia Fisheries Society, Indian Branch, Mangalore.

Juell, J.E. (1995) The behaviour of Atlantic salmon in relation to efficient cage rearing. *Reviews in Fish Biology and Fisheries*, **5**, 320–335.

Juell, J.E. & Westerberg, H. (1993) An ultrasonic telemetric system for automatic positioning of individual fish used to track salmon (*Salmo salar* L.) in a sea cage. *Aquacultural Engineering*, **12**, 1–18.

Juell, J.E., Furevik, D., & Bjordal, Å. (1993) Demand feeding in salmon farming by hydroacoustic food detection. *Aquacultural Engineering*, **12**, 155–167.

Juell, J.E., Bjordal, Å., Fernö, A. & Huse, I. (1994a) Effect of feeding intensity on food intake and growth of Atlantic salmon (*Salmo salar* L.) in sea cages. *Aquaculture and Fisheries Management*, **25**, 453–464.

Juell, J.E., Fernö, A., Furevik, D. & Huse, I. (1994b) Influence of hunger level and food availability on the spatial distribution of Atlantic salmon (*Salmo salar* L.) in sea cages. *Aquaculture and Fisheries Management*, **25**, 439–451.

Juell, J.E., Hemre, G.I., Holm, J.C., & Lie, Ø. (1998) Growth and feeding behaviour of caged Atlantic mackerel (*Scomber scombrus*). *Aquaculture Research*, **29**, 115–122.

Kadri, S., Metcalfe, N.B., Huntingford, F.A. and Thorpe, J.E. (1991) Daily feeding rhythms in Atlantic salmon in sea cages. *Aquaculture*, **92**, 219–224.

Kadri, S., Huntingford, F.A., Metcalfe, N.B. & Thorpe, J.E. (1996a) Social interactions and the distribution of food among one-sea-winter Atlantic salmon (*Salmo salar*) in a sea cage. *Aquaculture*, **139**, 1–10.

Kadri, S., Mitchell, D.F., Metcalfe, N.B., Huntingford, F.A. & Thorpe, J.E. (1996b) Differential patterns of feeding and resource accumulation in maturing and immature Atlantic salmon, *Salmo salar*. *Aquaculture*, **142**, 245–257.

Kamstra, A., Span, J. A. & Van Weerd, J. H. (1996) The acute toxicity and sublethal effects of nitrite on growth and feed utilization of European eel, *Anguilla anguilla* L. *Aquaculture Research*, **27**, 903–911.

Kerdchuen, N. & Legendre, M. (1992) Effet favorable des fortes densités pour l'adaptation d'un silure africain, *Heterobranchus longilis* (Pisces, Clariidae) en bacs de petit volume. *Revue d'Hydrobiologie Tropicale*, **25**, 63–67.

Kikuchi, K., Takeda, S., Honda, H. & Kiono, M. (1990) Oxygen consumption and nitrogenous excretion of starved Japanese flounder *Paralichthys olivaceus*. *Nippon Suisan Gakkaishi*, **56**, 1891.

Kils, U. (1989) Some aspects of schooling for aquaculture. *ICES C.M.* F:**12**, 10 pp.

King, D.L. & Garling, D.L. (1986) A state of the art overview of aquatic fertility with special reference to control exerted by chemical and physical factors. In: *Principles and Practices of Pond Aquaculture* (eds J.E Lannan, R.O. Smitherman & G. Tchobanoglous), pp. 53–65. Oregon State University Press, Corvallis, Oregon.

Knights, B. (1985) Energetics and fish farming. In: *Fish Energetics New Perspectives* (eds P. Tytler & P. Calow), pp. 309–340. The Johns Hopkins University Press, Maryland.

Labourg, P.J. (1979) Structure et évolution de la macrofaune invertébrée d'un écosystème lagunaire aménagé (réservoirs à poissons de Certes). In: *Colloque national « ecotron », Actes de Colloques 7* (eds P. Lasserre & D. Reys), pp. 591–614. Publications Scientifiques et Techniques du CNEXO, Paris.

Labourg, P.J. & Stequert, B. (1973) Régime alimentaire du bar *Dicentrarchus labrax* L. des réservoirs à poissons de la région d'Arcachon. *Bulletin d'Ecologie*, **4**, 187–194.

Lagardère, J.P., Bégout Anras, M.L. & Buchet, V. (1996) The acoustic positioning system as a valuable tool for estimating the well-being of fishes in aquaculture. In: *Underwater Biotelemetry. Proceedings of the first conference and workshop on fish telemetry in Europe, Liège, 4-6 April 1995* (eds E. Baras E. & J.C. Philippart), pp. 177–186. University of Liège, Liège, Belgium.

Lagardère, J.P., Favre, L., Mosneron Dupin, J. & Sureau, D. (1992) Orientation by olfactory perception in the common sole (*Solea solea* L.) evaluated using ultrasonic telemetry. In: *Wildlife Telemetry, remote monitoring and tracking of animals* (eds I.G. Priede & S.M. Swift), pp. 367–375. Ellis Horwood, New York.

Lagardère, J.P., Bégout, M.L., Lafaye, J.Y. & Villote, J.P. (1994) Influence of wind-produced noise on orientation in the sole (*Solea solea*). *Canadian Journal of Fisheries and Aquatic Sciences*, **51**, 1258–1264.

Lagardère, J.P., Bégout Anras, M.L., Breton, H. & Company I Claret, J. B. (1995) The effects of illumination, temperature and oxygen concentration on swimming activity of turbot *Psetta maxima* (Linné 1758). *Fisheries Research*, **24**, 167–173.

Lannan, J.E., Smitherman R.O. & Tchobanoglous, G. (eds) (1986) *Principles and Practices of Pond Aquaculture*. Oregon State University Press, Corvallis, Oregon.

Lemarie, G., Coves, D., Dutto, G., Gasset, E. & Person Le Ruyet, J. (1996) Chronic ammonia toxicity in European seabass (*Dicentrarchus labrax*) juveniles. In: *Applied environmental physiology of fishes,*

International congress on the biology of fishes, San Francisco State University, July 14-18, 1996. (eds C. Swanson, P. Young & D. MacKinlay), pp. 65–76. American Fisheries Society, Bethesda.

Lemly, A.D. & Smith, R.J.F. (1985) Effects of acute exposure to acidified water on the behavioral response of fathead minnows to chemical feeding stimuli. *Aquatic Toxicology*, **6**, 25–36.

Li, M. & Lovell, R.T. (1992a) Effect of dietary protein concentration on nitrogenous waste in intensively fed catfish ponds. *Journal of the World Aquaculture Society*, **23**, 122–127.

Li, M. & Lovell, R.T. (1992b) Comparison of satiate feeding and restricted feeding in channel catfish with various concentrations of dietary protein in production ponds. *Aquaculture*, **103**, 165–175.

Little, D. & Muir, J. (eds) (1987) *A Guide to Integrated Warm Water Aquaculture.* Institute of Aquaculture Publications, University of Stirling.

Losordo, T.M. & Piedrahita, R.H. (1991) Modelling temperature variation and thermal stratification in shallow aquaculture ponds. *Ecological Modelling*, **54**, 189–226.

Losordo, T.M. & Westers, H. (1994) System carrying capacity and flow estimation. In: *Aquaculture Water Systems: Engineering Design and Management* (eds M.B. Timmons & T.M. Losordo), *Development in Aquaculture and Fisheries Sciences*, **27**, 9–60.

Losordo, T.M., Piedrahita, R.H. & Ebeling, J.M. (1988) An automated water quality data acquisition system for the use in aquaculture ponds. *Aquacultural Engineering*, **7**, 265–278.

Mallekh, R., Lagardère, J.P., Bégout Anras, M.L. & Lafaye, J.Y. (1998) Variability in appetite of turbot, *Scophtalmus maximus*, under intensive rearing conditions: the role of environmental factors. *Aquaculture*, **165**, 123–138.

Meade, T.L. (1976) Closed system salmonid culture in the United States Marine Advisory Service. In: *Marine Memorandum* **40**. (ed. NOAA Sea Grant). University of Rhode Island, Kingston.

Milstein A. (1992) Ecological aspects of fish species interactions in polyculture ponds. *Hydrobiologia*, **231**, 177–186.

Mosneron Dupin, J. & Lagardère, J.P. (1990) Réactions comportementales du bar *Dicentrarchus labrax* (Linné, 1758) aux basses températures. Premières données recueillies en marais maritime par télémétrie acoustique. *Comptes Rendus de l'Académie des Sciences, Série III, Sciences de la Vie*, **310**, 279–284.

Muir, J. F. & Roberts, R.J. (eds) (1982) *Recent Advances In Aquaculture.* Croom Helm, London.

Munsiri, P. & Lovell, R.T. (1993) Comparison of satiate and restricted feeding of channel catfish with diets of varying protein quality in production ponds. *Journal of the World Aquaculture Society*, **24**, 459–465.

Nath, S.S., Bolte, J.P. & Ernst, D.G. (1995) Decision support for pond aquaculture planning and management. In: *Proceedings of the PACON Conference on Sustainable Aquaculture '95* (eds J. Bardach, J. Corbin & B. Duncan), pp. 270. PACON Conference on sustainable aquaculture.

Neji, H., Naimi, N., Lallier, R. & De la Noue, J. (1993) Relationship between feeding, hypoxia, digestibility and experimentally induced furunculosis in rainbow trout. In: *Fish Nutrition in Practice: 4th international symposium on fish nutrition and feeding* (eds S.J. Kaushik & P. Luquet), pp. 187–197. Editions INRA, Versailles, France.

Oppedal, F., Taranger, G.L., Juell, J.E., Fosseidengen, J.E. & Hansen, T. (1997) Light intensity affects growth and sexual maturation of Atlantic salmon (*Salmo salar* L.) postsmolts in sea cages. *Aquatic Living Resources*, **10**, 351–357.

Oppedal, F., Taranger, G.L., Juell, J.E. & Hansen, T. (1999) Growth, osmoregulation, and sexual maturation of underyearling Atlantic salmon smolt *Salmo salar* L. exposed to different intensities of continuous light in sea cages. *Aquaculture Research*, **30**, 491–499.

Perschbacher, P.W. & Strawn, K. (1984) Depth distribution and composition of macroinvertebrate communities in enriched, low-salinity fish culture ponds. *Journal of the World Mariculture Society*, **15**, 341–354.

Person Le Ruyet, J. & Bœuf, G. (1998) L'azote ammoniacal, un toxique potentiel en élevage de poissons: le cas du turbot. *Bulletin Français de la Pêche et de la Pisciculture*, **350-351**, 393–412.

Pfeiffer, T.J. & Lovell, R.T. (1990) Responses of grass carp, stocked intensively in earthen ponds, to various supplemental feeding regimes. *The Progressive Fish-Culturist*, **52**, 213–217.

Phelps, R.P. & Silva de Gomez A. (1992) Influence of shelter and feeding practices on channel catfish fingerling production. *The Progressive Fish-Culturist*, **54**, 21–24.

Pichavant, K., Person Le Ruyet, J., Severe, A., Le Roux, A. & Bœuf, G.(1998) Capacités adaptatives du turbot (*Psetta maxima*) juvénile à la photopériode. *Bulletin Français de la Pêche et de la Pisciculture*, **350-351**, 265–277.

Piedrahita, R.H. & Balchen, J.G. (1987) Sensitivity analysis for an aquaculture ponds model. *IFAC Progress Series*, **9**, 119–123.

Pierce, R.H., Weeks, J.M. & Prappas, J.M. (1993) Nitrate toxicity to five species of marine fish. *Journal of the World Aquaculture Society*, **24**, 105–107.

Pillay T.V.R. (ed.) (1990) *Aquaculture: Principle and Practices*. Fishing News Books, The University Press, Cambridge.

Poncin, P. & Ruwet, J.C. (1994) Applications to freshwater aquaculture of the methods used to measure the behaviour of fish: a brief review. In: *Measures for Success, Proceedings of the International Conference Bordeaux Aquaculture '94* (eds P. Kestemont, J. Muir, F. Sevilla & P. Williot), pp. 271–275. Editions Cemagref, Bordeaux France.

Pouliot, T. & De la Noue, J. (1989) Feed intake, digestibility and brain neurotransmitters of rainbow trout under hypoxia. *Aquaculture*, **79**, 317–327.

Poxton, M.G. (1991) Water quality fluctuations and monitoring in intensive fish culture. *European Aquaculture Society Special Publication*, **16**, 121–143.

Poxton, M.G. & Lloyd, N.J. (1989) Fluctuations in ammonia production by eels (*Anguilla anguilla* L.) as a result of feeding strategy. *European Aquaculture Society Special Publication*, **16**, 1125–1135.

Prein, M. (1985) *The influence of environmental factors on fish production in tropical ponds investigated with multiple regression and path analysis*. Diplom Arbeit, Christian-Albrechts University, Kiel, Germany.

Randolph, K.N. & Clemens, H.P. (1976a) Some factors influencing the feeding behaviour of channel catfish in culture ponds. *Transactions of the American Fisheries Society*, **105**, 718–724.

Randolph, K.N. & Clemens, H.P. (1976b) Home areas and swimways in channel catfish culture ponds. *Transactions of the American Fisheries Society*, **105**, 725–730.

Randolph, K.N. & Clemens, H.P. (1978) Effects of short-term food deprivation on channel catfish and implications for culture practices. *The Progressive Fish-Culturist*, **40**, 48–50.

Rasmussen, R.S. & Korsgaard, B. (1996) The effect of external ammonia on growth and food utilization of juvenile turbot (*Scophtalmus maximus*) L. *Journal of Experimental Marine Biology and Ecology*, **205**, 35–48.

Reymond, H. (1989a) Etude du régime alimentaire du bar *Dicentrarchus labrax* en élevage semi-intensif dans les bassins aquacoles de la station Aqualive. Rapport IFREMER, Contrat no. 88.5.524009, 25 pp.

Reymond, H. (1989b) Régime alimentaire du bar *Dicentrarchus labrax* en première année d'élevage semi-intensif en marais maritime: contribution relative des proies naturelles et de l'aliment à l'ingéré journalier. Rapport SEMDAC, Contrat no. 89.0001, 15 pp.

Reymond, H. (1991) Dynamique de la chaîne hétérotrophe benthique des marais maritimes en période estivale et son impact sur les productions aquacoles de carnivores: *Penaeus japonicus*, un modèle d'étude. DPhil thesis, University of Paris VI.

Robinson, E.H. & Li, M. (1995) Catfish nutrition part 2: Feeding. *Aquaculture Magazine*, **21**, 28–41.

Robinson, E.H., Jackson L.S., Li, M.H., Kingsbury, S.K. & Tucker, C.S. (1995) Effect of time of feeding on growth of channel catfish. *Journal of the World Aquaculture Society*, **26**, 320–322.

Romaire, R.P. & Boyd, C.E. (1979) Effects of solar radiation on the dynamics of dissolved oxygen in channel catfish ponds. *Transactions of the American Fisheries Society*, **108**, 473–478.

Ross, R.M. & Watten, B.J. (1998) Importance of rearing-unit design and stocking density to the behaviour, growth, and metabolism of lake trout (*Salvelinus namaycush*). *Aquacultural Engineering*, **19**, 41–56.

Ross, R.M., Watten, B.J., Krise, W.F. & DiLauro, M.N. (1995) Influence of tank design and hydraulic loading on the behaviour, growth, and metabolism of rainbow trout (*Oncorhynchus mykiss*). *Aquacultural Engineering*, **14**, 29–47.

Rottiers, D.V. (1992) Effects of day length and cleaning regiment on the growth of yearling parr Atlantic salmon. *The Progressive Fish-Culturist*, **54**, 69–72.

Russo, R.C. & Thurston, R.V. (1991) Toxicity of ammonia, nitrite and nitrate to fishes. *Journal of the World Aquaculture Society*, **3**, 58–59.

Sanchez-Vasquez, F.J., Madrid, J.A., Zamora, S. & Tabata, M. (1997) Feeding entrainment of locomotor activity rhythms in the goldfish is mediated by a feeding-entrainable circadian oscillator. *Journal of Comparative Physiology A.*, **181**, 121–132.

Sanni, S., Forsberg, O.I. & Bergheim, A. (1993) A dynamic model for fish metabolic production and water quality in landbased fish farms. In: *Fish Farming Technology* (eds Reinertsen, H., Dahle, L.A., Jorgensen, L. & Tvinnereim, K.), pp. 367–374. Balkema, Rotterdam.

Sarkar, S.K. & Konar, S.K. (1989) Influence of single superphosphate on the aquatic ecosystem in relation to fish growth. *Environmental Ecology*, **7**, 1042–1044.

Saroglia, M., Scarano, G. & Tibaldi, E. (1981) Acute toxicity of nitrite to seabass (*Dicentrarchus labrax*) and European eel (*Anguilla anguilla*). *Journal of the World Mariculture Society*, **12**, 121–126.

Schreck, C.B., Patino, C.K., Winton, J.R. & Holway, J.E. (1985) Effects of rearing density on indices of smoltification and performance of coho salmon, *Oncorhynchus kisutch*. *Aquaculture*, **104**, 37–50.

Scherer, E. (1992) Behavioural responses as indicators of environmental alterations: approach, results, developments. *Journal of Applied Ichthyology*, **8**, 122–131.

Schuster, C. (1994) The effect of fish meal content in trout food on water colour in a closed recirculating aquaculture system. *Aquaculture International*, **2**, 266–269.

Shang, Y.C. (1986) Pond production systems: stocking practices in pond fish culture. In: *Principles and Practices of Pond Aquaculture* (eds J.E Lannan, R.O. Smitherman & G. Tchobanoglous), pp. 85–96. Oregon State University Press, Corvallis, Oregon.

Skaar, A. & Bodvin, T. (1993) Full-scale production of salmon in floating enclosed systems. In: *Fish Farming Technology* (eds H. Reinertsen, L.A. Dahle, L. Jørgensen & K. Tvinnereim), pp. 325–328. Balkema, Rotterdam.

Smith, I.P., Metcalfe, N.B., Huntingford, F.A. & Kadri, S. (1993) Daily and seasonal patterns in the feeding behaviour of Atlantic salmon (*Salmo salar*) in a sea cage. *Aquaculture*, **117**, 165–178.

Smith, I.P., Metcalfe, N.B. & Huntingford, F.A. (1995) The effects of pellet dimensions on feeding responses by Atlantic salmon (*Salmo salar*) in a marine net pen. *Aquaculture*, **130**, 167–175.

Spataru, P. (1977) Gut contents of silver carp – *Hypophthalmichthys molitrix* (Val.) and some trophic relations to other fish species in a polyculture system. *Aquaculture*, **11**, 137–146.

Spataru, P., Hepher, B. & Halevy, A. (1980) The effect of the method of supplementary feed application on the feeding habits of carp (*Cyprinus carpio*) with regard to natural food in ponds. *Hydrobiologia*, **72**, 171–178.

Sutterlin, A.M. & Stevens, E.D. (1992) Thermal behaviour of rainbow trout and Arctic charr in cages moored in stratified waters. *Aquaculture*, **102**, 65–75.

Tackett, D.L., Carter, R.R. & Allen, K.O. (1987) Winter feeding of channel catfish based on maximum air temperature. *The Progressive Fish-Culturist*, **50**, 107–110.

Tackett, D.L., Carter, R.R. & Allen, K.O. (1988) Daily variation in feed consumption by channel catfish. *The Progressive Fish-Culturist*, **49**, 290–292.

Tacon, A.G.J. (1995) Application of nutrient requirement data under practical conditions: special problems of intensive and semi-intensive farming systems. *Journal of Applied Ichthyology*, **11**, 205–214.

Tacon, A.G.J. & De Silva, S.S. (1997) Feed preparation and feed management strategies within semi-intensive fish farming systems in the tropics. *Aquaculture*, **151**, 379–404.

Tacon, A.G.J., Rausin, N., Kadari, M. & Cornelis, P. (1991) The food and feeding of tropical marine fishes in floating net cages: Asian seabass, *Lates calcarifer* (Bloch), and brown spotted grouper, *Epinephelus tauvina* (Forskal). *Aquaculture and Fisheries Management*, **22**, 165–182.

Takeda, S. & Kiyono, M. (1990) Characterization of yellow substances accumulated in a closed recirculating system for culture. In: *The Second Asian Fisheries Forum* (ed. Asian Fisheries Society), pp. 991, Manila, Philippines.

Talbot, C., Corneillie, S. & Korsøen, Ø. (1999) Pattern of feed intake in four species of fish under commercial farming conditions: implications for feeding management. *Aquaculture Research*, **30**, 509–518.

Teichert-Coddington, D.R. (1996) Effect of stocking ratio on semi-intensive polyculture of *Colossoma macropomum* and *Oreochromis niloticus* in Honduras, Central America. *Aquaculture*, **143**, 291–302.

Teichert-Coddington, D.R. & Green, B.W. (1993) Influence of daylight and incubation interval on water column respiration in tropical fish ponds. *Hydrobiologia*, **250**, 159–165.

Thetmeyer, H., Waller, U., Black, K.D., Inselmann, S. & Rosenthal, H. (1999) Growth of European sea bass (*Dicentrarchus labrax* L.) under hypoxic and oscillating oxygen conditions. *Aquaculture*, **174**, 355–367.

Thorpe, J.E., Talbot, C., Miles, M.S., Rawlings, C. & Keay, D.S. (1990) Food consumption in 24 hours by Atlantic salmon (*Salmo salar* L.) in a sea cage. *Aquaculture*, **90**, 41–47.

Timmons, M.B., Summerfelt, S.T. & Vinci, B.J. (1998) Review of circular tank technology and management. *Aquacultural Engineering*, **18**, 51–69.

Torrans, E.L. (1986) Fish/plankton interactions. In: *Principles and Practices of Pond Aquaculture* (eds J.E Lannan, R.O. Smitherman & G. Tchobanoglous), pp. 67–81. Oregon State University Press, Corvallis, Oregon.

Tucker, C.S. (ed.) (1985) *Channel Catfish Culture*. Elsevier, Amsterdam, The Netherlands.

Tucker, C.S. (ed.) (1985) *Channel Catfish Culture*. Elsevier, Amsterdam, the Netherlands.

Tucker, C.S., Boyd C.E. & McCoy, E.W. (1979) Effects of feeding rate on water quality, production of channel catfish, and economic returns. *Transactions of the American Fisheries Society*, **108**, 389–396.

Udrea, V. (1978) Influence of some abiotic factors on the feeding rate of mullets in freshwater ponds. *Cercetäri Marine*, **11**, 173–180.

Van Ginneken, V.J.T., Van Eersel, R., Balm, P., Nieveen, M. & Van Den Thillard, G. (1997) Tilapia are able to withstand long-term exposure to low environmental pH, judged by their energy status, ionic balance and plasma cortisol. *Journal of Fish Biology*, **51**, 795–806.

Vijayan, M.M. & Leatherland, J.F. (1988) Effect of stocking density on the growth and stress-response in brook charr, *Salvelinus fontinalis*. *Aquaculture*, **75**, 159–170.

Villareal, C.A., Thorpe, J.E. & Miles, M.S. (1988) Influence of photoperiod on growth changes in juvenile Atlantic salmon, *Salmo salar* L. *Journal of Fish Biology*, **33**, 15–30.

Wallace, J.C., Kolbeinshavn, A.G. & Reinsnes, T.G. (1988) The effect of stocking density on early growth in Arctic charr, *Salvelinus alpinus*. *Aquaculture*, **73**, 101–110.

Watten, B.J. (1990) Comparative hydraulics and rearing trial performance of a product scale cross-flow rearing unit. *Aquacultural Engineering*, **9**, 245–266.

Webster, C.D, Tiu, L.G., Tidwell, J.H. Vanwyk, P. & Howerton, R.D. (1995) Effects of dietary-protein and lipid-levels on the growth and body-composition of sunshine bass (*Morone-chrysops × Morone-saxiatilis*) reared in cages. *Aquaculture*, **131**, 291–301.

Welch, H.E., Jorgenson, J.K. & Curtis, M.F. (1988) Emergence of Chironomidae (Diptera) in fertilized and natural lakes at Saqvaqjuac, N.W.T. *Canadian Journal of Fisheries and Aquatic Sciences*, **45**, 731–737.

Wieser, W. & Medgyesy, N. (1991) Metabolic rate and cost of growth in juvenile pike (*Esox lucius* L.) and perch (*Perca fluviatilis*): the use of energy budgets as indicators of environmental change. *Oecologia*, **87**, 500–505.

Wiggs, A.J., Henderson, E.B., Saunders, R.L. & Kutty, M.N. (1989) Activity, respiration, and excretion of ammonia by Atlantic salmon (*Salmo salar*) smolt and postsmolt. *Canadian Journal of Fisheries and Aquatic Sciences*, **46**, 790–795.

Wood, C.M. (1989) The physiological problems of fish in acid waters. In: *Acid Toxicity and Aquatic Animals* (eds R. Morris, D.J.A. Brown, E.W. Taylor & J.A. Brow), pp. 125–152. Society of Experimental Biology, Seminar Series, 34, Cambridge University Press, Cambridge.

Chapter 8
Feeding Rhythms

Juan Antonio Madrid, Thierry Boujard and F. Javier Sánchez-Vázquez

8.1 Introduction

A general property of living systems is that they function under conditions that are far from thermodynamic equilibrium. Each living organism must be considered as a different system depending on the time of the day, month or year when it is studied. For this reason, the homeostatic approach can be considered an oversimplified and unidimensional approach to the real complexity of life. Current physiological knowledge should aim to integrate a new source of information coded by the temporal organisation of cellular and systemic processes – the biological rhythms. Although the first evidence of endogenous periodicity in plants was obtained more than two hundred years ago, only in the past few decades has the endogenous nature of many periodic phenomena been recognised as a basic characteristic of living organisms.

In recent years an increasing number of studies have been devoted to biological rhythms in fish, and two books have been devoted to the subject (Thorpe 1978; Ali 1992). A chapter dealing with feeding rhythms in a book on feed intake in fish is justified by the growing importance of the topic in fish physiology. In addition to their basic interest, feeding rhythms in fish are of interest from an applied point of view because their knowledge may aid in deciding the feeding practices to be used under culture conditions to optimise feeding and improve production (see Chapter 14).

In the case of feeding rhythms in fish, it must be borne in mind that many exogenous factors in the aquatic environment exhibit periodic variation (Daan 1981). Thus, when fish exhibit rhythmic feeding behaviour it may be the result of fluctuations in:

(1) Abiotic parameters, such as light, temperature and oxygen concentration.
(2) Biotic factors, such as the relative abundance of prey and interactions with conspecifics and heterospecifics.
(3) Endogenous influences, such as pacemakers or hourglass processes.

The feeding rhythms, with a periodicity range from hours (prandial periodicity) to days and seasons, will be the output of the interaction of such factors.

Some of the forces acting on living organisms are unpredictable (e.g. confronting a competitor, infection, cloud cover, etc.), but others occur at times that may be quite predictably associated with certain phases of daylight, tides, lunar cycle, seasons, and so on. An ability to anticipate these events, to avoid their dangers and exploit their benefits, would be of adaptive advantage. It is, therefore, not surprising that most organisms have evolved various timekeeping strategies to adjust their lives to environmental cycles. For living organisms there are two main ways of measuring time: they can use the exogenous periodicity of the environment like a clock to activate different physiological or behavioural processes and/or, they can use some internal process to measure time (Aschoff 1981a,b; Pittendrigh 1981).

Although the appearance of certain biological rhythms is the result of an auto-organisational process, which is thermodynamically more efficient than the equivalent linear process, living organisms use the properties of physicochemical systems to increase their adaptive value and their rates of survival. Organisms can use their biological clocks to obtain three kinds of temporal information (Brady 1982):

(1)   Time-setting to set some processes so that they are 'on time' relative to environmental time, for example to be active during light hours or feed during the night.
(2)   Duration measurement to establish the duration of periodic events, such as the number of hours of light in 24 hours (i.e. photoperiod) to determine the season of the year.
(3)   Local time fixing to, for example, enable migration using celestial cues.

Using the above information, organisms can anticipate periodic events such as times of greatest predation risk or the time of food availability. Thus, physiological processes can be activated in advance of an external periodic event, allowing the organism to avoid risk or exploit the food available more efficiently. A biological rhythm can also be used to synchronise the activities of individuals within a population, such as in the synchronisation of reproductive cycles and the timing of spawning leading to increased survival rates of the offspring.

The periods of biological rhythms may range from a fraction of a second to several years. According to their frequency, three main categories have been established (Aschoff 1981a):

(1)   Ultradian rhythms, with a frequency of more than one cycle per 24 h, or a period of less than 20 h, e.g. tidal rhythms, or cardiac and respiratory cycles.
(2)   Circadian rhythms, with a period of about 24 h, are widespread in physiological and behavioural processes.
(3)   Infradian rhythms, with a frequency of less than one cycle per 24 h, such as annual and lunar rhythms.

In the case of feeding rhythms of fish, most examples reported have circadian periodicity, but a few studies have also been devoted to the study of tidal, lunar and annual rhythms of feeding.

The aims of the present chapter are not only to summarise the data concerning fish feeding rhythms, but also to integrate these data with ideas and concepts relating to biological rhythms. By adapting this approach we hope to highlight potential basic and applied research lines in fish chronobiology.

8.2 Feeding rhythms: descriptions and examples

8.2.1 Diel rhythms

Most animals are active either during the day or at night, but not throughout 24 hours, and species have acquired these behavioural patterns as a result of a long evolution under the influence of relatively stable selective forces (e.g. avoidance of predators, optimisation of feeding) (Daan 1981). Thus, in most species, diurnal or nocturnal feeding behaviour has been fixed genetically. Moreover, in certain cases, the temporal pattern of behaviour may also be conditioned by special sensory requirements, such as a dependence on vision for the capture of food. Examples of feeding patterns in fish are shown in Table 8.1.

Early work on feeding rhythms of fish under controlled conditions was carried out by Hoar (1942), who reported that during the summer two salmonids, the Atlantic salmon (*Salmo salar*) and the brook trout (*Salvelinus fontinalis*), preferred to feed during the daytime. Later, Rozin and Mayer (1961) found that goldfish (*Carassius auratus*) could learn to manipulate a rod in order to obtain food. This self-feeder device included a trigger, a food hopper and an event recorder to record feeding activity. Thus, the feeding rhythms of fish could be studied under a variety of self-feeding conditions.

Feeding rhythms in rainbow trout (*Oncorhynchus mykiss*) were also studied using a self-feeder (Adron *et al.* 1973; Landless 1976), and a peak of feeding activity was observed at dawn and a secondary peak 8 h later. However, this pattern – which was recorded in experiments undertaken during winter – was not always present. The peak of feeding activity associated with dawn was absent in October and November (Landless 1976), and the observed changes in feeding activity may have been provoked by differences in environmental conditions. Later, Grove *et al.* (1978) reported a strictly nocturnal pattern of feeding activity in rainbow trout during winter. In more recent studies on the same species under controlled laboratory conditions (Boujard & Leatherland 1992a; Sánchez-Vázquez & Tabata 1998), the main period of feeding has been reported to occur at dawn, regardless of the length of the photoperiod (Fig. 8.1a). These examples illustrate a general characteristic of feeding rhythms in fish: high intra-specific variability which is not completely understood. It has been hypothesised that when there is complete darkness during the night, trout feed exclusively during the daytime, but under natural conditions of moon- or starlight, trout will be able to operate the feeder and feed at night (Boujard & Leatherland 1992a). However, rainbow trout feeding rhythms are not always that simple, because when there was constant dim light during the night most trout decreased their feeding activity dramatically. Thus, factors other than the ability to detect the trigger of the self-feeder play a role in the control of feeding activity (Sanchez-Vázquez & Tabata 1998).

Feeding activity of Atlantic salmon has been studied in the laboratory and in fish held in sea cages. This species appears to be predominantly diurnal (Kadri *et al.* 1991, 1997; Jørgensen & Jobling 1992; Blyth *et al.* 1993; Smith *et al.* 1993; Paspatis & Boujard 1996). The fish feed most intensely in the water column during daylight hours (Jørgensen & Jobling 1992), although some authors have noted a change to nocturnal feeding at low temperature (Fraser *et al.* 1993, 1995, 1997).

The ability of fish to operate self-feeders has been exploited to study feeding rhythms in several species, such as the turbot (*Psetta maxima*) (Burel *et al.* 1997), which displayed

Table 8.1 Selected examples of feeding rhythms in fish held singly or in groups.

Species	Housing conditions	Main phase of feeding	Free-running rhythm	Source
Carassius auratus	Singly	Dual phasing	25.3 ± 1.8 h under DD; 24.4 ± 1.7 h under LD 0:45/0:45	Sánchez-Vázquez et al. (1996)
Dicentrarchus labrax	Groups	Seasonal shifts in feeding patterns	–	Anthouard et al. (1993)
Dicentrarchus labrax	Groups	Seasonal shifts in feeding patterns	–	Begout-Anras (1995)
Dicentrarchus labrax	Singly, Groups	Dual phasing	22.5–24.5 under LD 0:40/0:40	Sánchez-Vázquez et al. (1995a,b)
Dicentrarchus labrax	Groups	Seasonal shifts in feeding patterns	–	Sánchez-Vázquez et al. (1998)
Dicentrarchus labrax	Singly	Dual phasing	17.6–26.2 under DD	Aranda et al. (1999a)
Dicentrarchus labrax	Groups	Diurnal–nocturnal	–	Boujard et al. (1996)
Hoplosternum littorale	Groups	Nocturnal	–	Boujard et al. (1990)
Ictalurus nebulosus	Singly	Nocturnal	Circadian (LD 0:45/0:15)	Eriksson & Van Veen (1980)
Oncorhynchus mykiss	Singly	Dual phasing	–	Landless (1976)
Oncorhynchus mykiss	Groups	Nocturnal	6–8 h under LL	Grove et al. (1978)
Oncorhynchus mykiss	Groups	Diurnal	–	Boujard & Leatherland (1992a)
Oncorhynchus mykiss	Singly	Diurnal	Circadian	Cuenca & de la Higuera (1994)
Oncorhynchus mykiss	Singly	Diurnal	26.2 ± 0.3 h under LL; 21.9 ± 0.7 h under LD 0:45/0:45	Sánchez-Vázquez & Tabata (1998)
Salmo salar	Groups	Diurnal	–	Hoar (1942)
Salmo salar	Sea cages	Diurnal	–	Jørgensen & Jobling (1992)
Salmo salar	Singly	Seasonal shifts in feeding patterns	–	Smith et al. (1993)
Salmo salar	Groups	Dual phasing	–	Fraser et al. (1995)
Salmo salar	Field observations	Diurnal	–	Kadri et al. (1997)
Salmo trutta		Seasonal shifts in feeding patterns	–	Heggenes et al. (1993)
Salvelinus fontinalis		Diurnal	–	Hoar (1942)
Salvelinus alpinus	Groups	Seasonal shifts in feeding patterns	–	Jørgensen & Jobling (1989, 1990)
Silurus glanis	Groups	Nocturnal	–	Boujard (1995)

LD = light:dark cycle; LL = continuous light; DD = constant darkness.

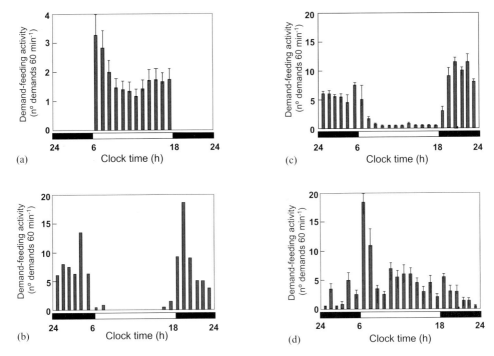

Fig. 8.1 Diel patterns of demand-feeding activity of three fish species held under laboratory conditions under an artificial photoperiod (L:D, 12:12). Horizontal open and solid bars at the bottom of each graph represent the light and dark phases, respectively. (a) Rainbow trout, *Oncorhynchus mykiss*, showing a diurnal pattern. (Redrawn from Sánchez-Vázquez & Tabata 1998.) (b) Atipa, *Hoplosternum littorale*, a nocturnal fish (redrawn from Boujard *et al.* 1990). (c, d) Sea bass, *Dicentrarchus labrax*, a dual phasing species which can show both diurnal and nocturnal feeding behaviour. (Redrawn from Sánchez-Vázquez *et al.* 1995a.)

diurnal rhythms of feeding activity. Tilapia (*Oreochromis niloticus*) also displays a predominantly diurnal rhythm of feeding activity, although a proportion of self-feeding activity may occur during the night (Toguyeni *et al.* 1997). This nocturnal feeding activity cannot be considered to be due to unintentional activation of the self-feeder because no food wastage was recorded.

Some species, such as several species of catfish, feed almost exclusively at night (Boujard & Luquet 1996) with an acrophase after dusk. This is the case for the atipa (*Hoplosternum littorale*) (Fig. 8.1b) (Boujard *et al.* 1990) and the European catfish (*Silurus glanis*) (Boujard 1995). In the latter case, such nocturnal behaviour is very strong because although the fish will feed during the light phase when food is only provided diurnally, nocturnal behaviour returns rapidly if the fish are allowed to feed during the whole 24-h cycle.

The presentation would be incomplete without mention of species that show a dual feeding pattern. According to Eriksson (1978), dualistic behaviour can be defined as a more or less instantaneous switching from diurnal to nocturnal activity and vice versa. However, in some fish species, switching seems to occur progressively, so dualistic behaviour might be considered simply as the ability of fish to shift from diurnal to nocturnal and vice versa. Dual species can invert, or shift, their feeding pattern either in response to environmental changes or spontaneously. For example, brown bullhead (*Ictalurus nebulosus*) exposed to low light intensity gradually shifted their locomotor activity from nocturnal to diurnal (Eriksson 1978; Eriksson & Alanärä 1992). In the case of feeding activity, dual behaviour under constant environmental conditions was first reported by Sanchez-Vázquez *et al.* (1995a,b) in European sea bass (*Dicentrarchus labrax*). In this species nocturnal and diurnal feeding patterns were observed simultaneously in different groups, or individuals, held under similar conditions (Fig. 8.1c and d). Moreover, some fish showed spontaneous and complete inversions in their feeding pattern within a few days.

It is clear that rhythmicity in feeding activity is widespread in fish although, in the light of the examples presented above, it is difficult to classify species as being strictly diurnal or nocturnal (Boujard 1999). It is also of interest to note that examples of arhythmicity in feeding activity are scarce. Rozin and Mayer (1961) did not find any rhythmicity in the feeding activity of goldfish, whereas others authors have done so (Sánchez-Vázquez *et al.* 1996). Arhythmic feeding may occur in fish larvae (Pérez & Buisson 1986), and also in fish that consume low-energy food sources; grass carp (*Ctenopharyngodon idella*) that eat plant material may need to feed almost continuously in order to cover their energy needs (Cui *et al.* 1993).

In nature, most fish may consume several types of food to obtain the nutrients they need. An interesting example of the effect of daily changes of food quality on foraging was demonstrated in two populations of blennies (*Parablennius sanguinolentus*), which fed mainly on algal turf and sea lettuce (*Ulva lactuca*), respectively. In fish of both populations, foraging paralleled changes in the energy content of the food source, although algal turf energy peaked in the afternoon, whereas *Ulva lactuca* energy peaked around noon (Zoufal & Taborsky 1991). The intake of different kinds of food may be partially (but not exclusively) regulated by changes in food availability and quality because specific nutritional needs may also play an important role in food selection (see also Chapter 11). In goldfish given the choice to feed from protein, fat and carbohydrate offered separately, no consistent daily pattern of nutrient preferences was obtained, but an opposite phasing for the selection of each macronutrient was observed in some individuals (e.g. carbohydrate during the light phase, protein during the dark phase, and fat during the transition phase) (Sánchez-Vázquez *et al.* 1998b). In rainbow trout, however, all animals showed a similar diurnal preference for all nutrients (Sánchez-Vázquez *et al.* 1999).

8.2.2 Tidal and lunar rhythms

Tidal rhythms in feeding activity are frequently related to vertical and horizontal movements of prey which use tidal currents to migrate or to synchronise their reproductive cycle. This tidal availability of prey results in rhythmic changes in feeding (Gibson 1992). A good example of tidally synchronised feeding rhythms is shown by the killifish (*Fundulus heteroclitus*), in which gut contents oscillate in parallel with the tidal cycle (Weisberg *et al.* 1981).

Brown (1946) reported that growth of brown trout (*Salmo trutta*) cycled with a periodicity of about three to four weeks, but no attempt was made to correlate growth rhythms with lunar periodicity. Since that time, lunar-related changes in food intake and body weight have been reported for several species, including coho salmon (*Oncorhynchus kisutch*) (parr and smolts) (Farbridge & Leatherland 1987a), Arctic charr (*Salvelinus alpinus*) (Dabrowski *et al.* 1992) and rainbow trout (Wagner & McKeown 1985; Farbridge and Leatherland 1987b;

Leatherland *et al.* 1992; Noël & Le Bail 1997). In the studies by Farbridge and Leatherland (1987a,b) on coho salmon and rainbow trout, fish were fed four times a day and held under constant 12:12 light:dark (LD) cycle at constant temperature. Under these conditions, rhythms of feed intake and growth had a period of fourteen to fifteen days, with peaks occurring between the new and full moon in coho salmon, and four to five days preceding the new or full moon in the rainbow trout. Growth in length also exhibited a semi-lunar rhythm in coho salmon, but was out of phase with the rhythm of body mass increase (Farbridge & Leatherland 1987a). Similar observations were made by Wagner & McKeown (1985) on rainbow trout, the rhythm for length increase being less pronounced and approximately 180° out of phase with that for increase in body mass.

8.2.3 Annual rhythms

In fish there may be a direct temperature-related effect on seasonal changes in feed intake, because fish are poikilothermic animals. Seasonal variations in feeding related to changes in photoperiod have also been demonstrated in a number of fish species (Swift 1955; Villarreal *et al.* 1988; Karas 1990; Palsson *et al.* 1992; Heggenes *et al.* 1993; Saether *et al.* 1996; Tveiten *et al.* 1996) (see also Chapters 6 and 15). In general, there is a marked increase in feed intake in spring, coinciding with the lengthening of the photoperiod, and a decrease in autumn, associated with a shortening of the photoperiod (Komourdjian *et al.* 1976; Higgins & Talbot 1985). In salmonid species this pattern may be influenced by the seasonal cycles of parr-smolt transformation and reproduction (Higgins & Talbot 1985; Jobling & Baardvik 1991; Tveiten *et al.* 1996).

In the wild, seasonal changes in the phasing of feeding rhythms may be related to seasonal changes in prey types. For example, sea bass have the ability to invert their feeding rhythms depending on the time of the year (Sánchez-Vázquez *et al.* 1998a), they consume a wide variety of prey organisms, and their dietary composition may vary seasonally (Pickett & Pawson 1994). Mediterranean sea bass consume mostly sardines (*Sardina pilchardus*) in summer and autumn, mullet (*Mugil cephalus*) in spring, and annelid worms and small fishes in winter (Kara & Derbal 1996). Given that these prey types may show different daily activity patterns, there is the (still unexplored) possibility that the dualism in sea bass is directed towards optimising the exploitation of a changing food source.

Seasonal variations in photoperiod and temperature may have a strong influence on the phasing of diel rhythms of feeding activity (Fig. 8.2). Early evidence of such seasonal changes came from studies on Arctic fishes, which are exposed to extremes of photoperiod and light intensity during the summer and winter solstices. For example, Müller (1978) reported complex annual variations in the phasing of diel activity rhythms of the burbot (*Lota lota*) and the sculpin (*Cottus peocilopus*) in a northern Swedish river, and Jørgensen and Jobling (1989, 1990) described seasonal phase inversions in daily feeding patterns of Arctic charr. According to Eriksson (1978), fish showing a seasonal phase inversion in their daily pattern of locomotor activity can be divided into two categories:

(1)   Crepuscular fishes, such as Atlantic salmon and rainbow trout, in which the apparent diurnalism or nocturnalism is the result of the fusion of dawn- and dusk-related activities, depending on the time of the year and the level of activity.

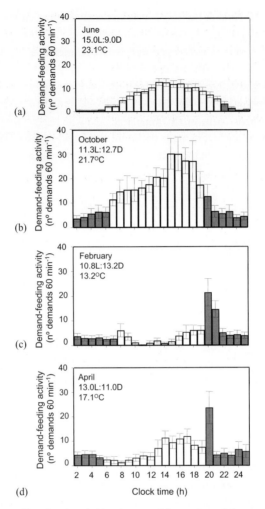

Fig. 8.2   Daily feeding profiles of sea bass held under natural fluctuations of photoperiod and water temperature throughout the year. Note that during winter (c), food demands made during daytime (represented by open bars) decreased and nocturnal food demands (represented by filled bars) increased sharply after dusk. (Modified from Sánchez-Vázquez *et al.* 1998.)

(2)   Biphasic fishes, such as burbot or sculpins, in which the seasonal inversion is more dramatic, and represents a true phase-inversion phenomenon.

These observations might result in a conclusion that seasonal phase inversions are a particular characteristic of species that live at high latitudes. However, species from temperate regions, such as sea bass, also display a seasonal phase inversion in their diel feeding rhythms (Anthouard *et al.* 1993; Sánchez-Vázquez *et al.* 1995a,b; Bégout Anras 1995; Boujard *et al.* 1996). This phenomenon was best exemplified by Sánchez-Vázquez *et al.* (1998a) who reported that sea bass underwent a double phase inversion, from being nocturnal in winter to diurnal during the rest of the year (Fig. 8.2). Inversions have also been observed under controlled laboratory conditions: the changes may appear spontaneously without any apparent relation to changes in the environment (Sánchez-Vázquez *et al.* 1995a,b, 1996; Aranda *et al.* 1999a). Figure 8.3 depicts examples of such spontaneous phase inversions in sea bass.

Why do fish show dualism? There is no clear answer; the ecological and physiological implications of dualistic behaviour in fish are uncertain. In the case of fish which live at high latitudes, and are subjected to extreme conditions, the seasonal changes in their behaviour might be justified by the wide variations in the photoperiod. However, this does not hold for species from temperate regions, such as sea bass and goldfish. Biotic factors (e.g. adaptation to the variable availability of food both in quantity and in quality, predation risk, etc.), in

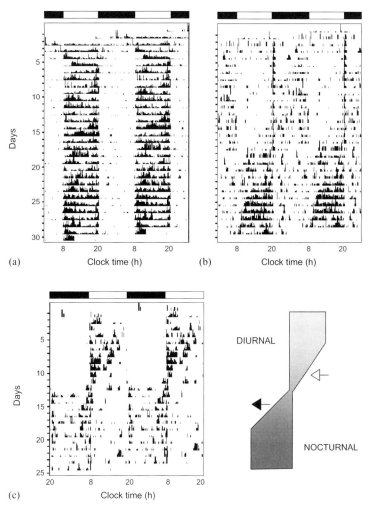

Fig. 8.3 Actograms of demand-feeding activity of groups of four sea bass maintained under constant 24°C and exposed to a 12:12 L:D cycle. Bars at the top of the graphs represent the photoperiod. For convenient visualisation, the activity records are double plotted (48-h horizontal scale) at a resolution of 10 min, the height of each point representing the percentage of food demands made in 24 h. Note that some fish exhibit a clear nocturnal feeding behaviour during the whole experimental period (a), while others were initially diurnal but spontaneously inverted to nocturnal in the middle of the experiment (b). The dynamics of such phase inversion, with contracting feeding by gradually advancing the offset of the feeding phase and expanding feeding by advancing the onset of feeding, is given in (c). (Redrawn from Sánchez-Vázquez *et al.* 1997.)

addition to the changes in abiotic factors, may be involved in the control of dualism. These hypotheses are based on the following situations in which dual phasing behaviour has been described: (i) light manipulation; (ii) temperature manipulation; and (iii) food availability restriction.

Since dual phasing behaviour is frequently synchronised to the seasons, some authors have tried to induce such inversions by manipulation of light and temperature. In the case of sea bass, photoperiod manipulation failed to bring about such inversions. The exposure of fish to photoperiods ranging from LD 2:22 to 22:2 cycles failed to induce inversions in feeding patterns (Aranda *et al.* 1999a). In brown bullheads, changes in light intensity seem to be effective in inducing feeding pattern inversions. Fish exposed to LD 12:12 cycles showed a nocturnal pattern at high light intensities and diurnal behaviour at low intensities (Eriksson 1978).

Diurnal and nocturnal feeding behaviour may not only be apparent under circadian LD cycles, but may also appear under ultradian cycles of illumination. For example, exposure of sea bass to 80-min LD pulses appeared to enhance phasing of feeding behaviour. Diurnal fish demanded feed only during the 40-min light periods, whereas nocturnal fish did so during the 40-min periods of darkness (Sánchez-Vázquez *et al.* 1995a). This suggests that fish can respond directly to illumination levels, and that the change in the synchronisation of the circadian rhythms of feeding activity to the LD 12:12 cycle is not the only factor responsible.

Under natural conditions, seasonal temperature changes often fluctuate more or less in parallel with changes in the photoperiod. In Atlantic salmon the proportion of nocturnal feeding activity increases at temperatures below 10°C, and it has been suggested that this external temperature stimulus is responsible for such changes (Fraser *et al.* 1993, 1995). However, these results should be interpreted with caution because in the studies described, a limited amount of food was continuously distributed. On the other hand, when sea bass were held individually, during spring and summer, and fed using self-feeders, gradual increases (from 22 to 28°C) or decreases (from 22 to 16°C) in water temperature were ineffective in inducing feeding pattern inversion (Aranda *et al.* 1999b). Further combinations of long (16L:8D) or short photoperiod (8L:16D) and warm (28°C) or cold temperature (16°C), simulating Mediterranean summer and winter conditions, failed to induce changes in the diurnal distribution of feeding (Aranda *et al.* 1999b). Taken together, these results indicate that the manipulation of water temperature may effectively influence the phasing of feeding rhythm only in some species, suggesting the existence of an internal mechanism which can be influenced by other factors in addition to water temperature.

Although the mechanisms regulating the inversion of feeding patterns remain unknown it is possible that, in addition to external cues, an endogenous annual clock may be involved. A similar process has been proposed to explain the seasonal reproductive cycling in sea bass (Carrillo *et al.* 1995; Prat *et al.* 1999), in which the endogenous clock could be synchronised by progressive changes in photoperiod, temperature and probably other unidentified external cues. In addition, the annual clock may require a specific entraining process involving gradual changes in both photoperiod and temperature. The existence of an endogenous annual clock may explain the failure of synchronisers to induce phase inversions at refractory times (Aranda *et al.* 1999a,b). The existence of refractory periods is suggested by the failure to induce gonad maturation several months after a reproductive period (Carrillo *et al.* 1995). In agreement with the endogenous clock hypothesis, Eriksson (1978) was able to construct an 'annual response curve' for brown trout following transfer from natural light to LD 12:12. The responses differed depending on season, with more diurnal behaviour being observed when fish were transferred in summer than in winter. However, to date, we have no firm direct evidence to confirm this hypothesis concerning the existence of an endogenous annual rhythm.

8.3 Other sources of variability in feeding rhythms

As indicated above, some of the discrepancies observed between studies conducted on a given species could be explained by differences in the season at which studies have been performed. Nevertheless, other factors, such as housing conditions (indoor or outdoor), self-feeder construction (rod position or protection) (Boujard *et al.* 1992a; Coves *et al.* 1998), reward level (Brännäs & Alanärä 1994; Alanärä & Kiessling 1996; Gélineau *et al.* 1998) and light intensity (Eriksson 1978), can also modify the diel patterns of feeding activity. Food intake varies from day to day both under controlled laboratory conditions and outdoor conditions (see also Chapters 6, 7 and 14). Such variations have been reported for many fish species including Atlantic salmon (Juell *et al.* 1993; Paspatis & Boujard 1996), rainbow trout (Alanärä 1992), turbot (Mallekh *et al.* 1998) and sea bass (Bégout-Anras 1995; Azzaydi *et al.* 1998) (Fig. 8.4). Mallekh *et al.* (1998) attempted to correlate feed intake of turbot with environmental factors, and found a significant correlation between feed intake and water temperature when data were treated as two-week running means, but failed to find any correlation on a daily basis. Thus it appears that food intake may be unpredictable on a daily basis, so growth and feed use may not be optimal if daily feed provision is predetermined to a fixed level (Juell *et al.* 1993; Azzaydi *et al.* 1998, 1999, 2000). Not only the amount but also the timing of food demand may change from day to day (see Chapter 10). In rainbow trout, for example, variability in feeding patterns may be observed on successive days within the same group of individuals even under fixed environmental conditions (Boujard & Leatherland 1992a): although feeding activity remained diurnal, the main peak was not always associated with dawn and the number of feeding bouts per day varied (Fig. 8.5).

Such differences may, in part, be the result of behavioural heterogeneity among individuals. For example, in an experiment designed to identify differences in feeding among rainbow trout, Brännäs and Alanärä (1997) reported that within a group most individuals demanded feed during daylight hours, but some were mainly active around dusk and during the early hours of the night. High inter-individual variability in feeding pattern was also observed in sea bass and goldfish housed individually, so the differences seen under group-housing conditions are not solely the result of social interaction with conspecifics (Sanchez-Vázquez *et al.* 1995a,b, 1996). It is tempting to speculate that a fish group might be regarded as a population of individual oscillators, although it should not be claimed that the expression of feeding rhythms in a group is the simple addition of individual feeding rhythms. Individual oscillators interact with each other to produce a single output (Lax *et al.* 1998). In the case of European catfish, for example, single individuals are often arrhythmic, but when they are maintained in pairs a high level of interaction occurs and growth is poor. When the size of the group is greater than two, the feeding rhythm becomes nocturnal and stable (Boujard 1995).

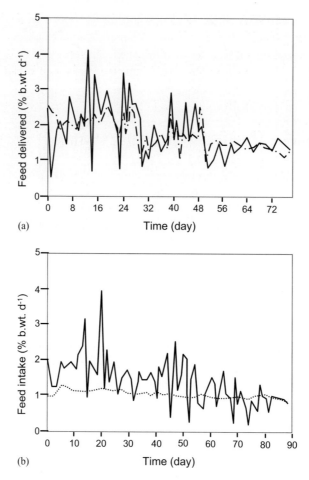

Fig. 8.4 Day-to-day variations in feeding activity. (a) Feed demanded by sea bass self-fed *ad libitum* (solid line) or time-restricted (dashed line). (Redrawn from Azzaydi *et al.* 1998.) (b) Feed intake in Atlantic salmon fed according to feeding tables (dotted line) or *ad libitum* using a hydroacoustic detection system (solid line). (Adapted from Juell *et al.* 1993.)

8.4 Regulation of feeding rhythms

8.4.1 *Endogenous control*

Hourglass mechanism

Many physiological processes are gated by external cues, continue for a certain time and then stop. These 'hourglass' mechanisms have been proposed as being responsible for biological rhythms, but their inability to self-oscillate without the external periodic cue by which they are gated constitutes an important argument against their being responsible for all biological rhythms. Nevertheless, hourglass mechanisms may be present in some periodic biological events. A typical hourglass process may be driven by the rate of gastric emptying, which may explain some of the rhythmicity in feeding: when the amount of food remaining in the

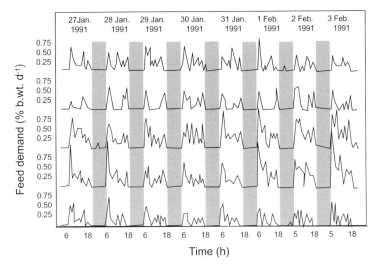

Fig. 8.5 Successive daily profiles of demand-feeding activity in five groups of thirty rainbow trout maintained under a 12:12 L:D cycle. Vertical bars represent the dark phase. (Redrawn from Boujard & Leatherland 1992a.)

stomach falls below a certain threshold, feeding can recommence if food is offered to the fish (Grove *et al*. 1978). Other biological variables related to feeding may also oscillate with a similar periodicity, but under fasting conditions these rhythms disappear. In addition to this non-self-sustaining capacity, hourglass-driven processes also show a wide variability in periodicity. In the above example, rates of digestion and evacuation, and metabolic rate – all of which are influenced by temperature – are factors that would modify prandial periodicity.

Pacemaker mechanism

The endogenous nature of a biological oscillation is demonstrated by the persistence of a rhythm under constant environmental conditions. Under such conditions the period of the rhythm usually deviates slightly from the environmental cycle to which it is normally syn-chronised, and it free-runs with its own 'natural' frequency. If such a free-running rhythm persists for many periods without attenuation, the rhythm belongs to the class of systems that are capable of self-sustaining oscillations. In contrast, rhythmic variations directly imposed upon the organism by exogenous factors (e.g. environmental temperature) disappear when the driving force is removed. In a strict sense, the term 'circa' in connection with the frequency of oscillation can only be applied to a rhythm in which a self-sustained endogenous oscilla-tion can be conclusively proved (Aschoff 1981a).

Although the fact that there is some internal control over several biological functions in fish has long been known (Spencer 1939; Harden Jones 1956), the endogenous origin of feed-ing rhythms remained uncertain until recently. Two types of evidence are used to test whether these daily rhythms are exogenously driven or are endogenously controlled:

(1)   The persistence of circadian rhythmicity under cycles of LD far from the circadian range, for example, in the ultradian range (e.g. LD 5:15 min or LD 40:40 min).

(2)   The persistence of rhythms in the absence of external cues in the circadian range.

Ultradian LD pulses were used by Eriksson and Van Veen (1980) to demonstrate free-running rhythms in the brown bullhead; the LD pulses were believed to act as stabilising factors preventing the rapid dissociation of the fish circadian system. The list of fish showing circadian feeding rhythms under LD pulses, or under constant illumination conditions (LL or DD), was later extended to include rainbow trout (Cuenca & De la Higuera 1994; Sánchez-Vázquez & Tabata 1998), European sea bass (Sánchez-Vázquez *et al.* 1995a) and goldfish (Sánchez-Vázquez *et al.* 1996). An example of fish showing free-running rhythms under constant laboratory conditions is given in Fig. 8.6.

Self-sustained oscillations can only be completely explained by assuming the existence of endogenous pacemakers. From a theoretical point of view, vertebrates have two different oscillators, the existence of which has been hypothesised in mammals. These oscillators are a light-entrained oscillator (LEO) located in the suprachiasmatic nucleus (SCN), and a food-entrained oscillator (FEO) which is anatomically and functionally independent of the LEO but for which the anatomical location has not been defined (for references, see Mistlberger 1994). Information on endogenous pacemakers in fish is scarce, and the location of the LEO remains uncertain. However, some characteristics of the retinorecipient region of the teleost hypothalamus show similarities with its counterpart in mammals (Holmqvist *et al.* 1992). Similarly, the existence – and therefore the location – of the FEO is also unknown. Some

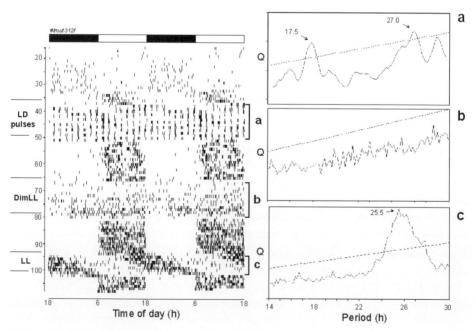

Fig. 8.6   Actogram of demand-feeding activity of an isolated trout maintained at constant 14°C and exposed to different free-running conditions: ultradian L:D pulses (45:45min light:dark), continuous dim light (2 lux) and continuous light (500 lux). Actogram descriptions as given in Fig. 8.3. Chi-square periodograms (dotted line represents a confidence level of 95%) during free-running conditions are given to the right. (Redrawn from Sánchez-Vázquez & Tabata 1998.)

results obtained in studies of locomotor rhythms in goldfish have revealed some properties of feeding entrainment that provide evidence of the existence of an FEO in fish (Spieler 1992; Sánchez-Vázquez *et al.* 1997) (see Chapter 9).

It is generally assumed that the vertebrate circadian system is composed of two structures in addition to the SCN: the pineal organ and the retina. The photosensitivity of the pineal organ, together with its ability to secrete melatonin into the bloodstream in a rhythmic manner, justifies the suggestion that the pineal organ plays a central role in the circadian system in fish (Kavaliers 1979; Tabata 1992; Ekström & Meissl 1997). However, pinealectomised fish may retain an ability to show free-running rhythms under constant illumination conditions, suggesting that other structures may act as central pacemakers (Kavaliers 1981; Tabata *et al.* 1991). Despite this, the pineal organ seems to act as an enhancing coupling factor (probably through its hormone melatonin) since 'splitting' of the activity rhythm, instability of period of circadian rhythms under constant conditions (tau) and variability in the circadian pattern under LD are described as consequences of pinealectomy (Kavaliers 1979, 1981). A schematic representation of possible components, and their interactions, of the fish circadian system is given in Fig. 8.7.

Despite the small effect that pinealectomy seems to have on behavioural circadian rhythms, the pineal organ of fish has been characterised as an endogenous pacemaker (see Falcon *et al.* 1992 for references). Cultured pineal tissue produces melatonin rhythmically under LD, and may show a free-running rhythm in continuous darkness (Iigo *et al.* 1991; Bolliet *et al.* 1994), although the rainbow trout pineal does not exhibit any circadian rhythm

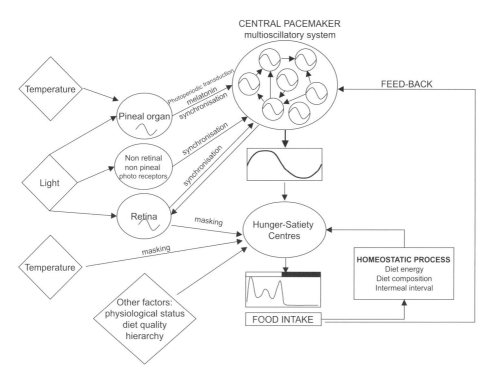

Fig. 8.7   A schematic representation of the circadian system of fish showing entraining pathways and interactions.

under DD (Gern & Greenhouse 1988). The retina of some fish also shows circadian rhythms in retinomotor movements, and in melatonin and serotonin production, both under LD and LL conditions (Zaunreiter *et al.* 1998a,b). Although its role in the fish circadian system is not completely understood, the neural connection of the retina to the hypothalamus suggests an involvement in circadian timekeeping (Tabata 1992).

Unlike in the case of circadian rhythms, no clear evidence of endogenous control has been obtained for infradian rhythms of lunar periodicity. For example, under constant light, coho salmon did not show semi-lunar rhythms of growth, although the rhythms were reinstated in phase with a semi-lunar periodicity after re-exposure to a LD 12:12 cycle (Leatherland *et al.* 1992).

There is some evidence to suggest that annual cycles of feeding and growth reflect endogenous circannual rhythms synchronised by photoperiodic changes (Eriksson & Lundqvist 1982; Villarreal *et al.* 1988; Smith *et al.* 1993; Saether *et al.* 1996). Such a conclusion has been drawn on the basis of the results obtained in long-term studies performed on fish exposed to constant photoperiod and temperature. Immature Arctic charr held at constant temperature (4°C) and photoperiod (LD 12:12) exhibited seasonal rhythms of feed intake and growth which did not differ significantly from those of fish exposed to a simulated natural photoperiod. In contrast, mature Arctic charr under LD 12:12 displayed changes in feed intake and growth, which were delayed in comparison to fish exposed to natural photoperiod. Effects related to gonadal maturation could have been the reason for these differences (Saether *et al.* 1996). To date, no data appear to be available concerning the endogenous character of annual rhythms of feeding and growth in non-salmonid fish species.

Synchronisers

Although many rhythms may be endogenously driven, their period, mean, amplitude and phase are modulated by periodic events in the environment. However, only a few environmental variables may act as entraining agents or 'zeitgebers' (Aschoff 1981b; Pittendrigh 1981). For most animals, light–dark alternation with a period of 24 h is the main zeitgeber, although other environmental cues such as periodic feeding, seasonal changes in photoperiod, tidal cycles and moonlight cycles may also act as zeitgebers.

For an oscillation in an environmental variable to be categorised as a zeitgeber, the following entrainment criteria must be met (Moore-Ede *et al.* 1982):

(1)   Absence of other time cues. The biological rhythm must free-run with a particular period before the time cue is imposed upon the animal and, conversely, must recover this free-running period after the time cue is removed.
(2)   Period control. Once the animal is exposed to the environmental cycle, the period of the biological rhythm must coincide with that of the environmental cycle.
(3)   Stable phase relationship. After a few days of exposure to the environmental cycle, a stable phase relationship between the rhythm and its zeitgeber must emerge. This phase relationship depends on the strength of the zeitgeber and on the endogenous period.
(4)   Phase control. When the zeitgeber is removed, the biological rhythm must free-run again starting from a phase determined by the environmental cycle, and not by the rhythm prior to entrainment.

The importance of the above criteria of entrainment should be apparent when it is appreciated that many environmental factors cycles modify biological rhythms by means of a direct effect on the waveform of the rhythm, and not by acting on the endogenous pacemaker. Such a direct action is called a masking effect, as opposed to the entraining effect of true zeitgebers. However, in practice, such a clear differentiation is extremely difficult to make because most zeitgebers exert a masking effect in addition to an entraining effect.

Although zeitgebers may differ according to the periodicity of biological rhythms, light is generally thought to be the main zeitgeber of biological rhythms in fish.

Ultradian rhythms exhibit a high degree of variability because they are not usually synchronised to any environmental cue but are driven by periodic endogenous processes (Peters & Veeneklaas 1992). There are exceptions to this, in that some fish show highly stable ultradian rhythms synchronised to the tides. Tidal rhythms with a period close to 12.5 h have been described for many biological functions, including feeding behaviour (Weisberg *et al.* 1981; see also Morgan 1991 and Gibson 1992 for references). Results of several studies provide evidence that the overt tidal rhythm may persist for a few periods in constant environmental conditions (see Morgan 1991 and Gibson 1992 for references). Such is the case in the rock goby (*Gobius paganellus*) and the common blenny (*Lipophrys pholis*), in which the tidal rhythm of locomotor activity persists for up to five days in constant conditions. However, the tidal factors that lead to adjustment of these rhythms in nature remain to be elucidated. Hydrostatic pressure, temperature, salinity, light, inundation and agitation have all been proposed as zeitgebers for some circatidal rhythms (Palmer 1973; Naylor 1982; Northcott *et al.* 1991).

In the case of biological rhythms of circadian periodicity, it is known that water temperature, O_2 and CO_2 content are abiotic factors that influence the pattern of feeding activity (Boujard & Leatherland 1992b) (see Chapters 6 and 7). However, to date it is not known whether such factors are zeitgebers. From evidence currently available it seems that only LD alternation and time-restricted feeding with a period of 24 hours act as true zeitgebers of fish circadian rhythms. However, we must be cautious in drawing this conclusion because in many experiments there was not differentiation between the role of light/dark alternation as a zeitgeber, acting on the endogenous pacemaker, and the masking effect of illumination conditions, not affecting the pacemaker but influencing the rhythmic variable (see section on Masking, p. 207).

In an attempt to assess the ability of LD cycles to act as zeitgebers, two types of experiments have been carried out:

(1)   Investigation of the number of transient cycles required to re-synchronise feeding rhythms following the shifting of the LD cycle by several hours.
(2)   Demonstration that free-running rhythms start to free-run from a phase determined by the last LD cycle.

As previously noted, sea bass, goldfish and rainbow trout have been shown to display a free-running circadian rhythm in feeding behaviour under continuous light. Thus, in these species LD alternation is a true zeitgeber, and feeding activity starts to free-run from the phase established by the previous LD cycle. This points to an entraining effect of LD alternation on the endogenous pacemaker (Sánchez-Vázquez *et al.* 1995a,1996; Sánchez-Vázquez & Tabata 1998).

An important aspect of the LD cycle in its entraining ability is related to the wavelength and intensity of the light. Hitherto, very little work has been carried out on light spectrum action. In rainbow trout, a LD cycle in which a red lamp (650 nm) was used as the light source synchronised the diel pattern of activity as well as a normal LD cycle (Molina Borja *et al.* 1990). Similarly, Boujard *et al.* (1992b) showed that feeding activity of atipa could be entrained with either red or blue light.

In addition to LD cycles, periodic food availability has also been reported to act as a zeitgeber in some fish (Sánchez-Vázquez *et al.* 1995b, 1997). When a fish has access to food only at certain times, many physiological and behavioural variables become synchronised to feeding time. This phenomenon is known as feed-anticipatory activity. The anticipation of feeding appears not only under LD cycles, but also in continuous light or darkness, suggesting an endogenous timing mechanism may be responsible for feeding entrainment (Mistlberger 1994).

Light and food may entrain different rhythms in the same individual. This effect has important metabolic implications because the phase relationship and the relative amplitudes of different rhythmic variables can shift as a consequence of relative shifts in LD and feeding schedules. This internal dissociation may be responsible for the differential effects of feeding time on growth, feed efficiency and flesh composition (see Chapter 10). The existence of internal dissociation also supports the hypothesis that the circadian system of fish is a multioscillatory system. Thus, it is interesting to consider that part of the circadian system of fish might be entrained by periodic feeding, whereas the remainder of the system is entrained to LD cycles, enabling the maintenance of other rhythms, such as those related to reproduction (Spieler 1992).

As previously noted, variability is one of the characteristics of circadian rhythms in fish. The appearance of circadian rhythms varies within and between species and even within a single individual. Thus, it is not surprising that a given variable may seem to be entrained preferentially by the LD cycle in some individuals and by feeding time in others of the same species. For example, when sea bass are exposed to two simultaneous zeitgebers with different periods, e.g. a LD 13:13 h cycle and a 24-h restricted feeding schedule (4 h of food availability each 24 h), demand feeding activity can be synchronised by the LD cycle in some fish, by restricted feeding in others, and by both zeitgebers in the remaining fish (Sánchez-Vázquez *et al.* 1995b).

As with circadian rhythms, circannual rhythms are also synchronised to environmental cues. In this case, annual changes in photoperiod seem to be the main zeitgeber. However, seasonal variations in water temperature are also a determining factor in some annual rhythms, such as reproductive cycling in some species (Zanuy *et al.* 1986). In some cases, the phase and amplitude of circadian rhythms are modified in accordance with an annual rhythm.

There is evidence to support the hypothesis that the circadian system is involved in the synchronisation of annual rhythms. According to Bunning's hypothesis (Bunning 1969), the combination of a narrow sensitivity window driven by the circadian system, together with a variation in photoperiod, can be used to monitor the number of hours of daylight or the length of the night. This type of monitoring would allow the appearance of certain photoperiod-induced processes such as reproduction or migrations. This model has been used to explain photoperiodicity in mammals, using the melatonin rhythm as the mediator of photoperiod and the pituitary-gonadal axis as the location of the sensitivity window. In some species, melatonin has an important gonadotrophic inhibitory effect at certain times of the day but not at others. In fish, some experiments have demonstrated the possibility of inducing gonadal maturation by simulating a complete photoperiod with skeleton photoperiods. These consist of a baseline non-stimulatory photoperiod accompanied by a short light pulse applied at certain times. Skeleton photoperiods allow one to differentiate the masking effect of light and darkness from the synchronising effect (Pittendrigh 1981). In the catfish *Heteropneustes fossilis* a 1-h light pulse given at different times during the dark phase may induce gonadal maturation when applied at certain times, but such a pulse is ineffective when applied at others (Sundararaj & Vasal 1976).

Masking

The passive influence of external factors may contribute in a large part to the overall feeding pattern of fish. Such influences, which do not modify the operation of endogenous oscillators, are called masking factors (Aschoff 1981b). Although no specific studies have been conducted to identify and demonstrate the quantitative influence of such factors in fish, most effects referred to by authors as an entraining process, should probably be considered as masking effects. For example, when the LD cycle is delayed or advanced by several hours, very few cycles are needed for circadian rhythms to be re-synchronised to the new LD cycle (Boujard *et al.* 1990; Sánchez-Vázquez *et al.* 1995a). Although these results could be considered as evidence for the weakness of the fish circadian system, they can also be explained by the existence of a strong masking effect of light conditions. In addition to acting as a zeitgeber, light may have a strong masking effect on circadian feeding rhythms. For example, if a fish is reliant upon the sense of vision to feed, complete darkness will remove this possibility, thus imposing restrictions on the output of the oscillator. In other cases the influence is not so clear. For example, in sea bass, light exerts a positive masking effect in diurnal fish but a negative effect in nocturnal fish (Sánchez-Vázquez *et al.* 1995a,b).

Daily changes in water temperature can also directly influence feeding patterns: favourable temperatures can increase feed intake, whereas temperature extremes will reduce it. The influence of temperature on locomotor activity and feeding activity varies according to species: the goldfish becomes active at water temperatures above 10°C (Hirata 1957), whereas salmonids are active at temperatures just above zero (see Chapter 6). In salmonids, however, a few days with water temperatures above 20°C leads to reduced food intake (Elliott 1975). Nevertheless, feed demands are not only driven by water temperature; other factors, such as previous history and photoperiod, modify such effects.

Social relations among fish are an important factor that influence feeding. Under natural conditions, it is thought that subordinate fish tend to feed mainly outside the feeding phase of dominant fish (Emery 1973; Brännäs & Alanärä 1997; see Chapters 3, 6 and 7).

Feeding rhythms in fish are often studied using self-feeding devices (see Chapters 3 and 7), and the characteristics of the devices can exert a strong influence on the results obtained. For example, trigger position, whether above or below the water surface, can affect the feeding pattern. A trigger positioned above the water surface can induce the appearance of a strictly diurnal pattern of feeding in rainbow trout, due to the difficulty of locating and operating the trigger in complete darkness (Boujard *et al.* 1992a). On the other hand, a trigger placed below the water surface could be activated accidentally in darkness by fish contacting the trigger when they swim close to the water surface. Such accidental demands are obvious when accompanied by large amounts of feed wastage. However, the installation of more or less complex protectors around the trigger to prevent accidental activation can impose difficulties in operating the feeders at night (Coves *et al.* 1998). In order to be sure that the construction of the self-feeding device does not markedly influence the feeding rhythm of the fish studied, several precautions should be taken:

(1)   Use a low light intensity at night in order to ensure that complete darkness does not have any masking effect.
(2)   Control that the feed demanded is eaten by the fish by installation of a feed waste collector.
(3)   Use triggering systems which reduce unintentional activation to a minimum.

8.5 Conclusions

There is increasing evidence that the properties of fish feeding rhythms have similarities with those of other groups of animals. However, it must be remembered that the relative homogeneity observed in mammals concerning the physiology of biological rhythms should probably not be expected in fish. Amongst other things, it could be argued that more that 400 million years of evolutionary history and over 25 000 fish species are not comparable with a history of 40 million years and 2000 species for mammals.

When food is continuously available, most fish feed during a specific phase of the diel light-dark cycle. According to their distribution of feeding preferences, fish were originally classified as being diurnal, nocturnal or crepuscular. However, there is considerable variability in feeding patterns between, and even within, individuals of the same species. This variability may partly be explained by a lack of homogeneity in experimental design, but it is also likely to be a characteristic of feeding rhythms in fish. Seasonal inversions in feeding rhythms are observed in some fish species, and this phenomenon might explain some of the contradictory results seen in the different studies.

As is the case for other vertebrates, fish feeding rhythms seem to result from the output of an internal pacemaker, probably composed of multiple oscillators synchronised by the daily alternation of light and darkness. Although the exact location of the pacemaker is not known, the retinorecipient hypothalamic area is a good candidate site. Light information can be gathered by the retina, pineal and by non-pineal, non-retinal brain receptors. In addition to the LD cycle, restricted feeding times can synchronise many biological rhythms. However, not all circadian variables are completely synchronised by feeding time, and some remain synchronised to the LD cycle.

More experimental work needs to be carried out to identify and understand the physiology of endogenous pacemakers and the specific zeitgebers that influence the biological rhythms of fish. The next few years will probably see the development of a new line in fish feeding research with a focus on the study of biological rhythms and their consequences for seasonal changes in nutritional demands and requirements.

8.6 References

Adron, J.W., Grant, P.T. & Cowey, C.B. (1973) A system for the quantitative study of the learning capacity of rainbow trout and its application to the study of food preferences and behaviour. *Journal of Fish Biology*, **5**, 625–636.

Alanärä, A. (1992) Demand feeding as a self-regulating feeding system for rainbow trout (*Oncorhynchus mykiss*) in net-pens. *Aquaculture*, **108**, 347–356.

Alanärä, A. & Kiessling, A. (1996) Changes in demand feeding behaviour in Arctic charr, *Salvelinus alpinus* L., caused by differences in dietary energy content and reward level. *Aquaculture Research*, **27**, 479–486.

Ali, M.A. (ed.) (1992) *Rhythms in Fishes*. Plenum Press, New York.

Anthouard, M., Divanach, P. & Kentouri, M. (1993) An analysis of feeding activities of sea bass (*Dicentrarchus labrax*, Moronidae) raised under different lighting conditions. *Ichtyophysiologica Acta*, **16**, 59–70.

Aranda, A., Madrid, J.A., Zamora, S. & Sánchez-Vázquez, F.J. (1999a) Synchronizing effect of photoperiod on the dual phasing of demand-feeding rhythms in sea bass. *Biological Rhythm Research*, **30**, 392–406.

Aranda, A., Sánchez-Vázquez, F.J. & Madrid, J.A. (1999b) Influence of temperature on demand-feeding rhythms in sea bass. *Journal of Fish Biology*, **55**, 1029–1039.

Aschoff, J. (1981a) A Survey on Biological Rhythms. In: *Handbook of Behavioral Neurobiology 4, Biological Rhythms* (ed. J. Aschoff), pp. 3–11. Plenum Press, New York.

Aschoff, J. (1981b) Freerunning and entrained circadian rhythms. In: *Handbook of Behavioral Neurobiology 4, Biological Rhythms* (ed. J. Aschoff), pp. 81–94. Plenum Press, New York.

Azzaydi, M., Madrid, J.A., Zamora, S., Sánchez-Vázquez, F.J. & Martínez, F.J. (1998) Effect of three feeding strategies (automatic, ad libitum demand-feeding and time-restricted demand-feeding) on feeding rhythms and growth in European sea bass (*Dicentrarchus labrax* L.). *Aquaculture*, **163**, 285–296.

Azzaydi, M., Martínez, F.J., Zamora, S., Sánchez-Vázquez, F.J. & Madrid, J.A. (1999) Effect of meal size modulation on growth performance and feeding rhythms in European sea bass (*Dicentrarchus labrax*, L). *Aquaculture*, **170**, 253–266.

Azzaydi, M., Martínez, F.J., Zamora, S., Sánchez-Vázquez, F.J. & Madrid, J.A (2000) The influence of nocturnal *versus* diurnal feeding under winter conditions on growth and feed conversion of European sea bass (*Dicentrarchus labrax*, L.). *Aquaculture*, **182**, 329–338.

Bégout-Anras, M.L. (1995) Demand-feeding behaviour of sea bass kept in ponds: diel and seasonal patterns, and influences of environmental factors. *Aquaculture International*, **3**, 186–195.

Blyth, P.J., Purser, G.J. & Russell, J.F. (1993) Detection of feeding rhythms in seacaged Atlantic salmon using new feeder technology. In: *Fish Farming Technology* (eds H. Reinertsen, L.A. Dahle, L. Jorgensen & K. Tvinnereim), pp. 209–216. Balkema, Rotterdam.

Bolliet, V., Bégay, V., Ravault, J-P., Ali, M.A., Collin, J.-P. & Falcón, J. (1994) Multiple circadian oscillators in the photosensitive pike pineal gland: a study using organ and cell culture. *Journal of Pineal Research*, **16**, 77–84.

Boujard, T. (1995) Diel rhythms of feeding activity in the European catfish, *Silurus glanis*. *Physiology and Behavior*, **58**, 641–645.

Boujard, T. (1999) Les rythmes circadiens d'alimentation chez les Téléostéens. In: *Comptes-rendus des XXIèmes journées de la Société Française d'Ichtyologie* (eds T. Boujard & J.Y. Sire), pp. 89–112. Cybium, 23 (suppl.).

Boujard, T. & Leatherland, J.F. (1992a) Demand-feeding behaviour and diel pattern activity in *Oncorhynchus mykiss* held under different photoperiod regimes. *Journal of Fish Biology*, **40**, 535–544.

Boujard, T. & Leatherland, J.F. (1992b) Circadian rhythms and feeding time in fish. *Environmental Biology of Fishes*, **35**, 109–131.

Boujard, T. & Luquet, P. (1996) Rythmes alimentaires et alimentation chez les siluroidei. In: *The Biology of Catfishes* (eds M. Legendre & J.P. Proteau), pp. 113–120. Aquatic Living Resources, **9** (suppl.).

Boujard, T., Dugy, X., Genner, D., Gosset, C. & Grig, G. (1992a) Description of a modular, low cost, eater meter for the study of feeding behavior and food preferences in fish. *Physiology & Behavior*, **52**, 1101–1106.

Boujard, T., Jourdan, M., Kentouri, M. & Divanach, P. (1996) Diel feeding activity and the effect of time-restricted self-feeding on growth and feed conversion in European sea bass. *Aquaculture*, **139**, 117–127.

Boujard, T., Keith, P. & Luquet, P. (1990) Diel cycle in *Hoplosternum littorale* (Teleostei): evidence for synchronization of locomotor, air breathing and feeding activity by circadian alternation of light and dark. *Journal of Fish Biology*, **36**, 133–140.

Boujard, T., Moreau, Y. & Luquet, P. (1992b) Diel cycles in *Hoplosternum littorale* (Teleostei): entrainment of feeding activity by low intensity coloured light. *Environmental Biology of Fishes*, **35**, 301–309.

Brady, J. (1982) Introduction to biological timekeeping. In: *Biological Timekeeping* (ed. J. Brady), pp. 1–7. Cambridge University Press, Cambridge.

Brännäs, E. & Alanärä, A. (1994) Effect of reward level on individual variability in demand feeding activity and growth rate in Arctic charr and rainbow trout. *Journal of Fish Biology*, **45**, 423–434.

Brännäs E. & Alanärä, A. (1997) Is diel dualism in feeding activity influenced by competition between individuals? *Canadian Journal of Zoology*, **75**, 661–669.

Brown, M.E. (1946) The growth of brown trout (*Salmo trutta, L.*) II. The growth of two-year-old trout at a constant temperature of 11.5°C. *Journal of Experimental Biology*, **22**, 130–144.

Bunning, F. (1969) Common features of photoperiodism in plants and animals. *Photochemistry and Photobiology*, **9**, 219–228.

Burel, C., Robin, J. & Boujard, T. (1997) Can turbot, *Psetta maxima*, be fed with self-feeders? *Aquatic Living Resources*, **10**, 381–384.

Carrillo, M., Zanuy, S., Prat, F., Cerdá, J., Ramos, J., Mañanós, E. & Bromage, N. (1995) Sea bass (*Dicentrarchus labrax*). In: *Broodstock Management and Egg and Larval Quality* (eds N.R. Bromage & R.J. Roberts), pp. 138–168. Blackwell Scientific Publications, Oxford.

Coves, D., Gasset, E., Lemarié, G. & Dutto, G. (1998) A simple way of avoiding wastage in European seabass, *Dicentrarchus labrax*, under self-feeding conditions. *Aquatic Living Resources*, **11**, 395–401.

Cuenca, E.M. & de la Higuera, M. (1994) Evidence for an endogenous circadian rhythm of feeding in the trout (*Oncorhynchus mykiss*). *Biological Rhythm Research*, **25**, 228–235.

Cui, Y., Chen, S., Wang S. & Liu, X. (1993) Laboratory observations on the circadian feeding pattern in the grass carp (*Ctenopharyngodon idella* Val.) fed three different diets. *Aquaculture*, **113**, 57–64.

Daan, S. (1981) Adaptive daily strategies in behavior. In: *Handbook of Behavioral Neurobiology 4, Biological Rhythms* (ed. J. Aschoff), pp. 275–298. Plenum Press, New York.

Dabrowski, K., Krumschnabel, G., Paukku, M. & Labanowski, J. (1992) Cyclic growth and activity of pancreatic enzymes in alevins of Arctic charr (*Salvelinus alpinus* L.). *Journal of Fish Biology*, **40**, 511–521.

Ekström, P. & Meissl, H. (1997) The pineal organ of teleost fishes. *Reviews in Fish Biology & Fisheries*, **7**, 199–284.

Elliott, J.M. (1975) Number of meals in a day, maximum weight of food consumed in a day and maximum rate of feeding for brown trout *Salmo trutta*, L. *Freshwater Biology*, **5**, 287–303.

Emery, A.R. (1973) Preliminary comparisons of day and night habits of freshwater fish in Ontario Lakes. *Journal of the Fisheries Research Board of Canada*, **30**, 761–774.

Eriksson, L.O. (1978) Nocturnalism versus diurnalism- dualism within individuals. In: *Rhythmic Activity of Fishes* (ed. J.E. Thorpe), pp. 69–89. Academic Press, London.

Eriksson, L.O. & Alanärä, A. (1992) Timing of feeding behaviour in Salmonids. In: *World Aquaculture Workshops* (eds J.E. Thorpe & F.A. Huntingford), pp. 41–48, vol. 2. The World Aquaculture Society, Halifax, Nova Scotia.

Eriksson, L.O. & Lundqvist, H. (1982) Circannual rhythms and photoperiod regulation of growth and smolting in Baltic salmon (*Salmo salar* L.) *Aquaculture*, **28**, 113–121.

Eriksson, L.O. & Van Veen, T. (1980) Circadian rhythms in the brown bullhead, *Ictalurus nebulosus* (Teleostei). Evidence for an endogenous rhythm in feeding, locomotor, and reaction time behaviour. *Canadian Journal of Zoology*, **58**, 1899–1907.

Falcon, J., Thibault, C. & Begay, V. (1992) Regulation of the rhythmic melatonin secretion by fish pineal photoreceptor cells. In: *Rhythms in Fishes* (ed. M.A. Ali), pp. 167–198. Plenum Press, New York.

Farbridge, K.J. & Leatherland, J.F. (1987a) Lunar cycles of coho salmon, *Oncorhynchus kisutch*. I. Growth and feeding. *Journal of Experimental Biology*, **128**, 165–178.

Farbridge, K.J. & Leatherland, J.F. (1987b) Lunar periodicity of growth cycles in rainbow trout, *Salmo gairdneri* Richardson. *Journal of Interdisciplinary Cycle Research*, **18**, 169–177.

Fraser, N.H.C. & Metcalfe, N.B. (1997) The cost of becoming nocturnal: feeding efficiency in relation to light intensity in juvenile Atlantic salmon. *Functional Ecology*, **11**, 760–767.

Fraser, N.H.C., Metcalfe, N.B. & Thorpe, J.E. (1993) Temperature-dependent switch between diurnal and nocturnal foraging in salmon. *Proceedings of the Royal Society of London B*, **252**, 135–139.

Fraser, N.H.C., Heggenes, J., Metcalfe, N.B. & Thorpe, J.E. (1995) Low summer temperatures cause juvenile Atlantic salmon to become nocturnal. *Canadian Journal of Zoology*, **73**, 446–451.

Gélineau, A., Corraze, G. & Boujard, T. (1998) Effects of restricted ration, time-restricted access and reward level on voluntary food intake, growth and growth heterogeneity of rainbow trout (*Oncorhynchus mykiss*) fed on demand with self-feeders. *Aquaculture*, **167**, 247–258.

Gern, W.A. & Greenhouse, S.S. (1988) Examination of in vitro melatonin secretion from superfused trout (*Salmo gairdneri*) pineal organs maintained under diel illumination or continuous darkness. *General and Comparative Endocrinology*, **71**, 163–174.

Gibson, R.N. (1992) Tidally-synchronised behaviour in marine fish. In: *Rhythms in Fishes* (ed. M.A. Ali), pp. 63–81. Plenum Press, New York.

Grove, D.J., Loizides, L.G. & Nott, J. (1978) Satiation amount, frequency of feeding and gastric emptying rate in *Salmo gairdneri*. *Journal of Fish Biology*, **12**, 507–516.

Harden Jones, F.R. (1956) The behaviour of minnows in relation to light intensity. *Journal of Experimental Biology*, **33**, 271–281.

Heggenes, J., Krog, O.M.W., Lindås, O.R., Dokk, J.G. & Bremnes, T. (1993) Homeostatic behavioural responses in a changing environment: Brown trout (*Salmo trutta*) become nocturnal during winter. *Journal of Animal Ecology*, **62**, 295–308.

Higgins, P.J. & Talbot, C. (1985) Growth and feeding in juvenile Atlantic salmon (*Salmo salar*, L.). In: *Nutrition and Feeding in Fish* (eds C.B. Cowey, A.M. Mackie & J.G. Bell), pp. 243–263. Academic Press, London.

Hirata, H. (1957) Diurnal rhythm of the feeding activity of goldfish in winter and early spring. *Bulletin of the Faculty of Fisheries, Hokkaido University*, **7**, 72–84.

Hoar, W.S. (1942) Diurnal variations in feeding activity of young salmon and trout. *Journal of the Fisheries Research Board of Canada*, **34**, 1655–1669.

Holmqvist, B.I., Östholm, T. & Ekström, P. (1992) Retinohypothalamic projections and the suprachiasmatic nucleus in the teleost brain. In: *Rhythms in Fishes* (ed. M.A. Ali), pp. 293–318. Plenum Press, New York.

Iigo, M., Kezuka, H., Aida, K. & Hanyu, I. (1991) Circadian rhythms of melatonin secretion from superfused goldfish (*Carassius auratus*) pineal glands in vitro. *General and Comparative Endocrinology*, **75**, 217–221.

Jobling, M. & Baardvik, B.M. (1991) Patterns of growth of maturing and immature Arctic charr, *Salvelinus alpinus*, in a hatchery population. *Aquaculture*, **94**, 343–354.

Jørgensen, E.H. & Jobling, M. (1989) Patterns of food intake in Arctic charr, *Salvelinus alpinus*, monitored by radiography. *Aquaculture*, **81**, 155–160.

Jørgensen, E.H. & Jobling, M. (1990) Feeding modes in Arctic charr, *Salvelinus alpinus* L: the importance of bottom feeding for the maintenance of growth. *Aquaculture*, **86**, 379–385.

Jørgensen, E.H. & Jobling, M. (1992) Feeding behaviour and effect of feeding regime on growth of Atlantic salmon, *Salmo salar*. *Aquaculture*, **101**, 135–146.

Juell, J.E., Furevik, D.M. & Bjordal, A. (1993) Demand feeding in salmon farming by hydroacoustic detection. *Aquacultural Engineering*, **12**, 155–167.

Kadri, S., Metcalfe, N.B., Huntingford, F.A. & Thorpe, J.E. (1991) Daily feeding rhythms of Atlantic salmon in sea cages. *Aquaculture*, **92**, 219–224.

Kadri, S., Metcalfe, N.B., Huntingford, F.A. & Thorpe, J.E. (1997) Daily feeding rhythms in Atlantic salmon II: size-related variation in feeding patterns of post-smolts under constant environmental conditions. *Journal of Fish Biology*, **50**, 273–279.

Kara, M.H. & Derbal, F. (1996) Régime alimentaire du loup *Dicentrarchus labrax* (poisson moronidé) du golfe d'Ánnaba, Algérie. *Annales de l'Institut Océanographique*, **72**, 185–194.

Karas, P. (1990) Seasonal changes in growth and standard metabolic rate of juvenile perch, *Perca fluviatilis* L. *Journal of Fish Biology*, **37**, 913–920.

Kavaliers, M. (1979) Pineal involvement in the control of circadian rhythmicity in the lake chub, *Couesius plumbeus*. *Journal of Experimental Zoology*, **209**, 33–40.

Kavaliers, M. (1981) Circadian organization in white suckers *Catostomus commersoni*: The role of the pineal organ. *Comparative Biochemistriy and Physiology*, **68A**, 127–129.

Komourdjian, M.P., Saunders, R.L. & Fenwick, J.C. (1976) Evidence for the role of growth hormone as a part of the 'light pituitary axis' in growth and smoltification of Atlantic salmon (*Salmo salar*). *Canadian Journal of Zoology*, **54**, 544–551.

Landless, P.J. (1976) Demand-feeding behaviour of rainbow trout. *Aquaculture*, **7**, 11–25.

Lax, P., Zamora, S. & Madrid, J.A. (1998) Coupling effect of locomotor activity on the rat's circadian system. *American Journal of Physiology*, **44**, R580–R587.

Leatherland, J.F., Farbridge, K.J., & Boujard, T. (1992) Lunar and semi-lunar rhythms in fishes. In: *Rhythms in Fishes* (ed. M.A. Ali), pp. 83–107. Plenum Press, New York.

Mallekh, R., Lagardère, J.P., Bégout-Anras, M.L. & Lafaye, J.Y. (1998) Variability in appetite of turbot, *Scophthalmus maximus* under intensive rearing conditions: the role of environmental factors. *Aquaculture*, **165**, 123–138.

Mistlberger, R.E. (1994) Circadian food-anticipatory activity: formal models and physiological mechanisms. *Neuroscience and Biobehavioral Reviews*, **18**, 171–195.

Molina Borja, M., Pérez, E., Pupier, R. & Buisson, B. (1990) Entrainment of circadian activity rhythm in the juvenile trout, *Salmo trutta* L., by red light. *Journal of Interdisciplinary Cycle Research*, **21**, 81–89.

Moore-Ede, M.C., Sulzman, F.M. & Fuller, C.A. (1982) *The Clocks that Time Us*. Harvard University Press, Cambridge.

Morgan., E. (1991) An appraisal of tidal activity rhythms. *Chronobiology International*, **8**, 283–306.

Müller, K. (1978) The flexibility of the circadian system of fish at different latitudes. In: *Rhythmic Activity of Fishes* (ed. J.E. Thorpe), pp. 91–104. Academic Press, London.

Naylor, E. (1982) Tidal and lunar rhythms in animals and plants. In: *Biological Timekeeping* (ed. J. Brady) pp. 33–48. Cambridge University Press, Cambridge.

Noël, O. & Le Bail, P.-Y. (1997) Does cyclicity of growth rate in rainbow trout exist? *Journal of Fish Biology*, **51**, 634–642.

Northcott, S.J., Gibson, R.N. & Morgan, E. (1991) On-shore entrainment of circatidal rhythmicity in *Lipophrys pholis* (Teleostei) by natural zeitgeber and the inhibitory effect of cageing. *Marine Behaviour and Physiology*, **19**, 63–73.

Palmer, J.D. (1973) Tidal rhythms: the clock control of the rhythmic physiology of marine organisms. *Biological Reviews*, **48**, 377–418.

Palsson, J.O., Jobling, M. & Jørgensen, E.H. (1992) Temporal changes in daily food intake of Arctic charr, *Salvenilus alpinus* L., of different sizes monitored by radiography. *Aquaculture*, **106**, 51–61.

Paspatis, M. & Boujard, T. (1996) A comparative study of automatic feeding and self-feeding in juvenile Atlantic salmon (*Salmo salar*) fed diets of different energy levels. *Aquaculture*, **145**, 245–257.

Pérez, E. & Buisson, B. (1986) Research on the origin of the circadian activities in the course of the ontogenesis of the trout, *Salmo trutta* L.: the activities of the eggs and vesiculed alevins in constant conditions. *Biology Zentralblatt*, **105**, 609–613.

Peters, R.C. & Veeneklaas, R.J. (1992) Ultradian rhythms in fishes. In: *Rhythms in Fishes* (ed. M.A. Ali), pp. 51–61. Plenum Press, New York.

Pickett, G.D. & Pawson, M.G. (eds) (1994) *Sea Bass: Biology, Exploitation and Conservation*. Chapman & Hall, London.

Pittendrigh, C.S. (1981) Circadian system: entrainment. In: *Handbook of Behavioral Neurobiology 4, Biological Rhythms* (ed. J. Aschoff), pp. 95–124. Plenum Press, New York.

Prat, F., Zanuy, S., Bromage, N. & Carrillo, M. (1999) Effect of constant short and long photoperiod regimes on the spawning performance and sex steroid levels of female and male sea bass. *Journal of Fish Biology*, **54**, 125–137.

Rozin, P. & Mayer, J. (1961) Regulation of food intake in the goldfish. *American Journal of Physiology*, **201**, 968–974.

Saether, B.S., Johnsen, H.K. & Jobling, M. (1996) Seasonal changes in food consumption and growth of Arctic charr exposed to either simulated natural or a 12, 12 LD photoperiod at constant water temperature. *Journal of Fish Biology*, **48**, 1113–1122.

Sánchez-Vázquez, F.J. & Tabata, M. (1998) Circadian rhythms of demand-feeding and locomotor activity in rainbow trout. *Journal of Fish Biology*, **52**, 255–267.

Sánchez-Vázquez, F.J., Madrid, J.A. & Zamora, S. (1995a) Circadian rhythms of feeding activity in sea bass, *Dicentrarchus labrax* L.: dual phasing capacity of diel demand-feeding pattern. *Journal of Biological Rhythms*, **10**, 256–266.

Sánchez-Vázquez, F.J., Zamora, S. & Madrid, J.A., (1995b) Light-dark and food restriction cycles in Sea bass: effect of conflicting zeitgebers on demand-feeding rhythms. *Physiology and Behavior*, **58**, 705–714.

Sánchez-Vázquez, F.J., Madrid, J.A., Zamora, S., Iigo, M. & Tabata, M., (1996) Demand feeding and locomotor circadian rhythms in the goldfish, *Carassius auratus*: dual and independent phasing. *Physiology and Behavior*, **60**, 665–674.

Sánchez-Vazquez, F.J., Madrid, J.A., Zamora, S. & Tabata, M. (1997) Feeding entrainment of locomotor activity rhythms in the goldfish is mediated by a feeding-entrainable circadian oscillator. *Journal of Comparative Physiology*, **181A**, 121–132.

Sánchez-Vázquez, F.J., Azzaydi, M., Martínez, F.J., Zamora, S. & Madrid, J.A. (1998a) Annual rhythms of demand-feeding activity in sea bass: evidence of a seasonal phase inversion of the diel feeding pattern. *Chronobiology International*, **15**, 607–622.

Sánchez-Vazquez, F.J., Yamamoto, T., Akiyama, T., Madrid, J.A. & Tabata, M. (1998b) Selection of macronutrients by goldfish operating self-feeders. *Physiology and Behavior*, **65**, 211–218.

Sánchez-Vázquez, F.J., Yamamoto, T., Akiyama, T., Madrid, J.A. & Tabata, M. (1999) Macronutrient self-selection through demand-feeders in rainbow trout. *Physiology and Behavior*, **66**, 45–51.

Smith, I.P., Metcalfe, N.B., Huntingford, F.A. & Kadri, S. (1993) Daily and seasonal patterns in the feeding behaviour of Atlantic salmon (*Salmo salar* L.) in a sea cage. *Aquaculture*, **117**, 165–178.

Spencer, W.P. (1939) Diurnal activity rhythms in fresh-water fishes. *Ohio Journal of Science*, **39**, 119–132.

Spieler, R.E. (1992) Feeding-entrained circadian rhythms in fishes. In: *Rhythms in Fishes* (ed. M.A. Ali), pp. 137–148. Plenum Press, New York.

Sundararaj, B.I. & Vasal, S. (1976) Photoperiod and temperature control in the regulation of reproduction in the female catfish *Heteropneustes fossilis*. *Journal of the Fisheries Research Board of Canada*, **33**, 959–973.

Swift, D.R. (1955) Seasonal variations in the growth rate, thyroid gland activity and food reserves of brown trout (*Salmo trutta* L.). *Journal of Experimental Biology*, **32**, 751–764.

Tabata, M. (1992) Photoreceptor organs and circadian locomotor activity in fishes. In: *Rhythms in Fishes* (ed. M.A. Ali), pp. 223–234. Plenum Press, New York.

Tabata, M., Minh-Nyo, M. & Oguri, M. (1991) The role of the eyes and the pineal organ in the circadian rhythmicity in the catfish *Silurus asotus*. *Nippon Suisan Gakkaishi*, **57**, 607–612.

Thorpe, J.E. (ed.) (1978) *Rhythmic Activity of Fishes*. Academic Press, London.

Toguyeni, A., Fauconneau, B., Boujard, T., *et al.* (1997) Feeding behaviour and food utilisation in tilapia, *Oreochromis niloticus*: effect of sex-ratio and relationship with the endocrine status. *Physiology and Behavior*, **62**, 273–279.

Tveiten, H., Johnsen, H.K. & Jobling, M. (1996) Influence of maturity status on the annual cycles of feeding and growth of Arctic charr reared at constant temperature. *Journal of Fish Biology*, **48**, 910–924.

Villarreal, C.A., Thorpe, J.E. & Miles, M.S. (1988) Influence of photoperiod on growth changes in juvenile Atlantic salmon, *Salmo salar* L. *Journal of Fish Biology*, **33**, 15–30.

Wagner, G.F. & McKeown, B.A. (1985) Cyclical growth in juvenile rainbow trout, *Salmo gairdneri*. *Canadian Journal of Zoology*, **63**, 2473–2474.

Weisberg, S.B., Whalen, R. & Lotrich, V.A. (1981) Tidal and diurnal influence on food consumption of a salt marsh killifish *Fundulus heteroclitus*. *Marine Biology*, **61**, 243–246.

Zanuy, S., Carrillo, M. & Ruiz, F. (1986) Delayed gametogenesis and spawning of sea bass (*Dicentrarchus labrax* L.) kept under different photoperiod and temperature regimes. *Fish Physiology and Biochemistry*, **2**, 53–63.

Zaunreiter, M., Brandstätter, R. & Goldschmid, A. (1998a) Evidence for an endogenous clock in the retina of rainbow trout: I. Retinomotor movements, dopamine and melatonin. *NeuroReport*, **9**, 1205–1209.

Zaunreiter, M., Brandstätter, R. & Goldschmid, A. (1998b) Evidence for an endogenous clock in the retina of rainbow trout: II. Circadian rhythmicity of serotonin metabolism. *NeuroReport*, **9**, 1475–1479.

Zoufal, R. & Taborsky, M. (1991) Fish foraging periodicity correlates with daily changes of diet quality. *Marine Biology*, **108**, 193–196.

Chapter 9
Feeding Anticipatory Activity

F. Javier Sánchez-Vázquez and Juan Antonio Madrid

9.1 Introduction

When considering feed intake in fish, one issue that merits attention is the timing of intake. As discussed in Chapter 8, fish do not usually feed continuously throughout 24 hours but consume meals at certain times of the day. Regular feeding produces important changes in many functions, including behaviour and physiology. For instance, the timing of meals may have profound effects on growth (Boujard & Leatherland 1992; also see Chapter 10). In this chapter, we will analyse particular patterns of activity displayed by fish in response to periodic meals, their nature and the mechanisms that permit feeding anticipatory behaviour.

9.1.1 What is feeding anticipatory activity?

When presented with food on a periodic basis (e.g. a daily feeding cycle), most animals develop some activity in anticipation of the forthcoming meal within a few days of the feeding cycle being established. Animals become active several hours prior to feeding and may display increasing activity which is three- to one hundred-fold greater than baseline, sustained for at least 30 min and not followed by inactivity for more than 1 hour (Stephan & Davidson 1998). This is known as 'feeding anticipatory activity'. The internal process, which couples biological rhythms and meal cycles, is called 'feeding entrainment'.

The ability of animals to 'anticipate' feeding time was demonstrated in rats by Richter (1922). In fish, the first evidence was provided by Davis (1963), who showed that bluegill (*Lepomis macrochirus*) and largemouth bass (*Micropterus salmoides*) exhibited prefeeding activity consisting of increasing swimming activity 1–3 h before the food was delivered. However, since feeding in this case coincided with the onset of light, the relative contribution of light and feeding to the expression of prefeeding activity was unclear. In a later study, Davis and Bardach (1965) reported that killifish (*Fundulus heteroclitus*) fed at mid-day showed prefeeding activity and this persisted during exposure to constant light (LL). These studies represented a landmark; they suggested the existence of an activity which was time-coordinated with feeding times, although the exact role of learning and endogenous clocks remained unexplored.

During the past two decades, several laboratory studies on fish have shown that activity rhythms may be strongly associated with a daily meal cycle (see Chapter 8). Typically, feeding-entrained fish exhibit a peak of activity a few hours prior to feeding, regardless of the

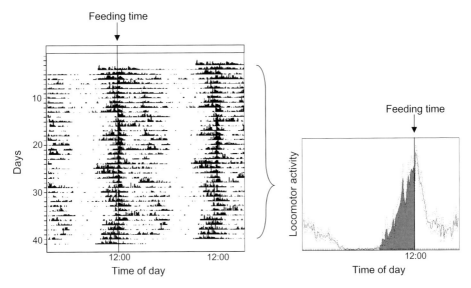

Fig. 9.1   Actogram of a locomotor activity record (left) and daily activity waveform (right) registered in an isolated goldfish (*Carassius auratus*) supplied with a single daily meal at 12:00 (about 1.5% of body weight) and exposed to constant light. For convenience, the activity record is double plotted, i.e. two consecutive days are shown side-by-side (48-h horizontal scale), at a resolution of 10 min. The height of each point represents the percentage of infrared light beam interruptions in 24 h. Note the increase of activity a few hours before feeding time: feeding anticipatory activity. (F.J. Sánchez-Vázquez, unpublished data.)

lighting conditions. For example, under continuous light and scheduled feeding the circadian pattern of activity may be modulated by feeding, with an active phase followed by a resting period (Fig. 9.1). Remarkably, the onset of the active phase precedes the time of feeding by 4 h. Such anticipatory activity is widely found among higher vertebrates (for a review, see Mistlberger, 1994) and has also been described in several fish species (for a review, see Spieler, 1992).

The first point we shall consider is why animals show anticipation to feeding and whether there is any adaptive value involved in anticipating a periodic meal.

9.1.2 Biological significance

Feeding-entrained rhythms, as other biological rhythms, provide fish with the ability to anticipate a recurrent event. The benefits of entraining to a regular meal cycle are basically similar to those of entraining to environment cycles such as day–night or temperature oscillations. If these events can be predicted, the animal can use this knowledge to its advantage by being prepared to cope with them.

Food is rarely constantly available in the wild, but has both spatial and temporal variation. Thus, foraging behaviour is commonly restricted to certain periods during which the abundance of prey is increased and the risk of predation is reduced. Since these factors often fluctuate periodically, the ability to predict a favourable period is clearly beneficial for the animal. Further, from a physiological standpoint, nutrient utilisation depends on the digestive and metabolic state of the animal (Potter *et al.* 1968; Krieger 1979; Suda & Saito 1979).

While a continuous active state is uneconomical, anticipation of an approaching meal will prepare the animal physiologically (e.g. circadian rhythms of duodenal motility entrain to feeding time and persist *in vitro*), which may improve food acquisition and utilisation (Comperatore & Stephan 1987).

9.1.3 Characteristics of feeding anticipatory activity

One interesting feature of feeding anticipatory activity is its gradual appearance. When fish are provided with a scheduled daily meal, they will (with time) exhibit increasing activity during the hours preceding feeding time. Feeding entrainment can be detected only after several feeding cycles, and when the meal is scheduled outside the active phase of the fish, feeding entrainment may take longer. For example, when a nocturnal fish is subjected to diurnal feed provision, fish usually require an initial period before the activity pattern shows any relation to feeding time. Such a relation is manifested by the progressive appearance of activity around feeding time (Fig. 9.2). In this case, the light/dark (LD) cycle is the dominant synchroniser of activity rhythms, and feeding seems to have little influence on the daily activity pattern. However, in the absence of an LD cycle, light entrainment disappears and activity rhythms are driven by feeding, with activity for some time before feeding and a peak im-

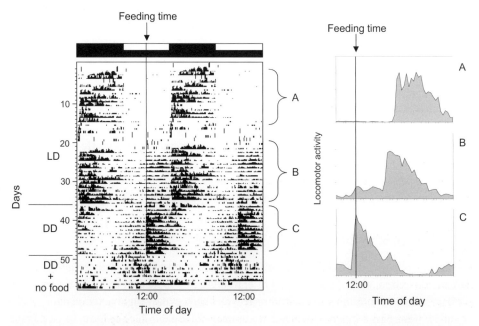

Fig. 9.2 Actogram of a locomotor activity record from an isolated goldfish exposed to a 12:12 LD cycle until day 36, and constant darkness (DD) thereafter. Food (about 1.5% of body weight) was supplied daily at 12:00 until day 49, and thereafter the fish was food-deprived to investigate feeding-entrained free-running rhythms. Actogram descriptions are as given in Fig. 9.1. Horizontal open and solid bars at the top of the graph represent the light and dark phase, respectively. Daily activity waveforms are given on the right side. Under LD conditions, light initially appeared to entrain activity rhythms more efficiently than meals (A), although some activity related to feeding gradually developed (B), gaining strength with time and being particularly evident under DD (C). Note that under DD + no food, the portion of activity that occurred directly after feeding (post-feeding activity) ceased and began to free-run. (Modified from Sánchez-Vázquez *et al.* 1997.)

mediately after. Activity prior to mealtime gradually increases from a few minutes to hours in the course of successive feeding cycles, and eventually becomes clearly distinguishable.

A further feature of feeding anticipatory activity is its persistence during food deprivation and its disappearance during *ad libitum* feeding (Sánchez-Vázquez *et al.* 1997). This indicates that restricted feeding is a prerequisite for an animal to express food-anticipatory bouts of activity.

9.1.4 Behavioural variables showing feeding anticipatory activity

Synchronisation to feeding has been reported to occur in several behaviours and metabolic activities of vertebrates (for a review, see Mistlberger 1994). Locomotor activity, feeding, drinking, body temperature and plasma corticosterone are among the behavioural and physiological variables that exhibit daily rhythms that can be entrained by a feeding cycle in higher vertebrates (Boulos & Terman 1980). In fish there is evidence for the synchronising effect of scheduled feeding (for review, see Spieler 1992), but the current list of studies on feeding-entrained behavioural rhythms is still short (Table 9.1).

Swimming activity is probably one of the most widely investigated behavioural variables (Kotrschal & Essler 1995), and the synchronisation of activity rhythms to food provision has been reported in several fish species such as loach (*Misgurnus anguillicaudatus*), mud-skipper (*Periopthalmus cantonensis*), killifish, bluegill, Atlantic salmon (*Salmo salar*), medaka (*Oryzias latipes*) and goldfish (*Carassius auratus*) (see Table 9.1). Despite the wide variety of experimental designs, fish are generally seen to respond to fixed meals by modulating their activity patterns in accordance to meal time. Goldfish, for instance, tend to be more diurnally active when fed during the day, and more nocturnally active when fed at night (Sánchez-Vázquez *et al.* 1997).

The timing and number of meals per day influences feeding entrainment; synchronisation to feeding is stronger with single daily meals scheduled in the active phase of fish, than with multiple daily meals. The time and place of food delivery also play an important role in feeding entrainment: animals must integrate this spatial and temporal information to optimise feeding. The ability to associate periodic food availability with particular feeding places has been demonstrated in insects, birds and mammals (Mistlberger 1994). Reebs (1993) tested time-place learning in a cichlid fish (*Cichlasoma nigrofasciatum*) fed four meals a day at different corners of the aquarium. The fish quickly learnt which corner gave food when food was provided in the same corner or in diagonally opposed corners; few visits to the other corners were recorded. When each of the four feeding sessions was associated with a different corner, fish failed to meet the criterion for time-place learning, suggesting that the experimental protocol and the signal-food association used may have interfered with each other.

Lever-pressing behaviour in demand-fed fish can also be entrained by limiting food availability to a few hours. Under such conditions, fish can only obtain food when their food demands coincide with the periods of food reward. For instance, when feeding is restricted to 1 h a day, goldfish significantly increase their level of response during the 30 min period immediately prior to the time feed becomes available (Gee *et al.* 1994). When goldfish were deprived of food for several days, the response rate declined rapidly but the temporal pattern of lever pressing remained during a few cycles. Demand-feeding behaviour can also be entrained by restricted feeding in sea bass; they quickly learn to activate the trigger during

Table 9.1 Examples of studies describing feeding-entrained, behavioural variables in fish.

Authors	Behavioural variable	Fish species	Feeding schedule
Davis (1963)	Locomotor activity	*Lepomis macrochirus* *Micropterus salmoides*	Single daily meal at the offset of different LD cycles
Davis & Bardach (1965)	Locomotor activity	*Fundulus heteroclitus* **Pseudopleuronectes americanus* **Microgadus tomcod* **Tautoglolabrus adespersus* **Stenotomus versicolor*	Single daily meal under LD and LL
Nishikawa & Ishibashi (1975)	Crawling activity	*Periophthalmus cantonensis*	One or two daily meals under LD and LL
Spieler & Noeske (1984)	Locomotor activity	*Carassius auratus*	Single daily meal at different times during LD
Spieler & Clougherty (1989)	Locomotor activity	*Carassius auratus*	Single daily meal under different LD, LL and DD
Sánchez-Vázquez et al. (1997)	Locomotor activity	*Carassius auratus*	Single daily meal at different times during LD and DD
Pradhan et al. (1989)	Phototactic behaviour	*Nemacheilys evezardi*	4 hours of restricted feeding
Weber & Spieler (1987)	Agonistic and reproductive behaviour	*Oryzias latipes*	Single daily meal at different times during LD
Boujard et al. (1993)	Demand-feeding	*Oncorhynchus mykiss*	Time-restricted feeding (4 h)
Reebs (1993)	Spatial activity distribution	*Cichlasoma nigrofasciatum*	Four daily meals delivered at different places
Naruse & Oishi (1994)	Locomotor activity	*Misgurnus anguillicaudatus*	Single daily meal at mid-L or mid-D
Juell et al. (1994)	Locomotor activity	*Salmo salar*	Two daily meals under natural LD
Bégout & Lagardère (1995)	Horizontal movements	*Sparus aurata*	Single daily meal under natural LD
Gee et al. (1994)	Operant feeding	*Carassius auratus*	1-h food availability under LD and LL
Sánchez-Vázquez et al. (1995)	Demand-feeding	*Dicentrarchus labrax*	Time-restricted feeding (4–12 h)
Boujard (1995)	Demand-feeding	*Silurus glanis*	Time-restricted feeding (12 h)
Sánchez-Vázquez et al. (1996)	Demand-feeding and locomotor activity	*Carassius auratus*	Scheduled and self-feeding regimes under LD
Azzaydi et al. (1998)	Demand-feeding and trigger activity	*Dicentrarchus labrax*	Three daily meals of 1-h duration each
Azzaydi et al. (1999)	Trigger activity	*Dicentrarchus labrax*	Three or two daily meals of 30-, 60-, 90- or 120-min duration

*Preliminary studies involving a very limited number of fish (one to three).

the periods of food reward, but fail to show sustained lever pressing during fasting (Sánchez-Vázquez *et al.* 1995).

Swimming and feeding activity rhythms are strongly influenced by feeding method (e.g. scheduled- versus self-feeding). For example, goldfish may show diurnal swimming activity when schedule-fed in the middle of the day, but change to nocturnal swimming and start to feed at night when given free access to food through self-feeders. However, swimming and feeding rhythms are not always in phase, as they seem to be controlled by a multioscillatory system (rather than a single oscillator) and they may entrain differently to the same stimuli (Sánchez-Vázquez *et al.* 1996).

Weber and Spieler (1987) investigated feeding entrainment of different types of behaviour in medaka (*Oryzias latipes*). They found that agonistic behaviour entrained to meal times, but courtship and egg-laying remained entrained to the LD cycle and were resistant to any phase-shifting effects of meal timing.

In summary, swimming and demand-feeding rhythms have been those recorded in most studies. Both rhythms quickly synchronise to restricted feeding, although demand-feeding is a more directed behaviour than swimming activity, which includes multiple behavioural components such as feeding, exploration, reproduction, etc. During fasting, on the other hand, trigger activations are rapidly extinguished – making the investigation of free-running rhythms difficult (Gee *et al.* 1994; Sánchez-Vázquez *et al.* 1995). In addition, if there is no restriction as regards the quantity of food during the period of food availability, the synchronising force of food may differ from that seen under scheduled feeding (in which a fixed amount of food is delivered) (Sánchez-Vázquez *et al.* 1995).

There have been very few in-depth studies into the nature of the endogenous timing mechanisms involved in feeding-entrainment in fish. In goldfish, there have been some attempts to investigate the mechanisms of feeding entrainment (Spieler & Noeske 1984; Spieler & Clougherty 1989; Sánchez-Vázquez *et al.* 1997), but the underlying system remains unresolved. There are several possible mechanisms to explain feeding anticipatory activity.

9.2 Models to explain anticipation to feeding

9.2.1 External versus internal origin

If animals used an external time cue associated with feeding to anticipate food access, they would quickly lose their capacity to predict the time of feeding on removal of the external cue. This does not occur however, so a hypothesis based solely on external cueing can be ruled out; animals maintained under a controlled environment, without any apparent variations that may serve to forewarn of food access, still show anticipation. As shown in Fig. 9.1, isolated fish, kept under constant light and temperature, become active several hours before mealtimes. This means that feeding anticipatory activity originates within the animal, and external signals are not required. Nevertheless, some environmental signals such as light, temperature or sound directly associated with food access, may influence the expression of anticipation to feeding. For example, the anticipatory activity of pigeons disappears if the pecking key is dimly lit during restricted feeding, but reappears when the key is lit continuously (Abe & Sugimoto 1987). In fish, feeding entrainment in self-fed trout (*Oncorhynchus mykiss*) disappeared when access to self-feeders was associated with dawn (Boujard *et al.* 1993).

As to internal models, those based on learning do not fully explain some functional properties of feeding entrainment. If animals anticipated food as part of an associate learning process, they would progressively tend to match their onset of activity with the start of meals. However, anticipation to feeding does not disappear with time. On the contrary, it develops with time, expands gradually and reaches a stable length (Fig. 9.2).

Current evidence supports the existence of an internal timing mechanism. There are basically two models: an hourglass or timer, and a self-sustained clock entrainable by feeding (Mistlberger 1994). The hourglass model has been suggested by physiologists and nutritionists, who propose that rhythmic behaviour related to feeding is driven by energy depletion and repletion cycles. This model, however, fails to explain reasonably why anticipation to feeding persists in fasted animals, which continue to become active around the previous feeding time. Thus, anticipation to feeding is not based on a timer mechanism triggered by cycles of nutrient uptake and metabolism. When food is removed and any external time cue associated with feeding disappears, anticipatory behaviour usually becomes slightly earlier or later as time progresses. Consequently, activity rhythms previously entrained by feeding start to free-run, and this points to the involvement of a circadian oscillator.

9.2.2 Self-sustained feeding-entrainable oscillator

The ability to anticipate a meal is thought to be mediated by an internal timing system or biological clock, which is synchronised to the environment (Aschoff 1986). Although the LD cycle is the most powerful environmental factor to entrain biological rhythms, periodic feeding can also act as a potent synchroniser or 'zeitgeber' (Edmonds 1977; Boulos & Terman 1980, see also Chapter 8). The entrainment of biological rhythms to light is mediated by a light-entrainable oscillator (LEO), which in rodents is localised in the suprachiasmatic nuclei (SCN) of the hypothalamus (Meijer & Rietveld 1989). Ablation of the SCN leads to the abolition, or serious disruption, of light-entrained free-running rhythms in rats (Moore & Eichler 1972; Stephan & Zucker 1972) and birds (Ebihara & Kawamura 1981). In contrast, feeding entrainment is not abolished in SCN-ablated rats, which suggests the existence of a feeding-entrainable oscillator (FEO) outside the SCN (Stephan *et al.* 1979). In fish, however, there is little information on the FEO and basic questions relating to the circadian nature and the mechanisms of feeding entrainment have not been answered (Spieler 1992; Sánchez-Vázquez *et al.* 1997). Two main experimental approaches can be used to investigate feeding-entrainment in fish: fasting to look at free-running rhythms, and altering feeding time to study resynchronisation.

Free-running rhythms under constant conditions

Experiments of the first type are designed to determine the endogenous nature of feeding entrainment. Feeding-entrained rhythms free-run during food deprivation and under constant conditions (e.g. continuous darkness and temperature), indicating that food anticipation is driven by a self-sustained FEO. To avoid the possible influence of any LEO, feeding entrainment is investigated under constant conditions before the removal of food. To date, the endogenous oscillatory nature of feeding entrainment has been shown in goldfish (Spieler & Clougherty 1989; Sánchez-Vázquez *et al.* 1997). It is noteworthy that feeding anticipatory activity usually free-runs during fasting; that is, when fish are deprived of food, the post-feeding activity observed after feeding ceases, whereas the pre-feeding activity (anticipatory activity) persists and begins to free-run (Fig. 9.2). Following food deprivation, resynchroni-sation to a new meal cycle is gradual, and usually involves some transient cycles (Fig. 9.3). This finding further supports the involvement of an endogenous clock in feeding anticipatory activity.

Resynchronisation after phase shifts of feeding time

The second technique involves advancing or delaying the timing of meals; this can be used to evaluate the contribution of endogenous mechanisms and the resetting properties of feeding entrainment. If meals have a strong influence on the FEO, the adaptation of the daily rhythm to the newly established meal cycle will be immediate. However, if a feeding-entrainable clock is involved, activity rhythms will resynchronise gradually, depending on the entraining strength of meals. Further there will be transient cycles to keep a stable phase relationship between the active phase and feeding time. Some attempts have been made to investigate this in fish, although the results reported are contradictory. Davis (1963) investigated the effect of

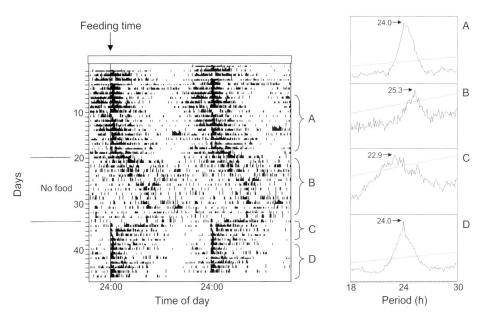

Fig. 9.3   Actogram (left) and periodogram analysis (right) of locomotor activity of an isolated goldfish exposed to continuous light (LL). Food (about 1.5% of body weight) was supplied daily at 24:00 until day 19 (A). From day 20 to day 34, no food was provided, to enable study of free-running rhythms (B). Thereafter, the meal cycle was re-established to investigate resynchronisation to feeding (C and D). Actogram descriptions are as given in Fig. 9.1. Chi-square periodograms (dotted line represents a confidence level of 95%, i.e. above the line is significant) during scheduled feeding (A and D), revealed a peak of 24 h, while during food deprivation activity rhythms free-run with a 25.3-h period (B). Note that during the first few days after the re-establishment of the meal cycle, fish resynchronise gradually and the periodogram reveals a peak of 22.9 h (C). (F.J. Sánchez-Vázquez, unpublished data.)

changing light and feeding time, but failed to observe shifts in the daily activity rhythm. This provided evidence that feeding time was not involved in the regulation of the pre-dawn peak of activity. When feeding was delayed or advanced by 6 h under LL, some fish were active at the time food was delivered on the first day, and in the succeeding days they showed a stepwise delay/advance to progressively compensate for the change in feeding time (Davis & Bardach 1965). In a more recent study, a shift in the feeding cycle by 9 h under DD resulted in gradual resynchronisation; fish either advanced or delayed their activity rhythms, and the previously observed phase relationships were re-established after some transient cycles (Sánchez-Vázquez *et al.* 1997). These results indicate that periodic feeding entrains a clock-controlled activity in fish, which keeps a stable phase angle of entrainment to feeding.

Higher vertebrates are believed to possess a FEO in addition to a LEO, which can be dissociated into independently oscillating units (Stephan 1986; Phillips *et al.* 1993; Mistlberger 1994; Rashotte & Stephan 1996). In fish, the properties of feeding entrainment provide support for the existence of a self-sustained FEO, although it is still uncertain whether or not fish possess a separate FEO, in addition to a LEO. Feeding-entrainment could be explained in mechanistic terms based on the properties of a complex single oscillator, which in turn would be entrained by both light and food. However, this is not the case for rats and birds, since the existence of a functionally and anatomically separate FEO has been demonstrated. However, at present it cannot be concluded that fish have an independent FEO, since it cannot be excluded that fish use other mechanisms to anticipate food based on the information supplied by a master LEO. Currently, information is compatible with the following two hypotheses: (i) that fish have separate but tightly coupled light- and food-entrainable oscillators; or (ii) that they have a single oscillator that is entrainable by both light and food (one synchroniser eventually being stronger than the other) (Sánchez-Vázquez *et al.* 1997).

9.3 Synchronising stimulus for feeding entrainment

Although many attempts have been made to identify the synchroniser, its transduction site and the afferent pathways responsible for feeding entrainment, matters remain unresolved (Spieler 1992; Mistlberger 1994). The contributions of pre-ingestive (visual, olfactory), ingestive (mechano- and chemoreceptors), digestive (enzymatic) and post-absorptive (hormonal, metabolic) factors are still uncertain.

Meal size and composition affect feeding entrainment in mammals. For example, rats only anticipate a palatable meal if it is large, and fail to anticipate non-nutritive meals (Mistlberger & Rusak 1987). The energy content of the diet seems to play an important role in feeding entrainment, since animals only respond to shifts in timing for diets above a certain energy threshold. The ingestion of bulky diets (diluted with cellulose) does not influence the resetting of the FEO, suggesting that caloric restriction rather than gastric distension acts as a synchroniser (Stephan 1997). The expression of feeding anticipatory activity requires an energy shortfall, to prime metabolism toward a catabolic state (Escobar *et al.* 1998). In other words, food restriction seems to be a prerequisite for the expression of feeding anticipatory activity, facilitating the assembly of the FEO, which may be dissociated under *ad libitum* conditions (when food is constantly available animals obviously do not need to anticipate feeding time). This may explain why feeding-entrainment appears and disappears when fish are transferred from self- to scheduled-feeding, and vice versa (Sánchez-Vázquez *et al.* 1996).

Food composition is a potential synchroniser for feeding entrainment, and macronutrients may differentially affect feeding entrainment: vegetable oil is far less effective than glucose in phase shifting the FEO in rats (Stephan & Davidson 1998). In fish, the role of some essential amino acids (tryptophan, tyrosine and phenylalanine) was investigated by Spieler *et al.* (1987). A diet deficient in these amino acids did not affect the ability of goldfish to resynchronise after a phase shift of mealtime. The authors concluded that the diet *per se* had little influence, and that other stimuli – such as pre-feeding processes (e.g. disturbance, social interaction) – may play a major role in resynchronisation. Some authors have suggested that there is a synchronising effect of group behaviour, so that feeding and circadian rhythms are socially facilitated (Kavaliers 1981; Spieler & Clougherty 1989). Nevertheless, since fish held in isolation are able to synchronise to meal times and express robust feeding anticipatory activity, social interactions may be considered as a reinforcing factor rather than an entraining stimulus (Spieler 1992).

There are many other potential candidates, apart from food itself, that may mediate feeding entrainment. Some of these are directly related to food intake. There is a relationship, for example, between food intake and the hormone melatonin. Melatonin is synthesised in several tissues including the enterochromaffin cells of the gastrointestinal tract. Thus, melatonin may transduce nutritional information into a chemical message (Huether 1994). The facts that food restriction increases circulating melatonin, the amount of melatonin in the digestive tract is higher than in any other organ including the pineal, and oral or portal, but not intraperitoneal or jugular administration of tryptophan (a melatonin precursor) causes an increase in melatonin, suggest that this hormone may be used as an internal synchroniser for the FEO. Although melatonin was not detected in the gut of rainbow trout (Gern *et al.* 1986), Bubenik and Pang (1997) reported melatonin in various segments of the gut (oesophagus, stomach, proximal and distal intestine) in seven species of lower vertebrates including a chondrostean (sturgeon, *Acipenser fulvescens*) and teleostean fish, carp, *Cyprinus carpio*, and trout. The function of the gut melatonin in fish, and its contribution, if any, to feeding entrainment is still unclear and needs further research.

9.4 Applications

9.4.1 Temporal integration

An examination of physiological functions in fish, as in other vertebrates, gives the impression of periodic changes, and animals typically exhibit biological rhythms synchronised to environment cycles. Periodic feeding is one of the most potent synchronisers, meaning that daily meals can be used to entrain circadian rhythms. This entraining ability of feeding cycles can be used to control the phase of a given rhythm. Feeding time acts as a time cue for temporal integration. For example, when food access is limited to certain times, most digestive and metabolic rhythms are reset and maintain a stable phase relation with feeding time. In contrast, if feeding is random, and food access cannot be predicted, behavioural and physiological rhythms may not be in phase. As a result, the internal temporal organisation may be deficient and some rhythms may change their phase relation or even free-run.

Scheduled feeding may be used to change and stabilise the phasing of some behavioural and hormonal rhythms in fish (Spieler 1992). Sea bass is a fish species characterised by dual phasing (i.e. the ability to shift from diurnal to nocturnal rhythms and vice versa). When the feeding of fish is restricted to the light or dark period contrary to that of their previous feeding phase (i.e. diurnal fish are allowed to feed only at night and nocturnal fish during the day), stable phase inversions from one type to the other may be induced (Sánchez-Vázquez *et al.* 1995). Such inversions can be so stable that, after restricted feeding, fish do not revert to their previous diurnal or nocturnal behaviour, or fish forced to feed out of phase may later return to their previous behaviour (Fig. 9.4). Summarising, we should not conclude that sea bass, or any other fish, simply respond to changing feeding schedules by phase shifting their rhythms because restricted feeding may not be effective in changing the diurnal/nocturnal behaviour.

A lack of temporal integration may result in an increase in feed wastage. During periodic feeding, fish can anticipate feeding and be prepared to deal with it. However, if the timing of

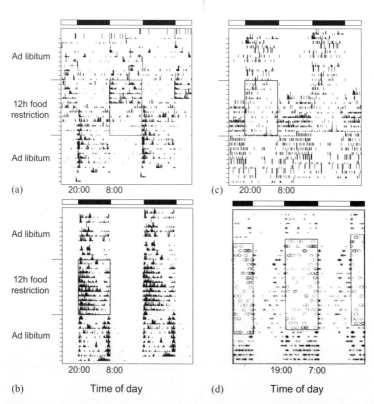

Fig. 9.4 Actograms of demand-feeding activity in sea bass (a–c) and European catfish (d) exposed to a 12:12 LD cycle and subjected to restricted feeding. Actogram descriptions are as given in Fig. 9.1. Fish were given free access to food through self-feeders, except in restrictive feeding periods (delimited by a rectangular box), during which food demands were rewarded only during the light or dark phase opposite to the previous feeding phase. Note that restricted food access induces diurnal fish to change to nocturnal, and vice versa (a, b), although some fish may not invert the phase of their rhythms (c) or may revert to their previous nocturnal behaviour (d). (Modified from Sánchez-Vázquez *et al.* 1995 and Boujard 1995.)

food delivery is random, they are unable to predict feeding and the wastage of food is likely to increase.

9.4.2 Anticipation of meals as an indicator of fish appetite

Farmed fish are usually fed by means of two feeding systems: demand-feeding, in which fish can decide the size and/or the time of their meal; and hand or automatic-feeding, in which both the size and the time of meals are fixed (see Chapters 3 and 14). In the latter system, food delivery is usually controlled by a clock and so feeding is periodic. In demand-feeding, access to food can also be time-restricted. Curiously, fish previously adapted to operating self-feeders, continue to activate a trigger when subjected to automatic-feeding (without a direct food reward), thus providing information that may be used as an 'appetite sensor'. Recent investigations on sea bass demonstrated that trigger activations may be useful for adjusting feeding schedules (Madrid *et al.* 1997; Azzaydi *et al.* 1998, 1999). Madrid *et al.* (1997) used a system that combined the recording of demand-feeding activity and detection of uneaten food pellets, and reported a hyperbolic relation between the two parameters. This means that above a certain level of trigger activations, no uneaten food is detected. When feeding in bass was restricted to 3 h each day (three meals of 1-h duration each), trigger activations were closely associated with these feeding periods, but fish activity increased 30 to 120 min before feeding in anticipation of the meal time. Further, anticipatory activity is not equally distributed among meals but is proportional to the number of rewarded food demands, the shortest anticipation (30–60 min) being associated with the least preferred meal time (Azzaydi *et al.* 1998). If food supply is divided into meals of different sizes, trigger activity differs according to feeding rhythms at different times of the day and season (Azzaydi *et al.* 1999). To summarise, anticipation of meals and trigger activity may have potential as a tool to test whether a given feeding schedule matches fish appetite.

9.4.3 Distribution of fish

Fish fed periodically tend to move horizontally and vertically to approach the feeder before feeding time. For example, Bégout and Lagardère (1995) reported that the distribution of sea bream (*Sparus aurata*) in an earthen pond was not uniform, fish usually gathering in areas near the inlet pipe. At about 1 h before the daily meal, fish approached the feeding area, and then returned to their previous positions after the meal (Fig. 9.5). As to vertical distributions, many fish exhibit diel changes in depth responding to variations in environmental factors such as light, predators and food availability (Neilson & Perry 1990; see Chapter 7). Despite the relatively large number of reports devoted to this problem, fundamental questions concerning the nature of vertical migrations remain (Kerfoot 1985). Daily changes in depth are often related to illumination conditions: fish ascend to the surface at night and descend to the bottom during the day. However, scheduled feeding seems to influence this spatial-temporal structure, modulating the activity of the fish. In Atlantic salmon, vertical distribution is influenced by feeding so that fish swim towards the surface and the feeding area during meals, but move downwards as they become sated (Juell *et al.* 1994; Ferno *et al.* 1995; see also Chapter 7).

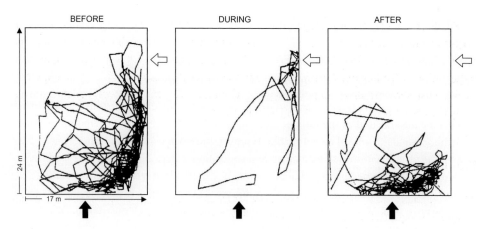

Fig. 9.5   Spatial distribution of sea bream fitted with ultrasonic transducers, and released in an earthen pond of 400 m² with 600 fish. Fish were fed daily during 5 min at the feeding areas (indicated by an open arrow). Note that the fish started to approach the feeding area about 1 h before feeding time (left panel), remained at this place during the meal (central panel), and went back to the their previous resting areas after the meal (right panel) (indicated by a solid arrow). (Modified from Bégout & Lagardère 1995.)

Under laboratory conditions, diel vertical movements can also be observed in small aquaria. For example, rainbow trout may swim close to the bottom during the day, but predominantly at the surface at night (Sánchez-Vázquez & Tabata 1998). Under a meal cycle, however, activity rhythms are modulated by feeding time, which concentrates most of the daily activity. Fish close to the surface may anticipate feeding, this consisting of a peak of activity prior to feeding, followed by a drop at meal times. In contrast, fish lower in the water column may fail to show anticipatory activity, but increase activity directly after food delivery. In other words, fish wait for food near the surface (where the food arrives) and so feeding anticipatory activity is displayed here, and then they descend to eat the food pellets. Immediately after feeding, the decrease in activity at the bottom could be seen as a passive response of fish to the feeding stimulus (i.e. a masking effect), while at the surface anticipation to feeding time is driven endogenously by a FEO (Sánchez-Vázquez *et al.* 1997).

9.5 Conclusions

There is strong evidence that periodic food availability synchronises many behavioural rhythms in fish. Fish typically respond to daily feeding schedules by increasing their activity a few hours prior to feeding; this is termed feeding anticipatory activity. Learning processes and hourglass or timer models fail to explain some properties of feeding entrainment and so they may be ruled out as sole contributor. The fact that anticipation of feeding persists and free-runs in fasting animals under constant conditions, and feeding-entrained rhythms resynchronise and keep a stable phase relation after shifting feeding times, suggests that feeding entrainment is endogenously driven by an internal timing system (the so-called feeding-entrainable oscillator). However, basic questions relating to the entraining stimulus, the transduction site and the afferent pathways of feeding entrainment remain unresolved and need further investigation.

Finally, scheduled feeding may have practical implications for the temporal integration of digestive and metabolic rhythms, which can be reset or phase-shifted by feeding cycles. The spatial distribution of fish, their approach to feeding areas, and the appearance of activity in anticipation of feeding may provide information about fish appetite. This may be important because fish showing strong anticipatory behaviour are likely to catch food pellets more eagerly than do conspecifics less interested in feeding.

9.6 References

Abe, H. & Sugimoto, S. (1987) Food-anticipatory response to restricted food access based on the pigeon's biological clock. *Animal Learning and Behaviour*, **15**, 353–359.

Aschoff, J. (1986) Anticipation of a meal: a process of 'learning' due to entrainment. *Monographia Zoologica Italia*, **20**, 195–219.

Azzaydi, M., Madrid, J.A., Zamora, S., Sánchez-Vázquez, F.J. & Martínez, F.J. (1998) Effect of three feeding strategies (automatic, ad libitum demand-feeding and time-restricted demand-feeding) on feeding rhythms and growth in European sea bass (*Dicentrarchus labrax*). *Aquaculture*, **163**, 285–296.

Azzaydi, M., Martínez, F.J., Zamora, S., Sánchez-Vázquez, F.J. & Madrid, J.A. (1999) Effect of meal size modulation on growth performance and feeding rhythms in European sea bass (*Dicentrarchus labrax*). *Aquaculture*, **170**, 253–266.

Bégout, M.L. & Lagardère, J.P. (1995) An acoustic telemetry study of seabream (*Sparus aurata* L.): first results on activity rhythm, effects of environmental variables and space utilization. *Hydrobiologia*, **300/301**, 417–423.

Boujard, T. (1995) Diel rhythms of feeding activity in the European catfish, *Silurus glanis. Physiology and Behaviour*, **58**, 641–645.

Boujard, T. & Leatherland, J.F. (1992) Circadian rhythms and feeding time in fishes. *Environmental Biology of Fish*, **35**, 109–131.

Boujard, T., Bett., S. Lin, L. & Leatherland, J.F. (1993) Effect of restricted access to demand-feeders on diurnal pattern of liver composition, plasma metabolites and hormone levels in *Oncorhynchus mykiss. Fish Physiology and Biochemistry*, **11**, 337–344.

Boulos, Z. & Terman, M. (1980) Food availability and daily biological rhythms. *Neuroscience and Biobehavioural Reviews*, **4**, 119–131.

Bubenik, G.A. & Pang, S.F. (1997) Melatonin levels in the gastrointestinal tissues of fish, amphibians, and a reptile. *General and Comparative Endocrinology*, **106**, 415–419.

Comperatore, C.A. & Stephan, F.K. (1987) Entrainment of duodenal activity to periodic feeding. *Journal of Biological Rhythms*, **2**, 227–242.

Davis, R.E. (1963) Daily 'predawn' peak of locomotor activity in fish. *Animal Behaviour*, **12**, 272–283.

Davis, R.E. & Bardach, E. (1965) Time-co-ordinated prefeeding activity in fish. *Animal Behaviour*, **13**, 154–162.

Ebihara, S. & Kawamura, H. (1981) The role of the pineal organ and the suprachiasmatic nucleus in the control of circadian locomotor rhythms in the Java Sparrow, *Padda oryzivora. Journal of Comparative Physiology*, **141 (A)**, 207–214.

Edmonds, S. (1977) Food and light as entrainers of circadian running activity in the rat. *Physiology and Behavior*, **18**, 915–919.

Escobar, C. Díaz-Muñoz, M. Encinas, F. & Aguilar-Roblero, R. (1998) Persistence of metabolic rhythmicity during fasting and its entrainment by restricted feeding schedules in rats. *American Journal of Physiology*, **274**, R1309–R1316.

Ferno, A., Huse, I., Juell, J.E. & Bjordal, A. (1995) Vertical distribution of Atlantic salmon (*Salmo salar* L.) in net pens: trade-off between surface light avoidance and food attraction. *Aquaculture*, **132**, 285–296.

Gee, P., Stephenson, D. & Wright, D.E. (1994) Temporal discrimination learning of operant feeding in goldfish (*Carassius auratus*). *Journal of Experimental Analytical Behaviour*, **62**, 1–13.

Gern, W.A., Duvall, D. & Nervina, J.M. (1986) Melatonin: a discussion of its evolution and actions in vertebrates. *American Zoologist*, **26**, 985–996.

Huether, G. (1994) Melatonin synthesis in the gastrointestinal tract and the impact of nutritional factors on circulating melatonin. *Annals of the New York Academy of Science*, **719**, 146–158.

Juell, J.E., Ferno, A., Furevik, D.M. & Huse, I. (1994) Influence of hunger level and food availability on the spatial distribution of Atlantic salmon (*Salmo salar* L.) in sea cages. *Aquaculture and Fisheries Management*, **25**, 439–451.

Kavaliers, M. (1981) Period lengthening and disruption of socially facilitated circadian activity rhythms of goldfish by lithium. *Physiology and Behavior*, **27**, 625–628.

Kerfoot, W.C. (1985) Adaptative value of vertical migration. In: *Migration: Mechanisms and Adaptative Significance* (ed. M.A. Rankin), *Contribution to Marine Science Supplement*, **27**, 92–113.

Kotrschal, K. & Essler, H. (1995) Goals and approaches in the analysis of locomotion in fish, with a focus on laboratory studies. *Review of Fisheries Science*, **3**, 171–200.

Krieger, D.T. (1979) Regulation of circadian periodicity of plasma corticosteroid concentrations and of body temperature by time of food presentation. In: *Biological Rhythms and their Central Mechanisms* (eds M. Suda, O. Hayaishi & H. Nakagawa), pp. 247–259. Elsevier, Amsterdam.

Madrid, J.A., Azzaydi, M., Zamora, S. & Sánchez-Vázquez (1997) Continuous recording of uneaten food pellets and demand-feeding activity: a new approach to studying feeding rhythms in fish. *Physiology and Behavior*, **62**, 689–695.

Meijer, J.H. & Rietveld, W.J. (1989) Neurophysiology of the SCN circadian pacemaker in rodents. *Physiological Reviews*, **69**, 671–707.

Mistlberger, R.E. (1994) Circadian food-anticipatory activity: formal models and physiological mechanisms. *Neuroscience and Biobehavioural Reviews*, **18**,171–195.

Mistlberger, R.E. & Rusak, B. (1987) Palatable daily meals entrain anticipatory activity rhythms in free-feeding rats: dependence on meal size and nutrient content. *Physiology and Behavior*, **41**, 219–226.

Moore, R.Y. & Eichler, V.B. (1972) Loss of circadian adrenal corticosterone rhythm following suprachiasmatic nucleus lesions in the rat. *Brain Research*, **71**, 17–33.

Naruse, M. & Oishi, T. (1994) Effects of light and food as zeitgebers on locomotor activity rhythms in the loach, *Misgurnus anguillicaudatus. Zoological Science*, **11**, 113–119.

Neilson, J.D. & Perry, R.I. (1990) Diel vertical migrations of marine fishes: an obligate or facultative process? *Advances in Marine Biology*, **26**, 115–168.

Nishikawa, M. & Ishibashi, T. (1975) Entrainment of the activity rhythm by the cycle of feeding in the mud-skipper, *Periopthalmus cantonensis. Zoological Magazine*, **84**, 184–189.

Phillips, D.L., Rautenberg, W., Rashotte, M.E. & Stephan, F.K. (1993) Evidence for a separate food-entrainable circadian oscillator in the pigeon. *Physiology and Behavior*, **53**, 1105–1113.

Potter, Van R., Baril, E.F., Watanabe, M. & Whittle, E.D. (1968) Systematic oscillations in metabolic functions in liver from rats adapted to controlled feeding schedules. *Federation Proceedings*, **27**, 1238–1245.

Pradhan, P.K., Pati, A.K. & Agarwal, S.M. (1989) Meal scheduling modulation of circadian rhythm of phototactic behaviour in cave dwelling fish. *Chronobiology International* **6**, 245–249.

Rashotte, M. & Stephan, F.K. (1996) Coupling between light- and food-entrainable circadian oscillators in pigeons. *Physiology and Behavior*, **59**, 1005–1010.

Reebs, S.G. (1993) A test of time-place learning in a cichlid fish. *Behavioural Processes*, **30**, 273–282.

Richter, C.P.A. (1922) A behavioristic study of the activity of the rat. *Comparative Psychological Monographs*, **1**, 1–55.

Sánchez-Vázquez, F.J. & Tabata, M. (1998) Circadian rhythms of demand-feeding and locomotor activity in rainbow trout. *Journal of Fish Biology*, **52**, 255–267.

Sánchez-Vázquez, F.J., Zamora, S. & Madrid, J.A. (1995) Light-dark and food restriction cycles in sea bass: effect of conflicting zeitgebers on demand-feeding rhythms. *Physiology and Behavior*, **58**, 705–714.

Sánchez-Vázquez, F.J., Madrid, J.A., Zamora, S., Iigo, M. & Tabata, M. (1996) Demand-feeding and locomotor circadian rhythms in the goldfish, *Carassius auratus*: dual and independent phasing. *Physiology and Behavior*, **60**, 665–674.

Sánchez-Vázquez, F.J., Madrid, J.A., Zamora, S. & Tabata, M. (1997) Feeding entrainment of locomotor activity rhythms in the goldfish is mediated by a feeding-entrainable circadian oscillator. *Journal of Comparative Physiology*, **181A**, 121–132.

Spieler, R.E. (1992) Feeding-entrained circadian rhythms in fishes. In: *Rhythms in Fishes* (ed. A. Ali), pp. 137–147. Plenum Press, New York.

Spieler, R.E. & Clougherty, J.J. (1989) Free-running locomotor rhythms of feeding-entrained goldfish. *Zoological Science*, **6**, 813–816.

Spieler, R.E. & Noeske, T. (1984) Effect of photoperiod and feeding schedule on diel variations of locomotor activity, cortisol and thyroxine in goldfish. *Transactions of the American Fisheries Society*, **113**, 528–539.

Spieler, R.E., Noeske-Hallin, T.A., DeRosier, T.A. & Poston, H.A. (1987) Some dietary amino acids and meal-feeding phase shifts of locomotor activity. *Medical Science Research*, **15**, 921–922.

Stephan, F.K. (1986) Coupling between feeding and light-entrainable circadian pacemakers in the rat. *Physiology and Behavior*, **38**, 537–544.

Stephan, F.K. (1997) Calories affect zeitgeber properties of the feeding entrainable circadian oscillator. *Physiology and Behavior*, **62**, 995–1002.

Stephan, F.K. & Davidson, A.J. (1998) Glucose, but not fat, phase shifts the feeding-entrainable circadian clock. *Physiology and Behavior*, **65**, 227–288.

Stephan, F.K. & Zucker, I. (1972) Circadian rhythms in drinking behaviour and locomotor activity of rats are eliminated by hypothalamic lesions. *Proceedings of the National Academy Science, USA*, **69**, 1583–1586.

Stephan, F.K., Swann, J.M. & Sisk, C.L. (1979) Entrainment of circadian rhythms by feeding schedules in rats with suprachiasmatic nucleus lesions. *Behavioral and Neural Biology*, **25**, 545–554.

Suda, M. & Saito, M. (1979) Coordinative regulation of feeding behaviour and metabolism by a circadian timing system. In: *Biological Rhythms and their Central Mechanisms* (eds M. Suda, O. Hayaishi & H. Nakagawa), pp. 263–271. Elsevier, Amsterdam.

Weber, D.N. & Spieler, R.E. (1987) Effects of the light-dark cycle and scheduled feeding on behavioural and reproductive rhythms of the cyprinodont fish, Medaka, *Oryzias latipes. Experientia*, **43**, 621–624.

Chapter 10

Effects of Feeding Time on Feed Intake and Growth

Valérie Bolliet, Mezian Azzaydi and Thierry Boujard

10.1 Introduction

Many fish species display rhythmic patterns of feeding when allowed free access to food. Fish have been generally classified as diurnal, nocturnal or crepuscular feeders, but some species change their feeding rhythms during the course of the year, depending on seasonal changes in photoperiod and/or temperature (Landless 1976; Eriksson and Alanärä 1992; Fraser *et al.* 1993, 1995; Sanchez-Vazquez *et al.* 1998; see Chapter 8).

Daily and seasonal changes in feeding activity may reflect adaptive responses to food availability and predators in the wild. This might also imply that nutrient partitioning and metabolism are not constant throughout the day, and that there is seasonal dependency. Further, it might be hypothesised that feeding activity could depend upon endogenous mechanisms, and occur when the fish is physiologically best prepared to use nutrients efficiently.

This raises the question of the effect of feeding time on growth and nutrient partitioning in cultivated fish – a question of prime interest to fish farmers with respect to the establishment of optimal feeding protocols. In traditional practice the quantities of feed delivered have usually been regulated according to temperature and fish size, but little attention has been given to the time of food delivery (see Chapter 14). Feed has often been distributed during normal working hours, regardless of any feeding rhythm of the cultivated species: thus, the timing of feed provision may not match the peak of appetite, and this could lead to poor growth, poor feed utilisation and feed wastage.

Feeding frequency is also known to affect growth and food conversion efficiency (Boujard & Leatherland 1992a; Jarboe & Grant 1997). For example, food conversion efficiency of channel catfish, *Ictalurus punctatus*, increased when the number of meals was increased from two to four at a fixed ration of 2% body weight per day (bw/day) (Greenland & Gill 1979). Further, rainbow trout, *Oncorhynchus mykiss*, fed continuously showed a better utilisation of carbohydrates than fish fed four meals per day at a restricted rate of 2% bw/day (Hung & Storebakken 1994).

Feeding time as used in this chapter is concerned with discrete meals delivered at different times of the light-dark cycle. We will focus on the effect of feeding time on feed intake, growth, nutrient utilisation and flesh quality, and will discuss possible mechanisms by which feeding time might affect growth.

10.2 Effect of feeding time on growth

Several studies dealing with the effect of feeding time on growth have been performed (Table 10.1). In siluriforms, which are considered to be nocturnal, feed provisioning during the hours of darkness may enhance growth. Kerdchuen and Legendre (1991) reported that African catfish, *Heterobranchus longifilis*, fed 3% bw/day during the night grew better than those fed the same amount of food during the day. A similar positive growth effect of feeding in darkness has been demonstrated in channel catfish, Indian catfish, *Heteropneustes fossilis,* and African catfish, *Clarias lazera* and *Heterobranchus longifilis* (Stickney & Andrews 1971; Hogendoorn 1981; Sundararaj *et al.* 1982; Baras *et al.* 1998). In the characin, *Piaractus brachyponus,* fish reared at stocking densities of 31–34 kg/m³ grew better when fed continuously during the night (20:30–08:30) than when fed during daylight hours (08:30–20:30) (Baras *et al.* 1996).

In contrast, feeding during daylight hours seems to enhance weight gain in rainbow trout. Although this species may change its feeding behaviour on a seasonal basis (see Eriksson & Alanärä 1992), it is generally believed that feeding in rainbow trout mostly occurs during the light phase, with a main peak of feeding activity at dawn and a secondary one around dusk (Boujard & Leatherland 1992a,b; Sanchez-Vazquez & Tabata 1998). In a study carried out during winter–spring, best growth was observed in rainbow trout fed at dawn, poorest growth in those fed at midnight, and growth rates of fish fed at mid-day or dusk were intermediate (Boujard *et al.* 1995). A later study performed on this species during the same period of the year confirmed that feeding fish at dawn resulted in better growth than feeding at midnight (Gélineau *et al.* 1996).

Taken together, the results suggest that the optimal time for feeding for promotion of good growth corresponds to the time of the natural daily peak of feeding activity. However, Zoccarato *et al.* (1993) reported better growth in rainbow trout fed at 16:00 than in those fed at dawn, whereas in another study fish fed at either mid-day or dusk were reported to display better growth than those fed at dawn (Reddy *et al.* 1994). Further, in several studies performed on carp, *Cyprinus carpio*, sea bass, *Dicentrarchus labrax* and channel catfish there was no marked effect of feeding time on growth (Noeske & Spieler 1984; Carillo *et al.* 1986; Perez *et al.* 1988; Robinson *et al.* 1995; Boujard *et al.* 1996; Jarboe & Grant 1996, 1997).

However, in none of these studies was food consumption monitored accurately. In addition, some of the discrepancies among studies may be attributable to the different conditions employed, for example with respect to size and age of the fish, rearing systems, stocking density, temperature, lighting conditions and ration levels. With such a variety of experimental conditions, one should not expect fish to be in the same behavioural or physiological state, and different responses to the time of feeding might be predicted.

In support of this are the results of a study carried out by Reddy *et al.* (1994), involving groups of rainbow trout fed 1.5, 2 or 2.5% bw at dawn, mid-day or dusk from August to January. The authors reported an effect of feeding time in the 1.5 and 2% bw groups, but not in the fish fed the 2.5% bw ration. Baras and colleagues (1998) investigated the effect of three different feeding schedules (during the day, during the night, or continuously) on the growth of African catfish of different ages and weights (31–101 days; 0.3–30 g). Fish fed at night had higher growth rates, lower mortality and were less heterogeneous than those fed either during the day or continuously. Although the same trends were observed in all sizes of fish,

Table 10.1 Examples of studies on the effect of feeding time on growth in fish.

Species	Season	Photoperiod	Feeding time (FT)	Optimum FT	Reference
Carassius auratus	March–May	12L:12D	0, 4, 8, 12, 16, 20 h after light onset	Photophase	Noeske et al. (1981)
	January	12L:12D	0, 6, 12, 18 h after light onset	18 h after light onset	Noeske & Spieler (1984)
Clarias lazera	?	?	Scotophase, photophase	Scotophase	Hogendoorn (1981)
Dicentrarchus labrax	June–September	15L:9D, light onset at 05:45	2, 7 h after light onset	No effect	Perez et al. (1988); Carrillo et al. (1986)
	April–June	Natural	Dawn, mid-day, dusk, midnight	No effect	Boujard et al. (1996)
	January–April	Natural	Three meal/photophase, three meal/scotophase	Scotophase	Azzaydi et al. (1999)
Heteropneustes fossilis	Winter	12L:12D	4, 10, 16, 22 h after light onset	16 h after light onset	Sundararaj et al. (1982)
Heterobranchus longifilis	July–August	12.5L:11.5D	2, 14 h after light onset	14 h after light onset	Kerdchuen & Legendre (1991)
	May, July, August, September	Natural	Scotophase, photophase	Scotophase	Baras et al. (1998)
Ictalurus punctatus	January–February.	12L:12D	1.5, 10 h after light onset	1.5 after light onset	Noeske-Hallin et al. (1985)
	May–September	Natural	08:30, 16:00, 20:00	No effect	Robinson et al. (1995)
	September–March	Natural	08:00, 12:00, 17:00	No effect	Jarboe & Grant (1996)
	Winter–spring Spring–summer		08:00, 17:00	No effect	Jarboe & Grant (1997)
Oncorhynchus mykiss	March–July, August–January	Natural	Postdawn, midlight, predusk	Midlight, predusk	Reddy et al. (1994)
	October–January	Natural	10:00, 17:00	17:00 (not significant)	Palmegiano et al. (1994)
	February–April January–March	Natural	Dawn, noon, midnight Dawn, midnight	Dawn	Boujard et al. (1995)
	January–March	Natural	Dawn, midnight	Dawn	Gélineau et al. (1996)
	September–November May–July	?	9:00, 16:00 10:00, 16:00, 22:00	16:00 No effect	Zoccarato et al. (1993)
Piaractus brachypomus	July–August	14L:10D	Scotophase, photophase	Scotophase	Baras et al. (1996)
Sparus auratus	?	Natural, Light onset at 05:00	5.25, 12.5, 17.25 22.25 h after light onset	22.25 h after light onset	Anthouard et al. (1996)
	June–September	15L:9D	2 h after onset +2 h after offset, 2 h after onset + 8 h before offset, 7 h after onset + 2 h before offset	2 h after onset +2 h after offset	Cerda Reverter et al. (1993)

the differences were only significant for the smallest fish. Also, in Arctic charr, *Salvelinus alpinus*, in which growth is depressed when fish are held at low stocking densities (Wallace *et al.* 1988; Baker & Ayles 1990; Jørgensen *et al.* 1993), Jørgensen and Jobling (1993) demonstrated that feeding fish during the hours of darkness eliminated density-dependent growth suppression.

10.2.1 *Finding the optimal feeding time: a problem of dualism?*

Fish may not rigidly confine their activities to the dark or the light phase, and fish of some species appear to shift from nocturnalism to diurnalism at different times of the year (see Chapter 8). This suggests that experiments carried out on given species at different times of the year might yield different results with respect to the effect of feeding time on growth. In support of this are the results of studies performed on goldfish, *Carassius auratus*, by Noeske and colleagues (1981) and Noeske and Spieler (1984). During spring, fish fed during the day (4–8 h after light-on) had a greater weight gain than those fed 4–8 h after light-off. In January, when goldfish were fed 1% bw in a single daily meal at different times of the light/dark cycle (0 h, 6 h, 12 h or 18 h after light onset), the fish fed 6 h after light-off were heavier and longer than those fed at other times of the day, and also had the highest condition factor.

Sea bass feed nocturnally in winter and switch to diurnalism in summer (Sanchez-Vazquez *et al.* 1998), and Azzaydi *et al.* (1999) reported that fish fed at night during the winter grew better than those fed during the day. At first sight, this latter result contrasts with previous findings for the same species in which no significant differences in growth were seen among fish allowed access to demand-feeders around dawn, mid-day, dusk, or midnight (Boujard *et al.* 1996). However, Boujard and colleagues (1996) carried out their experiment in spring – a period of transition – during which sea bass shift from being nocturnal to diurnal and are not rigid in their feeding behaviour (Sanchez-Vazquez *et al.* 1998).

It is now clear that the time of feeding affects the growth of fish, but finding the optimal time of feeding appears to be more difficult than initially thought. Holding conditions and season are liable to affect feeding behaviour. Further, in most studies, only two to four alternative meal times have been tested, and this is insufficient for determination of the best time for feeding. Investigations into optimal feeding times require experiments to be performed under controlled conditions combining the testing of multiple feeding times with study of the natural rhythm of feeding of the test species. Such investigations should assist in determining whether the optimal time of feeding corresponds to the timing of the natural daily peak of feeding activity.

There is some evidence that when a given amount of food is distributed in several equally sized meals, rather than a single meal, better growth may result (Greenland & Gill 1979; Jarboe & Grant 1997). In addition, the appetite of fish may fluctuate throughout the day, so that modulated feeding – in which the quantity of food delivered at each meal is varied in accordance with the natural daily pattern of feeding activity – might have potential for the enhancement of growth in fish. Support for this idea was provided by the results of a study carried out by Azzaydi *et al.* (1999): sea bass fed three meals daily, with amounts adjusted in accordance with the rhythm of self-feeding fish, had greater weight gain than fish fed three equally sized-meals each day.

10.3 Effect of feeding time on feed intake and nutrient utilisation

In most of the studies dealing with the effect of feeding time on growth, feed consumption has not been monitored accurately, so it is difficult to determine whether differences in weight gain were the result of differences in feed intake and/or differences in feed utilisation.

To the best of our knowledge, the first study suggesting an effect of meal timing on feed intake was performed on goldfish by Noeske and colleagues (1981). In this experiment, fish were fed a food patty for 40 min at one of six different times of the day, with the food being weighed before and after each meal. Under these conditions, the greatest amount of food was ingested by fish fed 8 h after light onset, whereas the lowest ingestion occurred in fish fed at night.

In the study carried out on rainbow trout by Boujard *et al.* (1995), approximately 9% of the food distributed at night was uneaten, whereas fish fed at dawn wasted little or none of the food provided. However, rainbow trout are predominantly visual feeders, so differences in feed intake might have resulted from the fish experiencing difficulties in capturing pellets during the night rather than from real fluctuations in appetite. Juvenile Atlantic salmon, *Salmo salar*, were unable to feed on drifting prey in complete darkness (Fraser & Metcalfe 1997), whereas Jørgensen and Jobling (1990, 1992) reported that Atlantic salmon and Arctic charr could feed in complete darkness when allowed access to pellets lying on the bottom of the tank. The fact that the fish in the study conducted by Boujard *et al.* (1995) may have had access to pellets for several minutes before they exited the tank gives support to the idea that feeding time had a direct effect on feed intake.

However, differences in growth of fish fed at different times may not be completely explained by differences in food consumption (Noeske *et al.* 1981; Boujard *et al.* 1995). Noeske and colleagues (1981) suggested that there may be an optimal time for feeding in terms of feed efficiency. Circumstantial support for this hypothesis may be provided by results of studies performed on channel catfish, African catfish and pirapatinga, *Piaractus brachypomus* (Noeske-Hallin *et al.* 1985; Carillo *et al.* 1986; Kerdchuen & Legendre 1991; Baras *et al.* 1996), but as feed intake was not measured accurately, any conclusions regarding feed efficiency are open to question.

Studies have been carried out on rainbow trout to investigate the effects of feeding time on both feed intake and nutrient utilisation (Boujard *et al.* 1995; Gélineau *et al.* 1996). In one experiment, fish fed 1.2% bw at dawn had a lower intake but similar final body weight to fish fed 1.5% bw at midnight (Fig. 10.1). In addition, fish fed at night had a 9% higher lipid concentration and a 16% poorer protein retention than fish fed at dawn. In a second study, the amount of feed distributed was adjusted each day to ensure an identical cumulative feed intake in all groups. Under these conditions, protein and lipid retention was most efficient in rainbow trout fed at dawn, i.e. in phase with their natural feeding rhythm (Gélineau *et al.* 1996).

10.4 Mechanisms involved in mediating the effects of feeding time on nutrient utilisation

In rainbow trout, the better feed efficiency and nutrient retention observed in fish fed in phase

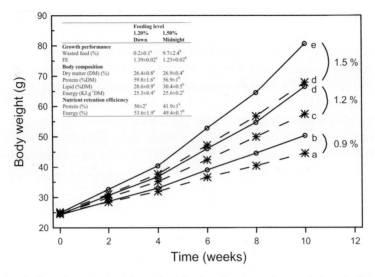

Fig. 10.1   Weight change over time in rainbow trout fed at dawn (sunrise + 1 h; open circles and solid curves) or at midnight (stars and broken curves) at 0.9%, 1.2% or 1.5% of their initial weight. There were twenty fish per replicate, and four replicates per treatment. Different letters indicate significantly different final weights. The standard deviation of mean final weight was 13, 5, 9.3 and 7.1 g for fish fed at dawn at a ration of 1.5%, 1.2 % and 0.9%, respectively. The standard deviation of mean final weight was 15.6, 14.8 and 13.3 g for fish fed at midnight at a ration of 1.5%, 1.2% and 0.9%, respectively. Fish fed 1.2% at dawn and those fed 1.5% at midnight had similar final body weights: their performances, proximate body compositions and nutrient retention efficiencies are shown in the table for comparison. Within lines, means with different superscripts are significantly different (ANOVA $P < 0.05$). Wasted feed = (feed distributed – feed intake)/(feed distributed); FE = gain/feed intake ratio (wet/dry). (Modified from Boujard *et al*. 1995.)

with their natural feeding activity appeared to be related to protein synthesis and retention. This was because ammonia excretion, thought to result from a rapid oxidation of exogenous amino acids (Brett & Zala 1975), was higher in trout fed at midnight than in those fed at dawn (Gélineau *et al*. 1998). Total energy expenditure assessed by indirect calorimetry measurements was similar in both groups. Trout fed at dawn seemed to have a higher capacity for protein synthesis (assessed as RNA:DNA ratio, cf. Bulow 1987) in the liver than those fed at night (Fig. 10.2a) (Gélineau *et al*. 1996). A reduced capacity for protein synthesis among fish fed at night would be expected to lead to amino acid deamination, thereby leading to a less efficient use of protein for growth.

Feeding activity fluctuates throughout the 24-h cycle and is thought to be under the control of endogenous oscillators synchronised by photoperiod and food availability (Boujard & Leatherland 1992b; Cuenca & de la Higuera 1994; Sanchez-Vazquez *et al*. 1995a,b, 1996; Boujard & Luquet 1996; Sanchez-Vazquez & Tabata 1998; see Chapter 8). There is a considerable body of evidence demonstrating that hormones or metabolites involved in feeding, growth and energy partitioning show significant daily fluctuations (for a review, see Spieler 1979; Boujard & Leatherland 1992a; LeBail & Boeuf 1997; MacKenzie *et al*. 1998; see also Chapter 12), suggesting that fish are in different physiological states at different times of the day. As such, they should respond differently to food depending on the time of feeding (Spieler 1979). Accordingly, feeding fish in phase with their internal rhythms might provide the best conditions for nutrient utilisation.

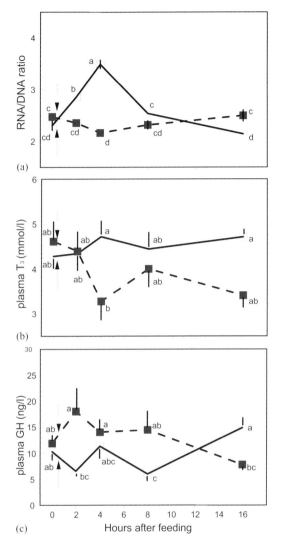

Fig. 10.2 Postprandial changes in liver RNA/DNA ratios (a), plasma tri-iodothyronine (T_3) concentrations (b) and plasma growth hormone (GH) concentrations (c) in rainbow trout fed at dawn (sunrise + 1 h; open squares and solid lines) or at midnight (solid squares and broken lines). Data are shown as mean ± SEM (RNA/DNA ratio, $n = 4$; Plasma T_3 and GH, $n = 20$). Means with the same letters are not significantly different at the 5% level. (Modified from Gélineau *et al.* 1996.)

On the other hand, it has been demonstrated that the act of feeding can entrain some circadian rhythms in fish (for review, see Boujard & Leatherland 1992a; Spieler 1992; Sanchez-Vazquez *et al.* 1997). Thus, feeding time might have an influence on the phase or amplitude of some of the endocrine cycles involved in the physiological regulation of feeding, thereby affecting processes involved in energy use, and in nutrient partitioning and storage.

10.4.1 Effect of feeding time on endocrine cycles

Data pertaining to the entraining effect of feeding time on endocrine cycles are scarce, and results are equivocal. Insulin and glucagon influence nutrient metabolism, and plasma concentrations are affected by feed intake (for a review, see Mommsen & Plisetskaya 1991; LeBail & Boeuf 1997; see also Chapter 12). An effect of feeding time on circulating insulin has been reported in sea bass fed 2 h or 7 h after the onset of light (05:45) (Perez *et al.* 1988). A peak in plasma insulin concentration was observed around 15:00, but fish fed in the morning had their lowest plasma insulin concentration around mid-day, and those fed in the afternoon exhibited their lowest plasma insulin concentration around midnight. In addition, the fish fed early in the photophase had significantly lower plasma insulin concentrations than those fed later. However, according to the authors, the differences in hormonal levels might have been the result of quantitative differences in feed intake rather than to a direct effect of feeding time.

A peak in plasma cortisol has been observed 4 h before feeding in the goldfish (Spieler & Noeske 1981, 1984), at feeding time in rainbow trout (Bry 1982; Rance *et al.* 1982; Laidley & Leatherland 1988; Boujard & Leatherland 1992c), and several hours after feeding in brown trout, *Salmo trutta* (Pickering & Pottinger 1983) and rainbow trout (Boujard *et al.* 1993). In goldfish and rainbow trout, it has been suggested that fluctuations in plasma cortisol concentrations might be entrained by both feeding and photoperiod (Spieler & Noeske 1984; Boujard *et al.* 1993). However, when synchrony is observed between the cortisol peak and feeding time it is difficult to know whether this is a response to feeding *per se* or whether it is an expression of stress resulting from competition for food (Boujard & Leatherland 1992b).

Thyroid hormones (T_3, T_4) are thought to play a permissive role in growth, by potentiating the effect of other anabolic hormones (Sumpter 1992; see Chapter 12). Spieler and Noeske (1981, 1984) reported that plasma T_4 in goldfish exhibited a peak during the photophase, and that time of feeding affected the amplitude of the oscillations. In one study, fish fed 8 h after light onset had lower plasma T_4 concentrations than those fed at light onset, and in another study, fish fed 4 or 6 h after light onset had lower plasma T_4 concentrations than those fed at other times of the day. However, in neither experiment did feeding time result in a phase shift of the thyroid hormone profile.

Studies performed on rainbow trout also failed to reveal a phase shift of the plasma T_4 profile related to time of feeding. With the exception of the study reported by Osborne *et al.* (1978), in which plasma T_4 peaked during the scotophase, there seems to be a diurnal acrophase of circulating hormone regardless of feeding time (Cook & Eales 1987; Boujard & Leatherland 1992c; Holloway *et al.* 1994; Reddy & Leatherland 1995). When fish were fed either four times during the photophase or once a day at mid-day, there was some (albeit not significant) increase in plasma T_4 at dawn (Leatherland *et al.* 1977; Boujard *et al.* 1993). More recently, Gomez *et al.* (1997) measured plasma profiles of T_4 in catheterised rainbow trout. No effect of meal timing on plasma T_4 could be observed, but there was high variability in individual plasma profiles. The peaks of circulating T_4 occurred at irregular intervals, presenting a picture of asynchrony among individuals.

An effect of feeding time on plasma T_3 was reported in the goldfish (Spieler & Noeske 1981). Fish fed in the afternoon had a highly significant rhythm of circulating hormone, the highest concentration occurring at 16:00, whereas fish fed in the morning did not show any significant rhythm. In rainbow trout, plasma T_3 concentrations tended to be stable in fish fed at dawn, but showed greater fluctuations in fish fed at midnight (Fig. 10.2b) (Gélineau *et al.* 1996). Others studies in which diel profiles of plasma T_3 have been investigated in rainbow trout have revealed either low or no fluctuations, and no effect of feeding time on the profile (Eales *et al.* 1981; Boujard & Leatherland 1992c; Boujard *et al.* 1993; Holloway *et al.* 1994; Reddy & Leatherland 1994, 1995; Gomez *et al.* 1997).

Growth hormone (GH) is considered to be a major hormone contributing to the regulation of somatic growth in teleosts (Björnsson 1997; see also Chapter 12). A phase shift of the postprandial peak of circulating GH related to meal timing has been reported in rainbow trout (Boujard *et al.* 1993; Reddy & Leatherland 1994, 1995; Gélineau *et al.* 1996). Plasma GH concentrations were higher in rainbow trout fed during the night than in those fed at dawn (Fig. 10.2c) (Gélineau *et al.* 1996). In another study conducted on the same species, no clear effects of feeding time on plasma hormonal profiles were demonstrated (Boujard *et al.* 1993). Gomez *et al.* (1997) investigated plasma profiles of GH in catheterised rainbow trout. The number of peaks of circulating GH varied between zero and three, and they occurred at irregular intervals, thus presenting an overall picture of asynchrony among individuals. In addition, no effect of meal timing on plasma GH concentrations could be observed.

Although it is believed that feeding is required for the maintenance of a rhythmic pattern of circulating hormones and metabolites (Cook and Eales 1987; Holloway *et al.* 1994; MacKenzie *et al.* 1998), the evidence for an effect of feeding time on the plasma profiles of hormones and metabolites involved in somatic growth is limited. Comparisons among studies are hampered by differences in experimental conditions: sampling interval, season, temperature, fish age and sex, reproductive stage and nutritional state. All are potentially confounding factors that may influence endocrine cycles (Leatherland *et al.* 1977; Osborne *et al.* 1978; Spieler 1979; Eales *et al.* 1981; Cook & Eales 1987; Perez-Sanchez *et al.* 1994). In addition, the daily fluctuations in plasma concentrations of GH or T_4 do not appear to be regular, and there seems to be great inter-individual heterogeneity (Gomez *et al.* 1996, 1997). This suggests that the examination of plasma hormone profiles may not be particularly suitable for gaining information about the mechanisms involved in the effect of feeding time on growth of fish. Investigation of hormonal receptors, gene expression or enzymatic activity might provide more pertinent information to elucidate how feeding time affects metabolism and nutrient utilisation.

10.5 Feeding time and flesh quality

Much aquaculture research has been devoted to the enhancement of growth in fish, but until recently far less attention was paid to the examination of factors that influence flesh quality (see Chapter 15). Flesh quality can be considered to be multifaceted, involving dressout percentage, nutrient composition and sensory attributes such as colour, texture, flavour and smell. Several of these attributes are influenced by lipid distribution, the lipid content of the flesh and the fatty acid composition of the lipid classes found in fish tissues (Bauvineau *et al.* 1993; Einen & Cheadle 1998; see Chapter 15).

The proximate composition of fish is known to depend on both the composition of the feed and ration levels (Takeuchi *et al.* 1978; Watanabe 1982; Greene & Selivonchick 1987; see also Chapter 15), but there may also be an effect of feeding time on body and fillet composition (Table 10.2). Rainbow trout fed 0.9, 1.2 or 1.5% bw at dawn had a higher percentage body lipid than fish fed during the night (Boujard *et al.* 1995). In the same species, body lipid concentration was higher in fish fed at 16:00 than in those fed at 09:00 (Zoccarato *et al.* 1993). However, in neither experiment was the tissue distribution of the lipids determined, so it is not known where the lipids were stored. This information is important because some species deposit large quantities of visceral fat, others deposit storage lipids in the liver, and many species have the swimming muscle as a major lipid storage depot (see Chapter 15). The pattern of lipid deposition would thus affect flesh sensory attributes, and probably also dressout percentage.

In the channel catfish, fish fed a daily ration of 2.5% bw, 3 h before dusk, had 36% more abdominal fat than those fed 1.5 h after dawn (Noeske-Hallin *et al.* 1985). In the same species, fish fed at the end of the afternoon had more visceral fat and a lower percentage of flesh lipid than fish fed in the morning (Robinson *et al.* 1995; Jarboe & Grant 1996). However, the differences were relatively small, and in a later study Jarboe and Grant (1997) were unable to detect any effect of feeding time on proximate composition. Kerdchuen and Legendre (1991) reported greater visceral fat deposition in the African catfish fed during the photophase than in those fed during the scotophase.

A higher percentage of perivisceral fat was recorded in rainbow trout fed in the afternoon (9.7 ± 2.4%) than in those fed in the morning (8.3 ± 1.3%) (Palmegiano *et al.* 1993). Reddy and Leatherland (1994) reported that visceral adipose tissue and skeletal muscle lipid tended to be higher in rainbow trout fed at the middle of the light phase than in fish fed just after dawn, and fish fed before dusk had intermediate values. Although differences were not significant, the trends were in accord with findings showing that trout fed at 16:00 had a higher proportion of fillet fat than fish fed at 09:00 (2.04 ± 0.79 and 0.91 ± 0.19, respectively) (Boccignone *et al.* 1993). In this latter study, the authors also reported significant differences in fillet fatty acid composition among fish fed at different times (09:00; 16:00; 21:00) (Table 10.3). Nevertheless, differences were small and none was observed between fish fed at 09:00 and 16:00. The absence of significant differences in fatty acid composition between fish fed 1.2% bw in either the morning or the afternoon was confirmed by Palmegiano *et al.* (1993, 1994).

Taken together, the results provide some evidence for an effect of feeding time on lipid deposition in fish. However, as with growth, there are discrepancies, possibly arising from the differences in experimental conditions employed in the various studies.

10.6 Conclusions

There is convincing evidence for an effect of feeding time on growth in fish, but more work is required to provide a better understanding of the phenomenon, and to provide grounds for an evaluation of the potential application of manipulating feeding time under farmed conditions. It has been suggested that the optimal feeding time to promote growth might correspond to the species' daily peak of feeding activity. However, results are equivocal – probably due to differences in the experimental conditions used – combined with a considerable degree of plasticity in fish.

The influences of feeding time on growth may result from effects on feed intake and

Table 10.2 Trials on the effect of feeding time on proximate composition of fish.

Species	Season	Photoperiod	Feeding time	Daily ration (bw/day)	Proximate component that differed between feeding treatments	Reference
Heterobranchus longifilis	July–August	12.5L:11.5D light onset 06:00	8:00, 20:00	3%	Visceral fat, body lipid and protein content	Kerdchuen & Legendre (1991)
Ictalurus punctatus	January–February	12L:12D light onset 06:00	7:30, 16:00	2.5%	Visceral fat	Noeske-Hallin et al. (1985)
	May–Sept.	Natural	8:30, 16:00, 20:00	Satiety	No effect	Robinson et al. (1995)
	Sept.–March	Natural	8:00, 12:00, 17:00	3%	No effect	Jarboe & Grant (1996)
	Winter–spring Spring–summer	Natural	8:00, 17:00	3%	No effect	Jarboe & Grant (1997)
Oncorhynchus mykiss	?	?	9:00, 16:00	?	Flesh lipid content	Boccignone et al. (1993)
	?	?	9:00, 16:00	1.2%	Visceral fat	Palmegiano et al. (1993)
	Sept.–Nov. May–July	?	9:00, 16:00; 10:00, 16:00, 22:00	1.4, 1.8%	Body lipid content	Zoccarato et al. (1993)
	Oct.–Jan.	Natural	10:00, 17:00	1.2%	No effect	Palmegiano et al. (1994)
	March–July August–Jan.	Natural	Post-dawn, mid-light, Pre-dusk	0.75%, 1.5%, 2.5%	No effect	Reddy et al. (1994)
	Jan.–March	Natural	Dawn, midnight	0.9%, 1.2%, 1.5%	Body lipid and protein content	Boujard et al. (1995)
Sparus auratus	June–Sept.	15L:9D	2 h after onset + 2 h after offset, 2 h after onset + 8 h before offset, 7 h after onset + 2 h before offset	ad lib.	Flesh lipid and protein content	Cerda Reverter et al. (1993)

Table 10.3   Effect of feeding time on the relative percentage of fatty acids (in % lipid) in the fillet of rainbow trout weighing ca. 200 g. Values within rows bearing the same letters are not significantly different. Data are given as mean ± SD. (Modified from Boccignone *et al.* 1993.)

Fatty acid	Feeding time		
	9:00	16:00	21:00
14:0	3.74 ± 0.2a	3.73 ± 0.31a	3.43 ± 0.26b
16:0	23.70 ± 1.06	23.83 ± 1.93	22.89 ± 1.16
16:1n7	7.44 ± 0.67	6.98 ± 1	7.41 ± 0.57
18:0	5.03 ± 0.52	4.96 ± 0.41	4.55 ± 0.33
18:1n9	26.31 ± 1.08	26.0 ± 1.9	27.62 ± 1.18
18:2n6	7.24 ± 0.5ab	7.4 ± 0.47a	6.66 ± 0.44b
18:3n6	0.74 ± 0.09	0.73 ± 0.14	0.64 ± 0.16
18:3n3	0.69 ± 0.07a	0.65 ± 0.09a	0.54 ± 0.12b
20:1n11	2.9 ± 0.2	3.02 ± 0.29	2.99 ± 0.26
22:1n13	4.33 ± 0.51	4.26 ± 0.71	4.27 ± 0.39
20:5n3	1.48 ± 0.51b	1.66 ± 0.17ab	1.82 ± 0.25a
22:6n3	13.39 ± 1.15	13.36 ± 2.11	13.5 ± 1.8
SFA	33.55 ± 0.84	33.72 ± 1.27	32.59 ± 1
MUFA	41.2 ± 1.58	41.02 ± 2.41	42.91 ± 1.63
PUFA	24.1 ± 1.41	24.26 ± 3.27	24.6 ± 1.51
n3/n6	1.98 ± 0.17ab	1.93 ± 0.24b	2.23 ± 0.32a
UFA/SFA	1.98 ± 0.07ab	1.94 ± 0.11b	2.07 ± 0.09a

nutrient utilisation. Proximate body composition also appears to be affected by the time of feeding, with variations in lipid content and distribution having been reported. Both are liable to affect flesh quality attributes and dressout percentage, which are of prime interest for the aquaculture and fish processing industries.

Studies involving the investigation of plasma concentrations of hormones and metabolites provide little evidence of an effect of feeding time on endocrine profiles. Experiments designed to test the effect of the timing of feeding on hormone receptors, allosteric regulation of enzymatic activity or gene expression might be more valuable for the elucidation of the mechanisms involved in mediating the effects of feeding time on nutrient utilisation.

10.7 References

Anthouard, M., Dermoncourt, E., Divanach, P., Paspatis, M. & Kentouri, M. (1996) Les rythmes d'activité trophique chez la daurade (*Sparus aurata*, L.) en situation de libre accès alimentaire total ou temporellement limité. *Ichtyophysiologica Acta*, **19**, 91–113.

Azzaydi M., Martinez, F.J., Zamora, S., Sanchez-Vazquez, F.J. & Madrid J.A. (1999) Effect of meal size modulation on growth performance and feeding rhythms in European sea bass (*Dicentrarchus labrax*, L.). *Aquaculture*, **170**, 253–266.

Baker, R.F. & Ayles, G.B. (1990) The effect of varying density and loading level on the growth of Arctic charr (*Salvelinus alpinus* L.) and rainbow trout (*Oncorhynchus mykiss* W.). *World Aquaculture*, **21**, 58–62.

Baras, E., Melard, C., Grignard, J.C. & Thoreau, X. (1996) Comparison of food conversion by pirapatinga, *Piaractus brachypomus* under different feeding time. *The Progressive Fish-Culturist*, **58**, 59–61.

Baras, E., Tissier, F., Westerloppe, L., Mélard, C. & Philippart, J.C. (1998) Feeding in darkness alleviates density-dependent growth of juvenile vundu catfish *Heterobranchus longifilis* (Claridae). *Aquatic Living Resources*, **11**, 335–340.

Bauvineau, C., Laroche, M., Heil, F., *et al.* (1993) Incidences de la vitesse de croissance sur les caractéristiques de la chair de le truite fario (*Salmo trutta*) élevée en mer. *Sciences des aliments*, **13**, 201–211.

Björnsson B.Th. (1997) The biology of salmon growth hormone: from daylight to dominance. *Fish Physiology and Biochemistry*, **17**, 9–24.

Boccignone, M., Forneris, G., Salvo, F., Ziino, M. & Leuzzi, U. (1993) Size and meal timing: effect on body composition in rainbow trout. In: *Fish Nutrition in Practice* (eds S.J. Kaushik & P. Luquet), pp. 293–296. INRA editions, Paris, France.

Boujard, T. & Leatherland, J.F. (1992a) Circadian rhythms and feeding time in fishes. *Environmental Biology of Fishes*, **35**, 109–131.

Boujard, T. & Leatherland, J.F. (1992b) Demand-feeding behaviour and diel pattern of feeding activity in *Oncorhynchus mykiss* held under different photoperiod regimes. *Journal of Fish Biology*, **40**, 535–544.

Boujard, T. & Leatherland, J.F. (1992c) Circadian pattern of hepatosomatic index, liver glycogen and lipid content, plasma non-esterified fatty acids, glucose, T3, T4, growth hormone and cortisol concentrations in *Oncorhynchus mykiss* held under different photoperiod regimes and fed using demand feeders. *Fish Physiology and Biochemistry*, **10**, 111–122.

Boujard T. & Luquet P. (1996) Rythmes alimentaires et alimentation chez les siluriformes. *Aquatic Living Resources*, **11**, 335–340.

Boujard, T., Bett, S., Lin, L. & Leatherland J.F. (1993) Effect of restricted access to demand-feeders on diurnal pattern of liver composition, plasma metabolites and hormone levels in *Oncorhynchus mykiss*. *Fish Physiology and Biochemistry*, **11**, 337–344.

Boujard, T., Gélineau, A. & Corraze, G. (1995) Time of a single daily meal influences growth performance in rainbow trout (*Oncorhynchus mykiss*). *Aquaculture Research*, **26**, 341–349.

Boujard, T., Jourdan, M., Kentouri, M. & Divanach, P. (1996) Diel feeding activity and the effect of time-restricted self-feeding on growth and feed conversion in European sea bass. *Aquaculture*, **139**, 117–127.

Brett, J.R. & Zala C.A. (1975) Daily pattern of nitrogen excretion and oxygen consumption of sockeye salmon (*Oncorhynchus nerka*) under controlled conditions. *Journal of the Fisheries Research Board of Canada*, **32**, 2479–2486.

Bry, C. (1982) Daily variations in plasma cortisol levels of individual female rainbow trout, *Salmo gairdneri*: evidence for a post feeding peak in well adapted fish. *General and Comparative Endocrinology*, **48**, 462–468.

Bulow F.J. (1987) RNA-DNA ratios as indicators of growth in fish: A review. In: *The Age and Growth in Fish* (eds R.C. Summerfelt & G.E. Hall), pp. 45–54. The Iowa State University Press, Ames, Iowa.

Carrillo, M., Perez, J. & Zanuy, S. (1986) Efecto de la hora de ingesta y de la naturaleza de la dieta sobre el crecimiento de la lubina (*Dicentrarchus labrax* L.). *Investigaciones Pesqueras*, **50**, 83–95.

Cerda Reverter, J.M., Pineda, J., Zanuy, S., Carrillo, M., Ramos, J. & Gutierrez, J. (1993) Efecto de la hora de alimentacion sobre las variaciones diarias de los niveles plasmaticos de glucosa e insulina y sobre el crecimiento de juveniles de dorada (*Sparus aurata*). In: *Actas del IV Congreso Nacional de Acuicultura.* (eds A. Cerviño, A. Landín, A. de Coo, A. Guerra & M. Torre), pp. 103–108. Pondevadra University, Pontevedra, Spain.

Cook, R.F. & Eales, J.G. (1987) Effects of feeding and photocycle on diel changes in plasma thyroid hormone levels in rainbow trout, *Salmo gairdneri*. *Journal of Experimental Zoology*, **242**, 161–169.

Cuenca, E.M. & de la Higuera, M. (1994) Evidence for an endogenous circadian rhythm of feeding in the trout (*Oncorhynchus mykiss*). *Biological Rhythm Research*, **25**, 336–337.

Eales, J.G., Hughes, M. & Uin, L. (1981) Effect of food intake on diel variation in plasma thyroid hormone levels in rainbow trout, *Salmo gairdneri*. *General and Comparative Endocrinology*, **45**, 167–174.

Einen, O. & Cheadle, G. (1998) Quality characteristics in raw and smoked fillets of Atlantic salmon, *Salmo salar*, fed high energy diets. *Aquaculture Nutrition*, **4**, 99–108.

Eriksson L.O. & Alanärä A. (1992) Timing of feeding behaviour in salmonids. In: *The Importance of Feeding Behavior for the Efficient Culture of Salmonid Fishes* (eds J.E. Thorpe & F.A. Huntingford), pp. 41–48. World Aquaculture Society, Baton Rouge, Louisiana.

Fraser, N.H. & Metcalfe, N.B. (1997) The cost of becoming nocturnal: feeding efficiency in relation to light intensity in juvenile Atlantic salmon. *Functional Ecology*, **11**, 385–391.

Fraser, N.H., Metcalfe, N.B. & Thorpe, J.E. (1993) Temperature-dependent switch between diurnal and nocturnal foraging in salmon. *Proceedings of the Royal Society of London*, **252B**, 135–139.

Fraser, N.H., Heggenes, J., Metcalfe, N.B. & Thorpe, J.E. (1995) Low summer temperatures cause juvenile Atlantic salmon to become nocturnal. *Canadian Journal of Zoology*, **73**, 446–451.

Gélineau, A., Mambrini, M., Leatherland, J.F. & Boujard, T. (1996) Effect of feeding time on hepatic nucleic acid, plasma T_3, T_4 and GH concentrations in rainbow trout. *Physiology and Behavior*, **59**, 1061–1067.

Gélineau, A., Médale, F. & Boujard, T. (1998) Effect of feeding time on postprandial nitrogen excretion and energy expenditure in rainbow trout. *Journal of Fish Biology*, **52**, 655–664.

Gomez, J.M., Boujard, T., Fostier, A. & Le Bail, P.-Y. (1996) Characterization of growth hormone nycthemeral plasma profiles in catheterized rainbow trout (*Oncorhynchus mykiss*). *Journal of Experimental Zoology*, **274**, 171–180.

Gomez, J.M., Boujard, T., Boeuf, G., Solari, A. & Le Bail, P.-Y. (1997) Individual diurnal plasma profiles of thyroid hormones in rainbow trout (*Oncorhynchus mykiss*) in relation to cortisol, growth hormone and growth rate. *General and Comparative Endocrinology*, **107**, 74–83.

Greene, D.H.S. & Selivonchick, D.P. (1987) Lipid metabolism in fish. *Progress in Lipid Research*, **26**, 53–85.

Greenland, D.C. & Gill R.L. (1979) Multiple daily feedings with automatic feeders improve growth and feed conversion rates of channel catfish. *The Progressive Fish-Culturist*, **41**, 151–153.

Hogendoorn, H. (1981) Controlled propagation of the African catfish, *Clarias lazera* (C. & V.). IV. Effect of feeding regime in fingerling culture. *Aquaculture*, **24**, 123–131.

Holloway, A.C., Reddy, P.K., Sheridan, M.A. & Leatherland, J.F. (1994) Diurnal rhythms of plasma growth hormone, somatostatin, thyroid hormones, cortisol and glucose concentrations in rainbow trout, *Oncorhynchus mykiss*, during progressive food deprivation. *Biological Rhythm Research*, **25**, 415–432.

Hung, S.S.O. & Storebakken, T. (1994) Carbohydrate utilization by rainbow trout is affected by feeding strategy. *Journal of Nutrition*, **124**, 223–230.

Jarboe, H.H. & Grant, W.J. (1996) Effects of feeding time and frequency on growth of channel catfish, *Ictalurus punctatus*, in closed recirculating raceway systems. *Journal of the World Aquaculture Society*, **27**, 235–239.

Jarboe, H.H. & Grant, W.J. (1997) The influence of feeding time and frequency on the growth, survival, feed conversion and body composition of channel catfish, *Ictalurus punctatus,* cultured in a three-tier, closed, recirculating raceway system. *Journal of Applied Aquaculture,* **7**, 43–52.

Jørgensen, E.H. & Jobling, M. (1990) Feeding modes in Arctic charr, *Salvelinus alpinus* L.: the importance of bottom feeding for the maintenance of growth. *Aquaculture,* **86**, 379–386.

Jørgensen, E.H. & Jobling, M. (1992) Feeding behaviour and effect of feeding regime on growth of Atlantic salmon, *Salmo salar. Aquaculture,* **101**, 135–146.

Jørgensen, E.H. & Jobling, M. (1993) Feeding in darkness eliminates density-dependent growth suppression in Arctic charr. *Aquaculture International,* **1**, 90–93.

Jørgensen, E.H., Christiansen, J.S. & Jobling, M. (1993) Effects of stocking density on food intake, growth performance and oxygen consumption in Arctic charr (*Salvelinus alpinus*). *Aquaculture,* **110**, 190–204.

Kerdchuen N. & Legendre M. (1991) Influence de la fréquence et de la période de nourrissage sur la croissance et l'efficacité alimentaire d'un silure africain, *Heterobranchus longifilis. Aquatic Living Resources,* **4**, 241–248.

Laidley, C.W. & Leatherland, J.F. (1988) Circadian studies of plasma cortisol, thyroid hormone, protein, glucose and ion concentration, liver glycogen concentration and liver and spleen weight in rainbow trout, *Salmo gairdneri* Richardson. *Comparative Biochemistry and Physiology,* **89A**, 495–502.

Landless, P.J. (1976) Demand-feeding behaviour of rainbow trout. *Aquaculture,* **7**, 11–25.

Leatherland, J. F., Cho, C.Y. & Slinger, S.J. (1977) Effects of diet, ambient temperature and holding conditions on plasma thyroxine levels in rainbow trout (*Salmo gairdneri*). *Journal of the Fisheries Research Board of Canada,* **34**, 677–682.

LeBail, P.Y. & Boeuf, G. (1997) What hormones may regulate food intake in fish? *Aquatic Living Resources,* **10**, 371–379.

MacKenzie, D.S., VandenPutte, C.M. & Leiner, K.A. (1998) Nutrient regulation of endocrine function in fish. *Aquaculture,* **161**, 3–25.

Mommsen, T.P. & Plisetskaya, E.M. (1991) Insulin in fishes and agnathans: history, structure and metabolic regulation. *Reviews in Aquatic Sciences,* **4**, 225–259.

Noeske, T.A. & Spieler, R.E. (1984) Circadian feeding time affects growth of fish. *Transactions of the American Fisheries Society,* **113**, 540–544.

Noeske, T.A., Erickson, D. & Spieler, R.E. (1981) The time of day goldfish receive a single daily meal affects growth. *Journal of the World Mariculture Society,* **12**, 73–77.

Noeske-Hallin, T.A., Spieler, R.E., Parker, N.C. & Suttle, M.A. (1985) Feeding time differentially affects fattening and growth of channel catfish. *Journal of Nutrition,* **115**, 1228–1232.

Osborne, R.H., Simpson, T.H. & Youngson, A.F. (1978) Seasonal and diurnal rhythms of thyroidal status in the rainbow trout, *Salmo gairdneri* R. *Journal of Fish Biology,* **12**, 531–540.

Palmegiano, G.B., Bianchini, M.L., Boccignone, M., Forneris, G., Sicuro, B. & Zoccarato, I. (1993) Effect of starvation and meal timing on fatty acid composition in rainbow trout (*Oncorhynchus mykiss*). *Rivista Italiana Acquacoltura,* **28**, 5–11.

Palmegiano, G.B., Bianchini, M.L., Boccignone, M., *et al.* (1994) Effect of timing of meals on meat composition in rainbow trout (*Oncorhynchus mykiss*). *Rivista Italiana Acquacoltura,* **29**, 67–72.

Perez, J., Zanuy, S. & Carrillo, M. (1988) Effects of diet and feeding time on daily variations in plasma insulin, hepatic c-AMP and other metabolites in a teleost fish, *Dicentrarchus labrax. Fish Physiology and Biochemistry*, **5**, 191–197.

Perez-Sanchez, J., Marti-Palanca, H. & LeBail, P.Y. (1994) Seasonal changes in circulating growth hormone (GH), hepatic GH-binding and plasma insulin-like growth factor-I immunoreactivity in a marine fish, gilthead sea bream, *Sparus aurata. Fish Physiology and Biochemistry*, **13**, 199–208.

Pickering, A.D. & Pottinger, T.G. (1983) Seasonal and diel changes in plasma cortisol levels of the brown trout, *Salmo trutta. General and Comparative Endocrinology*, **49**, 232–239.

Rance, T.A., Baker, B.I. & Webley, G. (1982) Variations in plasma cortisol concentrations over a 24-hour period in the rainbow trout *Salmo gairdneri. General and Comparative Endocrinology*, **48**, 269–274.

Reddy, P.K. & Leatherland, J.F. (1994) Does the time of feeding affect the diurnal rhythms of plasma hormone and glucose concentration and hepatic glycogen content of rainbow trout? *Fish Physiology and Biochemistry*, **13**, 133–140.

Reddy, P.K. & Leatherland, J.F. (1995) Influence of the combination of time of feeding and ration level on the diurnal hormone rhythms in rainbow trout. *Fish Physiology and Biochemistry*, **14**, 25–36.

Reddy, P.K., Leatherland, J.F., Khan, M.N. & Boujard, T. (1994) Effect of the daily meal time on the growth of rainbow trout fed different ration levels. *Aquaculture International*, **2**, 1–15.

Robinson, E.H., Jackson, L.S., Li, M.H., Kingsbury, S.K. & Tucker, C.S. (1995) Effect of time of feeding on growth of channel catfish. *Journal of the World Aquaculture Society*, **26**, 320–322.

Sanchez-Vazquez, F.J. & Tabata, M. (1998) Circadian rhythms of demand-feeding and locomotor activity in rainbow trout. *Journal of Fish Biology*, **52**, 255–267.

Sanchez-Vazquez, F.J., Madrid, J.A. & Zamora S. (1995a) Circadian rhythms of feeding activity in sea bass, *Dicentrarchus labrax* L: dual phasing capacity of diel demand-feeding pattern. *Journal of Biological Rhythms*, **10**, 256–266.

Sanchez-Vazquez, F.J., Zamora, S. & Madrid, J.A. (1995b) Light-dark and food restriction cycles in sea bass: effect of conflicting zeitgebers on demand-feeding rhythms. *Physiology and Behavior*, **58**, 705–714.

Sanchez-Vazquez, F.J., Madrid, J.A., Zamora, S., Iigo M. & Tabata, M. (1996) Demand feeding and locomotor circadian rhythms in the goldfish, *Carassius auratus*: dual and independent phasing. *Physiology and Behavior*, **60**, 665–674.

Sanchez-Vazquez, F.J., Madrid, J.A., Zamora, S. & Tabata, M. (1997) Feeding entrainment of locomotor activity rhythms in the goldfish is mediated by a feeding-entrainable circadian oscillator. *Journal of Comparative Physiology*, **181A**, 121–132.

Sanchez-Vazquez, F.J., Azzaydi, M., Martinez, F.J., Zamora, S. & Madrid, J.A. (1998) Annual rhythms of demand-feeding activity in sea-bass: evidence of a seasonal phase inversion of the diel feeding pattern. *Chronobiology International*, **15**, 607–622.

Spieler, R.E. (1979) Diel rhythms of circulating prolactin, cortisol, thyroxine, and triiodothyronine levels in fishes. A review. *Revue Canadienne de Biologie*, **38**, 301–315.

Spieler, R.E. (1992) Feeding-entrained circadian rhythms in fishes. In: *Rhythms in Fishes* (ed. M.A. Ali), pp. 137–148. Plenum Press, New York.

Spieler, R.E. & Noeske, T.A. (1981) Timing of a single daily meal and diel variations of serum thyroxine, triiodothyronine and cortisol in goldfish, *Carassius auratus. Life Science*, **28**, 2939–2944.

Spieler, R.E. & Noeske, T.A. (1984) Effects of photoperiod and feeding schedule on diel variations of locomotor activity, cortisol, and thyroxine in goldfish. *Transactions of the American Fisheries Society*, **113**, 528–539.

Stickney, R.R. & Andrews, J.W. (1971) The influence of photoperiod on growth and food conversion of channel catfish. *The Progressive Fish-Culturist*, **33**, 204–205.

Sumpter, J.P. (1992) Control of growth of rainbow trout (*Oncorhynchus mykiss*). *Aquaculture*, **100**, 299–320.

Sundararaj, B.I., Nath, P. & Halberg, F. (1982) Circadian meal timing in relation to lighting schedule optimizes catfish body weight gain. *Journal of Nutrition*, **112**, 1085–1097.

Takeuchi, T., Watanabe, T. & Ogino, C. (1978) Supplementary effect of lipids in a high protein diet for rainbow trout. *Bulletin of the Japanese Society for Scientific Fisheries*, **44**, 677–681.

Wallace, J.C., Kolbeinshavn, A.G. & Reinsnes, T.G. (1988) The effect of stocking density on early growth in Arctic charr, *Salvelinus alpinus* (L.). *Aquaculture*, **73**, 101–110.

Watanabe, T. (1982) Lipid nutrition in fish. *Comparative Biochemistry and Physiology*, **73B**, 3–15.

Zoccarato, I., Boccignone, M., Palmegiano, G.B., Anselmino, M., Benatti, G. & Leveroni Calvi, S. (1993) Meal timing and feeding level: effect on performances in rainbow trout. In: *Fish Nutrition in Practice* (eds S.J. Kaushik & P. Luquet), pp. 297–300. INRA Editions, Paris, France.

Chapter 11

Effects of Nutritional Factors and Feed Characteristics on Feed Intake

Manuel de la Higuera

11.1 Introduction

Feed selection must be directed to satisfy nutrient needs, and for that reason feed quality selection is even more important than feed quantity. Setting up feed preferences, as a selecting process, depends not only on the chemical composition of feed but also on the sensitivity of fish species to certain of its components. It may be postulated that chemical images exist at a central level – similar to visual patterns of identification – such that feed search and discrimination would be directed to accept those feeds in which nutrient contents are similar to the central needs image. Similarly, when given a choice, fish will be expected to reject feeds with compositions distant from the central image of reference. It may be difficult to prove that animals select a diet to meet their nutrient requirements when given a choice of foods (Forbes 1998), but it is clear that there are many situations in which animals show considerable 'nutritional wisdom'. Furthermore, it is far from clear what internal mechanisms are involved in controlling diet selection; in fact no study has been made in this area with fish.

The self-feeding capacity of fish gives us the opportunity to investigate feed preferences, not only as a direct function of organoleptic properties of feeds but also as influenced by the physiological status of the fish and environmental conditions. In any case, a diet should not only cover nutrient needs – it should also be accepted and ingested in quantities to ensure adequate growth and reproduction. In this sense, the presence of certain non-nutrient components in the feed can influence palatability and intake. The main objective of this chapter is to provide a summary overview of the effects of feed composition on intake.

11.2 Physical characteristics and feed intake

To cover the nutritional requirements of an animal, feeds must be formulated in accordance with the animal's needs, and they must be ingested in adequate quantities. Feed acceptance depends on a variety of chemical and nutritional factors and on physical characteristics. The choice of equipment, processing and system variables used in the feed manufacturing process not only affects nutrient availability – it can also have influences on the acceptability of the finished feed. The ways in which the various characteristics of feeds can affect acceptability and ingestion have been discussed in some detail in Chapters 2 and 6, and so only a brief summary, with some examples, will be given here.

The intake of a pelleted feed can be affected by physical characteristics such as shape, size (diameter and length), density (floating or sinking rate), as well as grinding and other processing procedures that can affect nutrient availability and the elimination of antinutritional factors and deterrent compounds (see Chapters 2 and 6). For example, although Atlantic halibut, *Hippoglossus hippoglossus*, did not seem to show any preference for pellets of a particular size (Helland *et al.* 1997), it has been reported that Atlantic salmon, *Salmo salar*, shows clear preferences. When salmon were held in a net pen, pellet diameter and length influenced particle capture, with salmon taking longer to capture small pellets (Smith *et al.* 1995). In the same study it was observed that pellet length, but not diameter, was positively correlated to pellet rejection. Stradmeyer *et al.* (1988) reported that juvenile Atlantic salmon show a preference for long pellets over round ones. Further, it was found that although pellet texture did not affect capture, it did affect feed intake, with fish preferring soft pellets (with gelatine added) to hard pellets. More studies are, however, needed to elucidate the effects of the appearance and texture of feed pellets on feed intake, growth and feed efficiency.

11.3 Dietary nutrients and sources affecting feed intake

Dietary energy (from protein, fat and carbohydrate) is one of the main factors postulated to influence feed intake (in g/100 g body weight) in fish (Grove *et al.* 1978; Boujard & Médale 1994; see also Chapter 14). The relationships between dietary energy and feed intake are closer when digestible (energy in food minus the energy in faeces) or metabolisable energy (energy in the food less the energy lost in faeces, urine and through excretion from the gills), rather than gross energy (heat of combustion measured using a bomb calorimeter; see Chapter 1), are used as the basis for comparison (Kaushik & Médale 1994; Morales *et al.* 1994). However, many of the studies indicating that feed intake is regulated in relation to dietary energy have employed an experimental design in which food has been diluted with indigestible bulk (e.g. Rozin & Mayer 1961; Grove *et al.* 1978; Bromley & Adkins 1984). Under such circumstances it is not known whether the feeding responses are truly regulated by energy concentration, or whether they may be nutrient specific, i.e. animals provided with a range of diluted feeds would need to vary the amounts consumed to regulate both energy intake and amounts of specific nutrients. It is important to know whether fish regulate their feed intake primarily on the basis of energy concentration or whether there is a separate regulation of individual nutrients or nutrient groups, i.e. macronutrients (proteins, lipids and carbohydrates) and micronutrients (vitamins and minerals). For example, the assumption that a fish is regulating energy intake irrespective of macronutrient ratios, when it is really attempting to regulate its intake and use of specific nutrients, could have major negative consequences with regard to the formulation of feeds for farmed fish. Consequently, more research is needed on the influence of dietary energy on feed intake, especially when considering the way in which the energy available to the fish from the different macronutrients influences feeding frequency and preferences.

Although energy is usually considered to be the main factor governing feed intake, availability of specific essential nutrients such as protein (or amino acids), vitamins and minerals in the food is also of importance. Specific appetites for essential nutrients have been reported for terrestrial animal species (Forbes 1998), and essential nutrient deficiency is known to reduce feed intake in fish (see Fig. 11.1 for some essential amino acids). When animals are fed deficient diets, two opposite responses can be induced: an increase in feed intake to reach the absolute requirement level when the deficiency is mild – as observed in poultry and pigs for essential amino acids by Boorman (1979) and Henry (1985), respectively – or an inhibitory response when the deficiency is greater (Fig. 11.1). A possible explanation for the latter type of response is that, in terms of metabolic economy, animals prefer to reduce feed intake instead of forcing their metabolism by eating more of the unbalanced diet. The reduction in intake would prevent or delay the onset of metabolic disorders. Among the first symptoms of an essential-nutrient deficiency is appetite loss and consequent cessation of growth. Data from experiments to determine essential amino acid (EAA) requirements of rainbow trout, *Oncorhynchus mykiss*, have been analysed and plotted in Fig. 11.1 to show the relationship between the concentration of an EAA and feed intake. The figure shows that increasing EAA availability stimulates intake until dietary EAA concentrations reach requirement levels, while higher EAA concentrations do not markedly alter feed intake. For any given protein source, one EAA will eventually become limiting and some amino acids will also be used to provide energy (see Chapter 13). Unless a form of non-protein energy is provided at an adequate level, an animal will use amino acids for energy metabolism, thereby decreasing amino acid utilisation for protein synthesis and growth.

Increasing the non-protein dietary energy to meet the caloric needs of fish reduces amino acid oxidation and, consequently, improves the utilisation of dietary protein for growth and reduces nitrogen excretion (Kaushik & Cowey 1991; Cho 1992). This protein-sparing effect of dietary lipid and/or carbohydrates has been widely demonstrated in fish (Wilson 1989; Cho & Kaushik 1990; Sanz *et al.* 1993; Vergara *et al.* 1996). Lipids contain more energy per unit weight than protein or carbohydrate. Lipids are not only used efficiently by fish but also

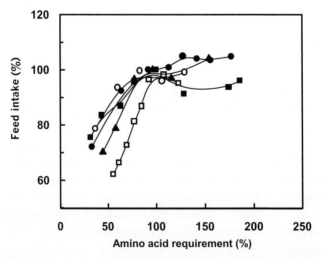

Fig. 11.1   Relationship between dietary essential amino acid level (as % requirement) and feed intake. Feed intake was calculated from growth and feed efficiency data of fish fed to satiation from experiments to evaluate requirement of : ● arginine (Kim *et al.* 1992b); ○ arginine (Walton *et al.* 1986); ■ leucine (Choo *et al.* 1991); □ lysine (Kim *et al.* 1992b); ▲ methionine (Kim *et al.* 1992a). Intake of the feed supplying the essential amino acid at the level closest to requirement, was considered as 100%.

increase the palatability of feeds, reduce dust and stabilise pellets. Although carbohydrate is a cheaper source of energy than protein or lipid it is less efficiently utilised, and the maximal permissible levels in the diet differ among fish species (Wilson 1994). Increasing the proportion of digestible non-protein energy sources in commercial feeds has been successful in terms of improving fish growth and feed efficiency (Cho & Kaushik 1990; Cho 1992; Sanz *et al.* 1993; Vergara *et al.* 1996), and the use of high-energy feeds, especially for salmonid culture, is becoming common practice (Cho 1992; Austreng 1994). Few experiments have been conducted in which protein and non-protein components have been fed separately in order to evaluate intake of each (Kaushik & Luquet 1983), but there is evidence to suggest that fish control feed intake in order to meet energy needs, and that control is related to the digestible energy of the diet (Boujard & Médale 1994; Morales *et al.* 1994).

As fish regulate their energy intake, the use of high-energy diets may have a direct influence on feed intake. Some studies have used self-feeding techniques to test the ability of fish to regulate their energy intake. For example, Rozin and Mayer (1961) with the goldfish, *Carassius auratus*, and Grove *et al.* (1978) with the rainbow trout, have demonstrated that intake of feeds with kaolin added as an inert material is increased to compensate for the lower feed-energy content. According to Bromley and Adkins (1984), rainbow trout can compensate for an α-cellulose (inert bulk) content of up to 30% in the diet. Further, dietary bulk agents (at 20% level) resulted in increased intake by sea bass, *Dicentrarchus labrax*, so fish maintained a similar energy intake to that of bass fed a diet without the inert bulking agent (Dias *et al.* 1998). When rainbow trout self-fed on diets of similar protein concentration but differing in fat content – and hence total energy content – they ate more of the low-energy diets, but maintained total energy intake regardless of dietary energy content (Boujard & Médale 1994). Similar results were obtained by Paspatis and Boujard (1996), who studied Atlantic salmon self-fed on diets with different energy densities. However, under commercial production conditions trout were unable to adjust feed demands to dietary energy content (Alanärä 1994). Similarly, no compensatory response to the dietary energy content was observed in Arctic charr, *Salvelinus alpinus*, reared under comparable conditions (Alanärä & Kiessling 1996). Given the discrepancies in results between small- and large-scale experiments, the relationship between dietary energy content and the responses of fish allowed to self-feed deserves further study.

With the advent of high-energy feeds the question arises as to whether fish respond by reducing intake or whether lipid-rich feeds induce hyperphagia, as observed in the rat (Horn *et al.* 1996). On the other hand, there is also evidence that energy intake is regulated in relation to body energy stores, in order to maintain a particular set-point (e.g. Liebelt *et al.* 1965). In the same way, studies on Arctic charr provide additional evidence that some factors associated with monitoring of the size of the energy reserves may contribute to appetite control and the modulation of feed intake (Jobling & Miglavs 1993; see also Chapters 12 and 15). These authors suggest that appetite declines as body fat accumulates and energy depots become replete, so that energy reserves are maintained close to some set-point.

The intake of feeds containing different protein sources and levels has been reported to be inversely related to their digestible, or metabolisable, energy content (Morales *et al.* 1994). Further, when feeds were reformulated to be isocaloric in terms of digestible and metabolisable energy, feed intake was similar, regardless of the protein source and level. However, this has not always been reported to be the case, and the decrease in feed intake induced by feeds containing protein sources other than fishmeal might be attributed to (apart from palatability differences; see Chapter 2) deficiencies in EAA. For example, European eel, *Anguilla anguilla*, fed on a feed containing sunflower-protein, increased feed intake when the feed was supplemented with deficient amino acids (lysine, methionine, histidine and threonine). The overall result was a feed intake and efficiency, and muscle protein synthesis and deposition, similar to that of fish fed a fishmeal-based feed (García-Gallego *et al.* 1998; de la Higuera *et al.* 1999). Similarly, a partial restoration of feed intake was noted by Médale *et al.* (1998) when a soy-protein-concentrate feed was supplemented with methionine, although intake did not reach that of fish given a feed based on fishmeal (control).

An adequate postprandial amino acid pattern is not only necessary to ensure amino acid utilisation for protein synthesis and growth (see Chapter 13), but might also determine feed acceptability. In this sense, the manner of supplementing a protein-deficient diet could alter feed intake by determining the availability of the supplemented amino acid, not only as a function of time but also in relation to the remaining amino acids available for protein synthesis and growth (Sierra 1995). Free amino acids may be rapidly absorbed compared with protein-bound amino acids (Murai *et al.* 1982), thereby inducing oxidation. As such, a relative deficiency may persist when the postprandial amino acid pattern is considered, and when supplemented amino acids are provided in coated form the amino acid pattern for protein synthesis is improved (Cowey & Walton 1988; de la Higuera *et al.* 1998). As a consequence, the availability of the supplemented amino acid, as well as feed intake and growth, are enhanced (de la Higuera *et al.* 1998). An alternative to achieve an adequate postprandial amino acid profile for growth might be to increase feeding frequency, as demonstrated in carp, *Cyprinus carpio* (Yamada *et al.* 1981) and pig (Batterham & Bayley 1989). Experiments involving self-feeding would be informative in showing the way by which changes in the postprandial timing of supplemented and protein-bound amino acids might influence protein synthesis through alterations in feeding frequency.

When a substantial amount of fishmeal is replaced with plant and other protein sources, feed intake generally declines (Morales *et al.* 1994; Gomes *et al.* 1995), sometimes due to poorer palatability (i.e. the overall sensory impression the animal receives from its food) (see Chapters 2 and 5). Replacement of fishmeal by soybean meal in feeds given to largemouth bass, *Micropterus salmoides*, resulted in reduced feed intake in proportion to the amount of fishmeal replaced (Kubitza *et al.* 1997). Supplementation with inosine-5'-monophosphate, but not betaine or free amino acids, enhanced feed intake of a soybean-protein-based feed, but inclusion of 10% fishmeal was most effective in increasing intake. A low intake of feeds containing alternative protein sources might be a consequence of EAA deficiency (see Fig. 11.1), and EAA supplementation leads to restoration of feed intake in some fish species (Sierra 1995; de la Higuera *et al.* 1998, 1999).

The removal or inactivation of antinutritional factors present in some plant-protein sources enhances their nutritive quality (see Chapter 2). Some compounds are not directly responsible for a decreased nutritive utilisation, but affect feed acceptability. For example, heat-stable alcohol-soluble factors present in soybean meals appear to be responsible for reduced feed intake in some fish species (van den Ingh *et al.* 1996), and Bureau *et al.* (1998) suggested that saponins cause alcohol extracts of soybean meal to be highly unpalatable for chinook salmon *Oncorhynchus tshawytscha*. The replacement of up to 100% of the fishmeal with a soy-protein concentrate in feeds for rainbow trout did not affect growth or feed utilisation, but negative effects were observed when soy flour (up to 50% replacement) was included in the feed (Kaushik *et al.* 1995). Some animal-protein sources can also contain compounds that reduce feed intake. For example, a high protein content and a balanced amino acid profile appear to make the earthworm *Eisenia foetida* a useful alternative protein source for fish. However, this invertebrate contains unpalatable compounds in the coelomic fluid. The removal of these compounds resulted in an increase in feed intake, and the use of squid extract as a feeding stimulant further improved feed intake of fish fed formulations containing earthworm meal (Cardenete *et al.* 1993).

The time-course of availability of nutrients affects feed utilisation (see Chapter 10), and feed composition influences the rate of gastric evacuation and, hence, feeding frequency (Grove *et al.* 1978; Jobling 1981, 1987; Rösch 1987). Feeding of multiple meals results in higher stomach evacuation rates, as compared with single-meal feeding (Fletcher *et al.* 1984; Persson 1984; Grove *et al.* 1985). Rates of gastric evacuation and intestinal transit are also related to body size and water temperature, both of which affect frequency of feeding (Grove *et al.* 1978; Schade 1982). A close relationship between evacuation rate and feeding has been reported (Bromley 1987), pointing to an influence of gastrointestinal factors in the regulation of feeding (see Chapters 12 and 13).

11.4 Nutrient selection

The ability of fish to operate self-feeding devices (see Chapters 3 and 14), together with their chemoreceptive sensitivity (see Chapter 5), can be applied to discrimination studies where fish show feed preferences as a function of taste and/or content and availability of dietary nutrients. Feed selection in animals should be directed to satisfy nutrient requirements, although palatability also plays a role. The acquisition of a balanced intake of nutrients is the main objective of feed selection. Animals able to select from a range of available feeds will try to ingest different proportions of each to ensure an adequate intake of essential nutrients and energy for normal body functions and growth (Simpson & Raubenheimer 1999). In higher vertebrates, feed selection does not appear to be random or purposeless; it seems to be directed towards attaining a certain level of nutrient intake (Kyriazakis 1997).

When fish make a choice between two or more feeds it is implied that they can distinguish physicochemical properties and nutritional characteristics. Animals will select feeds with positive post-ingestion consequences and will avoid those with negative or less positive ones (Provenza & Cincotta 1993; Kyriazakis 1997; Simpson & Raubenheimer 1999). Animals rapidly associate the sensory properties of a feed with its metabolic and physiological effects, and may remember the image for a long time (Ralphs 1997). Day *et al.* (1998) consider that animals assess foods in order to identify whether they are nutritionally beneficial or harmful, and propose that food choice and intake centre upon the concept of information gathering through intrinsic and extrinsic exploration.

The ability of goldfish to select a balanced diet from three single-macronutrient feeds (50% protein or fat or carbohydrates, the rest being cellulose, vitamins, minerals and binder) has been studied (Sánchez-Vázquez *et al.* 1998). The fish showed a complex pattern of feed selection with different daily demands for single macronutrients, but the goldfish maintained their energy intake by selecting feeds with different energy values. The ability to regulate energy intake when provided with feeds differing in energy concentration has been documented previously for goldfish (Rozin & Mayer 1961) and other fish species (Boujard & Médale 1994; Paspatis & Boujard 1996). In the goldfish, the self-selection from the three single-macronutrient feeds resulted in an overall diet containing approximately 22% protein, 32% fat and 46% carbohydrates. A comparable study was conducted on the rainbow trout (Sánchez-Vázquez *et al.* 1999). Three feeds, each containing protein, fat or carbohydrates as the only macronutrient, were offered to trout in three different self-feeders, so the fish could choose among them to compose their diet. After six days, the trout showed a clear pattern of preferences that remained stable for two months. The intake pattern corresponded to a dietary regimen containing 64% protein, 19% fat and 18% carbohydrate. Both species, goldfish and trout, composed a diet according to their preferences and regulated feed intake to balance energy intake. Further, the selected mixture seemed to reflect the feeding habits of the species. Goldfish, an omnivorous species, selected a lower proportion of protein than did rainbow trout, which is considered to be carnivorous. It should be emphasised that preferences need not coincide with a balanced practical diet formulated to give maximum growth and efficient feed conversion. According to Harper and Peters (1989) the amount of protein selected varies greatly among animals and is usually above the requirement level, defined as the lowest level for maximum growth.

Excessively high concentrations of protein in feeds have been shown to decrease feed intake and growth in some fish species (e.g. Gurure *et al.* 1995; Santinha *et al.* 1996). Such negative effects have also been observed in higher vertebrates fed diets containing high levels of protein and certain amino acids (Li & Anderson 1983). The reduction in feed intake might be a protective response to reduce the possible harmful effects caused by an excessive intake of these dietary components.

The selection of a diet is a problem that an animal must to solve when faced with a variety of feeds that differ in nutritional value. Under such conditions the animal must select a blend which meets its requirement for essential nutrients [for principles of diet selection, see Emmans (1991) and Simpson & Raubenheimer (1999)]. This means that the animal must have some knowledge of the nutritional properties of the feeds available, and Day *et al.* (1998) proposed that animals are motivated to sample food items in order to evaluate nutritional benefit or harm. Thus, animals acquire and retain relevant nutritional information as a result of interacting with feeds. This makes sense, because if an animal has a choice between a deficient feed and one containing adequate levels of an essential nutrient, selection of the appropriate feed is vital. For example, when rainbow trout were given the choice between zinc-deficient and zinc-sufficient feeds (1.9 and 26.0 ppm, respectively), the fish showed a clear preference for the feed that met their zinc requirements (Fig. 11.2). When the position of the self-feeder providing the zinc-sufficient feed was changed, in order to test for positional influences on feed choice, the trout soon readjusted their preference to ensure an adequate intake of this essential element (Expósito 1999).

A feed preference associated with the availability of vitamin C has been demonstrated in the sea bream, *Sparus aurata* (Paspatis *et al.* 1997). Fish fed a commercial feed enriched with vitamin C displayed better growth than those fed on a non-supplemented feed, implying that the amount of vitamin C in the commercial feed was inadequate to sustain maximum growth. When given the choice between these two feeds, the fish showed a net preference for the supplemented feed. Discrimination was clear after three weeks, probably after the fish detected

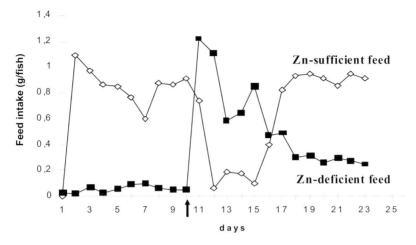

Fig. 11.2   Self-selection of zinc-sufficient (◇, 26.0 ppm) and zinc-deficient (■, 1.9 ppm) feeds by rainbow trout. Results are means for two groups of forty fish. The arrow indicates a change of position of the feeds in the self-feeders (Expósito 1999).

the beneficial effects of increased vitamin C intake. Thus, contrary to the results from the studies on dietary zinc preference in trout (Fig. 11.2), the feed choice in the sea bream occurred after some time. The fish may have required time to associate the positive effects of an increased vitamin C intake with the properties of the feed.

The utility of self-feeding experiments to determine essential nutrient requirements in fish remains to be demonstrated. Hidalgo *et al.* (1988) attempted to use a self-feeding protocol to establish the methionine requirements of the sea bass. Five feeds differing in methionine concentration (0.30, 0.65, 1.00, 1.35 and 1.7 g/100 g feed) were simultaneously offered to sea bass of different sizes. After eight days exploring the self-feeders, the fish showed a net preference for the feed containing 1.35% methionine; the optimum dietary level according to the authors, and close to the requirement range established by Thebault *et al.* (1985). Nevertheless, larger fish were more influenced by spatial preferences and, at the same time, exhibited a certain aversion for the feed containing the highest methionine concentration. Although toxicity associated with high dietary levels of methionine has been reported from studies on mammals (Harper *et al.* 1970) and rainbow trout (Kaushik & Luquet 1980), negative influences on intake do not seem to have occurred in other experiments carried out to evaluate methionine requirements of fish species (calculated from Kim *et al.* 1992a; Sierra 1995).

The selection of a diet balanced for a particular nutrient, from a choice of two or more feeds, is considered to indicate that there is a specific appetite for that nutrient. An internal, or metabolic, deficiency will increase an animal's preference for the feed containing the nutrient, while intake of feeds containing concentrations that exceed requirements will not necessarily decrease unless a high concentration of the nutrient has detrimental effects.

11.5 Feeding stimulants

The influence of chemical stimuli on feeding behaviour is well known, and there several reviews of this subject have been made (Carr 1982; Mackie 1982; Mackie & Mitchell 1985; Takeda and Takii 1992; Hara 1994; see also Chapter 5). Nevertheless, a classification of chemicals that influence feeding behaviour is somewhat confusing, although a terminology has been proposed (Lindstedt 1971): arrestant; attractant; repellent; incitant; suppressant; stimulant; and deterrent. The first three terms are used for substances detectable in very low concentrations, and are related to positive or negative orientation responses to the feed. The remaining terms apply to positive or negative responses relating to initiation or continuation of feeding, and direct contact with the feed is usually required for the response. Sometimes a single chemical fits two different definitions when activating chemoreceptors at markedly different concentrations (e.g. attractant and stimulant). Given that in this chapter we are concerned mainly with the practical feeding of fish, the influence of chemical stimuli on the continuation of feeding will be our focus. In this sense, after contact with the feed, two opposite responses could be promoted: rejection of the feed due to the presence of deterrent compounds, or initiation and continuation of feeding induced by stimulants.

Electrophysiological and behavioural studies on fish (mainly salmonids) have shown that olfactory and gustatory senses are especially sensitive to components which are naturally familiar to the fish or directly related to their food (Hara 1994; see Chapter 5). In general terms, it appears that while olfactory cues contribute to food detection, taste is responsible for the final intake or rejection of the food (see Chapter 5). As feeding behaviour is mediated by mixtures of chemicals, a number of chemosensory cells must be stimulated to induce a feeding response (Mackie *et al.* 1980). Chemical stimuli have received attention in terms of the control of feed intake, and valuable information is available on appetite suppressants and stimulants in normally feeding fish. According to Mackie and Mitchell (1985), fish have a specific demand with regard to feeding stimulants, but may respond to a wide variety of feeding deterrents.

The method used to identify feeding stimulants is based on omission tests, where the effects of mixtures are compared with those of the individual components. Omission as well as supplementation tests would be the basis for adequate designs to determine the capacity of essential nutrients and stimulants to induce feeding behaviour. An experimental protocol to determine the chemical basis of feeding behaviour might be as follows:

(1)   Behavioural and electrophysiological studies to test responses of fish to tissue extracts of natural prey.
(2)   Physicochemical analysis of active extracts.
(3)   Additional behavioural and electrophysiological tests to determine the activity of fractions from extracts or, even better, of individual compounds.
(4)   Examination of effectiveness of identified mixtures or compounds in practical feeding trials.

Although a variety of feeding stimulants have been identified, the mixtures that cause the greatest behavioural responses are composed of free amino acids, nucleotides and nucleosides, and quaternary ammonium bases (Mackie & Michell 1985; see Chapters 2 and 5). The most common physicochemical properties of feeding stimulants for fish are the following: non-volatile; low molecular weight; nitrogen-containing; amphoteric; water-soluble; stable to heat treatment; and broad biological distribution. These properties are consistent with those of free amino acids and related nitrogenous substances of low molecular weight (e.g. nucleotides, nucleosides and quaternary ammonium bases).

Based on electrophysiological studies, two types of fish have been identified (see review by Hara 1994; see Chapter 5): (i) those that respond to a wide range of amino acids; and (ii) those that respond to a limited range. Of the twenty-seven species tested, ten (channel catfish, *Ictalurus punctatus*, and some marine species) proved to be in the wide response category, whereas others (eleven salmonids, two cyprinids and several marine species) showed a limited response range. Further, in general terms, carnivorous fish appear to be sensitive to alkaline and neutral amino acids (glycine, proline, taurine, valine) while herbivorous species seem to respond more to acidic amino acids (aspartic and glutamic acids). Alanine, glycine, proline, valine, tryptophan, tyrosine, phenylalanine, lysine and histidine appear to be major components of feeding stimulants for many fish species, although the composition of the active amino acid mixtures varies among species. A feed containing D-forms of amino acids, rather than L-forms, proved repellent to rainbow trout (Adron & Mackie 1978). D-Amino acids seemed to be without effect on plaice, *Pleuronectes platessa*, and dab, *Limanda limanda* (Mackie 1982), indicating a possible stereospecificity at amino acid receptor sites. A mixture of proline, glycine, taurine, alanine and arginine was highly stimulatory for plaice and dab, while in rainbow trout these amino acids were inactive or repellent. In general terms, individual amino acids are either without effect or repellent.

Mackie *et al.* (1980) showed that feed intake of Dover sole, *Solea solea*, is mediated by chemicals that diffuse from the feed into the water to stimulate chemosensory cells. Difficulties are encountered in getting this species to accept artificial feed. Dover sole accept feeds containing mussel, *Mytilus edulis*, but refuse to eat those which include fishmeal in the formula. Further, Dover sole ate a casein-based feed containing a chemical mixture based on the low molecular fraction of mussel flesh. Neither the amino acid fraction, nor the mixture minus betaine, had stimulatory activity, and betaine was the most effective feeding stimulant. The authors proposed that betaine is a chemoattractant, and that sole has an absolute requirement for it to initiate feeding. The reason for such specificity may be that the natural food of sole consists primarily of small crustaceans and molluscs, which contain large amounts of betaine. Betaine has been shown to be a feeding stimulant for several species, although the stimulatory properties are usually associated with the presence of certain amino acids. For example, the best feeding stimulant identified for Japanese eel, *Anguilla japonica*, by Takeda *et al.* (1984) was a mixture of the amino acids glycine, alanine, proline and histidine, along with betaine; these are all compounds found in the natural prey of this species.

Feeding stimulants are species-specific. For example, an L-amino acid mixture that induced a positive response in trout was ineffective for turbot, *Scophthalmus maximus*, whereas non-amino acid components of a synthetic squid mixture were highly stimulatory for the turbot (Adron & Mackie 1978; Mackie & Adron 1978). Of over forty nucleotides tested, only inosine and its derivatives were highly stimulatory. Nucleotides and nucleosides [inosine, inosine-5'-monophosphate (IMP), adenosine-5'-diphosphate (ADP), guanosine-5'-monophosphate (GMP), uridine-5'-monophosphate (UMP)], have been identified as feeding stimulants for fish. IMP has feeding stimulant activity for several fish species (Takeda & Takii 1992). The chemical structure of nucleotides and nucleosides also influence their stimulatory activity (e.g. having a purine base and a phosphatidic base at the 5'-position). Quaternary ammonium bases (glycine-betaine and sulphur analogue dimethylthetin, trimethylglycine, dimethyl-β-propiothetin) have also been reported to act as feeding stimulants in several fish species (Takeda & Takii 1992).

Among non-nitrogenous compounds, studies on the rainbow trout have shown that glucose (an aldose) is better accepted than ketoses. Also, monosaccharides seem to be more palatable than their disaccharide derivatives, e.g. glucose and fructose versus sucrose (Jones 1990). Other substances with feeding stimulant capacity include lecithin derivatives, lactic acid, alcohols and peptides. Lecithin-type compounds having trimethyl in the α-N-moiety and higher saturated fatty acids in α- and β-residue groups may be attractants (Harada 1987). Compounds such as lactic acid, the tripeptide glutathione (GSH), and various extracts from plants and animals have also shown a certain capacity to stimulate feed intake. However, it remains to be established which of these compounds, and their relative proportions, induce an optimal feeding response.

11.5.1 Feeding stimulants and fish nutrition

The feeding activity of fish changes with feed acceptability and palatability, both of which are associated with the chemical and physical properties of the feed (see Chapters 2 and 6). Environmental conditions will also influence feeding activity (Fletcher 1984; see Chapter 6). An increase in dietary palatability results in a higher intake and a shorter feeding time in the rat (Davis & Smith 1988), and such responses to palatable feeds by fish would have the advantage of decreasing the time that pellets spend in contact with water, and hence reduce nutrient leaching. Feeding stimulants can be used to increase the palatability of formulated feeds (e.g. Dias *et al.* 1997), and the acceptability of alternative protein sources can often be improved by reducing concentrations of deterrent substances (Cardenete *et al.* 1993; van den Ingh *et al.* 1996; Bureau *et al.* 1998; see also Chapter 2).

A major practical objective in marine fish culture is the replacement of 'natural', live food by formulated feeds during the rearing of larvae and juveniles. Weaning of the young fish on to formulated feeds may be a problem if they have been previously fed on live foods. There is evidence (Métailler *et al.* 1983; Takii *et al.* 1986a; Kumai *et al.* 1989) that feeding stimulants can be used to stimulate feeding of larval and juvenile fish, leading to increased survival and growth, and improved feed conversion. For example, juvenile goldfish and rainbow trout fed diets containing the sulphonium compound dimethyl-β-propiothetin, a feeding stimulant, not only exhibited better growth than controls, but also had greater resistance against environmental stressors, such as oxygen deficiency and raised water temperature (Nakajima 1992).

The physiological mechanisms underlying the stimulatory effects of feeding stimulants on growth are not clear. Improved growth may be attributed to enhanced digestion and metabolic utilisation of nutrients, as well as to increased food intake. For example, when Japanese eels were fed on a feed containing a feeding stimulant mixture (L-alanine, glycine, L-proline, L-histidine and UMP) they grew better than fish fed an unsupplemented feed despite daily feeding rates being almost identical (Takii *et al.* 1986a). This suggests that chemical signals induced by feeding stimulants may enhance a cephalic reflex response, resulting in improved nutrient utilisation. An alternative explanation might be that the nutritional values of feeds are improved by stimulant supplementation. However, most stimulants are not essential nutrients and they are present at such low concentrations that they are unlikely to influence the nutritional balance of feeds. Thus, the former hypothesis is more plausible, i.e. the chemical cues may stimulate a cephalic reflex response and promote digestive and metabolic functions to improve food utilisation, as has been demonstrated in mammals (Giduck *et al.* 1987). The hypothesis has been tested in fish: Japanese eels given a feed supplemented with a feeding stimulant mixture had better growth, feed efficiency, protein efficiency ratio and protein and energy retention than controls (Takii *et al.* 1986a), and digestive functions (pepsin-like and trypsin-like enzyme activities, as well as protein and carbohydrate digestibility) were augmented by the feeding stimulant. The greater growth of eels reared on the supplemented feed could also be indirectly attributed to more efficient metabolism, i.e. enhanced carbohydrate utilisation and increased synthesis of body protein and fat (Takii *et al.* 1986b). In view of the significance of feed development in aquaculture, more research on the influence of feeding stimulants on feed utilisation is needed.

11.6 Deterrent compounds

There are many examples of reduced appetite, poor growth and disease being associated with nutritional deficiency. Also, certain compounds directly influence feed intake by acting as feeding deterrents (see Chapter 2). For example, this may be the case for single D-amino acids (e.g. proline), certain combinations of amino acids (e.g. taurine, alanine and arginine) (Adron & Mackie 1978), trimethylamine and its oxidation products produced in decaying fish flesh (Mackie 1982; Hughes 1991), highly oxidised oils (Murai *et al.* 1988; Ketola *et al.* 1989), non-nutrient and antinutritional factors present in soybean (Bureau *et al.* 1998), aflatoxins and T-2 toxins produced by moulds, and many other compounds present in feedstuffs (Black *et al.* 1988). The elimination of antinutritional factors, such as those present in soybean meal, exerts a positive effect on feed intake. There is also evidence that artificial contaminants, such as industrial chemicals, pesticides and herbicides, sometimes depress feed intake and trigger metabolic disorders (see Chapter 6).

In addition to certain components present in protein sources alternative to fishmeal, some of the antibiotics included in medicated feeds (erythromycin, fluoroquinolone, ormetoprim, oxytetracycline, sulfadimethoxine) have been shown to act as feeding deterrents (Schreck & Moffitt 1987; Poe & Wilson 1989; Robinson *et al.* 1990; Robinson & Tucker 1992; Toften *et al.* 1995; Boujard & Le Gouvello 1997; Toften & Jobling 1997). For example, when Atlantic salmon were given a choice between medicated (oxytetracycline) and medicated + squid-extract feeds, the fish showed a preference for the feed containing the feeding stimulant (squid extract), at least during the first three weeks of the experiment (Toften *et al.* 1995). However, in a longer-term experiment, the fish seemed to become accustomed to the medicated feed and showed no difference in feed intake compared with the control group after 44–65 days (Toften & Jobling 1997). In this case, the use of feeding stimulants resulted in a temporary increase in feed intake, and the authors concluded that short-term studies may be inadequate to test whether deterrent or stimulant properties of feed ingredients are of practical importance in feed formulation (Toften & Jobling 1997). The possibility of avoiding the negative influences of medicated feeds on feed intake may reside in adequate technological treatment, such as microencapsulation of the medicament. For example, rainbow trout can discriminate between medicated feeds containing microgranulated or free fluoroquinolone (Boujard & Le Gouvello 1997), the latter being detected as a deterrent compound.

11.7 Conclusions: feed acceptance and palatability studies

In order to avoid the misinterpretation of results in feed preference studies, it is important that the control feed meets the essential nutrient requirements of the species, or ontogenetic stage, being investigated. It should be remembered that quantitative nutrient requirements are species- and life-stage-specific, and a feed must fulfil these requirements if normal feed intake and growth are to be maintained. A deficiency in one or more essential nutrient will probably alter the feeding behaviour of the fish, making it difficult to assess the significance of the variable under examination. High-energy feed formulation should take into account the relative proportions of essential nutrients per unit of digestible energy, because an excess of dietary energy might curb consumption before sufficient essential nutrients are ingested.

According to Mackie (1982) the ideal assay system for testing feeding stimulants consists of determining whether the fish will eat a fully defined test feed, followed by omission of components until the fish no longer eats significant amounts of the feed. An alternative approach is to add potential feeding stimulants to an otherwise unacceptable or tasteless feed (Jones 1989). Either method is time-consuming, but there is no real alternative. Mackie (1982) proposes vitamin-free casein as the preferred protein source, since it should not contribute to the taste of the final feed. The unflavoured casein-based feed should be unacceptable to the fish. After the identification of a chemical mixture with feeding stimulant properties, the next stage would be to test the amino acid and the non-amino acid fractions separately. If either of these fractions is effective, other components can be omitted until the activity is lost. To establish how many amino acid mixtures are active is almost impossible to ascertain because of, among other things, the great number of possible combinations. Irrespective of the approach adopted, once the feeding stimulant is identified it is essential to confirm the results by carrying out a growth experiment conducted over a sufficiently long time to assess practical relevance.

11.8 References

Adron, J.W. & Mackie, A.M. (1978) Studies on the chemical nature of feeding stimulants for rainbow trout, *Salmo gairdneri* Richardson. *Journal of Fish Biology*, **12**, 303–310.

Alanärä, A. (1994) The effect of temperature, dietary energy content and reward level on the demand feeding activity of rainbow trout (*Oncorhynchus mykiss*). *Aquaculture*, **126**, 349–359.

Alanärä, A. & Kiessling, A. (1996) Changes in demand feeding behaviour in Arctic charr, *Salvelinus alpinus* L., caused by differences in dietary energy content and reward level. *Aquaculture Research*, **27**, 479–486.

Austreng, E. (1994) Historical development of salmon feed. Annual report 1993, *Institute of Aquaculture Research (AKVAFORSK)*, As-NLH, Norway, 32 pp.

Batterham, E.S. & Bayley, H.S. (1989) Effect of frequency of feeding of diets containing free or protein-bound lysine on the oxidation of (14C)phenylalanine by growing pigs. *British Journal of Nutrition*, **62**, 647–655.

Black, J.J., Maccubbin, A.E., Myers, H.K. & Zeigel, R.F. (1988) Aflatoxin B1 induced hepatic neoplasia in Great Lakes coho salmon. *Bulletin of Environmental Contamination and Toxicology*, **41**, 742–745.

Boorman, K.N. (1979) Regulation of protein and amino acid intake. In: *Food Intake Regulation in Poultry* (eds K.N. Boorman & B.M. Freeman), pp. 87–126. Longman, Edinburgh.

Boujard, T. & Le Gouvello, R. (1997) Voluntary feed intake and discrimination of diets containing a novel fluoroquinolone in self-fed rainbow trout. *Aquatic Living Resources*, **10**, 343–350.

Boujard, T. & Médale, F. (1994) Regulation of voluntary feed intake in juvenile rainbow trout fed by hand or by self-feeders with diets containing two different protein/energy ratios. *Aquatic Living Resources*, **7**, 211–215.

Bromley, P.J. (1987) The effects of food type, meal size and body weight on digestion and gastric evacuation in turbot, *Scophthalmus maximus* L. *Journal of Fish Biology*, **30**, 501–512.

Bromley, P.J. & Adkins, T.C. (1984) The influence of cellulose filler on feeding, growth and utilization of energy in rainbow trout, *Salmo gairdneri* Richardson. *Journal of Fish Biology*, **24**, 235–244.

Bureau, D.P., Harris, A.M. & Cho, C.Y. (1998) The effects of purified alcohol extracts from soy products on feed intake and growth of chinook salmon (*Oncorhynchus tshawytscha*) and rainbow trout (*Oncorhynchus mykiss*). *Aquaculture*, **161**, 27–43.

Cardenete, G., Garzón, A., Moyano, F. & de la Higuera, M. (1993) Nutritive utilization of earthworm protein by fingerling rainbow trout (*Oncorhynchus mykiss*). In: *Fish Nutrition in Practice* (eds S.J. Kaushik and P. Luquet), pp. 923–926. INRA Editions, Paris.

Carr, W.E.S. (1982) Chemical stimulation of feeding behavior. In: *Chemoreception in Fishes* (ed. T.J. Hara), pp. 259–273. Elsevier, Amsterdam.

Cho, C.Y. (1992) Feeding systems for rainbow trout and other salmonids with reference to current estimates of energy and protein requirements. *Aquaculture*, **100**, 107–123.

Cho, C.Y. & Kaushik, S.J. (1990) Nutritional energetics in fish: energy and protein utilization in rainbow trout (*Salmo gairdneri*). *World Review of Nutrition and Dietetics*, **61**, 132–172.

Choo, P., Smith, T.K., Cho, C.Y. & Ferguson, H.W. (1991) Dietary excesses of leucine influence growth and body composition of rainbow trout. *Journal of Nutrition*, **121**, 1932–1939.

Cowey, C.B. & Walton, M.J. (1988) Studies on the uptake of (14C)-amino acids derived from both dietary (14C) protein and dietary (14C) amino acids by rainbow trout, *Salmo gairdneri* Richardson. *Journal of Fish Biology*, **33**, 293–305.

Davis, J.D. & Smith, G.P. (1998) Analysis of lick rate measures the positive and negative feedback effects of carbohydrate on eating. *Appetite*, **11**, 229–238.

Day, J.E.L., Kyriazakis, I. & Rogers, P.J. (1998) Food choice and intake: towards a unifying framework of learning and feeding motivation. *Nutrition Research Reviews*, **11**, 25–43.

Dias, J., Gomes, E.F. & Kaushik, S.J. (1997) Improvement of feed intake through supplementation with an attractant mix in European seabass fed plant-protein rich diets. *Aquatic Living Resources*, **10**, 385–389.

Dias, J., Huelvan, C., Dinis, M.T. & Métailler, R. (1998) Influence of dietary bulk agents (silica, cellulose and a natural zeolite) on protein digestibility, growth, feed intake and feed transit time in European seabass (*Dicentrarchus labrax*) juveniles. *Aquatic Living Resources*, **11**, 219–226.

de la Higuera, M., Garzón, A., Hidalgo, M.C., Peragón, J., Cardenete, G. & Lupiáñez, J.A. (1998) Influence of temperature and dietary-protein supplementation either with free or coated lysine on the fractional protein-turnover rates in the white muscle of carp. *Fish Physiology and Biochemistry*, **18**, 85–95.

de la Higuera, M., Akharbach, H., Hidalgo, M.C., Peragón, J., Lupiáñez, J.A. & García-Gallego, M. (1999) Liver and white muscle protein turnover rates in the European eel (*Anguilla anguilla*): effects of dietary protein quality. *Aquaculture*, **179**, 203–216.

Emmans, G.C. (1991) Diet selection by animals: theory and experimental design. *Proceedings of the Nutrition Society*, **50**, 59–64.

Expósito, A. (1999) *Disponibilidad de cinc en dietas para truchas (*Oncorhynchus mykiss*): selección específica del alimento, distribución y recambio tisular de cinc, síntesis proteica y actividades enzimáticas cinc-dependientes.* PhD Thesis, University of Granada, Spain.

Fletcher, D.J. (1984) The physiological control of appetite in fish. *Comparative Biochemistry and Physiology*, **78A**, 617–628.

Fletcher, D.J., Grove, D.J., Basimi, R.A. & Ghaddaf, A. (1984) Emptying rates of single and double meals of different quality from the stomach of the dab, *Limanda limanda* (L.). *Journal of Fish Biology*, **25**, 435–444.

Forbes, J.M. (1998) *Voluntary Food Intake and Diet Selection in Farm Animals.* CAB International, Oxford.

García-Gallego, M., Akharbach, H. & de la Higuera, M. (1998) Use of protein sources alternative to fish meal in diets with amino acids supplementation for the European eel (*Anguilla anguilla*). *Animal Science*, **66**, 285–292.

Giduck, S.A., Threatte, R.M. & Kare M.R. (1987) Cephalic reflexes: their role in digestion and possible roles in absorption and metabolism. *Journal of Nutrition*, **117**, 1191–1196.

Gomes, E.F., Rema, P & Kaushik, S.J. (1995) Replacement of fish meal by plant proteins in the diet of rainbow trout (*Oncorhynchus mykiss*): digestibility and growth performance. *Aquaculture*, **130**, 177–186.

Gurure, R.M., Moccia, R.D. & Atkinson, J.L. (1995) Optimal protein requirements of young Arctic charr (*Salvelinus alpinus*) fed practical diets. *Aquaculture Nutrition*, **1**, 227–234.

Grove, D.J., Loizides, L. & Nott, J. (1978) Satiation amount, frequency of feeding and gastric emptying rate in *Salmo gairdneri*. *Journal of Fish Biology*, **12**, 507–516.

Grove, D.J., Moctezuma, M.A., Flett, H.R.J., Foott, J.S., Watson, T. & Flowerdew, M.W. (1985) Gastric emptying and return of appetite in juvenile turbot, *Scophthalmus maximus* L., fed on artificial diets. *Journal of Fish Biology*, **26**, 339–354.

Hara, T.J. (1994) The diversity of chemical stimulation in fish olfaction and gustation. *Reviews in Fish Biology and Fisheries*, **4**, 1–35.

Harada, K. (1987) Relationships between structure and feeding attraction activity of certain L-amino acids and lecithin in aquatic animals. *Nippon Suisan Gakkaishi*, **53**, 2243–2247.

Harper, A.E. & Peters, J.C. (1989) Protein intake, brain amino acid and serotonin concentrations and protein self-selection. *Journal of Nutrition*, **119**, 677–689.

Harper, A.E., Benevenga, N.J. & Wohlhueter, R.M. (1970) Effects of ingestion of disproportionate amounts of amino acids. *Physiological Review*, **50**, 428–558.

Helland, S.J., Grisdale-Helland, B. & Berge, G.M. (1997) Feed intake and growth of Atlantic halibut (*Hippoglossus hippoglossus* L.) fed combinations of pellet sizes. *Aquaculture*, **156**, 1–8.

Henry, Y. (1985) Dietary factors involved in feed intake regulation in growing pigs: a review. *Livestock Production Science*, **12**, 339–354.

Hidalgo, F., Kentouri, M. & Divanach, P. (1988) Sur l'utilisation du self-feeder comme outil d'epreuve nutritionnelle du loup, *Dicentrarchus labrax*. Résultats préliminaires avec la méthionine. *Aquaculture*, **68**, 177–190.

Horn, C.C., Tordoff, M.G. & Friedman, M.I. (1996) Does ingested fat produce satiety? *American Journal of Physiology (RICP)*, **39**, R761–R765.

Hughes, S.G. (1991) Response of first-feeding spring chinook salmon to four potential chemical modifiers of feed intake modifiers. *The Progressive Fish-Culturist*, **52**, 15–17.

Jobling, M. (1981) Dietary digestibility and the influence of food components on gastric evacuation in plaice, *Pleuronectes platessa* L. *Journal of Fish Biology*, **19**, 29–36.

Jobling, M. (1987) Influences of food particle size and dietary energy content on patterns of gastric evacuation in fish: test of a physiological model of gastric emptying. *Journal of Fish Biology*, **30**, 299–314.

Jobling, M. & Miglavs, I. (1993) The size of lipid depots – a factor contributing to the control of food intake in Arctic charr, *Salvelinus alpinus*? *Journal of Fish Biology*, **43**, 487–489.

Jones, K.A. (1989) The palatability of amino acids and related compounds to rainbow trout, *Salmo gairdneri* Richardson. *Journal of Fish Biology*, **34**, 149–160.

Jones, K.A. (1990) Chemical requirements of feeding in rainbow trout, *Oncorhynchus mykiss* (Walbaum); palatability studies on amino acids, amides, alcohols, aldehydes, saccharides and other compounds. *Journal of Fish Biology*, **37**, 413–423.

Kaushik, S.J. & Cowey, C.B. (1991) Ammoniogenesis and dietary factors affecting nitrogen excretion. In: *Nutritional Strategies & Aquaculture Waste* (eds C.B. Cowey & C.Y. Cho), pp. 3–19. University of Guelph, Guelph, Canada.

Kaushik, S.J. & Luquet, P. (1980) Influence of bacterial protein incorporation and of sulphur amino acid supplementation to such diets on growth of rainbow trout, *Salmo gairdneri* Richardson. *Aquaculture*, **19**, 163–175.

Kaushik, S.J. & Luquet, P. (1983) Relationship between protein intake and voluntary energy intake as affected by body weight with an estimation of maintenance needs in rainbow trout. *Zeitschrift für Tierphysiologie, Tierernährung und Futtermittelkunde*, **51**, 57–69.

Kaushik, S.J. & Médale, F. (1994) Energy requirements, utilization and supply to salmonids. *Aquaculture*, **124**, 81–97.

Kaushik, S.J., Cravedi, J.P., Lalles, J.P., Sumpter, J., Fauconneau, B. & Laroche, M. (1995) Partial or total replacement of fish meal by soybean protein on growth, protein utilization, potential estrogenic or antigenic effects, cholesterolemia and flesh quality in rainbow trout, *Oncorhynchus mykiss*. *Aquaculture*, **133**, 257–274.

Ketola, H.G., Smith, C.E. & Kindschi, G.A. (1989) Influence of diet and oxidative rancidity on fry of Atlantic and coho salmon. *Aquaculture*, **79**, 417–423.

Kim, K., Kayes, T.B. & Amundson, C.H. (1992a) Requirements for sulfur amino acids and utilization of D-methionine by rainbow trout (*Oncorhynchus mykiss*). *Aquaculture*, **101**, 95–103.

Kim, K., Kayes, T.B. & Amundson, C.H. (1992b) Requirements for lysine and arginine by rainbow trout (*Oncorhynchus mykiss*). *Aquaculture*, **106**, 333–344.

Kubitza, F., Lovshin, L.L. & Lovell, R.T. (1997) Identification of feed enhancers for juvenile largemouth bass *Micropterus salmoides*. *Aquaculture*, **148**, 191–200.

Kumai, H., Kimura, I., Nakamura, M., Takii, K. & Ishida, H (1989) Studies on digestive system and assimilation of a flavoured diet in ocellate puffer. *Nippon Suisan Gakkaishi*, **55**, 1035–1043.

Kyriazakis, I. (1997). The nutritional choices of farm animals: to eat or what to eat? In: *Animal Choices* (eds J.M. Forbes, T.L.J. Lawrence, R.G. Rodway & M.A. Varley), pp. 55–65. British Society of Animal Science, Edinburgh.

Li, E.T.S. & Anderson, G.H. (1983) Amino acids in the regulation of food intake. *Nutrition Abstract and Reviews in Clinical Nutrition*, **53**, 169–178.

Liebelt, R.A., Ichinoe, S. & Nicholson, N. (1965) Regulatory influences of adipose tissue on food intake and body weight. *Annals of the New York Academy of Sciences*, **131**, 559–582.

Lindstedt, K.J. (1971) Chemical control of feeding behaviour. *Comparative Biochemistry and Physiology*, **39A**, 553–581.

Mackie, A.M. (1982) Identification of the gustatory feeding stimulants. In: *Chemoreception in Fishes* (ed. T.J. Hara), pp. 275–291. Elsevier, Amsterdam.

Mackie, A.M. & Adron, J.W. (1978) Identification of inosine and inosine-5'-monophosphate as the gustatory feeding stimulants for the turbot, *Scophthalmus maximus*. *Comparative Biochemistry and Physiology*, **60A**, 79–83.

Mackie, A.M. & Mitchell, A.I. (1985) Identification of gustatory feeding stimulants for fish – Application in aquaculture. In: *Nutrition and Feeding in Fish* (eds C.B. Cowey, A.M. Mackie & J.G. Bell), pp. 177–189. Academic Press, London.

Mackie, A.M., Adron, J.W. & Grant, P.T. (1980) Chemical nature of feeding stimulants for the juvenile Dover sole, *Solea solea* (L.). *Journal of Fish Biology*, **16**, 701–708.

Médale, F., Boujard, T., Vallée, F., *et al.* (1998) Voluntary feed intake, nitrogen and phosphorus losses in rainbow trout (*Oncorhynchus mykiss)* fed increasing dietary levels of soy protein concentrate. *Aquatic Living Resources*, **11**, 239–246.

Métailler, R., Cadena-Roa, M. & Person-Le Ruyet, J. (1983) Attractive chemical substances for the weaning of Dover sole (*Solea vulgaris*): qualitative and quantitative approach. *Journal of the World Mariculture Society*, **14**, 679–684.

Morales, A.E., Cardenete, G., de la Higuera, M. & Sanz, A. (1994) Effects of dietary protein source on growth, feed conversion and energy utilization in rainbow trout, *Oncorhynchus mykiss. Aquaculture*, **124**, 117–126.

Murai, T., Akiyama, T., Ogata, H., Hirasawa, Y. & Nose, T. (1982) Effect of coating amino acids with casein supplemented to gelatin diet on plasma free amino acids of carp. *Bulletin of the Japanese Society of Scientific Fisheries*, **48**, 703–710.

Murai, T., Akiyama, T., Ogata, H. & Suzuki, T. (1988) Interaction of dietary oxidized fish oil and glutathione on fingerling yellowtail *Seriola quinqueradiata*. *Bulletin of the Japanese Society of Scientific Fisheries*, **54**, 145–149.

Nakajima, K. (1992) Activation effect of a short term of dimethyl-β-propiothetin supplementation on goldfish and rainbow trout. *Nippon Suisan Gakkaishi*, **58**, 1453–1458.

Paspatis, M. & Boujard, T. (1996) A comparative study of automatic feeding and self-feeding in juvenile Atlantic salmon (*Salmo salar*) fed diets of different energy levels. *Aquaculture*, **145**, 245–257.

Paspatis, M., Kentouri, M & Krystalakis, N. (1997) Vitamin C: a factor on feed preference of sea bream (*Sparus aurata*). In: *Proceedings of the 5th Hellenic Symposium on Oceanography and Fisheries*, Kavala, Greece. Vol. II, pp. 169–172.

Persson, L. (1984) Food evacuation and models for multiple meals in fishes. *Environmental Biology of Fishes*, **10**, 305–309.

Poe, W.E. & Wilson, R.P. (1989) Palatability of diets containing sulfadimethoxine, ormetoprim, and Romet 30 to channel catfish fingerlings. *The Progressive Fish-Culturist*, **51**, 226–228.

Provenza, F.D. & Cincotta, R.P. (1993) Foraging as a self-organizational learning process: accepting adaptability at the expense of predictability. In: *Diet Selection* (ed. R.N. Hughes), pp. 78–101. Blackwell Scientific Publications, Oxford.

Ralphs, M.H. (1997) Persistence of aversions to larkspur in naive and native cattle. *Journal of Range Management*, **50**, 367–370.

Robinson, E.H. & Tucker, C.S. (1992) Palatability of sarafloxin HCl-medicated feed to channel catfish. *Journal of Applied Aquaculture*, **1**, 81–87.

Robinson, E.H., Brent, J.R., Crabtree, J.T. & Tucker, C.S. (1990) Improved palatability of channel catfish feeds containing Romet-30. *Journal of Aquatic Animal Health*, **2**, 43–48.

Rösch, R. (1987) Effect of experimental conditions on the stomach evacuation of *Coregonus lavaretus* L. *Journal of Fish Biology*, **30**, 521–531.

Rozin, P. & Mayer, J. (1961) Regulation of food intake in the goldfish. *American Journal of Physiology*, **201**, 968–974.

Sánchez-Vázquez, F.J., Yamamoto, T., Akiyama, T., Madrid, J.A. & Tabata, M. (1998) Selection of macronutrients by goldfish operating self-feeders. *Physiology and Behavior*, **65**, 211–218.

Sánchez-Vázquez, F.J., Yamamoto, T., Akiyama, T., Madrid, J.A. & Tabata, M. (1999) Macronutrient self-selection through demand-feeders in rainbow trout. *Physiology and Behavior*, **66**, 45–51.

Santinha, P.J.M., Gomes, E.F.S. & Coimbra, J.O. (1996) Effects of protein level of the diet on digestibility and growth of gilthead sea bream, *Sparus aurata* L. *Aquaculture Nutrition*, **2**, 81–87.

Sanz, A., Suarez, M.D., Hidalgo, M.C., García-Gallego, M. & de la Higuera, M. (1993) Feeding of the European eel *Anguilla anguilla*. III. Influence of the relative proportions of the energy yielding nutrients. *Comparative Biochemistry and Physiology*, **105A**, 177–182.

Schade, R. (1982) Untersuchungen zur Nahrungsausnutzung im Darm von Karpfen (*Cyprinus carpio* L.). *Archiv für Hydrobiologia*, **59** (suppl.), 377–415.

Schreck, J.A. & Moffitt, C.M. (1987) Palatability of feed containing different concentrations of erythromycin thiocyanate to Chinook salmon. *The Progressive Fish-Culturist*, **49**, 241–247.

Sierra, M.A. (1995) *La encapsulación como estrategia para establecer las necesidades de metionina y la suplementación de proteina de soja. Consecuencias sobre el recambio proteico tisular y el crecimiento de la dorada* (Sparus aurata). PhD Thesis, University of Granada, Spain.

Simpson, S.J. & Raubenheimer, D. (1999) Assuaging nutritional complexity: a geometric approach. *Proceedings of the Nutrition Society*, **58**, 779–789.

Smith, I.P., Metcalfe, N.B. & Huntingford, F.A. (1995) The effects of food pellet dimensions on feeding responses by Atlantic salmon (*Salmo salar* L.) in a marine net pen. *Aquaculture*, **130**, 167–175.

Stradmeyer, L., Metcalfe, N.B. & Thorpe, J.E. (1988) Effect of food pellet shape and texture on the feeding response of juvenile Atlantic salmon. *Aquaculture*, **73**, 217–228.

Takeda, M. & Takii, K. (1992) Gustation and nutrition in fishes: application to aquaculture. In: *Fish Chemoreception* (ed. T.J. Hara), pp. 271–287. Chapman & Hall, London.

Takeda, M., Takii, K. & Matsui, K. (1984) Identification of feeding stimulants for juvenile eel. *Bulletin of the Japanese Society of Scientific Fisheries*, **59**, 645–651.

Takii, K., Shimeno, S., Takeda, M. & Kamekawa, S. (1986a) The effect of feeding stimulants in diet on digestive enzyme activities of eel. *Bulletin of the Japanese Society of Scientific Fisheries*, **52**, 1449–1454.

Takii, K., Shimeno, S. & Takeda, M. (1986b) The effect of feeding stimulants in diet on some hepatic enzyme activities of eel. *Bulletin of the Japanese Society of Scientific Fisheries* **52**, 2131–2134.

Thebault, H., Alliot, E. & Pastoureaud, A. (1985) Quantitative methionine requirement of juvenile seabass (*Dicentrarchus labrax*). *Aquaculture*, **50**, 75–87.

Toften, H. & Jobling, M. (1997) Feed intake and growth of Atlantic salmon, *Salmo salar* L., fed diets supplemented with oxytetracycline and squid extract. *Aquaculture Nutrition*, **3**, 145–151.

Toften, H., Jørgensen, E.H. & Jobling, M. (1995) The study of feeding preferences using radiography: oxytetracycline as a feeding deterrent and squid extract as a stimulant in diets for Atlantic salmon. *Aquaculture Nutrition*, **1**, 145–149.

van den Ingh, T.S.G.A.M., Olli, J. & Krogdahl, Å. (1996) Alcohol-soluble components in soybeans cause morphological changes in the distal intestine of Atlantic salmon, *Salmo salar* L. *Journal of Fish Diseases*, **19**, 47–53.

Vergara, J.M., Robaina, L., Izquierdo, M. & de la Higuera, M. (1996) Protein sparing effect of lipids in diets for fingerlings of gilthead sea bream. *Fisheries Science*, **62**, 624–628.

Walton, M.J., Cowey, C.B., Coloso, R.M. & Adron, J.W. (1986) Dietary requirements of rainbow trout for tryptophan, lysine and arginine determined by growth and biochemical measurements. *Fish Physiology and Biochemistry*, **2**, 161–169.

Wilson, R.P. (1989) Amino acids and proteins. In: *Fish Nutrition* (ed. J.E. Halver), pp. 111–151. Academic Press, London.

Wilson, R.P. (1994) Utilization of dietary carbohydrate by fish. *Aquaculture*, **124**, 67–80.

Yamada, S., Tanaka, Y. & Katayama, T. (1981) Feeding experiments with carp fry fed an amino acid diet by increasing the number of feedings per day. *Bulletin of the Japanese Society of Scientific Fisheries*, **47**, 1247–1251.

Chapter 12

Regulation of Food Intake by Neuropeptides and Hormones

Nuria de Pedro and Björn Thrandur Björnsson

12.1 Introduction

The mechanisms which control feeding and satiation are complex and multifactorial. Even in mammals, where the mechanisms have been studied for decades, they are not yet clearly defined. It is generally accepted that food intake is under the control of a central feeding system in combination with a peripheral satiation system. The balance between neuropeptides and monoamines constitutes the central feeding system, while the peripheral satiation system involves some gastrointestinal peptides and hormones. Most peptides that influence feeding exert an inhibitory effect, with only a few stimulating food intake (for review, see Morley 1987). Although most information stems from studies on mammals – particularly rodents – some investigations have been carried out on fish. These are sufficient to warrant a review of the neuroendocrine regulation of food intake in fish, using mammalian data to speculate on possible mechanisms of action.

12.2 Inhibitory peptides

12.2.1 Corticotrophin-releasing factor

Corticotrophin-releasing factor (CRF) is a 41-amino acid polypeptide that was first isolated and characterised from the ovine hypothalamus. It is considered to be the major physiological regulator of the secretion of adrenocorticotrophic hormone (ACTH), β-endorphin and other pro-opiomelanocortin-derived peptides from the adenohypophysis (anterior pituitary) (Vale *et al.* 1981). As is the case for a number of other hypothalamic 'factors', its 'status' has sometimes been raised to that of a 'hormone' by calling it corticotrophin-releasing hormone (CRH). The primary structure of CRF is similar in all mammalian and fish species studied (for reviews, see Lederis *et al.* 1990; Owens & Nemeroff 1991). CRF from the sucker, *Catostomus commersoni* (Okawara *et al.* 1988), shows greater homology to porcine, rat and human CRF (it differs in only one or two residues) than it does to CRF from ruminants (seven to eight substituted amino acids). Recently, the amino acid sequence of sockeye salmon, *Oncorhynchus nerka*, CRF has been characterised: it exhibits between 66 and 80% homology with mammalian and sucker CRFs, respectively (Ando *et al.* 1999).

The distribution of neuronal bodies and fibres containing immunoreactive CRF (CRF-ir) has been investigated in the brain and pituitary of several fish species (Fryer & Lederis 1986; Olivereau & Olivereau 1988; González *et al.* 1992; Coto-Montes *et al.* 1994; Mancera & Fernández-Llébrez 1995). In general, most CRF-ir neurones have been localised in the preoptic and lateralis tuberis nuclei, their axons projecting towards the pituitary. In addition, immunopositive neurones and fibres may occur in extrahypothalamic regions, such as the telencephalon and tegmentum mesencephali (González *et al.* 1992; Coto-Montes *et al.* 1994).

Cloning of cDNAs from different tissues and species (rat, mouse, human, chicken and *Xenopus*) has revealed two receptor subtypes, CRF_1 and CRF_2 (for review, see Spiess *et al.* 1998). Although CRF receptors are yet to be characterised in fish, binding sites for ovine CRF have been identified in the pituitary of goldfish, *Carassius auratus* (Fryer & Lederis 1986), and it has been suggested that receptors equivalent to those described in mammals, birds and amphibians are likely to exist in fish (Coto-Montes *et al.* 1994).

The most consistently observed action of CRF in fish is stimulation of the pituitary-adrenal axis (Fryer & Lederis 1986; Tran *et al.* 1990). CRF has also been demonstrated to influence the hypothalamic-pituitary-thyroid axis and food consumption in cyprinids (De Pedro *et al.* 1995a,b). In addition, due to the extensive distribution of CRF within the central nervous system (CNS), it has been suggested that CRF can act as both a neuromodulator and neuro-transmitter in the CNS (González *et al.* 1992; Coto-Montes *et al.* 1994).

Food intake is reduced in response to central administration of CRF in fish: in goldfish at 2 h post-injection (De Pedro *et al.* 1993) and tench, *Tinca tinca*, at 8 h post-injection (De Pedro *et al.* 1995b). CRF-induced decreases in feeding have been described in mammals (rat, mouse, rabbit, pig and monkey; for reviews, see Morley 1987; Glowa *et al.* 1992) and amphibians (*Rana perezi* tadpoles; Gancedo *et al.* 1992), suggesting that the anorectic action of CRF is highly conserved in vertebrates. A central mediation of the anorectic action of CRF is suggested for fish, because peripheral CRF administration does not suppress food intake (De Pedro *et al.* 1993, 1995b). In addition, the intracerebroventricular (i.c.v.) CRF-induced feeding reduction can be blocked by the administration of the antagonist, α-helical $CRF_{[9-41]}$ (De Pedro *et al.* 1997), a finding that is in accord with mammalian data (Glowa *et al.* 1992). However, the exact mechanism by which CRF induces anorexia in fish has not been eluci-dated. Cortisol and/or catecholamines may mediate the CRF-induced feeding reduction in fish, because the decrease in food intake parallels increases in plasma cortisol and reductions in hypothalamic content of noradrenaline (NA) and dopamine (DA) in CRF-treated goldfish; both responses are reversed by the antagonist α-helical $CRF_{[9-41]}$ (Fig. 12.1; De Pedro *et al.* 1997). Different routes of CRF-administration in goldfish have shown that cortisol increases after both i.c.v. and intraperitoneal (i.p.) CRF injections. Decreased feeding was only ob-served after i.c.v. injection (De Pedro *et al.* 1993, 1997), suggesting that the central anorectic action of CRF is independent of pituitary adrenal activation. As for the catecholamines, the blockade of the CRF-induced suppression of food intake by pre-treatment with specific D_1- and D_2-dopaminergic and α_1-adrenergic antagonists, but not α_2-adrenergic antagonists, at 2 h post-injection, suggests that the anorectic effect of CRF may be mediated by α_1-adrenergic and dopaminergic receptors in goldfish (De Pedro *et al.* 1998a). There may also be additional interactions between CRF and the monoaminergic system in the regulation of food intake in

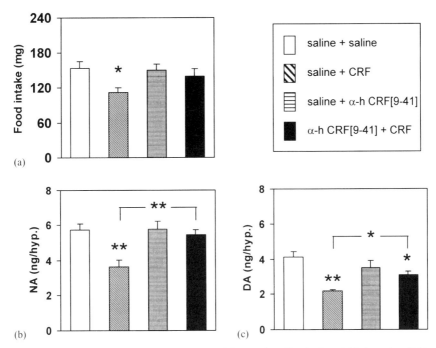

Fig. 12.1   (a) Food intake and hypothalamic content of (b) noradrenaline (NA) and (c) dopamine (DA) at 2 h post-injection after intracerebroventricular (i.c.v.) administration of saline, corticotrophin-releasing factor (CRF) (1 µg), the antagonist α-helical CRF$_{[9-41]}$ (10 µg), and CRF and antagonist in goldfish. (Duncan test, $P < 0.05$, $P < 0.01$ compared with control group). (From De Pedro *et al.* 1997.)

fish. CRF and serotonin may act together to inhibit feeding, with CRF mediating the anorectic action of serotonin in goldfish (De Pedro *et al.* 1998b).

12.2.2 Bombesin

The tetrapeptide bombesin (BBS) was originally isolated from the skin of the frog *Bombina bombina* (Anastasi *et al.* 1971). This peptide belongs to a family of BBS-like peptides present throughout all vertebrate classes (Vigna & Thorndyke 1989). The carboxyl terminal hepta-peptide is strongly conserved in all members of this peptide family (for review, see Dietrich 1994). The family of BBS-like peptides has been divided into three (bombesins, ranatensins and phylitorins) and two subfamilies [neuromedins and gastrin-releasing peptides (GRPs)] in amphibians and mammals, respectively (Dietrich 1994). BBS-like peptides have also been described in fish, although to date the only known GRP sequence is that from rainbow trout, *Oncorhynchus mykiss* (Jensen & Conlon 1992).

BBS/GRP-like immunoreactive (ir) peptides have been found in the CNS of the lesser spotted dogfish, *Scyliorhinus canicula* (Vallarino *et al.* 1990), and goldfish (Himick & Peter 1995a). In both species, a prominent BBS-like ir fibre system was detected in the ventral tel-encephalon, diencephalon (inferior lobes and ventroposterior regions of the hypothalamus) and the hypothalamo-pituitary neurosecretory system. BBS-like peptides have also been observed in endocrine cells and nerve fibres of the gastrointestinal and cardiovascular system of several fish species (Kiliaan *et al.* 1992; Holmgren *et al.* 1994).

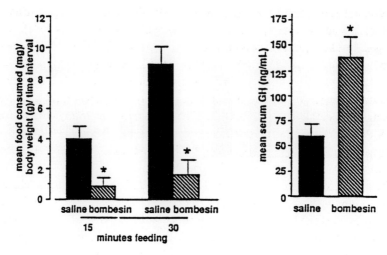

Fig. 12.2   Effects of i.c.v. injection of bombesin (60 ng/g) on cumulative food intake and plasma growth hormone (GH) levels in goldfish. (Student's *t*-test, $P < 0.05$). (From Himick & Peter 1994a.)

Binding sites for BBS/GRP-like peptides have been demonstrated in the goldfish brain and pituitary (Himick *et al.* 1995). High densities of binding sites were localised in the ventral and medial regions of the hypothalamus, and in the neurointermediate lobe of the pituitary. Such BBS/GRP binding sites display high affinity for members of the BBS/GRP subfamily of peptides, but exhibit lower affinity for neuromedins. This is in accordance with findings from mammals, where two classes of receptors have been identified; one mainly interacts with BBS and GRP peptides, and the other has a high affinity for neuromedins (for review, see Dietrich 1994). It is likely that a high-affinity neuromedin receptor also exists in fish, although it has not yet been characterised.

BBS-related peptides may have roles in gastrointestinal and cardiovascular regulation (Kiliaan *et al.* 1992; Holmgren *et al.* 1994), pituitary hormone release (Himick & Peter 1994a, 1995a) and feeding (Himick & Peter 1994a) in teleosts. Himick and Peter (1994a) demonstrated that bombesin acutely (at 45 min post-injection) suppresses food intake in the goldfish following either central (Fig. 12.2) or peripheral administration, as is the case in mammals and birds (for reviews, see Morley 1987, 1995; Rowland & Kalra 1997). Concomitant increases in plasma growth hormone (GH) levels were detected in goldfish (Himick & Peter 1994a), but the possible relationship between bombesin and GH in the suppression of food intake remains to be clarified. BBS influences the gastrointestinal tract in fish (Kiliaan *et al.* 1992; Holmgren *et al.* 1994), so the peptide could exert its anorectic effect through signals arising at the gastrointestinal level (gastrointestinal contractility, gut emptying, etc.) (Himick & Peter 1994a).

12.2.3 Cholecystokinin

Cholecystokinin (CCK) is a peptide which occurs in two main forms: the primary gastrointestinal form which contains 33 or 39 amino acids, and the carboxy-terminal octapeptide (CCK-8) which is found both in the CNS and in the gut (for a review, see Baile *et al.* 1986).

Within vertebrates, the members of the CCK/gastrin family of peptides appear to have a common biologically active carboxy-terminal pentapeptide sequence (Norris 1985).

CCK/gastrin-like peptides have been detected in the CNS of numerous fish species (Vigna *et al.* 1985; Moons *et al.* 1992; Himick & Peter 1994b). There is a widespread distribution of CCK/gastrin-like ir fibres and cell bodies in the brain, with the greatest concentration of CCK/gastrin ir being in the ventrocaudal regions of the hypothalamus. CCK/gastrin-like immunoreactivity has also been localised in the fish gut (Vigna *et al.* 1985; Kiliaan *et al.* 1992; Himick & Peter 1994b) and pancreas (Eilertson *et al.* 1996).

CCK/gastrin binding sites have been reported in the brain and pituitary of sea bass, *Dicentrarchus labrax* (Moons *et al.* 1992) and goldfish (Himick *et al.* 1996), with high densities within the telencephalon and hypothalamus. Himick *et al.* (1996) have also characterised CCK/gastrin binding sites in the CNS of goldfish. There appears to be a single, saturable, high-affinity binding site for the sulphated forms of CCK and gastrin. These findings provide support for the existence of a single CCK/gastrin receptor in ectotherms. This might represent the origin of the two CCK receptor subtypes of endotherms: the CCK-A and CCK-B receptors (Vigna *et al.* 1986).

CCK in fish is strongly identified with regulation of gastrointestinal function: secretion of gastric acid and pancreatic enzymes, gallbladder contraction and gut motility (for a review, see Vigna 1983). CCK also modulates the release of pancreatic somatostatins in rainbow trout (Eilertson *et al.* 1996), and plays a role in the central regulation of food intake and pituitary hormone release in goldfish (Himick & Peter 1994b). CCK has potent satiation effects in goldfish, with either i.c.v. or i.p. (Fig. 12.3) injections of the sulphated octapeptide CCK (CCK-8s) leading to reduced food intake at 45 min post-injection (Himick & Peter 1994b). CCK also decreases feeding in a broad range of species, including mammals, birds and molluscs (for reviews, see Morley 1995; Le Bail & Boeuf 1997). The effects of CCK on food intake seem to involve peripheral satiation mechanisms in some species, whereas in others the major site of action appears to be the CNS.

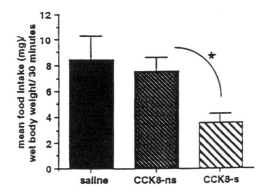

Fig. 12.3   Food intake of goldfish after i.p. injection of either non-sulphated CCK-8 (CCK-8ns; 95.6 ng/g) or CCK-8s (100 ng/g). (Duncan test, $P < 0.05$). (From Himick & Peter 1994b.)

12.3 Stimulatory peptides

12.3.1 Galanin

Galanin is a 29-amino acid peptide which was first isolated from the porcine small intestine. The name galanin arises from the amino-terminal residues glycine and amidated carboxy-terminal alanine residues (Tatemoto *et al.* 1983). The amino acid sequence of galanin has been determined for representatives of different vertebrate groups, including mammals (rat, dog, cow, sheep, pig, human), birds (chicken) and reptiles (alligator), and the first fifteen amino-terminal residues have been found to be identical (for a review, see Crawley 1995). Fish galanin has also been isolated and sequenced, and a highly conserved amino-terminal region has been described. The galanin isolated from rainbow trout, lesser spotted dogfish and bowfin, *Amia calva*, stomach extracts is identical to that of other vertebrates with respect to the first fourteen (trout, Anglade *et al.* 1994) or fifteen residues (bowfin and dogfish, Wang & Conlon 1994). Habu *et al.* (1994) isolated and sequenced galanin from the pituitary of yellowfin tuna, *Thunnus albacares*, and found two substitutions in the amino-terminal 1–15 sequence.

Immunohistochemical studies have provided evidence for an extensive system of galanin-ir neurones in the brain of several species of teleosts (Olivereau & Olivereau 1991; for review, see Merchenthaler *et al.* 1993) and in the brain of an elasmobranch, the lesser spotted dogfish (Vallarino *et al.* 1991). Galanin-ir is similarly distributed, with neuronal populations in the rostral preoptic area and ventral hypothalamus, and galanin-ir fibres in several brain areas. The hypothalamus-pituitary region, particularly the preoptic periventricularis, preoptic and lateralis tuberis nuclei, has the most dense concentration of galanin-ir fibres and cell bodies of the CNS. Galanin-like peptides have also been detected in the peripheral nervous system of fish, there being dense populations of galanin-ir in nerve fibres in the gastrointestinal tract, heart and lung (Kiliaan *et al.* 1993; Holmgren *et al.* 1994).

Autoradiographic studies have indicated that there are galanin receptors in several brain regions and in the pituitary of teleosts (Moons *et al.* 1991; Holmqvist & Carlberg 1992). There are similar distributions in Atlantic salmon, *Salmo salar*, and sea bass; ^{125}I-galanin binding sites occur in the telencephalon, tuberal and posterior hypothalamus, thalamus, ventral medulla oblongata and tectum opticum. The most dense galanin binding sites were found in the posterior hypothalamus, which contains an abundant galanin-like ir fibre innervation (Moons *et al.* 1991). Although this provides anatomical evidence of galanin-specific binding sites in the teleost brain and pituitary, it remains to be demonstrated that these binding sites are functional receptors.

The widespread and correlated distribution of galanin and its putative receptors throughout the peripheral and central nervous system suggests that galanin has physiological functions in fish. This is substantiated by findings showing that galanin is involved in the neuroendocrine control of the pituitary pars distalis (Merchenthaler *et al.* 1993; Power *et al.* 1996), the release of pancreatic somatostatin (Eilertson *et al.* 1996), cardiovascular and gastrointestinal nervous control (Kiliaan *et al.* 1993; Holmgren *et al.* 1994; Le Mével *et al.* 1998), and the regulation of food intake (De Pedro *et al.* 1995c; Guijarro *et al.* 1999). Intracerebroventricular administration of galanin results in increased food intake in two species of cyprinids, goldfish (De Pedro *et al.* 1995c) and tench (Guijarro *et al.* 1999), at 2 and 8 h after the injections.

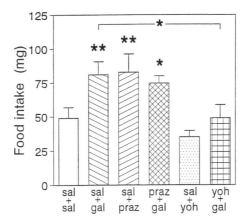

Fig. 12.4   Food intake at 2 h post-injection after i.c.v. administration of saline (group 1); 1 μg galanin (group 2); 10 μg prazosin (α_1-adregenergic antagonist, group 3); 10 μg prazosin + 1 μg galanin (group 4); 10 μg yohimbine (α_2-adrenergic antagonist, group 5); 10 μg yohimbine + 1 μg galanin (group 6) in goldfish. (Duncan test, $P < 0.05$, $P < 0.01$ compared with control group). (From De Pedro *et al.* 1995c.)

This is similar to the findings reported for mammals (for reviews, see Morley 1987; Leibowitz 1995; Rowland & Kalra 1997). Peripherally injected galanin does not stimulate feeding in either goldfish (De Pedro *et al.* 1995c) or tench (Guijarro *et al.* 1999), suggesting a central role for this peptide in the regulation of feeding in cyprinids. It is possible that the differences seen in the responses to exogenous galanin could be due to a more rapid metabolism of the molecule in the periphery than in cerebrospinal fluid. In mammals, the half-life of galanin is 100 min in the hypothalamus compared with 4.6 min in plasma, indicating that galanin is unstable in the peripheral circulation (Merchenthaler *et al.* 1993).

Central administration of galantide, a potent antagonist of galanin receptors, blocked the orexigenic effects of exogenous galanin in goldfish at 2 h post-injection (De Pedro *et al.* 1995c). Similar findings have been obtained with other antagonists in the rat (for review, see Leibowitz 1995). The ability of galanin to stimulate feeding appears to be dependent on α_2-adrenergic activation following the release of NA. An α_2-adrenergic receptor antagonist, but not an α_1-receptor antagonist, counteracted the stimulatory effect of galanin in goldfish (Fig. 12.4; De Pedro *et al.* 1995c), and there is a strong positive correlation between galanin-stimulated NA levels and the magnitude of the feeding response in rats (Kyrkouli *et al.* 1992).

Repeated central administration of galanin leads to prolonged hyperphagia and weight gain in mammals (Morley 1987; Leibowitz 1995). However, chronic i.c.v. treatment (nine days) with this peptide in goldfish only resulted in increased food intake at 8 h post-injection during the first day of treatment, and did not have any effect on body weight (N. De Pedro & A.I. Guijarro, unpublished data). Thus, it seems possible that goldfish may develop tolerance to the orexigenic effects of galanin, but further studies are needed to confirm this.

12.3.2 Opioid peptides

Endogenous opioid peptides have been classified into three major categories: endorphins, enkephalins and dynorphins. The representatives of each category are derived from a different precursor protein, pro-opiomelanocortin (POMC), proenkephalin and prodynorphin, respectively (for a review, see Simon & Hiller 1994). β-Endorphin, the predominant opioid peptide derived from POMC, is the only endogenous opioid peptide that has been shown to play a role in the control of food intake in fish. β-Endorphin is a 29- to 31-amino acid polypeptide, and its sequence has been determined in different vertebrate groups, including mammals, birds, amphibians and fish (Dores *et al.* 1990). Although β-endorphin is highly conserved throughout phylogeny, sequence comparison reveals some differences between tetrapods and fish, mainly in the carboxy-terminal region (for reviews, see Dores *et al.* 1990; Arends *et al.* 1998).

β-Endorphin-like immunoreactivity has been observed in the brain and pituitary of fish (Dores *et al.* 1990, 1993). Cell bodies and fibres staining for β-endorphin are mainly localised in the hypothalamus, particularly in the preoptic and lateralis tuberis nuclei. Similar observations have been made on mammals and amphibians (Vallarino 1985).

In mammals, opioid receptors have been pharmacologically classified into mu (mu_1, mu_2), delta ($delta_1$, $delta_2$) and kappa ($kappa_1$, $kappa_2$, $kappa_3$) subtypes (for a review, see Simon & Hiller 1994). Binding studies using selective radioligands have identified mu and kappa receptors in the brain of goldfish (Brooks *et al.* 1994) and coho salmon, *Oncorhynchus kisutch* (Ebbesson *et al.* 1996). Darlison *et al.* (1997) identified a family of opioid receptor-like proteins in the sucker. The sequences of these are related to the rat mu-, kappa- and delta-opioid receptors. To date, pharmacological and functional properties have only been investigated with respect to the mu-opioid receptors. The mu-opioid receptors have been highly conserved during the course of vertebrate evolution, suggesting important physiological roles (Darlison *et al.* 1997).

In fish, opioid neuropeptides have been implicated in the control of a variety of behavioural and physiological processes, including fear habituation (Olson *et al.* 1978), nociception (Enrensing & Mitchell 1982), thermoregulation (Kavaliers 1983), shoaling behaviour (Kavaliers 1989), gonadotrophin secretion (Rosenblum & Peter 1989), parr-smolt transformation (Ebbesson *et al.* 1996) and the regulation of food intake (De Pedro *et al.* 1995d; Guijarro *et al.* 1999).

Treatment with exogenous β-endorphin leads to increased food intake in goldfish and tench shortly (2 and 8 h) after i.c.v. administration, whereas i.p. treatment is without effect (De Pedro *et al.* 1995d; Guijarro *et al.* 1999). These results are in agreement with findings for endotherms, where β-endorphin exerts a stimulatory effect on food ingestion by acting centrally (Baile *et al.* 1986; Morley 1987, 1995).

The involvement of endogenous opioids in the regulation of feeding in goldfish has been demonstrated using selective agonists (kappa, U-50488; mu, DAMGO; delta, DPEN) and antagonists (general, naloxone; kappa, nor-BNI; mu, β-FNA; mu_1, naloxonazine; $delta_1$, BNTX; $delta_2$, naltriben) of opioid receptors (De Pedro *et al.* 1995d, 1996). The results are in general agreement with those obtained in studies on endotherms (Olson *et al.* 1998): CNS-administration of opioid receptor agonists stimulates feeding, whereas opioid antagonists either reduce ingestion or are without effect (Fig. 12.5). The ability of kappa, mu and delta antagonists to block the stimulatory effect of β-endorphin on feeding at 2 h post-injection (Fig. 12.5) indicates that such stimulation is mediated via mu-opioidergic receptors, involving, at least, the mu_1 subtype (De Pedro *et al.* 1996).

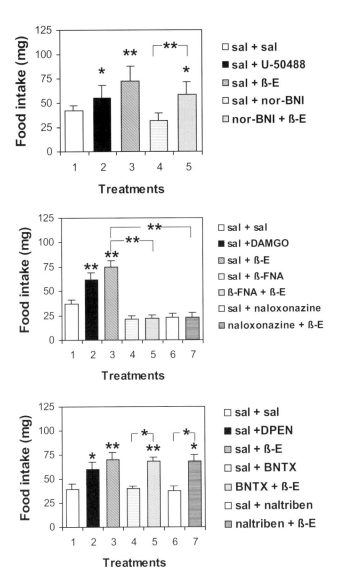

Fig. 12.5   Effects of i.c.v. administration of agonists (1 µg) and antagonists (5 µg) of opioidergic receptors on feeding at 2 h post-injection in goldfish. Top (kappa): 1 µl saline alone (group 1); 1 µg U-50488 (kappa agonist, group 2); 1 µg β-endorphin (group 3); 5 µg nor-binaltorphamine (kappa antagonist, group 4); 5 µg nor-binaltorphamine + 1 µg β-endorphin (group 5). Middle (mu): 1 µl saline alone (group 1); 1 µg DAMGO (mu agonist, group 2); 1 µg β-endorphin (group 3); 5 µg β-funaltrexamine (mu antagonist, group 4); 5 µg β-funaltrexamine + 1 µg β-endorphin (group 5); 5 µg naloxonazine (mu$_1$ antagonist, group 6); 5 µg naloxonazine + 1 µg β-endorphin (group 7). Bottom (delta): 1 µl saline alone (group 1); 1 µg DPEN (delta agonist, group 2); 1 µg β-endorphin (group 3); 5 µg BNTX (delta$_1$ antagonist, group 4); 5 µg BNTX + 1 µg β-endorphin (group 5); 5 µg naltriben (delta$_2$ antagonist, group 6); 5 µg naltriben + 1 µg β-endorphin (group 7). (Duncan test, $P < 0.05$, $P < 0.01$ compared with control group). (From De Pedro *et al.* 1996.)

Prolonged administration of β-endorphin to central sites over nine days decreased food intake (8 h) and body weight in goldfish (López-Patiño *et al.* 1999a). These findings parallel those from mammalian studies (for reviews, see Baile *et al.* 1986; Olson *et al.* 1998), and indicate that β-endorphin may exert different effects depending on the pattern of administration. Although tolerance to β-endorphin could have been developed, other factors may also be involved. Opioid agonists delay gastrointestinal emptying in mammals (Olson *et al.* 1998), so it is possible that chronic β-endorphin administration reduces gastrointestinal emptying, leading to the generation of a satiation signal that inhibits food intake and weight gain.

12.3.3 Neuropeptide Y

Neuropeptide Y (NPY) is an amidated 36-amino acid peptide belonging to a family that also includes peptide YY (PYY) and pancreatic polypeptide (PP) in tetrapods, and peptide Y (PY) in certain fish (for a review, see Larhammar 1996). NPY was first isolated (and characterised) from porcine brain (Tatemoto 1982). The structure of NPY has remained extremely well conserved throughout vertebrate phylogeny: of the thirty-six amino acid residues, twenty-two are identical for the eighteen known sequences of NPY from different mammals, birds, reptiles, amphibians and fish (for reviews, see Larhammar 1996; Larhammar *et al.* 1997).

NPY-like immunoreactivity occurs widely in the brain of several fish species (Vecino & Ekström 1992; Chiba 1997), with the highest density in the diencephalon. Silverstein *et al.* (1998a) demonstrated that the optic tectum, caudoventral telencephalon, thalamus, pre-optic area and caudal hypothalamus of Pacific salmon species (*Onchorhynchus kisutch* and *Onchorhynchus tshawytscha*) express NPY-like mRNA. Immunohistochemical analysis of peripheral organs has shown NPY to be present in the endocrine pancreas and intestine of the sea lamprey, *Petromyzon marinus* (Cheung *et al.* 1991).

There is pharmacological evidence for the existence of six NPY receptor family subtypes, and five subtypes (Y_1, Y_2, Y_4, Y_5 and Y_6) have been cloned in mammals (for a review, see Larhammar *et al.* 1997). In fish, three NPY receptor genes (zYa, zYb and zYc) have been identified in the zebrafish, *Danio rerio* (Larhammar *et al.* 1997), and the sequence of the cod, *Gadus morhua*, Yb gene has been elucidated (Arvidsson *et al.* 1998). The receptors described in fish are more similar to the mammalian Y_1-Y_4-Y_6 subfamily than to Y_2 or Y_5. However, it remains to be resolved whether additional subtypes are still to be discovered in each lineage, or whether fish and mammals have distinct repertoires of NPY receptor subtypes.

The widespread and abundant expression of NPY in the nervous system of all vertebrates investigated is reflected in a multitude of physiological effects. Stimulation of secretion of pituitary hormones, such as GH and gonadotrophins (Peng *et al.* 1990), and the stimulation of feeding (López-Patiño *et al.* 1999b) have been described in fish. Studies on goldfish provide evidence for the involvement of NPY in the control of food intake. López-Patiño *et al.* (1999b) found that i.c.v. administration of NPY stimulated feeding in goldfish at 2 h after the injections, whereas no significant effects on food intake were observed after peripheral injections. This suggests a central stimulatory action of this peptide, as is the case in other vertebrates [mammals (Morley 1987; Leibowitz 1995; Rowland & Kalra 1997), birds (Richardson *et al.* 1995) and reptiles (Morris & Crews 1990)].

The NPY receptor subtype involved in food intake regulation has been described as 'Y_1-like', because agonists selective for this receptor subtype, such as [Leu[31], Pro[34]]-NPY, but not Y_2 agonists, stimulate feeding in goldfish at 2 h after central administration (De Pedro *et al.* 2000). Further, administration of [D-Tyr[27,36], D-Thr[32]]-NPY (27-36), a general antagonist for

NPY receptors, counteracts the increased feeding induced by NPY and the Y_1 agonist (López-Patiño *et al.* 1999b). These findings are in agreement with data for mammals, although the Y_5 receptor also seems to be involved in the regulation of feeding in rats (for a review, see Rowland & Kalra 1997). However, it is not clear whether the Y_1 and Y_5 receptor subtypes mediate the NPY-induced stimulation of feeding, or if a novel receptor subtype with general characteristics of both Y_1 and Y_5 receptors exists.

The mechanisms by which NPY stimulates feeding are still not known. Central administration of naloxone decreases NPY-induced feeding in goldfish at 2 h post-injection (De Pedro *et al.* 2000), as in mammals (for a review, see Levine & Billington 1997). This is indicative of an interaction between NPY and opioid systems in food intake regulation in vertebrates. NPY may lead to initiation of food intake, and opioids then maintain feeding.

NPY is involved in the feeding evoked following food deprivation in mammals, implying that hypothalamic NPY released during a fast generates the feeding stimulation (Yoshihara *et al.* 1996; Levine & Billington 1997). A similar mechanism has been suggested in fish; an increase in NPY-like gene expression was observed in the preoptic area of the hypothalamus after three weeks of fasting in Pacific salmon (Silverstein *et al.* 1998a). Further, following a 24- or 72-h period of food deprivation, goldfish increased feeding to a similar degree to that after central injection of NPY. The administration of a general NPY antagonist counteracted the stimulation of food intake induced by a 24-h fast (Fig. 12.6) (López-Patiño *et al.* 1999b).

The body weight of goldfish remained unchanged following chronic i.c.v. treatment (eight days) with NPY (M.A. López-Patiño & N. De Pedro, unpublished data): tolerance may have developed because food intake was increased only during the first two days of treatment.

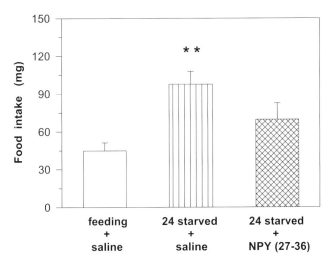

Fig. 12.6   Food intake of goldfish at 2 h post-injection. Feeding + saline: fish daily fed and injected i.c.v. with 1 µl saline; 24 h-starved + saline: fish injected i.c.v. with 1 µl saline after 24 h starvation; 24 h-starved + NPY (27-36): fish injected i.c.v. with 5 µg neuropeptide Y receptor antagonist [D-Tyr[27-36], D-Thr[32]]-NPY (27-36) after 24 h starvation. (Student-Newman-Keuls test, $P < 0.01$ compared with feeding + saline group). (From López-Patiño *et al.* 1999b.)

12.4 Hormones

12.4.1 Growth hormone and insulin-like growth factor I

Since the mid-1980s, the amino acid sequence of growth hormone (GH) – also termed soma-totrophin – has been elucidated for over thirty fish species, either by direct isolation and amino acid sequencing, or by deduction from the encoding gene sequence. The GH molecule is a single-chain polypeptide, usually consisting of 190 amino acid residues, shaped into a four-helix bundle, with both species-variant and -invariant regions. Phylogenetic comparison of teleost GH, encompassing sequences from twenty-six species (Rubin & Dores 1995), reveals relatively large species divergence. This indicates that several bursts of GH evolution have taken place in the teleosts, possibly reflecting changes in the functions of the hormone that are additional to the basic growth-promoting actions (Wallis 1996). The divergence also explains why fish GH immunoassays are usually valid only for single species or groups of closely related species.

GH is produced in, and secreted by, endocrine cells – the somatotrophs – located in the proximal pars distalis of the anterior pituitary. The control of GH secretion in mammals is thought to be under the dual regulation of hypothalamic GH-releasing factor (GRF) and so-matotrophin release-inhibiting factor (SRIF, somatostatin). These two factors act in concert to control GH secretion, where SRIF controls the baseline secretion and GRF regulates the GH secretion pulses (Arimura & Culler 1985; Giustina & Wehrenberg 1995). These factors are secreted into the portal blood vessels running from the hypothalamus to the pituitary via the pituitary stalk.

The regulation of pituitary GH secretion in teleosts and mammals has certain similarities, although brain structure differs between teleosts and other vertebrates. The pituitary portal system is absent in teleosts, and it appears that hypothalamic neurones enter the pars distalis and either directly innervate the hormone-producing cells or release regulatory peptides in their proximity (Peter *et al.* 1990). While GRF and SRIF seem to have similar GH-regulatory functions in fish and mammals (Luo & McKeown 1991), additional hypothalamic GH in-hibitors [NA and serotonin (5-HT)], and secretagogues [gonadotrophin-releasing hormone (GnRH), DA, NPY, thyrotrophin-releasing hormone (TRH) and CCK] have been identified in the goldfish (for a review, see Peter & Marchant 1995). In salmonids, both DA (Agustsson *et al.* 2000) and GnRH have been recognised as GH secretagogues (for a review, see Holloway & Leatherland 1998).

The molecular structure of the GH receptor has not yet been characterised in fish. How-ever, it appears that the receptor-binding characteristics of the GH molecule are highly con-served, as interspecific biological activity is commonly found. Mammalian GHs (e.g. ovine and bovine) have been widely used to investigate the roles of GH in fish, because homologous hormones have usually not been available. Comparisons between mammalian and piscine GHs are limited (Bolton *et al.* 1987; Le Gac *et al.* 1992; MacLatchy *et al.* 1992; Fine *et al.* 1993; O'Connor *et al.* 1993), but the GHs appear to have similar physiological effects when tested on fish, even though potencies may differ. Further, recombinant carp GH has been tested on human lymphoma cells with lactogenic (prolactin; PRL) receptors, and on human preadipocyte cells having somatogenic (GH) receptors. The bioactivity of the carp GH was found to be 0.01 and 6–10% that of human GH, on the respective preparations (Fine *et al.*

1993). Thus, it appears likely that GH receptors in fish are similar to the mammalian somatogenic GH receptor, which is a single membrane-spanning protein of about 620 amino acids (Kelly *et al.* 1991, 1994).

Radioreceptor studies on a number of teleost species have demonstrated the presence of a single class of high-affinity, low-capacity binding sites for GH in various tissues, including liver, ovaries, testis, brain, gill, intestine, muscle, cartilage, head kidney and posterior body kidney (Gray *et al.* 1990; Hirano 1991; Sakamoto & Hirano 1991; Yao *et al.* 1991; Le Gac *et al.* 1992; Ng *et al.* 1992; Pérez-Sánchez *et al.* 1994a; Calduch-Giner *et al.* 1995; Zhang & Marchant 1996). The greatest density of GH receptors is in the liver, where GH stimulates the production and secretion of insulin-like growth factor I (IGF-I) (Funkenstein *et al.* 1989; Duan & Hirano 1992; Moriyama 1995).

As the widespread tissue distribution of putative GH receptors indicates, GH is a multifunctional hormone in fish. It is a major regulator of parr-smolt transformation in anadromous salmonids, and has a seawater-adapting hypo-osmoregulatory role (for a review, see Björnsson 1997). A similar osmoregulatory role has recently been indicated in tilapia, *Oreochromis mossambicus* (Morgan *et al.* 1997; Sakamoto *et al.* 1997). There are increasing indications that GH has roles in teleost sexual maturation (for reviews, see Björnsson *et al.* 1994; Holloway & Leatherland 1998), and protein and lipid metabolism (Sheridan 1986; Foster *et al.* 1991; O'Connor *et al.* 1993; Fauconneau *et al.* 1996), as well as having effects on behaviour, e.g. increasing dominance, aggression and food intake, and decreasing predator avoidance (Johnsson & Björnsson 1994; Johnsson *et al.* 1996; Jönsson *et al.* 1996, 1998).

The best documented function of GH in fish, as in other vertebrates, is that of growth promotion. In-vivo treatment using either piscine or mammalian GH increases fish growth, including that of salmonid species (for a review, see Björnsson 1997), tilapia (Clarke *et al.* 1977; Shepherd *et al.* 1997), carp, *Cyprinus carpio* (Fu *et al.* 1998) and sea bream, *Sparus aurata* (Ben-Atia *et al.* 1999). Furthermore, the overexpression of the GH gene in GH-transgenic fish results in greatly increased growth (Devlin *et al.* 1994, 1995; Martínez *et al.* 1996).

GH can induce skeletal growth under sub-maintenance feeding conditions (Johnsson & Björnsson 1994), illustrating the fundamental stimulatory effect of GH on skeletal growth. This, together with the effects on nutrient partitioning, favouring protein accretion and lipid utilisation, leads to a lower weight-to-length ratio (condition factor) in GH-treated fish (for a review, see Björnsson 1997). Although the effects of GH on skeletal growth and weight gain have been thoroughly documented, the mechanisms by which GH exerts its effects have been much less studied. At the whole-animal level, a sustained growth increase can only be achieved by increased food intake and/or improved food utilisation. GH treatment results in improved feed conversion efficiency (Markert *et al.* 1977; Gill *et al.* 1985; Garber *et al.* 1995), to the extent of improving weight gain without any significant increase in food intake (Markert *et al.* 1977). However, GH-treated and GH-transgenic fish may also have increased food intake relative to controls (Johnsson & Björnsson 1994; Devlin *et al.* 1999), although the mechanisms by which GH stimulates food intake are so far unknown. Two major pathways can be envisaged: a direct stimulation at the level of the CNS, or an indirect effect resulting from stimulation of peripheral tissue growth or metabolic changes, leading to feed-back signals to the brain. While the latter is likely, the former should not be dismissed because GH receptors have been identified in the brain, and GH can be detected in the cerebrospinal fluid of salmonids (V. Johansson & B.Th. Björnsson, unpublished data). The origin of this GH, as well as the biological properties of the blood–brain barrier in fish, are still unclear.

The biological effects of GH are to some extent mediated by IGF-I. This is a 70-amino acid-long polypeptide, with a highly conserved sequence among fish species and from fish to mammalian species (Cao *et al.* 1989; Reinecke *et al.* 1997). The 'dual effector' integration of the GH-IGF-I system (Green *et al.* 1985) has been examined in greatest detail in mammals, but data on the regulation of cartilage growth in fish indicate that the mechanisms may be similar (Gray & Kelley 1991; Tsai *et al.* 1994; Cheng & Chen 1995). According to the dual effector hypothesis, GH both stimulates IGF-I secretion and increases tissue sensitivity to IGF-I; IGF-I, in turn, stimulates cell differentiation, growth, and proliferation.

IGF-I levels in fish increase during periods of increased GH levels, such as during salmonid smoltification (McCormick *et al.* 2000) or following GH treatment (Moriyama 1995). However, during prolonged feed deprivation, there is a separation of the GH-IGF-I axis. GH levels increase (Marchelidon *et al.* 1996; Björnsson 1997), whereas hepatic GH-binding sites (Pérez-Sánchez *et al.* 1994b), IGF-I mRNA expression (Duan & Plisetskaya 1993) and circulating IGF-I levels decrease (Moriyama *et al.* 1994; Pérez-Sánchez *et al.* 1994b). It is tempting to speculate that the function of the GH increase during feed deprivation is to mobilise energy reserves, while the IGF-I part of the axis is down-regulated to minimise stimulation of cellular growth.

In rainbow trout, GH levels have been found to increase following a meal, and to be generally higher the better-fed the fish (Reddy & Leatherland 1995); data collected from goldfish suggest that changes in BBS and CCK provide links to the neuroendocrine system for the postprandial regulation of GH (Himick & Peter 1995b).

12.4.2 *Hormones regulating metabolism*

Hormones that regulate metabolism may not necessarily be directly involved in the regulation of food intake, but it is likely that they affect food intake indirectly through their effects on energy balance and nutrient status. In fish, the evidence linking thyroid hormones, insulin, glucagon, and glucagon-like peptide to regulation of food intake is, however, too sparse to justify anything more than a brief summary of these endocrine systems.

Thyroid hormones

The main iodothyronine hormone produced in and secreted by the teleost thyroid gland is usually thyroxine (T_4). The more biologically active form, triiodothyronine (T_3), is therefore largely produced peripherally by deiodinase activity. The thyroid hormones are recognised as having important roles at various ontogenetic developmental stages such as during larval growth, flatfish metamorphosis and salmonid smoltification. However, they also participate in regulating growth and metabolism of fish (for reviews, see Eales 1985; Leatherland 1994). Nutritional status influences the thyroid hormone system. During feed deprivation, both T_3 and T_4 levels decreased in rainbow trout (Flood & Eales 1983; Leatherland & Farbridge 1992), but in the Nile tilapia, *Oreochromis niloticus*, only T_3 levels decreased (Toguyeni *et al.* 1996). Following refeeding of fasted trout, a rapid increase in T_4, but not T_3, was seen

(Himick & Eales 1990). This may be coupled to a postprandial increase in glucose, because glucose injections induced similar effects. Thyroid hormone treatment has been reported to increase food intake and growth (Fagerlund *et al.* 1980; Refstie 1982): these studies were carried out on juvenile salmonids which makes it difficult to separate effects on growth and food intake from possible smoltification-related effects. There is no convincing evidence to strongly tie thyroid hormones with the short-term regulation of food intake.

Insulin, glucagon and glucagon-like peptides (GLPs)

These peptide hormones are secreted by the endocrine pancreas, which is organised as a discrete organ, the Brockmann bodies, in many teleost species (for a review, see Mommsen & Plisetskaya 1991). All these hormones exert either an anabolic or a catabolic influence on the intermediary metabolism of fish. Insulin consists of two chains, usually 21 and 31 amino acids, joined by two disulphide bridges; it is an important anabolic regulator of carbohydrate, protein and lipid metabolism, increasing tissue stores of glycogen, protein and lipids. Glucagon is a 29 amino acid-long peptide; it opposes the actions of insulin, in fish as in mammals, mainly through glycogenolytic action. Glucagon-like peptide 1 (GLP-1) has an insulinotropic, catabolic, function in mammals, but exerts gluconeogenic, glycogenolytic and lipolytic actions on the fish liver, making it a 'superglucagon' (for a review, see Plisetskaya 1995).

As regulators of intermediate metabolism, these hormones respond to the nutritional balance of the fish, e.g. postprandial changes in glucose levels (for a review, see Le Bail & Boeuf 1997). Any influence on the regulation of food intake may be indirect via influences on the peripheral satiation system, but the presence of insulin and insulin receptors in the brain (Mommsen & Plisetskaya 1991; Plisetskaya *et al.* 1993) makes central regulatory influences a possibility.

Leptin

Leptin is a 167-amino acid protein, which after signal peptide cleavage (21 amino acids) circulates as a 16 kDa peptide. The leptin gene has been cloned from several mammalian species, but not yet from fish. It appears that adipose tissue is the major site for expression of the gene, and that leptin is produced in proportion to adipose tissue mass (for reviews, see Blum 1997; Friedman 1998).

A group of high-affinity, membrane-bound, leptin receptors (OB-R) have been characterised in mammals: five isoforms with a short cytoplasmic tail, and a single isoform with a long (303 amino acids) tail. The short forms of the OB-R are found in various tissues, whereas the long form is found predominantly in the regions of the hypothalamus which have been implicated in the regulation of feeding and body weight. Leptin is now proposed to have a number of physiological roles in mammalian species, including several related to the regulation of food intake and energy balance, the timing of sexual maturation (puberty), and fertility in mature females (Buchanan *et al.* 1998; Houseknecht *et al.* 1998; Rosenbaum & Leibel 1998; Vuagnat *et al.* 1998). These effects of leptin were initially believed to be mediated solely by interaction of leptin with hypothalamic receptors in areas that regulate feeding, metabolism and reproduction, but several direct effects on peripheral target tissues outside the CNS have now been demonstrated.

The ways in which the brain is able to monitor the size of the fat stores (energy reserves) are still not completely understood, but signalling mechanisms involving leptin may be important. A variety of signalling mechanisms and pathways have been hypothesised, primarily based on the 'lipostatic model' (Kennedy 1953) (for reviews, see Weigle 1994; Inui 1999; and for discussion as to how this relates to fish species, see Rowe *et al.* 1991; Silverstein *et al.* 1998b, 1999; Jobling & Johansen 1999). Although the presence of leptin has not been confirmed in fish, there is evidence that the size of the fat reserves may exert some control over feeding (Metcalfe & Thorpe 1992; Jobling & Miglavs 1993; Silverstein *et al.* 1998b, 1999). Leptin probably exerts control on food intake and body weight homeostasis via NPY transmission, but several hypothalamic neuropeptides may also be involved in the mediation of leptin action (Inui 1999).

12.5 Conclusions

The information presented in this chapter supports the contention that many of the neuropeptides involved in the regulation of feeding have conserved structures throughout vertebrate evolution and the same neuropeptides seem to function across the vertebrate series in the modulation of food intake.

All the peptides known to modify food intake in fish appear to act in the short term, and most act centrally, with the hypothalamus as the probable site of action. Corticotrophin-releasing factor, bombesin and cholecystokinin inhibit food intake, whereas galanin, β-endorphin and neuropeptide Y elicit feeding (Fig. 12.7). It should be noted that although this has not been discussed in detail, several peptides and monoamines may act in concert to influence food intake, giving a highly complex regulatory system.

There are also links between the neuropeptides that act at the hypothalamic-hypophyseal level and the GH-IGF-I regulatory system, and together, these provide for regulation of food intake and growth.

The actions of neuropeptides and monoamines are generated in response to visceral and metabolic information, which reflect nutritional status and the immediate past history of feeding. Peripheral hormones, such as insulin, may signal nutritive state, while a further such candidate, leptin, may influence several neuroendocrine axes leading to the modulation of the pituitary secretion of pituitary hormones such as GH.

12.6 Acknowledgements

The research carried out by N. De Pedro and co-workers was supported by the Spanish Ministry of Education and Science, and the Community of Madrid, while the studies of B.Th. Björnsson and co-workers were supported by the Swedish Council for Forestry and Agriculture Research. N. De Pedro is grateful to Professor M. Alonso Bedate and all members of her research group for invaluable support, and to M. J. Delgado for careful reading of the manuscript and constructive comments and discussions.

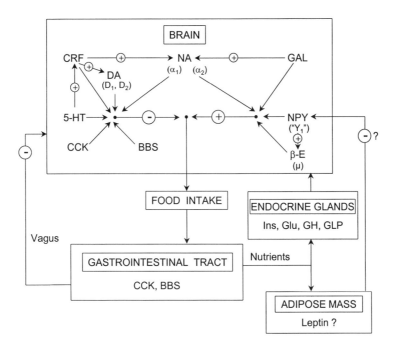

Fig. 12.7   A model for food intake regulation by neuropeptides and hormones in fish. α_1, α_1-adrenergic receptors; α_2, α_2-adrenergic receptors; β-E, β-endorphin; BBS, bombesin; CCK, cholecystokinin; CRF, corticotrophin-releasing factor; D_1, D_1-dopaminergic receptors; D_2, D_2-dopaminergic receptors; DA, dopamine; GAL, galanin; GH, growth hormone; GLP, glucagon-like peptide; Glu, glucagon; 5-HT, serotonin; Ins, insulin; NA, noradrenaline; NPY, neuropeptide Y; Y_1, Y_1 NPY receptors; μ, μ opioidergic receptors; +, stimulatory input; –, inhibitory input.

12.7 References

Agustsson, T., Ebbesson, L.O.E. & Björnsson, B.Th. (2000) Dopaminergic innervation of the rainbow trout pituitary and stimulatory effect of dopamine on pituitary growth hormone secretion in vitro. *Comparative Biochemistry and Physiology* A, **127**, 355–364.

Anastasi, A., Erspamer, V. & Bucchi, M. (1971) Isolation and structure of bombesin and alytensin, two analogous active peptides from skin of the European amphibians, *Bombina* and *Alytes*. *Experientia*, **27**, 166–167.

Ando, H., Hasegawa, M., Ando, J. & Urano, A. (1999) Expression of salmon corticotropin-releasing hormone precursor gene in the preoptic nucleus in stressed rainbow trout. *General and Comparative Endocrinology*, **113**, 87–95.

Anglade, I., Wang, Y., Jensen, J., Tramu, G., Kah, O. & Conlon, J.M. (1994) Characterization of trout galanin and its distribution in trout brain and pituitary. *The Journal of Comparative Neurology*, **350**, 63–74.

Arends, R.J., Vermeer, H., Martens, G.J.M., Leunissen, J.A.M., Wendelaar Bonga, S.E. & Flik, G. (1998) Cloning and expression of two proopiomelanocortin mRNAs in the common carp (*Cyprinus carpio* L.). *Molecular and Cellular Endocrinology*, **143**, 23–31.

Arimura, A. & Culler, M.C. (1985) Regulation of growth hormone secretion. In: *The Pituitary Gland* (ed. H. Imura), pp. 221–246. Raven Press, New York.

Arvidsson, A.K., Wraith, A., Jönsson-Rylander, A.-C. & Larhammar, D. (1998) Cloning of a neuro-peptide Y/peptide YY receptor from the Atlantic cod: the Yb receptor. *Regulatory Peptides*, **75-76**, 39–43.

Baile, C.A., McLaughlin, C.L. & Della-Fera, M.A. (1986) Role of cholecystokinin and opioid peptides in control of food intake. *Physiological Reviews*, **66**, 172–234.

Ben-Atia, I., Fine, M., Tandler, A., *et al.* (1999) Preparation of recombinant gilthead seabream (*Sparus aurata*) growth hormone and its use for stimulation of larvae growth by oral administration. *General and Comparative Endocrinology*, **113**, 155–164.

Björnsson, B.Th. (1997) The biology of salmon growth hormone: from daylight to dominance. *Fish Physiology and Biochemistry*, **17**, 9–24.

Björnsson, B.Th., Taranger, G.L., Hansen, T., Stefansson, S.O. & Haux, C. (1994) The interrelation be-tween photoperiod, growth hormone, and sexual maturation of adult Atlantic salmon (*Salmo salar*). *General and Comparative Endocrinology*, **93**, 70–81.

Blum, W.F. (1997) Leptin: the voice of the adipose tissue. *Hormone Research*, **48**, 2–8.

Bolton, J.P., Collie, N.L., Kawauchi, H. & Hirano, T. (1987) Osmoregulatory actions of growth hor-mone in rainbow trout (*Salmo gairdneri*). *Journal of Endocrinology*, **112**, 63–68.

Brooks, A.I., Standifer, K.M., Cheng, J., Ciszewska, G. & Pasternak, G.W. (1994) Opioid binding in giant toad and goldfish brain. *Receptor*, **4**, 55–62.

Buchanan, C., Mahesh, V., Zamorano, P. & Brann, D. (1998) Central nervous effects of leptin. *Trends in Endocrinology and Metabolism*, **9**, 146–150.

Calduch-Giner, J.A., Sitja-Bobadilla, A., Alvárez-Pellitero, P. & Pérez-Sánchez, J. (1995) Evidence for a direct action of GH on haemopoietic cells of a marine fish, the gilthead sea bream (*Sparus aurata*). *Journal of Endocrinology*, **146**, 459–467.

Cao, Q.P., Duguay, S.J., Plisetskaya, E.M., Steiner, D.F. & Chan, S.J. (1989) Nucleotide sequence and growth hormone-regulated expression of salmon insulin-like growth factor I mRNA. *Molecular Endocrinology*, **3**, 2005–2010.

Cheng, C.M. & Chen, T.T. (1995) Synergism of GH and IGF-I in stimulation of sulphate uptake by teleostean branchial cartilage *in vitro*. *Journal of Endocrinology*, **147**, 67–73.

Cheung, R., Andrews, P.C., Plisetskaya, E.M. & Youson, J.H. (1991) Immunoreactivity to peptides belonging to the pancreatic polypeptide family (NPY, aPY, PP, PYY) and to glucagon-like peptide in the endocrine pancreas and anterior intestine of adult lampreys, *Petromyzon marinus*: immuno-histochemical study. *General and Comparative Endocrinology*, **81**, 51–63.

Chiba, A. (1997) Distribution of neuropeptide Y-like immunoreactivity in the brain of the bichir, *Polypterus senegalus*, with special regard to the terminal nerve. *Cell and Tissue Research*, **289**, 275–284.

Clarke, W.C., Farmer, S.W. & Hartwell, K.M. (1977) Effect of teleost pituitary growth hormone on growth of *Tilapia mossambica* and on growth and seawater adaptation of sockeye salmon (*Onco-rhynchus nerka*). *General and Comparative Endocrinology*, **33**, 174–178.

Coto-Montes, A., García-Fernández, J.M., Del Brío, M.A. & Riera, P. (1994) The distribution of corti-cotropin-releasing factor immunoreactive neurons and nerve fibres in the brain of *Gambusia affinis* and *Salmo trutta*. *Histology and Histopathology*, **9**, 233–241.

Crawley, J.N. (1995) Biological actions of galanin. *Regulatory Peptides*, **59**, 1–16.

Darlison, M.G., Greten, F.R., Harvey, R.J., *et al.* (1997) Opioid receptors from a lower vertebrate (*Catostomus commersoni*): sequence, pharmacology, coupling a G-protein-gated inward-rectifying potassium channel (GIRK1) and evolution. *Proceedings of the National Academy of Sciences of USA*, **94**, 8214–8219.

De Pedro, N., Alonso-Gómez, A.L., Gancedo, B., Delgado, M.J. & Alonso-Bedate, M. (1993) Role of corticotropin-releasing factor (CRF) as a food intake regulator in goldfish. *Physiology and Behavior*, **53**, 517–520.

De Pedro, N., Gancedo, B., Alonso-Gómez, A.L., Delgado, M.J. & Alonso-Bedate, M. (1995a) CRF effect on thyroid function is not mediated by feeding behavior in goldfish. *Pharmacology, Biochemistry and Behavior*, **51**, 885–890.

De Pedro, N., Gancedo, B., Alonso-Gómez, A.L., Delgado, M.J. & Alonso-Bedate, M. (1995b) Alterations in food intake and thyroid tissue content by corticotropin-releasing factor in *Tinca tinca*. *Journal of Physiology and Biochemistry*, **51**, 71–76.

De Pedro, N., Céspedes, M.V., Delgado, M.J. & Alonso-Bedate, M. (1995c) The galanin-induced feeding stimulation is mediated via α_2-adrenergic receptors in goldfish. *Regulatory Peptides*, **57**, 77–84.

De Pedro, N., Delgado, M.J. & Alonso-Bedate, M. (1995d) Central administration of β-endorphin increases food intake in goldfish: pretreatment with the opioid antagonist naloxone. *Regulatory Peptides*, **55**, 189–195.

De Pedro, N., Céspedes, M.V., Delgado, M.J. & Alonso-Bedate, M. (1996) Mu-opioid receptor is involved in β-endorphin-induced feeding in goldfish. *Peptides*, **17**, 421–424.

De Pedro, N., Alonso-Gómez, A.L., Gancedo, B., Valenciano, A.I., Delgado, M.J. & Alonso-Bedate, M. (1997) The effect of α-Helical-CRF$_{[9-41]}$ on feeding in goldfish: involvement of cortisol and catecholamines. *Behavioral Neuroscience*, **3**, 398–403.

De Pedro, N., Delgado, M.J., Pinillos, M.L. & Alonso-Bedate, M. (1998a) The anorectic effect of CRF is mediated by α_1-adrenergic and dopaminergic receptors in goldfish. *Life Sciences*, **62**, 1801–1808.

De Pedro, N., Pinillos, M.L., Valenciano, A.I., Alonso-Bedate, M. & Delgado, M.J. (1998b) Inhibitory effect of serotonin on feeding behavior in goldfish: involvement of CRF. *Peptides*, **19**, 505–511.

De Pedro, N., López-Patiño, M.A., Guijarro, A.I., Pinillos, M.L., Delgado, M.J. & Alonso-Bedate, M. (2000) NPY receptors and opioidergic system are involved in NPY-induced feeding in goldfish. *Peptides*, **21**, 1495–1502.

Devlin, R.H., Yesaki, T.Y., Biagi, C.A., Donaldson, E.M., Swanson, P. & Chan, W.-K. (1994) Extraordinary salmon growth. *Nature*, **371**, 209–210.

Devlin, R.H., Yesaki, T.Y., Donaldson, E.M., Du, S.J. & Hew, C.-L. (1995) Production of germline transgenic Pacific salmonids with dramatically increased growth performance. *Canadian Journal of Fisheries and Aquatic Sciences*, **52**, 1376–1384.

Devlin, R.H., Johnsson, J.I., Smailus, D.E., Biagi, C.A., Jönsson, E. & Björnsson, B.Th. (1999) Increased ability to compete for food by growth hormone-transgenic coho salmon (*Oncorhynchus kisutch* Walbaum). *Aquaculture Research*, **30**, 479–482.

Dietrich, J.B. (1994) Neuropeptides, antagonists and cell proliferation: bombesin as an example. *Cellular and Molecular Biology*, **40**, 731–746.

Dores, R.M., McDonald, L.K., Stevenson, T.C. & Sei, C.A. (1990) The molecular evolution of neuropeptides: prospects for the '90s. *Brain, Behavior and Evolution*, **36**, 80–99.

Dores, R.M., Kaneko, D.J. & Sandoval, F. (1993) An anatomical and biochemical study of the pituitary proopiomelanocortin systems in the polypteriform fish *Calamoichthys calabaricus*. *General and Comparative Endocrinology*, **90**, 87–99.

Duan, C. & Hirano, T. (1992) Effects of insulin-like growth factor-I and insulin on the *in vitro* uptake of sulphate by eel branchial cartilage: evidence for the presence of independent hepatic and pancreatic sulphation factors. *Journal of Endocrinology*, **133**, 211–219.

Duan, C. & Plisetskaya, E.M. (1993) Nutritional regulation of insulin-like growth factor-I mRNA expression in salmon tissues. *Journal of Endocrinology*, **139**, 243–252.

Eales, J.G. (1985) The peripheral metabolism of thyroid hormones and regulation of thyroidal status in poikilotherms. *Canadian Journal of Zoology*, **63**, 1217–1231.

Ebbesson, L.O.E., Deviche, P. & Ebbesson, S.O.E. (1996) Distribution and changes in μ- and κ-opiate receptors during the midlife neurodevelopmental period of coho salmon, *Oncorhynchus kisutch*. *The Journal of Comparative Neurology*, **364**, 448–464.

Eilertson, C.D., Carneiro, N.M., Kittilson, J.D., Comley, C. & Sheridan, M.A. (1996) Cholecystokinin, neuropeptide Y and galanin modulate the release of pancreatic somatostatin-25 and somatostatin-14 *in vitro*. *Regulatory Peptides*, **63**, 105–112.

Enrensing, R.H. & Mitchell, G.F. (1982) Similar antagonism of morphine analgesia by MIF-1 and naloxone in *Carassius auratus*. *Pharmacology, Biochemistry and Behavior*, **17**, 757–761.

Fagerlund, U.H.M., Higgs, D.A., McBride, J.R., Plotnikoff, M.D. & Dosanjh, B.S. (1980) The potential for using the anabolic hormones 17-alpha-methyltestosterone and/or 3,5,3'-triiodo-L-thyronine in the fresh water rearing of coho salmon (*Oncorhynchus kisutch*) and the effects on subsequent seawater performance. *Canadian Journal of Zoology*, **58**, 1424–1432.

Fauconneau, B., Mady, M.P. & Le Bail, P.Y. (1996) Effect of growth hormone on muscle protein synthesis in rainbow trout (*Oncorhynchus mykiss*) and Atlantic salmon (*Salmo salar*). *Fish Physiology and Biochemistry*, **15**, 49–56.

Fine, M., Sakal, E., Vashdi, D., *et al.* (1993) Recombinant carp (*Cyprinus carpio*) growth hormone: Expression, purification, and determination of biological activity *in vitro* and *in vivo*. *General and Comparative Endocrinology*, **89**, 51–61.

Flood, C.G. & Eales, J.G. (1983) Effects of starvation and refeeding on plasma thyroxine and triiodothyronine levels and thyroxine deiodination in rainbow trout, *Salmo gairdneri*. *Canadian Journal of Zoology*, **61**, 1949–1953.

Foster, A.R., Houlihan, D.F., Gray, C., *et al.* (1991) The effects of ovine growth hormone on protein turnover in rainbow trout. *General and Comparative Endocrinology*, **82**, 111–120.

Friedman, J.M. (1998) Leptin, leptin receptors and the control of body weight. *Nutrition Reviews*, **56**, S38–S46.

Fryer, J.N. & Lederis, K. (1986) Control of corticotropin secretion in teleost fishes. *American Zoologist*, **26**, 1017–1026.

Fu, C., Cui, Y., Hung, S.S.O. & Zhu, Z. (1998) Growth and feed utilisation by F4 human growth hormone transgenic carp fed diets with different protein levels. *Journal of Fish Biology*, **53**, 115–129.

Funkenstein, B., Silbergeld, A., Cavari, B. & Laron, Z. (1989) Growth hormone increases plasma levels of insulin-like growth factor (IGF-I) in a teleost, the gilthead seabream (*Sparus aurata*). *Journal of Endocrinology*, **120**, R19–R21.

Gancedo, B., Corpas, I., Alonso-Gómez, A.L., Delgado, M.J., Morreale de Escobar, G. & Alonso-Bedate, M. (1992) Corticotropin-releasing factor stimulates metamorphosis and increases thyroid hormone concentration in prometamorphic *Rana perezi* larvae. *General and Comparative Endocrinology*, **87**, 6–13.

Garber, M.J., DeYonge, K.G., Byatt, J.C., *et al.* (1995) Dose-response effects of recombinant bovine somatotropin (Posilac™) on growth performance and body composition of two-year-old rainbow trout (*Oncorhynchus mykiss*). *Journal of Animal Science*, **73**, 3216–3222.

Gill, J.A., Sumpter, J.P., Donaldson, E.M., *et al.* (1985) Recombinant chicken and bovine growth hormones accelerate growth in aquacultured juvenile Pacific salmon *Oncorhynchus kisutch*. *Bio/technology*, **3**, 643–646.

Giustina, A. & Wehrenberg, W.B. (1995) Influence of thyroid hormones on the regulation of growth hormone secretion. *European Journal of Endocrinology*, **133**, 646–653.

Glowa, J.R., Barret, J.E., Russell, J. & Gold, P.W. (1992) Effects of corticotropin releasing hormone on appetitive behaviors. *Peptides*, **13**, 609–621.

González, G.C., Belenky, M.A., Polenov, A.L. & Ledereis, K. (1992) Comparative localization of corticotropin and corticotropin releasing factor-like peptides in the brain and pituitary of a primitive vertebrate, the sturgeon *Acipenser ruthenus* L. *Journal of Neurocytology*, **21**, 885–896.

Gray, E.S. & Kelley, K.M. (1991) Growth regulation in the gobiid teleost, *Gillichthys mirabilis*: roles of growth hormone, hepatic growth hormone receptors and insulin-like growth factor-I. *Journal of Endocrinology*, **131**, 57–66.

Gray, E.S., Young, G. & Bern, H.A. (1990) Radioreceptor assay for growth hormone in coho salmon (*Oncorhynchus kisutch*) and its application to the study of stunting. *The Journal of Experimental Zoology*, **256**, 290–296.

Green, H., Morikawa, M. & Nixon, T. (1985) A dual effector theory of growth-hormone action. *Differentiation*, **29**, 195–198.

Guijarro, A.I., Delgado, M.J., Pinillos, M.L., López-Patiño, M.A., Alonso-Bedate, M. & De Pedro, N. (1999) Galanin and β-endorphin as feeding regulators in cyprinids: effect of temperature. *Aquaculture Research*, **30**, 483–489.

Habu, A., Ohishi, T., Mihara, S., *et al.* (1994) Isolation and sequence determination of galanin from the pituitary of yellowfin tuna. *Biomedical Research*, **15**, 357–362.

Himick, B.A. & Eales, J.G. (1990) The acute effects of food and glucose challenge on plasma thyroxine and triiodothyronine levels in previously starved rainbow trout (*Oncorhynchus mykiss*). *General and Comparative Endocrinology*, **78**, 34–41.

Himick, B.A. & Peter, R.E. (1994a) Bombesin acts to suppress feeding behavior and alter serum growth hormone in goldfish. *Physiology and Behavior*, **55**, 65–72.

Himick, B.A. & Peter R.E. (1994b) CCK/gastrin-like immunoreactivity in brain and gut, and CCK suppression of feeding in goldfish. *American Journal of Physiology*, **267**, R841–R851.

Himick, B.A. & Peter, R.E. (1995a) Bombesin like immunoreactivity in the forebrain and pituitary and regulation of anterior pituitary hormone release by bombesin in goldfish. *Neuroendocrinology*, **61**, 365–376.

Himick, B.A. & Peter, R.E. (1995b) Neuropeptide regulation of feeding and growth hormone secretion in fish. *Netherlands Journal of Zoology*, **45**, 3–9.

Himick, B.A., Vigna, S.R. & Peter, R.E. (1995) Characterization and distribution of bombesin binding sites in the goldfish hypothalamic feeding center and pituitary. *Regulatory Peptides*, **60**, 167–176.

Himick, B.A., Vigna, S.R. & Peter, R.E. (1996) Characterization of cholecystokinin binding sites in goldfish brain and pituitary. *American Journal of Physiology*, **271**, R137–R143.

Hirano, T. (1991) Hepatic receptors for homologous growth hormone in the eel. *General and Comparative Endocrinology*, **81**, 383–390.

Holloway, A.C. & Leatherland, J.F. (1998) Neuroendocrine regulation of growth hormone secretion in teleost fishes with emphasis on the involvement of gonadal steroids. *Reviews in Fish Biology and Fisheries*, **8**, 409–429.

Holmgren, S., Fritsche, R., Karila, P., *et al.* (1994) Neuropeptides in the Australian lungfish *Neoceratodus forsteri*: effects *in vivo* and presence in autonomic nerves. *American Journal of Physiology*, **266**, R1568–R1577.

Holmqvist, B.I. & Carlberg, M. (1992) Galanin receptors in the brain of a teleost: autoradiographic distribution of binding sites in the Atlantic salmon. *The Journal of Comparative Neurology*, **326**, 44–60.

Houseknecht, K., Baile, C.A., Matteri, R.L. & Spurlock, M.E. (1998) The biology of leptin: a review. *Journal of Animal Science*, **76**, 1405–1420.

Inui, A. (1999) Feeding and body-weight regulation by hypothalamic neuropeptides – mediation of the actions of leptin. *Trends in Neuroscience*, **22**, 62–67.

Jensen, J. & Conlon, M. (1992) Isolation and primary structure of gastrin-releasing peptide from a teleost fish, the trout (*Oncorhynchus mykiss*). *Peptides*, **15**, 995–999.

Jobling, M. & Johansen, S.J.S. (1999) The lipostat, hyperphagia and catch-up growth. *Aquaculture Research*, **30**, 473–478.

Jobling, M. & Miglavs, I. (1993) The size of lipid depots – a factor contributing to the control of food intake in Arctic charr, *Salvelinus alpinus* (L.). *Journal of Fish Biology*, **43**, 487–489.

Johnsson, J.I. & Björnsson, B.Th. (1994) Growth hormone increases growth rate, appetite and dominance in juvenile rainbow trout, *Oncorhynchus mykiss*. *Animal Behavior*, **48**, 177–186.

Johnsson, J.I., Petersson, E., Jönsson, E., Björnsson, B.Th. & Järvi, T. (1996) Domestication and growth hormone alter anti-predator behaviour and growth patterns in juvenile brown trout, *Salmo trutta*. *Canadian Journal of Fisheries and Aquatic Sciences*, **53**, 1546–1554.

Jönsson, E., Johnsson, J.I. & Björnsson, B.Th. (1996) Growth hormone increases predation exposure of rainbow trout. *Proceedings of the Royal Society of London B*, **263**, 647–651.

Jönsson, E., Johnsson, J.I. & Björnsson, B.Th. (1998) Growth hormone increases aggressive behavior in juvenile rainbow trout. *Hormones and Behavior*, **33**, 9–15.

Kavaliers, M. (1983) Pineal mediation of the thermoregulatory and behavioral activating effects of β-endorphin. *Peptides*, **3**, 679–685.

Kavaliers, M. (1989) Day-night rhythms of shoaling behavior in goldfish: opioid and pineal involvement. *Physiology and Behavior*, **46**, 167–172.

Kelly, P.A., Djiane, J., Postel-Vinay, M.-C. & Edery, M. (1991) The prolactin/growth hormone receptor family. *Endocrine Reviews*, **12**, 235–251.

Kelly, P.A., Goujon, L., Sotiropoulos, A., *et al.* (1994) The GH receptor and signal transduction. *Hormone Research*, **42**, 133–139.

Kennedy, G.C. (1953) The role of depot fat in hypothalamic control of food intake in the rat. *Proceedings of the Royal Society of London B*, **140**, 578–592.

Kiliaan, A.J., Holmgren, S., Jönsson, A.-C., Dekker, K. & Groot, J. (1992) Neurotensin, substance P, gastrin/cholecystokinin, and bombesin in the intestine of the tilapia (*Oreochromis mossambicus*) and the goldfish (*Carassius auratus*): immunochemical detection and effects on electrophysiological characteristics. *General and Comparative Endocrinology*, **88**, 351–363.

Kiliaan, A.J., Holmgren, S., Jönsson, A.-C., Dekker, K. & Groot, J.A. (1993) Neuropeptides in the intestine of two teleost species (*Oreochromis mossambicus*, *Carassius auratus*): localization and electrophysiological effects on the epithelium. *Cell and Tissue Research*, **271**, 123–134.

Kyrkouli, S.E., Stanley, B.G. & Leibowitz, S.F. (1992) Differential effects of galanin and neuropeptide Y on extracellular norepinephrine levels in the paraventricular hypothalamic nucleus of the rat: a microdialysis study. *Life Sciences*, **51**, 203–210.

Larhammar, D. (1996) Evolution of neuropeptide Y, peptide YY and pancreatic polypeptide. *Regulatory Peptides*, **62**, 1–11.

Larhammar, D., Arvidsson, A.-K., Berglund, M.M., *et al.* (1997) Evolution of neuropeptide Y and its receptors. In: *Advances in Comparative Endocrinology* (eds S. Kawashima & S. Kikuyama), pp. 551–557. Monduzzi Editore, Bologna.

Le Bail, P.Y. & Boeuf, G. (1997) What hormones may regulate food intake in fish? *Aquatic Living Resources*, **10**, 371–379.

Le Gac, F., Ollitrault, M., Loir, M. & Le Bail, P.Y. (1992) Evidence for binding and action of growth hormone in trout testis. *Biology of Reproduction*, **46**, 949–957.

Le Mével, J.-C., Mabin, D., Hanley, A.M. & Conlon, J.M. (1998) Contrasting cardiovascular effects following central and peripheral injections of trout galanin in trout. *American Journal of Physiology*, **275**, R1118–R1126.

Leatherland, J.F. (1994) Reflections on the thyroidology of fishes: from molecules to humankind. *Guelph Ichthyology Reviews*, **2**, 1–67.

Leatherland, J.F. & Farbridge, K.J. (1992) Chronic fasting reduces the response of the thyroid to growth hormone and TSH, and alters the growth hormone-related changes in hepatic 5'-monodeiodinase activity in rainbow trout, *Oncorhynchus mykiss*. *General and Comparative Endocrinology*, **87**, 342–353.

Lederis, K.P., Okawara, Y., Richter, D. & Morley, S.D. (1990) Evolutionary aspects of corticotropin releasing hormone. In: *Progress in Comparative Endocrinology. Progress in Clinical and Biological Research*, vol. **342** (eds A. Epple, C.G. Scanes & M.H. Stetson), pp. 467–472. Wiley-Liss, New York.

Leibowitz, S.F. (1995) Brain peptides and obesity: pharmacologic treatment. *Obesity Research*, **3**, S573–S589.

Levine, A.S. & Billington, C.J. (1997) Why do we eat? A neural systems approach. *Annual Reviews in Nutrition*, **17**, 597–619.

López-Patiño, M.A., Guijarro, A.I., Pinillos, M.L., *et al.* (1999a) Is the opioid system involved in the regulation of feeding behavior and body weight in fish? *Dolor*, **14**, 55.

López-Patiño, M.A., Guijarro, A.I., Delgado, M.J., Alonso-Bedate, M. & De Pedro, N. (1999b) Neuropeptide Y has a stimulatory action on feeding behavior in goldfish (*Carassius auratus*). *European Journal of Pharmacology*, **377**, 147–153.

Luo, D. & McKeown, B.A. (1991) Interaction of carp growth hormone-releasing factor and somatostatin on *in vitro* release of growth hormone in rainbow trout (*Oncorhynchus mykiss*). *Neuroendocrinology*, **54**, 359–364.

MacLatchy, D.L., Kawauchi, H. & Eales, J.G. (1992) Stimulation of hepatic thyroxine 5-deiodinase activity in rainbow trout (*Oncorhynchus mykiss*) by Pacific salmon growth hormone. *Comparative Biochemistry and Physiology*, **101A**, 689–691.

Mancera, J.M. & Fernández-Llébrez, P. (1995) Localization of corticotropin-releasing factor immuno-reactivity in the brain of the teleost *Sparus aurata*. *Cell and Tissue Research*, **264**, 539–548.

Marchelidon, J., Schmitz, M., Houdebine, L.M., Vidal, B., Le, B.N. & Dufour, S. (1996) Development of a radioimmunoassay for European eel growth hormone and application to the study of silvering and experimental fasting. *General and Comparative Endocrinology*, **102**, 360–369.

Markert, J.R., Higgs, D.A., Dye, H.M. & MacQuarrie, D.W. (1977) Influence of bovine growth hormone on growth rate, appetite, and food conversion of yearling coho salmon (*Oncorhynchus kisutch*) fed two diets of different composition. *Canadian Journal of Zoology*, **55**, 74–83.

Martínez, R., Estrada, M.P., Berlanga, J., *et al.* (1996) Growth enhancement in transgenic tilapia by ectopic expression of tilapia growth hormone. *Molecular and Marine Biology and Biotechnology*, **5**, 62–70.

McCormick, S.D., Moriyama, S. & Björnsson, B.Th. (2000) Low temperature limits the photoperiod control of smolting in Atlantic salmon through endocrine mechanisms. *American Journal of Physiology*, **278**, R1352–R1361.

Merchenthaler, I., López, F.J. & Negro-Vilar, A. (1993) Anatomy and physiology of central galanin-containing pathways. *Progress in Neurobiology*, **40**, 711–769.

Metcalfe, N.B. & Thorpe, J.E. (1992) Anorexia and defended energy levels in over-wintering juvenile salmon. *Journal of Animal Ecology*, **61**, 175–181.

Mommsen, T.P. & Plisetskaya, E.M. (1991) Insulin in fishes and agnathans: history, structure and metabolic regulation. *Reviews in Aquatic Sciences*, **4**, 225–259.

Moons, L., Batten, T.F.C. & Vandesande, F. (1991) Autoradiographic distribution of galanin binding sites in the brain and pituitary of the sea bass (*Dicentrarchus labrax*). *Neuroscience Letters*, **123**, 49–52.

Moons, L., Vandesande, F. & Batten, T.F.C. (1992) Comparative distribution of substance P (SP) and cholecystokinin (CCK) binding sites and immunoreactivity in the brain of the sea bass (*Dicentrarchus labrax*). *Peptides*, **13**, 37–46.

Morgan, J.D., Sakamoto, T., Grau, E.G. & Iwama, G.K. (1997) Physiological and respiratory responses of the Mozambique tilapia (*Oreochromis mossambicus*) to salinity acclimation. *Comparative Biochemistry and Physiology*, **117A**, 391–398.

Moriyama, S. (1995) Increased plasma insulin-like growth factor-I (IGF-I) following oral and intraperitoneal administration of growth hormone to rainbow trout, *Oncorhynchus mykiss*. *Growth Regulation*, **5**, 164–167.

Moriyama, S., Swanson, P., Nishii, M., *et al.* (1994) Development of homologous radioimmunoassay for coho salmon insulin-like growth factor-I. *General and Comparative Endocrinology*, **96**, 149–161.

Morley, J.E. (1987) Neuropeptide regulation of appetite and weight. *Endocrine Reviews*, **8**, 256–287.

Morley, J.E. (1995) The role of peptides in appetite regulation across species. *American Zoologist*, **35**, 437–445.

Morris, Y.A. & Crews, D. (1990) The effects of exogenous neuropeptide Y on feeding and sexual behavior in the red-sided garter snake (*Thamnophis sirtalis parietalis*). *Brain Research*, **530**, 339–341.

Ng, T.B., Leung, T.C., Cheng, C.H.K. & Woo, N.Y.S. (1992) Growth hormone binding sites in tilapia (*Oreochromis mossambicus*) liver. *General and Comparative Endocrinology*, **86**, 111–118.

Norris, D.O. (ed.) (1985) The gastrointestinal peptides. In: *Vertebrate Endocrinology*, 2nd edition, pp. 364–379. Lea & Febiger, Philadelphia.

O'Connor, P.K., Reich, B. & Sheridan, M.A. (1993) Growth hormone stimulates hepatic lipid mobilization in rainbow trout, *Oncorhynchus mykiss*. *Journal of Comparative Physiology*, **163B**, 427–431.

Okawara, Y., Morley, S.D., Burzio, L.O., Zwiers, H., Lederis, K. & Richter, D. (1988) Cloning and sequence analysis of cDNAs for corticotropin releasing factor precursor from teleost fish *Catostomus commersoni*. *Proceedings of the National Academy of Sciences of USA*, **85**, 8439–8443.

Olivereau, M. & Olivereau, J.M. (1988) Localization of CRF-like immunoreactivity in the brain and pituitary of teleost fish. *Peptides*, **9**, 13–21.

Olivereau, M. & Olivereau, J.M. (1991) Immunocytochemical localization of a galanin-like peptidergic system in the brain and pituitary of some teleost fish. *Histochemistry*, **96**, 343–354.

Olson, G.A., Olson, R.D., Vaccarino, A.L. & Kastin, A.J. (1998) Endogenous opiates: 1997. *Peptides*, **19**, 1791–1843.

Olson, R.D., Kastin, A.J., Mitchel, G.F., Olson, G.A., Coy, D.H. & Montalbano, D. (1978) Effects of endorphin and enkephalin analogs on fear habituation in goldfish. *Pharmacology, Biochemistry and Behavior*, **9**, 111–114.

Owens, M.J. & Nemeroff, C.B. (1991). Physiology and pharmacology of corticotropin-releasing factor. *Pharmacological Reviews*, **43**, 425–470.

Peng, C., Huang, Y. & Peter, R.E. (1990) Neuropeptide Y stimulates growth hormone and gonadotropin release from the goldfish pituitary *in vitro*. *Neuroendocrinology*, **52**, 28–34.

Pérez-Sánchez, J., Marti-Palanca, H. & Le Bail, P.Y. (1994a) Seasonal changes in circulating growth hormone (GH), hepatic GH-binding and plasma insulin-like growth factor-I immunoreactivity in a marine fish, gilthead sea bream, *Sparus aurata*. *Fish Physiology and Biochemistry*, **13**, 199–208.

Pérez-Sánchez, J., Marti-Palanca, H. & Le Bail, P.Y. (1994b) Homologous growth hormone (GH) binding in gilthead sea bream (*Sparus aurata*). Effect of fasting and refeeding on hepatic GH-binding and plasma somatomedin-like immunoreactivity. *Journal of Fish Biology*, **44**, 287–301.

Peter, R.E. & Marchant, T.A. (1995) The endocrinology of growth in carp and related species. *Aquaculture*, **129**, 299–321.

Peter, R.E., Yu, K.L., Marchant, T.A. & Rosenblum, P.M. (1990) Direct neural regulation of the teleost adenohypophysis. *The Journal of Experimental Zoology Supplement*, **4**, 84–89.

Plisetskaya, E.M. (1995) Peptides of insulin and glucagon superfamilies in fish. *Netherlands Journal of Zoology*, **45**, 181–188.

Plisetskaya, E.M., Bondareva, V.M., Duan, C. & Duguay, S.J. (1993) Does salmon brain produce insulin? *General and Comparative Endocrinology*, **91**, 74–80.

Power, D.M., Canario, A.V.M. & Ingleton, P.M. (1996) Somatotropin release-inhibiting factor and galanin innervation in the hypothalamus and pituitary of seabream (*Sparus aurata*). *General and Comparative Endocrinology*, **101**, 264–274.

Reddy, P.K. & Leatherland, J.F. (1995) Influence of the combination of time of feeding and ration level on the diurnal hormone rhythms in rainbow trout. *Fish Physiology and Biochemistry*, **14**, 25–36.

Refstie, T. (1982) The effect of feeding thyroid hormones on saltwater tolerance and growth rate of Atlantic salmon (*Salmo salar*). *Canadian Journal of Zoology*, **60**, 2706–2712.

Reinecke, M., Schmid, A., Ermatinger, R. & Loffing, C.D. (1997) Insulin-like growth factor I in the teleost *Oreochromis mossambicus*, the tilapia: Gene sequence, tissue expression, and cellular localization. *Endocrinology*, **138**, 3613–3619.

Richardson, R.D., Boswell, T., Raffety, B.D., Seeley, R.J., Wingfield, J.C. & Woods, S.C. (1995) NPY increases food intake in white-crowned sparrows: effect in short and long photoperiods. *American Journal of Physiology*, **268**, R1418–R1422.

Rosenbaum, M. & Leibel, R.L. (1998) Leptin: a molecule integrating somatic energy stores, energy expenditure and fertility. *Trends in Endocrinology and Metabolism*, **9**, 117–124.

Rosenblum, P.M. & Peter, R.E. (1989) Evidence for the involvement of endogenous opioids in the regulation of gonadotropin secretion in male goldfish, *Carassius auratus*. *General and Comparative Endocrinology*, **73**, 17–21.

Rowe, D.K., Thorpe, J.E. & Shanks, A.M. (1991) Role of fat stores in the maturation of male Atlantic salmon (*Salmo salar* L.) parr. *Canadian Journal of Fisheries and Aquatic Sciences*, **48**, 405–413.

Rowland, N.E. & Kalra, S.P. (1997) Potential role of neuropeptide ligands in the treatment of overeating. *CNS Drugs*, **7**, 419–426.

Rubin, D.A. & Dores, R.M. (1995) Obtaining a more resolute teleost growth hormone phylogeny by the introduction of gaps in sequence alignment. *Molecular Phylogenetics and Evolution*, **4**, 129–138.

Sakamoto, T. & Hirano, T. (1991) Growth hormone receptors in the liver and osmoregulatory organs of rainbow trout: characterization and dynamics during adaptation to seawater. *Journal of Endocrinology*, **130**, 425–434.

Sakamoto, T., Shepherd, B.S., Madsen, S.S., *et al.* (1997) Osmoregulatory actions of growth hormone and prolactin in an advanced teleost. *General and Comparative Endocrinology*, **106**, 95–101.

Shepherd, B.S., Sakamoto, T., Nishioka, R.S., *et al.* (1997) Somatotropic actions of the homologous growth hormone and prolactins in the euryhaline teleost, the tilapia, *Oreochromis mossambicus*. *Proceedings of the National Academy of Sciences of USA*, **94**, 2068–2072.

Sheridan, M.A. (1986) Effects of thyroxin, cortisol, growth hormone, and prolactin on lipid metabolism of coho salmon, *Oncorhynchus kisutch*, during smoltification. *General and Comparative Endocrinology*, **64**, 220–238.

Silverstein, J.T., Breininger, J., Baskin, D.G. & Plisetskaya, E.M. (1998a) Neuropeptide Y-like gene expression in the salmon brain increases with fasting. *General and Comparative Endocrinology*, **110**, 157–165.

Silverstein, J.T., Shearer, K.D., Dickhoff, W.W. & Plisetskaya, E.M. (1998b) Effects of growth and fatness on sexual development of chinook salmon (*Oncorhynchus tshawytscha*) parr. *Canadian Journal of Fisheries and Aquatic Sciences*, **55**, 2376–2382.

Silverstein, J.T., Shearer, K.D., Dickhoff, W.W. & Plisetskaya, E.M. (1999) Regulation of nutrient intake and energy balance in salmon. *Aquaculture*, **177**, 161–169.

Simon, E.J. & Hiller, J.M. (1994) Opioid peptides and opioid receptors. In: *Basic Neurochemistry: Molecular, Cellular, and Medical Aspects*, 5th edition (eds G.J. Siegel & B.W. Agranoff), pp. 321–339. Raven Press, New York.

Spiess, J., Dautzenberg, F.M., Sydow, S., *et al.* (1998) Molecular properties of the CRF receptor. *Trends in Endocrinology and Metabolism*, **9**, 140–145.

Tatemoto, K. (1982) Neuropeptide Y: complete amino acid sequence of the brain peptide. *Proceedings of the National Academy of Sciences of USA*, **79**, 5485–5489.

Tatemoto, K., Rökaeus, A., Jöruvall, H., MacDonald, J. & Mutt, V. (1983) Galanin – a novel biologically active peptide from porcine intestine. *Federation of Biochemical Societies Letters*, **164**, 124–128.

Toguyeni, A., Baroiller, J.F., Fostier, A., *et al.* (1996) Consequences of food restriction on short-term growth variation and on plasma circulating hormones in *Oreochromis niloticus* in relation to sex. *General and Comparative Endocrinology*, **103**, 167–175.

Tran, T.N., Fryer, J.N., Lederis, K. & Vaudry, H. (1990) CRF, urotensin I, and sauvagine stimulate the release of POMC-derived peptides from goldfish neurointermediate lobe cells. *General and Comparative Endocrinology*, **78**, 351–360.

Tsai, P.I., Madsen, S.S., McCormick, S.D. & Bern, H.A. (1994) Endocrine control of cartilage growth in coho salmon: GH influence *in vivo* on the response to IGF-I *in vitro*. *Zoological Sciences (Tokyo)*, **11**, 299–303.

Vale, W., Spiess, J., Rivier, C. & Rivier, J. (1981) Characterization of a 41-residue ovine hypothalamic peptide that stimulates secretion of corticotropin and β-endorphin. *Science*, **213**, 1394–1397.

Vallarino, M. (1985) Occurrence of β-endorphin-like immunoreactivity in the brain of the teleost, *Boops boops*. *General and Comparative Endocrinology*, **60**, 63–69.

Vallarino, M., D'Este, L., Negri, L., Ottonello, I. & Renda, T. (1990) Occurrence of bombesin-like immunoreactivity in the brain of the cartilaginous fish, *Scyliorhinus canicula*. *Cell and Tissue Research*, **259**, 177–181.

Vallarino, M., Feuilloley, M., Vandesande, F. & Vaudry, H. (1991) Immunohistochemical mapping of galanin-like immunoreactivity in the brain of the dogfish *Scyliorhinus canicula*. *Peptides*, **12**, 351–357.

Vecino, E. & Ekström, P. (1992) Colocalization of neuropeptide Y (NPY)-like and FMRF amide-like immunoreactivities in the brain of the Atlantic salmon (*Salmo salar*). *Cell and Tissue Research*, **270**, 435–442.

Vigna, S.R. (1983) Evolution of endocrine regulation of gastrointestinal function in lower vertebrates. *American Zoologist* **23**, 729–738.

Vigna, S.R. & Thorndyke, M.C. (1989) Bombesin. In: *The Comparative Physiology of Regulatory Peptides* (ed. S. Holmgren), pp. 34–60. Chapman & Hall, London.

Vigna, S.R., Fisher, B.L., Morgan, J.L.M. & Rosenquist, G.L. (1985) Distribution and molecular heterogeneity of cholecystokinin-like immunoreactive peptides in the brain and gut of the rainbow trout, *Salmo gairdneri*. *Comparative Biochemistry and Physiology*, **82C**, 143–146.

Vigna, S.R., Thorndyke, M.C. & Williams, J.A. (1986) Evidence for a common evolutionary origin of brain and pancreas cholecystokinin receptors. *Proceedings of the National Academy of Sciences of USA*, **83**, 4355–4359.

Vuagnat, B.A.M., Pierroz, D.D., Lalaoui, M., *et al.* (1998) Evidence for a leptin-neuropeptide Y axis for the regulation of growth hormone secretion in the rat. *Neuroendocrinology*, **67**, 291–300.

Wallis, M. (1996) The molecular evolution of vertebrate growth hormones: a pattern of near-stasis interrupted by sustained bursts of rapid change. *Journal of Molecular Evolution*, **43**, 93–100.

Wang, Y. & Conlon, J.M. (1994) Purification and characterization of galanin from the phylogenetically ancient fish, the bowfin (*Amia calva*) and dogfish (*Scyliorhinus canicula*). *Peptides*, **15**, 981–986.

Weigle, D.S. (1994) Appetite and the regulation of body composition. *Federation of American Society of Experimental Biology Journal*, **8**, 302–310.

Yao, K., Niu, P., Weigle D., Le Gac, F. & Le Bail, P.Y. (1991) Presence of specific growth hormone binding sites in rainbow trout (*Oncorhynchus mykiss*) tissues: characterization of the hepatic receptor. *General and Comparative Endocrinology*, **81**, 72–82.

Yoshihara, T., Honma, S. & Honma, K.I. (1996) Effects of restricted daily feeding on neuropeptide Y release in the rat paraventricular nucleus. *Endocrinology and Metabolism*, **33**, E589–E595.

Zhang, Y. & Marchant, T.A. (1996) Characterization of growth hormone binding sites in the goldfish, *Carassius auratus*: effects of hypophysectomy and hormone injection. *Fish Physiology and Biochemistry*, **15**, 157–165.

Chapter 13
Physiological Effects of Feeding

*Chris Carter, Dominic Houlihan, Anders Kiessling, Francoise Médale
and Malcolm Jobling*

13.1 Introduction

Physiology is the study of how animals function. When we think of physiology in fish, the experiments that usually spring to mind involve analysis of tissues or blood, the flow of substances in or out of the animal, or the ways in which particular fish species seem to be adapted to their environment. Measurements may be based on individuals or groups of fish. A characteristic feature of physiological measurements in fish has been the use of unfed animals, i.e. in a postabsorptive state (Schreck & Moyle 1990). This is usually done in an attempt to reduce the variability that exists within a group of fish held under normal feeding and rearing conditions. Thus, we know quite a lot about the physiology of unfed fish, but relatively little about the physiology of fish whose food intake has been carefully measured before experiments have been started.

Exceptions to this generalisation are the studies on bioenergetics and the physiological changes following a single meal (see reviews by Brett 1979, 1995; Brett & Groves 1979; Elliott 1979; Cho *et al.* 1982; Schreck & Moyle 1990; Jobling 1993, 1994, 1997) and on food processing (Fänge & Grove 1979; Grove 1986). In such studies, individuals or groups of fish have been fed a known amount of food, and measurements such as oxygen consumption made (e.g. Solomon & Brafield 1972; Beamish 1974). It has long been recognised that feeding introduces a dynamic into individual physiological systems as the animals deal with the consequences of the meal. Fish nutrition is a particular branch of physiology that has largely been concerned with establishing relationships between ration and growth, comparisons between feed ingredients and the determination of nutritional requirements. In the latter, emphasis has been on feed composition and assessment of mean response. The quantity of food eaten by the fish has usually not been measured, and this has resulted in requirements being expressed as a relative proportion or percentage of the diet (e.g. NRC 1993). Several books and monographs are available that provide discussions of feed composition and nutritional requirements (Halver 1972, 1989; Cowey 1982; Cowey *et al.* 1985; Wilson 1991; Higgs *et al.* 1995; see also Chapters 2 and 11).

The focus of this chapter is on studies where food intake has been measured and related to physiological effects. By pointing to the studies where food intake has been measured, we aim to demonstrate that in such cases there is a marked improvement in the interpretation of the results. In particular, we have highlighted the effects of the quantification of protein intake on amino acid metabolism, working from the rationale that growth has a close relationship with protein accretion.

13.2 Different methods of feeding

The physiological responses to feeding in fish are dependent upon the amount of food eaten. If the physiological question that is being posed demands that the amount of food being consumed is known, a method for the measurement of food intake must be incorporated into the experimental design (see Chapter 3). Decisions must also be made concerning the use of single or groups of fish. Food intake of single fish or of individuals in a small group may be measured over the duration of an experiment, but the monitoring of uptake by individuals within larger groups is problematic. The measurement of food intake of fish using hand-feeding techniques is extremely time-consuming, and this places limits on the number of treatments and replicates. Hand-feeding can, however, yield accurate information for relating food intake to physiological effects (Elliott 1976; Houlihan *et al.* 1989; Lyndon *et al.* 1992). Observations (direct or from a video recording) made on an individual's food intake (see Chapter 3) can be related to physiological effects measured at a later time. This approach has often formed the basis for the study of energy and nitrogen budgets because it allows investigation of relationships between known food intake and measures of metabolic physiology and growth (e.g. Solomon & Brafield 1972; Cui & Wootton 1988; Carter & Brafield 1991). The measurement of day-to-day variation in food intake is also possible, but only relatively few studies have related individual patterns (day-to-day variation) of food intake to a physiological response (Rozin & Mayer 1961; Jobling & Baardvik 1994; Ali *et al.* 1998; Shelverton & Carter 1998).

Manual or automatic feeding may be used for larger groups of fish and the food intake of individual animals determined by X-radiography (Talbot & Higgins 1983; see Chapter 3). This approach has been used to relate food intake to a variety of physiological effects, particularly in relation to the performance of individual fish within groups (see reviews by Jobling *et al.* 1990, 1995; McCarthy *et al.* 1993; Carter *et al.* 1995b; see also Chapter 3). There are several approaches to measuring food intake by groups of fish via combining different methods of delivery (e.g. manual feeding, automatic feeders, demand feeders) with assessment of uneaten food (e.g. collection of uneaten food from the tank outflow or use of hydroacoustic transducers; see Chapter 3). Such methods relate to physiological effects in different ways and do not allow investigation of physiological effects at the level of the individual. The collection of uneaten food allows the calculation of food intake and this, in turn, enables assessment of utilisation efficiencies to be made (see Chapter 1).

For all of these methods, long-term (long enough for measurable growth to occur) physiological effects depend on measuring or predicting total food intake over weeks. Also fish should be subject to a stable nutritional regime (fed the same amount in the same way each day), and if terminal measurements are to be made the nutritional regime used on the final day should reflect the normal regime. In the following account, short-term (within 24 hours) physiological effects relate to the food intake from the meal immediately preceding the measurements.

13.3 Short-term effects of a meal

Fish on a stable nutritional regime exhibit repeatable physiological events that cycle over 24 hours. Daily growth cannot easily be measured, but is the result of the links between these cycles and food intake on, for example, oxygen consumption, nitrogenous excretion and protein metabolism. In studies in which care has been taken to isolate the effects of food intake from time of feeding it has been shown that a given ration may be used differently when consumed at different times of the day (Boujard *et al.* 1995; Gelineau *et al.* 1998; Verbeeten *et al.* 1999; see Chapter 10).

The main purpose of this section is to detail short-term physiological effects of feeding in order to provide a link to long-term growth. Food processing by the gastrointestinal tract (GIT) may have major effects on both the absorption of nutrients and on food intake (Fänge & Grove 1979; Grove 1986; Holmgren *et al.* 1986). Investigation of food processing has usually been focused on relating major parameters (species, fish size, temperature) and variations in meal size, meal frequency and the composition of the meal to gastrointestinal emptying (for reviews see Fänge & Grove 1979; Jobling 1986, 1987; dos Santos & Jobling 1991; Bromley 1994). The overriding impression is of flexibility in patterns of emptying (for a selected species) depending on the characteristics of the feeding regime and of the food (Grove 1986; Jobling 1986, 1987; dos Santos & Jobling 1991). For example, Fletcher *et al.* (1984) showed that when two meals were fed to dab (*Limanda limanda*) 3 h apart, the inclusion of a binder in the feed meant that the two meals did not mix in the stomach. Without the binder, the two meals became mixed in the stomach and this slowed evacuation. Such studies provide ways of predicting the rate of gastrointestinal emptying and, in turn, the rate of appetite return (Grove *et al.* 1978). Regulation of the GIT at a pharmacological level is extremely complex, and few detailed descriptions are available (see Fänge & Grove 1979; Holmgren *et al.* 1986; Chapter 12). However, it is thought that the physical and chemical (nutritional) characteristics of the feed will have short-term effects on satiety through responses elicited in the GIT via chemo- and mechanoreceptors (Fletcher 1984).

A large part of the increase in oxygen consumption that follows the ingestion of a meal (Heat Increment of Feeding, HIF) is thought to be due to amino acid fluxes and the turnover of protein (Jobling 1981, 1993; Houlihan *et al.* 1988; Houlihan 1991). The balance between protein and energy utilisation is obviously central to growth. Associated with the rise in oxygen consumption following a meal is an increase in ammonia production from the oxidation of amino acids. The working hypothesis is that the flow of ingested amino acids to the tissues causes an increase in both amino acid oxidation and protein synthesis, but there is no single study that tests this hypothesis. Thus, data from different experiments need to be combined to elaborate the relative changes in short-term physiological effects that occur over the 24 h following feeding (Fig. 13.1). Protein synthesis increases by the largest relative amount and appears to show two peaks, the first at 6 h and a larger peak at 18 h (Fig. 13.1). Part of the explanation for this may be that tissues have different temporal patterns of protein synthesis; for example, liver protein synthesis activity tends to peak before that of white muscle (Lyndon *et al.* 1992). There are smaller relative increases in both oxygen consumption (Lyndon *et al.* 1992) and ammonia excretion (Ramnarine *et al.* 1987), but these are not synchronised (c.f. Fig. 13.4). These two experiments provide a useful comparison (data selected for similarity in protein intake, fish weight and temperature) and raise questions about the degree

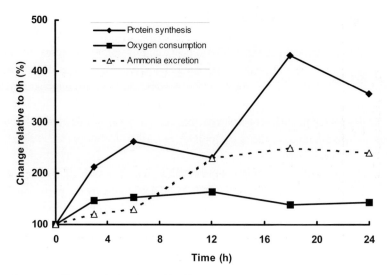

Fig. 13.1   Short-term effects (% of 0-h values) of food intake on protein synthesis and oxygen consumption (Lyndon *et al.* 1992), and on ammonia excretion (Ramnarine *et al.* 1987) in Atlantic cod.

of synchronisation between physiological effects. They also highlight a direction for future research.

13.4 Tissue metabolic physiology

Amino acid flux and temporal changes in free amino acid concentrations in different tissues highlight the dynamic relationships between protein intake, tissue metabolism and the utilisation of amino acids for protein synthesis. An ability to regulate a large influx of amino acids, presumably to maintain tissue homeostasis as well as to optimise the utilisation of dietary protein and energy, is central to these relationships. Feeding results in an influx of dietary amino acids that is usually greater than the total amount of free amino acids in the whole fish (Carter *et al.* 1995a; Houlihan *et al.* 1995a). It is therefore not unexpected that tissue free amino acid concentrations change following feeding. Changes are sequential between tissues and, to some extent, reflect a route through the tissues concerned with digestion, absorption, metabolism and then growth. However, these changes appear relatively small (Walton & Wilson 1986; Espe *et al.* 1993; Lyndon *et al.* 1993; Carter *et al.* 1995a), especially when seen in relation to relative changes in events such as protein synthesis (Fig. 13.2). This supports the hypothesis that protein synthesis has a key role in the regulation of free amino acid concentrations.

The liver has a central role in amino acid metabolism and in the regulation of amino acid transport to 'downstream' tissues. Its free amino acid pool concentration is influenced by the import and export of amino acids as free amino acids, peptides or proteins as well as by the use of amino acids within the tissue (Jürss & Bastrop 1995). There is, therefore, a link between cycles of protein turnover, free amino acid fluxes and nitrogenous excretion. In the livers of unfed salmonids total free amino acid concentrations are typically between 30 and 50 µmol/g,

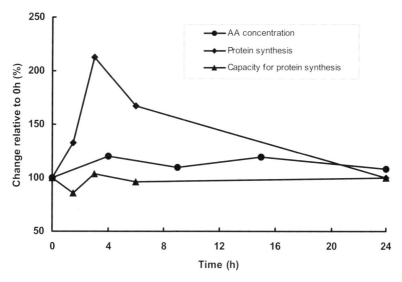

Fig. 13.2   Short-term effects (% of 0-h values) of food intake on free amino acid pool concentrations (Carter *et al.* 1995a), rates of tissue protein synthesis and the capacity for protein synthesis (RNA:protein) (McMillan & Houlihan 1992) in the liver of rainbow trout following feeding at 0 h.

with essential amino acids (EAAs) accounting for from 10 to 15% of the total (Walton & Wilson 1986; Médale *et al.* 1987; Carter *et al.* 1995a). Total free amino acid concentrations increase over a few hours following feeding (Walton & Wilson 1986; Carter *et al.* 1995a), and EAA show the largest peaks. Thus, it could be hypothesised that EAAs are not metabolised as rapidly as the non-essential amino acids (Walton & Wilson 1986). One additional feature may be differences in patterns shown by specific EAAs. Arginine, phenylalanine and threonine concentrations peak first, and a second peak was due to valine, isoleucine, leucine and methionine, although the latter amino acids may also have increased during the first peak (Walton & Wilson 1986). This pattern implies an initial response to dietary amino acids, and a second peak that may have been related to increases in endogenous amino acid metabolism and rates of protein synthesis (Lyndon *et al.* 1992). RNA concentration expressed as the capacity for protein synthesis (mg RNA per g protein; Sugden & Fuller 1991) remains relatively stable (McMillan & Houlihan 1992), which suggests that most of the increased protein synthesis in the liver is due to an increase in ribosomal activity (Fig. 13.2). These studies refer to the effects induced by a single meal, following several days without food; comparable studies on fish fed daily would be of great interest.

Free amino acid pools in white muscle are relatively unaffected by feeding (Fig. 13.3) (Ogata 1986; Espe *et al.* 1993; Lyndon *et al.* 1993; Carter *et al.* 1995a), although there are some differences between studies that are not readily explained. Before feeding, total free amino acid pool concentrations in the white muscle in salmonids are around 30 µmol/g, with EAAs accounting for 20% of the total (Medale *et al.* 1987; Espe *et al.* 1993; Carter *et al.* 1995a). Following feeding, concentrations rarely exceed 150% of the pre-feeding levels, and the proportion of EAAs does not change significantly (Kaushik & Luquet 1977; Espe *et al.* 1993; Carter *et al.* 1995a). As in the liver, there is considerable variation in the time course of the changes in different EAAs. It is also difficult to find consistent trends among the results

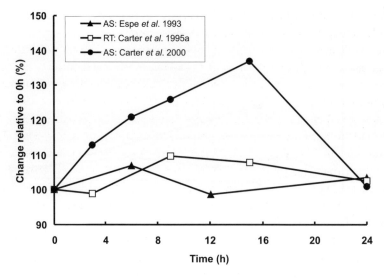

Fig. 13.3   Short-term effects (% of 0-h values) of food intake on free amino acid pools in the white muscle for rainbow trout (RT) and Atlantic salmon (AS) following feeding at 0 h. (Espe *et al.* 1993; Carter *et al.* 1995a, 2000).

from different studies. However, although there are differences between individual amino acids, the total free amino acid pool concentrations and composition show considerable stability and there is, presumably, some form of regulation (Lyndon *et al.* 1993; Carter *et al.* 1995a; Houlihan *et al.* 1995a,b).

Attempts have been made to relate amino acid profiles in tissues, particularly the blood, to the amino acid composition of the diet (Nose 1972; Walton & Wilson 1986). If food intake is measured, then quantification of the amount of amino acid consumed is possible. Amino acid intake was found to correlate with the total and EAA free pool concentrations in the pylorus 4 h following feeding, illustrating the rapid uptake of dietary amino acids into this tissue (Carter *et al.* 1995a). There were few significant correlations between amino acid intake and amino acids in the liver or white muscle free pools (Walton & Wilson 1986; Lyndon *et al.* 1993; Carter *et al.* 1995a). This is not surprising given the large number of factors involved in regulating the concentration and composition of the tissue free amino acid pools (Fuller & Garlick 1994; Arzel *et al.* 1995). Recently, the importance of the gastrointestinal tract in metabolising amino acids derived from recycling (rather than the food) has been recognised in mammals (Reeds *et al.* 1999), and this may also be of significance in fish.

We know less about tissue concentrations of other metabolites or metabolic enzyme activities (see p. 309, Biochemical correlates of food intake). The data on diurnal variations of liver glycogen content do not demonstrate a definite effect of food intake, and the delay between mealtime and the highest liver glycogen concentrations varies with season and the age of the fish (Delahunty *et al.* 1978; Laidley & Leatherland 1988). The time-course of the appearance of plasma glucose (and of other simple sugars) following feeding will depend on dietary factors, such as total carbohydrate content, feed ingredients and their processing, as well as on the species (feeding habits and trophic niche), water temperature and nutritional history (Bergot & Breque 1983; Hung 1991). Omnivorous and herbivorous species utilise diets containing higher carbohydrate contents than do carnivorous species, but some carnivorous fish are able to adapt to dietary carbohydrate (Wilson 1994).

Ingestion of a diet containing relatively high levels of digestible carbohydrate and lipid leads to an increase in plasma glucose and triacylglycerol concentrations in rainbow trout (*Oncorhynchus mykiss*) (Brauge *et al.* 1994, 1995). The arrival of such nutrients in the plasma generally occurs within 2 h after feeding (Brauge *et al.* 1994; Medale *et al.* 1999). Higher temperatures speed up the appearance of digestion products in the plasma (Brauge *et al.* 1995), presumably due to an increase in digestion and absorption rates. Consequently, in addition to diet composition, the profile of plasma metabolite concentrations mainly depends upon water temperature that also affects energy demands. The magnitude of the plasma glucose increase following a meal is influenced by the amount of digestible carbohydrates supplied by the diet (Brauge *et al.* 1994; Hemre *et al.* 1995). Increasing dietary wheat starch was positively correlated with plasma glucose, but negatively correlated with plasma triacylglycerols in Atlantic salmon (*Salmo salar*) (Hemre *et al.* 1995). A wide range of dietary lipid (31 to 47%) had no effect on plasma triacylglycerol concentration. When the highest dietary levels of lipid were combined with the lowest carbohydrate levels, the result was the highest plasma glucose levels; these may have been due to unknown stresses associated with the very high level of fat (Hemre & Sandnes 1999). Such studies emphasise the importance of measuring food intake, and point to problems in interpreting the effects of dietary manipulation when more than one variable is changed.

Insulin secretion, which acts on amino acid and glucose regulation, appears to be driven by food intake (Plisetskaya *et al.* 1991; Rungruangsak-Torrissen *et al.* 1999). Individual food intake was positively correlated with plasma insulin and growth rate in Atlantic salmon (Rungruangsak-Torrissen *et al.* 1999), and the links to feeding may partly explain the correlations between plasma insulin, body weight and growth rate found in an earlier study (Plisetskaya *et al.* 1991). Plasma glucose concentrations do not seem to provide an index of glucose utilisation (Wilson 1994). Fish, especially carnivores, are best able to utilise carbohydrate when rates of absorption and appearance in the blood are low (Meton *et al.* 1999).

Generally, body lipid reflects dietary lipid both in terms of amount and fatty acid composition (see Chapter 15). As food intake increases, so does lipid deposition, and the influence of the phospholipid fatty acid fraction associated primarily with the cell membranes becomes less (Pickova 1997). Lipids can also be synthesised from carbohydrates and amino acids. When the effects of feeding time are separated from the amount of food consumed it has been demonstrated that there is a depressive effect of feeding time on non-esterified fatty acids (NEFA) concentrations in plasma (Boujard & Leatherland 1992). These results are consistent with data obtained from rats and humans where plasma NEFA levels produced by lipolysis are inversely correlated with time after food intake (Aschoff 1979).

13.5 Whole-animal metabolic physiology

13.5.1 Respiration and excretion

The rate of oxygen consumption increases after feeding in fish. Usually, there is a peak in oxygen consumption two to three times above the pre-feeding level within a few hours after the end of the meal; the metabolic rate then declines to the pre-feeding level. Linear relationships between feed intake and HIF have been established (Jobling 1981, 1985; Carter & Brafield 1992). Carter and Brafield (1992) measured daily food intake and the associated HIF of individual grass carp (*Ctenopharyngodon idella*) held in respirometers for 20–30 days. They found that, depending on feed composition, HIF increased between 12.7 and 18.5 mg O$_2$/g per day for each kJ of energy absorbed. Patterns of ammonia and carbon dioxide excretion have also been measured. These have been linked to food intake, diet composition and the utilisation of metabolic substrates through the calculation of respiratory and ammonia (nitrogen) quotients (Brafield 1985; Gelineau *et al.* 1998; Owen *et al.* 1998; Médale *et al.* 1999). There appears to be synchronisation between postprandial increases in metabolism of common carp (*Cyprinus carpio*) measured in these ways (Fig. 13.4), but relative differences between the pathways may reveal changes in the balance between the use of substrates for aerobic respiration. In some cases the physiological responses to feeding have been observed to return to pre-feeding levels within 24 h (e.g. Owen *et al.* 1998; Médale *et al.* 1999), but responses may be longer-lasting in fish such as in Atlantic cod (*Gadus morhua*), that take over 24 h to process a meal (e.g. Saunders 1963; Ramnarine *et al.* 1987). The duration of the various physiological responses to a meal may be expected to depend upon such factors as the size of the meal, the ease with which it is digested, its nutrient composition and temperature (Jobling 1981).

The design of the respirometer chamber used in such studies may affect the results obtained due to influences on the behaviour and physiology of the fish. Oxygen consumption is often high for as long as 24 h after a fish is introduced into a respirometer, so there needs to be an adequate period of acclimatisation before the recording of metabolic rates. Other problems relate to establishing the feeding regime. One approach is to hold animals individually in a respirometer for several days, or longer (e.g. Solomon & Brafield 1972; Carter & Brafield

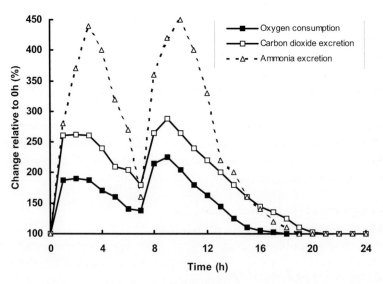

Fig. 13.4 Short-term effects (% of 0-h values) of oxygen consumption, carbon dioxide excretion and ammonia excretion for common carp following feeding at 0 and 7 h (fed 1.5% body weight distributed as two equal rations). (Adapted from Médale *et al.* 1999.)

1991). In such experiments feeding behaviour can be observed, the amount of food consumed can be measured, and consumption can then be related to metabolic responses and growth. An advantage when working with individuals is that behaviour can be recorded and, if it appears abnormal, then the results can be excluded. Under such circumstances increases in oxygen consumption can be ascribed to the effects of the meal and not confounded with increases in activity. Experiments may also be carried out on groups of fish where the amount of food consumed by the group is determined, and the oxygen consumption and other metabolic parameters of the animals recorded (Heinsbroek *et al.* 1993; Owen *et al.* 1998).

The quantity (absolute and relative to other macronutrients) and quality (amino acid composition) of absorbed protein will influence the HIF, rates of protein synthesis, ammonia production and the relationships between them. Analysis of diurnal cycles in amino acid metabolism suggest that the surge in protein synthesis following feeding occurs at approximately the same time as the oxidation of amino acids and the excretion of ammonia (P. Campbell *et al.* unpublished data). A link between the postprandial profiles of ammonia excretion and the amount of food fed has been found in rainbow trout (Médale *et al.* 1995). The time-course of protein breakdown is not known, but may be of considerable importance (Millward & Rivers 1988). For example, it is possible that, depending upon the size of the meal and the speed of digestion, protein accretion may be controlled by the rapidity of the rise and fall not only of protein synthesis but also protein breakdown.

13.5.2 Protein turnover

Stimulation of protein synthesis following feeding (McMillan & Houlihan 1988; Lyndon *et al.* 1992) and the incorporation of infused amino acids into proteins (Brown & Cameron 1991) are both associated with an increase in oxygen consumption (Brown & Cameron 1991; Lyndon *et al.* 1992). The extent of this stimulation in protein synthesis seems to be closely related to the size of the meal. Estimates made from results obtained from studies carried out on cod indicated that each gram of protein consumed resulted in a gram of protein being synthesised on the day of the meal (Houlihan *et al.* 1988). Subsequent measurements undertaken on Atlantic salmon and rainbow trout indicate that for every gram of protein consumed, 0.33–0.89 g of protein is synthesised (Houlihan *et al.* 1995a; Owen *et al.* 1999). It is to be expected that the amount of protein synthesised will depend not only on the amount of absorbed protein but also on the amino acid balance of the protein and the digestible energy intake (see above). Only a proportion of the proteins that are synthesised following a meal are retained as growth. For example, in salmonids the retention efficiency of synthesised proteins has been reported to fall within the range 23 to 62% (Houlihan *et al.* 1995a; Owen *et al.* 1999).

13.6 Long-term effects of food intake

Fish may be broadly classified according to their feeding habits, e.g. plant eaters and detritus feeders, planktivores, benthophages and piscivores, and this is reflected in feeding behaviours, gut morphology and digestive physiology (Kapoor *et al.* 1975; Fänge & Grove 1979; Gerking 1994; Hidalgo *et al.* 1999). However, many species may display considerable intraspecific variation in trophic habits, with distinct phenotypes being associated with particular patterns of prey or habitat use. Thus, certain individuals, or groups of individuals, may exploit only a small proportion of the array of prey types exploited by the species as a whole, and this trophic polymorphism is usually associated with a range of behavioural, morphological and physiological adaptations (Bryan & Larkin 1972; Meyer 1987, 1990a,b; Wainwright *et al.* 1991; Mittelbach *et al.* 1992, 1999). Most of these adaptations seem to reflect a phenotypic plasticity, rather than being genetic in nature, and they develop as a consequence of – rather than a prerequisite to – dietary specialisation. However, once developed, such dietary specialisations may be self-perpetuating if increased specialisation on particular prey types leads to reduced capture efficiencies for others (Bence 1986; Ehlinger 1990; Schluter 1995; Robinson *et al.* 1996). These specialisations may involve changes in the morphology of the mouthparts and gastrointestinal tract, and in digestive and absorptive capacities (Kapoor *et al.* 1975; Hofer 1979a,b; Buddington *et al.* 1987, 1997; Wainwright *et al.* 1991; Magnan & Stevens 1993; Mittelbach *et al.* 1999).

A phenotypic plasticity that enables individuals within a species to modify feeding behaviours and regulate digestive functions in relation to changes in prey availability and dietary composition has obvious ecological implications; those species that are capable of making such modifications will be able to exploit a wider range of food resources. For example, there is considerable variation in jaw morphology and the size of the pharyngeal jaw muscles among populations of pumpkinseed sunfish (*Lepomis gibbosus*), much of this variation seeming to be associated with the proportions of snails making up the diet of the fish inhabiting different lakes or habitats (Wainwright *et al.* 1991; Mittelbach *et al.* 1992). Large differences in jaw morphology and muscle size are inducible by providing pumpkinseeds with diets with or without snails, indicating that the differences among populations are predominantly the result of a developmental plasticity that allows the fish to exploit a range of prey types (Mittelbach *et al.* 1999). Fish may also have some capacity to adapt digestive processes, e.g. enzyme profile and secretion, and nutrient transport and absorption, to match changes in diet (Kapoor *et al.* 1975; Hofer 1979a,b; Buddington *et al.* 1987, 1997), but this ability seems to vary among species. Carnivores, for example, appear to have a limited capacity to alter digestive and nutrient transport functions to match changes in dietary composition, whereas omnivores display a much greater ability to modulate their digestive and absorptive physiology (Buddington *et al.* 1987, 1997).

Diet-induced changes in gastrointestinal morphology are reasonably well documented in fish (Kapoor *et al.* 1975; Buddington *et al.* 1997), with high-roughage or nutrient-dilute diets tending to induce hypertrophy of the stomach and enlargement of the intestine. This enlargement of the GIT following feeding on foods of low nutrient density (Hilton *et al.* 1983; Ruohonen & Grove 1996) may represent one facet of the suite of adaptations that enable fish to maintain rates of nutrient, and energy, intake when provided with diluted feeds (Rozin & Mayer 1961; Grove *et al.* 1978; see also Chapter 11). Similarly, a common response to reductions in feeding frequency or time-restricted feeding is an increase in meal size, presumably brought about by an increase in gastric capacity and hypertrophy of gut tissues. For example, plaice (*Pleuronectes platessa*) that were fed every other day developed larger and more bag-like stomachs than those fed more frequently, and the fish that were fed infrequently also increased their meal sizes to a greater extent than predicted (Jobling 1982). Hypertrophy of the GIT may commence shortly after the imposition of such feeding regimes, but some days or weeks may be required before the changes in the relative size of the GIT are complete, so that meal size adaptations may occur gradually over time. This is what appears to occur when fish are exposed to time-restricted feeding regimes; feeding activity and/or feed intake of fish fed according to time-restricted regimes increases with the passage of time and gradually approaches that of conspecifics allowed continuous access to food (Alanärä 1992; Boujard *et al.* 1996; Koskela *et al.* 1997).

The measurement of physiological effects over long periods allows identification of changes that occur within an experimental population or in individuals. This provides a basis for the investigation of life history strategies, such as whether low or high protein turnover impact on long-term growth (e.g. Carter *et al.* 1998), or how feeding and growth link to the maturation of different individuals and strains (e.g. Sæther *et al.* 1996; Tveiten *et al.* 1996; Damsgård *et al.* 1999). Measurements can be made several times during the course of a study, and these can either be related to food intake directly preceding the measurements or to estimates of cumulative food intake. For example, flounder (*Pleuronectes flesus*) were held for 212 days and nitrogen flux measured three times over this period; significant relationships between protein intake and protein growth, protein synthesis and ammonia excretion were demonstrated (Carter *et al.* 1998). Studies of this type allow stable feeding regimes to be established, measurable growth occurs and repeated measurements of metabolic parameters can be made (Carter *et al.* 1993a,b; McCarthy *et al.* 1994; Owen *et al.* 1998). Results of such studies have, for example, supported the conclusion that individual differences in protein turnover have a marked effect on growth performance.

13.6.1 *Protein intake, synthesis and growth*

Tissue accretion (growth) is related to food intake, and when growth is defined as protein growth (Houlihan *et al.* 1995a,b) rates of protein growth will be determined by rates of protein synthesis and protein degradation. Measurements of protein metabolism can be made on individual animals whose food consumption has been determined (see reviews by Houlihan *et al.* 1995a,b), and such investigations have revealed possible reasons for the wide differences between individual fish in the efficiency of conversion of food into growth.

Protein growth is related to the amount of protein consumed through the stimulation in protein synthesis and efficiency of retention of synthesised proteins, so analyses of rates of protein synthesis and protein growth should be combined with data on food intake (Houlihan *et al.* 1988, 1989; Carter *et al.* 1993a,b, 1998; McCarthy *et al.* 1994). Food intake appears to stimulate RNA synthesis, and this is indicated by an increase in the RNA:protein ratio. This probably provides part of the mechanism facilitating increased protein synthesis, so net growth occurs when rates of protein synthesis are higher than protein degradation (see Fig. 13.7). Protein growth is the difference between protein synthesis and degradation (see Fig. 13.7). A difference in protein growth between fish with the same food intake could be due to higher synthesis, lower degradation, or a combination of both. Individuals with high growth efficiency have been compared to individuals with the same food intake but which grow more slowly. Improved efficiency of conversion of food into growth seems to relate to lower rates of protein degradation (turnover) in the efficient fish rather than to higher rates of protein synthesis (Carter *et al.* 1993b; McCarthy *et al.* 1994).

The influence of diet composition on protein synthesis and growth has also been investigated. Not only protein synthesis but also protein degradation (increased protein turnover)

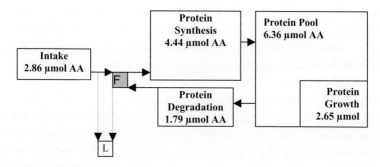

Fig. 13.5   Amino acid flux diagram for turbot larvae (at 17 days post hatch) fed with natural zooplankton, based on the model of Millward & Rivers (1988) and showing relative sizes for each component (redrawn from Conceicao *et al.* 1997). The values for the free AA (F = 0.24 µmol AA) and protein pools and for the rates of protein intake, synthesis, degradation and growth are from Conceicao *et al.* (1997). Values assume an average amino acid profile where 1 g of protein was calculated to be equivalent to 8.1 mmol free amino acid. Amino acid losses (L) of 0.21 µmol AA were calculated as intake minus protein growth.

can be correlated with increased protein intake (Houlihan *et al.* 1989; Meyer-Burgdorff & Rosenow 1995). Different methods, the short-term using flooding dose (Houlihan *et al.* 1989) and the medium-term end-product method (Meyer-Burgdorff & Rosenow 1995), have been used to measure protein synthesis. P. Campbell (unpublished data) found that as protein consumption increased (with no increase in ration or energy consumption), both protein synthesis and protein degradation increased in rainbow trout, and Meyer-Burgdorff and Rosenow (1995) reported that when common carp were fed high-protein feeds, both protein turnover and amino acid deamination increased. However, when the fish were fed low-protein feeds (sub-optimal protein concentrations), protein turnover was even higher, presumably due to increased recycling of amino acids (released from protein degradation) into protein synthesis. This result has also been found in rainbow trout fed diets with a poor amino acid balance (Perera 1995), and supports the hypothesis that protein synthesis has a major role in the regulation of amino acid concentrations (see below, Amino acid flux model).

13.7 Amino acid flux model: food intake and amino acid flux

One aim of incorporating food intake data into an analysis of the physiological response to feeding is to construct flux diagrams. This is done in an attempt to analyse amino acid flux from the food into the free pools and hence into protein synthesis, protein growth and amino acid losses (Fig. 13.5). Such flux models are at an early stage of development, and represent an oversimplification. Current models present a single free amino acid pool (ignoring differences in free amino acid composition between tissues), and are based on estimates of whole body rates of protein synthesis. They also suffer from problems relating to differences in time scale between protein synthesis measurements and protein growth measurements (Houlihan *et al.* 1995b,c). Nevertheless, amino acid flux models are useful as they give an indication of the proportion of ingested protein which is retained as growth, and demonstrate the effects a meal has on the size of the tissue free amino acid pools and protein synthesis.

An amino acid flux model for turbot (*Scophthalmus maximus*) larvae has been constructed based upon intake values calculated from the amount of amino acids offered to the larvae each day (Conceicao *et al.* 1997). The model (Fig. 13.5) provides indications that a very high proportion (over 90%) of the ingested protein was retained as growth compared with values reported previously for larval herring (*Clupea herengus*) (60–65%) (Houlihan *et al.* 1995c) and juvenile fish (ca. 50%) (Houlihan *et al.* 1995a; Grisdale-Helland & Helland 1997). Protein conversion efficiencies for larval fish are not generally available because of the difficulty of measuring food intake. Also, the estimates of amino acid intake are based upon the amino acid composition of the food offered. The flux diagram (Fig. 13.5) indicates that, on a daily basis, the dietary amino acid supply may be as much as ten times the free amino acid pool. Daily removal by protein synthesis was estimated to be twenty times the free amino acid pool, and protein breakdown was estimated to return ten times the free pool. Thus, the free amino acid pool of larval turbot is extremely dynamic, and it is possible that the amino acid profile of the absorbed amino acids should match the requirements very closely if excessive amino acid oxidation is to be avoided.

An amino acid flux model constructed for larger fish (a 250-g rainbow trout consuming 2% body weight ration) revealed that consumption of amino acids was equivalent to approximately twice the free amino acid pool (Houlihan *et al.* 1995c). Free amino acid pools in the majority of tissues are relatively stable following a meal (Carter *et al.* 1995a; see p.300, Tissue metabolic physiology), and they are regulated through anabolic and catabolic pathways. In the trout model it was estimated that protein synthesis removed 1.3 times the free pool; recycling of amino acids into the free pool through protein breakdown was equivalent to less than half its size, and losses of amino acid nitrogen above the maintenance rate were similar to the amount retained as growth (Houlihan *et al.* 1995c).

13.8 Biochemical correlates of food intake

It is to be expected that the greater amount of food eaten by fish, the greater will be the activities and concentrations of enzymes involved in cellular metabolism (Houlihan *et al.* 1993). However, there are only a limited number of studies where such relationships have been investigated at the level of the individual animal. More frequently the food intake of a group of fish has been measured and, because of the relationship between food intake and growth, the growth rate of individual animals has been correlated with the biochemical correlate under consideration. Thus, food intake of the group has been assumed to have driven both growth rate and the metabolic, or biochemical, correlate. Several metabolic variables have been used in an attempt to obtain some indirect assessment of the condition and recent growth history of fish (Bulow 1987; Busacker *et al.* 1990; Houlihan *et al.* 1993; Couture *et al.* 1998; Dutil *et al.* 1998). The metabolic indicators monitored include concentrations of nucleic acids, and enzymatic indicators of aerobic and glycolytic capacities of muscle, liver and intestine.

One frequently measured cellular component used as a growth correlate is RNA. RNA concentration is frequently expressed as a RNA:protein ratio on the grounds that this ratio correlates well with rates of protein synthesis, and the RNA:protein ratio has been described as the capacity for protein synthesis (mg RNA per gram protein; Sugden & Fuller 1991). Some 85% of total RNA is believed to be ribosomal, and it seems that the concentration of the ribosomes has a major influence on rates of protein synthesis. Ornithine decarboxylase (ODC) is the first and rate-limiting enzyme in the biosynthesis of nucleic acids and proteins (Benfey 1992; Benfey *et al.* 1994). Thus, it might be expected that changes in ODC activity would precede changes in other biochemical indices, e.g. protein synthesis, tissue amino acid incorporation and RNA:DNA ratios, used for the assessment of condition and short-term changes in fish growth. In brook trout (*Salvelinus fontinalis*) there was a positive correlation between hepatic ODC activity and gut contents (Benfey 1992).

The tissue RNA:DNA ratio has often been used as an indirect measure of recent growth in fish. It has been argued that because the DNA content of a cell is relatively constant, and RNA content varies with the rate of protein synthesis, the ratio of RNA to DNA provides an index of protein synthetic activity and hence growth (Bulow 1987); it is not, however, clear whether this argument is true for multinuclear muscle cells. In many studies with fish larvae it is the ratio of RNA to DNA that has been used as the growth correlate. For example, results of a recent study on cod larvae indicate that the analysis of nucleic acids may provide valuable information about the recent growth and condition of individual larvae (McNamara *et al.* 1999). Concentrations of RNA, DNA, 18S ribosomal RNA (rRNA), poly(A) messenger RNA (mRNA) and two mRNAs coding for myofibrillar proteins were examined in cod larvae held in the laboratory under conditions of feeding and starvation. At the time of yolk exhaustion there were significant differences between fed and starved larvae for all measured components. In an experiment with older larvae (3–4 weeks), 18S rRNA was significantly reduced following three days of food deprivation, and mRNAs and total RNA responded in a similar manner (McNamara *et al.* 1999).

In a study carried out on juvenile cod where the animals were fed individually, the rate of food consumption was found to be positively correlated with white muscle RNA:protein ratio (Houlihan *et al.* 1989) and also to growth, protein synthesis and protein degradation rates (Houlihan *et al.* 1988). Data for Atlantic salmon gave similar relationships, and the strength of the relationships with food intake provided indications that the concentration of RNA (and by inference ribosomal numbers) was more important than ribosomal activity (i.e. grams of protein synthesised per gram RNA per day) (Carter *et al.* 1993b). Broadly similar results have been obtained with grass carp (Carter *et al.* 1993a) and rainbow trout (McCarthy *et al.* 1994).

The strength of the positive relationships between RNA concentration and growth rate reported in many studies has led to various attempts to estimate the growth rates of wild animals from the RNA concentrations in their tissues (e.g. Mathers *et al.* 1992a,b; Pelletier *et al.* 1994). However, RNA concentrations in fish tissues may be affected by temperature and are also influenced by body size (Houlihan *et al.* 1993), and it is by no means clear that RNA concentrations always provide a good predictor of growth rate (Miglavs & Jobling 1989; Foster *et al.* 1993a; Pelletier *et al.* 1995). An additional complication is the time-course of the change in RNA concentration in response to the change in rate of food intake (Foster *et al.* 1993b).

During periods of intense feeding the activity of white muscle enzymes, such as phosphofructokinase (PFK) and 3-hydroxyacyl-CoA (HAD) appear to be positively correlated with feeding and growth rates (Kiessling *et al.* 1991a). The effect was not observed in the PFK activity of the red muscle. In small rainbow trout, as food intake increased, HAD and citrate synthase activities increased in red muscle but not in white muscle (Kiessling *et al.* 1991a).

Activities of mitochondrial enzymes, such as HAD, citrate synthase (citric acid cycle) and cytochrome oxidase (respiratory chain) in the white muscle may also be positively correlated with growth rates, but mitochondrial enzymes in red muscle appear to be poor predictors of growth. This is not surprising given the fact that during fasting it is the white muscle proteins that are mobilised, whereas the aerobic, red muscle seems to be preferentially conserved (Love 1988; see Chapter 15). However, there is not necessarily a strong link between enzyme activity, growth rate and food intake. No long-term effects were observed in white or red muscle enzyme activity of Chinook salmon (*Oncorhynchus tshawytscha*) despite a 20% increase in food intake when fish were exposed to increased water velocities (Kiessling *et al.* 1994), indicating that food intake had no significant effect on muscle metabolism *per se*. Both Walzem *et al.* (1991) and Bastrop *et al.* (1992) reported positive correlations between liver metabolism, growth and allotted ration level, so the liver seems to be an organ that is very responsive in terms of changes in enzyme levels in response to feeding (e.g. Bastrop *et al.* 1992).

Enzyme systems that may be influenced by intensity of feeding are also to be found in the digestive system. For example, intestinal mitochondrial enzyme activity appears to correlate with feeding and growth (Couture *et al.* 1998; Dutil *et al.* 1998). Nutrient absorption by the enterocytes of the gut involves active transport (Buddington *et al.* 1987, 1997; Hirst 1993); this relates to Na^+K^+ ATPase activity, and the enterocytes are mitochondria-rich cells in which ATP is produced aerobically. Consequently, links between feeding intensity, intestinal mitochondrial enzyme activity, nutrient absorptive capacity and growth might be expected. In the cod, chymotrypsin activity in the digestive caeca and alkaline phosphatase and glutamyl-transferase in the intestine have been found to correlate with the ingestion of food (Lemieux *et al.* 1999). The authors suggested that trypsin could be limiting growth rate in cod, and this has also been postulated for Atlantic salmon strains with different trypsin isozymes. The different trypsin isozymes may produce differences in free amino acid concentrations in the plasma (Torrissen *et al.* 1994; Rungruangsak-Torrissen *et al.* 1999).

13.9 Effects on body composition and growth efficiency

Although the previous sections have emphasised the link between food intake and physiological responses, a more frequent approach is to relate food intake to whole animal growth. The relationship between food intake and wet weight growth has been described as being curvilinear. This means that above a certain level of food intake (defined as the optimum ration) the rate of increase in growth declines and efficiency decreases from its maximum value (Brett 1979; Elliot 1994; Jobling 1997). In many studies there has been a failure to differentiate between food provided and food consumed, leading to an overestimation of food intake and making interpretation of the data difficult (see Chapter 1 for discussion). In some studies in which food intake has been measured there is evidence for the curvilinear relationship (e.g. Jobling *et al.* 1989; Christiansen & Jobling 1990; Carter *et al.* 1992, 1994), and the same has been found in some studies in which cumulative food intake has been measured for fish held individually (e.g. Allen & Wootton 1982; Cui & Wootton 1988). On the other hand, a linear relationship between food intake and growth (Fig. 13.6) (e.g. Cui *et al.* 1994, 1996; Jobling 1995; Kiessling *et al.* 1995; Xie *et al.* 1997) or between protein intake and protein growth

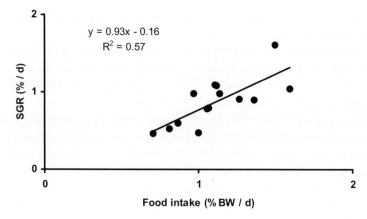

Fig. 13.6   The relationship between mean daily feed intake (% bw per day) and specific growth rate (% per day) of individually fed Atlantic salmon (700–1100 g) held at 9–14°C. (Adapted from Kiessling *et al.* 1995.)

(Fig. 13.7) is apparent in some instances. Thus, the form of the relationship between food consumption and growth is far from being resolved. There may be real species differences, confusion relating to the units used (absolute versus relative rates, wet versus dry weight, etc.) and temporal effects (see Jobling 1997 for discussion). The discrepancies may also reflect

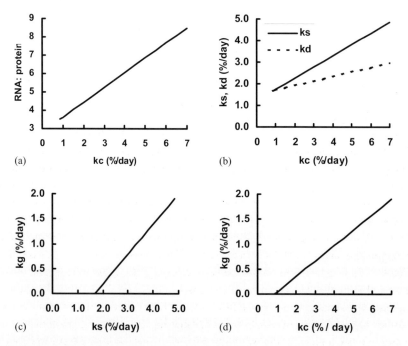

Fig. 13.7   Summary figure of protein turnover (% per day) in fish as represented by relationships between: (a) protein intake (kc) and the capacity for protein synthesis (RNA:protein ratio); (b) protein intake and protein synthesis (ks) and protein degradation (kd); (c) protein synthesis and protein growth; and (d) protein consumption and protein growth (kg). Based on Atlantic cod. (Redrawn from Houlihan *et al.* 1988, 1989.)

problems encountered in measuring the relationship over the entire range of food intake, and under different environmental conditions.

In terrestrial mammals, where individual food intake measurements are easier to perform, fast-growing individuals have been shown to utilise feed more efficiently than slow grow-ing animals (Tomas *et al.* 1991; Oddy 1999). In fish, this relationship has been far less well studied, partly due to the difficulty in measuring food intake accurately (see Thodesen *et al.* 1999). However, in most breeding programmes for fish it is assumed that the fastest-growing fish are also the most efficient at utilising the feed (see Gjedrem 1983), and there is evidence that fish of fast-growing strains may have lower maintenance requirements than those of slower-growing strains (Kolok 1989).

There are perceived to be several benefits to be gained from the production of 'single-sex' populations or sterile triploid fish for farming purposes (Donaldson & Devlin 1996; Maclean 1998). For example, sterile triploids are expected to grow better than diploids because of the allocation of resources to somatic weight gain rather than in the production of gonads. Further, the induction of sterility is envisaged to be a method that can be used to circumvent the problem of deterioration of flesh quality that accompanies sexual maturation in several fish species. The expected differences between diploids and triploids have not always been observed, and the results of growth trials have been equivocal (Henken *et al.* 1987; Carter *et al.* 1994; Galbreath *et al.* 1994; Habicht *et al.* 1994; Galbreath & Thorgaard 1995; Ojolick *et al.* 1995; O'Keefe & Benfey 1999). Thus, in some cases triploids have been reported to outperform diploids, in others little difference has been observed, and in some studies the performance of triploid fish has been inferior to that of diploid conspecifics. In cases where food intake has been measured only minor differences between triploids and diploids have been recorded (Carter *et al.* 1994; O'Keefe & Benfey 1999), although there is some evidence that triploids may be competitively inferior when reared together with diploids.

Developments within the field of molecular genetics are playing an increasing role in ag-ricultural production through the application of transgenic technology to enhance commer-cially important traits. Organisms into which new genetic material has been introduced are termed 'genetically modified' or 'transgenic', and such organisms express proteins encoded by the introduced genes (Steele & Pursel 1990; Chen *et al.* 1995; Donaldson & Devlin 1996; Sin 1997; Maclean 1998; Prieto *et al.* 1999). Expression of peptide hormones, such as growth hormone (GH), in transgenic animals induces a series of physiological alterations relating to metabolism and growth, and it is transgenic fish for GH that have been the focus of most atten-tion (Chen *et al.* 1995; Donaldson & Devlin 1996; Maclean 1998; Cook *et al.* 2000a,b,c).

Fish transgenic for GH have accelerated growth and development relative to non-trans-genic conspecifics, results that are in line with findings that application of exogenous GH also results in growth enhancement (Donaldson *et al.* 1979; Chen *et al.* 1995; Björnsson 1997; see also Chapter 12). Changes seen in fish transgenic for GH also seem to resemble those seen in fish treated with exogenous hormone (Björnsson 1997; Devlin *et al.* 1999; Cook *et al.* 2000a). Thus, food intake of transgenic fish is greater than among non-transgenic conspecif-ics, and there are also differences in nutrient partitioning and body composition parameters, with transgenic fish tending to deposit less body lipid (Cook *et al.* 2000a). However, when large transgenic coho salmon were compared with their conspecifics (average body weight 4.5 kg), no differences in muscle structure were evident (A. Kiessling & R.H. Devlin, unpub-lished data), underlining the fact that the effects of very long-term exposure to elevated GH levels are not well understood and may well be different from short-term exposure effects. Other changes seen in fish transgenic for GH are increased hyperplasia (Hill *et al*. 2000). Transgenic fish also display a suite of behavioural changes, including increased foraging activity (Abrahams & Sutterlin 1999) and increased competitiveness under conditions of restricted food supply (Devlin *et al*. 1999), that may enhance feed acquisition.

Metabolic rates of transgenic fish also tend to be higher than those of non-transgenic conspecifics (Stevens *et al*. 1998; Cook *et al*. 2000b,c). The increased metabolic rate may, in part, be a result of increased activity (Stevens *et al*. 1998; Abrahams & Sutterlin 1999) and increased feeding by the transgenics (Cook *et al*. 2000a), because an increase in food intake would result in a high metabolic rate due to HIF. This is, however, not the full explanation because transgenic fish maintain elevated metabolic rates even when food-deprived, and this results in a more rapid depletion of energy reserves than seen in non-transgenic conspecifics (Cook *et al*. 2000c). There are, for example, marked differences in rates of depletion in lipid reserves, possibly reflecting the central role of GH in the catabolism of lipid during periods of food shortage. In other words, under conditions of food deprivation there appears to be an uncoupling of GH from the insulin-like growth factor system, and GH is directed away from growth promotion towards the regulation of lipolysis and catabolism (see Chapter 12). However, in GH transgenic coho salmon there were no differences in the β-oxidation enzyme HAD in the white muscle, but significantly higher levels of the mitochondrial respiratory chain enzyme cytochrome oxidase and the glycolytic enzyme PFK – results that agree with the higher metabolic rates of these fish reported by others (Stevens *et al*. 1998; Cook *et al*. 2000c).

Increased ration leads to increased growth (Brett 1979; Elliott 1994; Jobling 1997), with growth being a function of food intake in combination with the capacity for the fish to utilise the ingested food. The metabolism rates of ectotherms, such as fish, are low compared with those of endotherms of similar size, and are some five to ten times lower even at the same body temperature. This means that maintenance requirements of fish are much lower than those of ectothermic vertebrates, and fish have considerable potential for the efficient conversion of ingested food into somatic growth. Thus, when fish are held at temperatures close to the optimum, the maintenance ration is often found to be 15–25% of the maximum ration (e.g. Brett 1979; Cui *et al*. 1994, 1996; Elliott 1994; Jobling 1995; Mélard *et al*. 1996; Xie *et al*. 1997), and conversion efficiencies in excess of 30% on an energy basis are frequently recorded. However, the energetic costs of swimming are relatively high, and for fish consuming fixed submaintenance rations there is an increased depletion of energy reserves at progressively higher swimming speeds. Similarly, the energetic costs of swimming may result in reduced growth in fish fed at fixed rations that are above maintenance (White & Li 1985). On the other hand, increased swimming speed may lead to a stimulation of feed intake in fish allowed to feed to satiation, with the result that there is no negative effect upon growth rate but feed efficiency is reduced (Fig. 13.8; Kiessling *et al*. 1994).

A variety of nutritional factors such as the amount and composition of the diet as well as the feeding regime can affect body composition, particularly body lipid (see Chapter 15). In long-term growth trials body size becomes a complicating factor (Shearer 1994). For example, it is well known that lipid deposition differed according to digestible energy intake, whereas body protein content (expressed as percent wet weight) was dependent on fish weight and

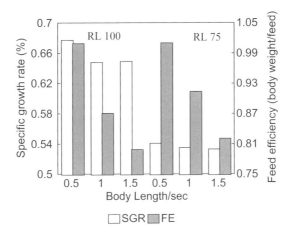

Fig. 13.8   Specific growth rate (% per day) and feed efficiency ratio (g/g) in Chinook salmon exposed to different levels of sustained exercise (swimming speed, body lengths/second) and ration levels (RL100 and RL75). (Adapted from Kiessling *et al.* 1994.)

appeared not to be affected by food intake (Storebakken & Austreng 1987a,b; Kiessling *et al.* 1991b; Shearer 1994).

The amount of food consumed is likely to affect a range of physiological systems in addition to those already mentioned. The measurement of food intake of individuals in large groups allows the description of social hierarchies in terms of the proportion of the food consumed by each individual (see Chapter 3). For example, in *Tilapia rendalli* there was a correlation between feeding rank and dominance index when fish were fed from a point source (McCarthy *et al.* 1999). The effects of feeding and/or social hierarchy rank on brain serotonergic and dopaminergic activity (Winberg *et al.* 1993; Overli *et al.* 1998), fin damage (Moutou *et al.* 1998) and stress (Christiansen *et al.* 1991) have also been examined. The emerging conclusion is that there are higher costs associated with being dominant or subordinate than being in the middle of a hierarchy (Moutou *et al.* 1998). Different facets of the abiotic and biotic environment can be manipulated to control the formation and strength of hierarchies, thereby restricting their physiological effects (see Chapter 6). For example, brain serotonergic activity was elevated in subordinate Arctic charr compared with dominants (Overli *et al.* 1998), but during subsequent rearing in isolation, food intake gradually increased in previously subordinate fish, while serotonergic activity fell to be close to that of dominants. The total amount of food available (McCarthy *et al.* 1992), as well as its spatial distribution and predictability of supply (Grand & Grant 1994; McCarthy *et al.* 1999), influence feeding, inter-individual variation in food intake and therefore physiological effects and growth.

Water speed can affect schooling behaviour in fish, and this may affect growth and growth efficiency (see Chapter 6). Higher growth efficiency of groups of Arctic charr subjected to sustained swimming was explained by lower energy expenditure associated with agonistic activity than for groups in static water (Christiansen & Jobling 1990). Differences between individual food intake (measured by X-radiography) and growth demonstrated that higher maintenance costs (and lower growth for a given food intake) were associated with static water (Christiansen & Jobling 1990). Interestingly, plasma cortisol concentrations were higher in fish forced to swim, and this was thought to reflect general changes in metabolism rather than higher levels of stress (Christiansen *et al.* 1991). Some credence to this is given by the finding that the maintenance requirements of rainbow trout treated with cortisol implants (leading to chronic elevation of plasma cortisol concentrations) are higher than those of sham-treated and control individuals. In other words, elevated cortisol concentrations seem to result in 'higher costs of living' for the fish (Gregory & Wood 1999). Furthermore, chronic elevation of plasma cortisol concentration had negative effects on feed intake, growth and feed conversion in fish fed to satiation. Negative effects on growth and feed conversion were even more pronounced in fish provided with sub-maximum rations (Gregory & Wood 1999), which would be expected if treatment with cortisol resulted in increased metabolic costs to the fish.

One interesting area is the possible relationship between food intake and immunology. Lysozyme activity, which is thought to play a role in disease defence, was found to be reduced in rainbow trout fed a protein-deficient diet (Kiron *et al.* 1995). Unfortunately this study, like many others, used dietary manipulations but did not control rates of food intake, beyond feeding to satiation. In another study, in which food intake of trout was determined, there were found to be significant negative correlations between the amount of food consumed and lysozyme and antiprotease activity (Thompson 1993). This surprising result was also accompanied by a slight immunostimulatory effect of feed deprivation. The hypothesis that high feeding rates can result in immunosuppression, possibly as a consequence of pressure on metabolic scope, deserves to be further investigated.

13.10 Physiological effects and the regulation of food intake

Regulation of food intake is highly complex. It involves positive and negative feedback mechanisms acting over a hierarchy of time scales to determine the size of a single meal, the balance of essential nutrients and, in the long term, body composition (see reviews by Forbes 1988, 1992, 1999; Langhans & Scharrer 1992; Langhans 1999). Positive feedback determines the initiation and continuation of feeding, and results from relationships between the sensory properties of foods, experience concerning nutrient availability and the physiological status of the animal. Negative feedback can be divided into gastrointestinal and metabolic, or pre- and post-absorptive, phases (Langhans 1999). Forbes (1999) has argued that since the natural condition of an animal is to eat, only systems involved in the cessation of feeding need to be considered. Few reviews that concern the factors involved in regulating food intake in fish are available (see Peter 1979; Vahl 1979; Fletcher 1984; LeBail & Boeuf 1997; Silverstein *et al.* 1999), but the broad levels of organisation and integrated models proposed for mammals also appear to apply to fish (see Chapter 12).

We have attempted to integrate models of nutrient flux with models of food intake (Forbes 1999; Langhans 1999) in order to discuss physiological effects (Fig. 13.9). Thus, we adopt part of the 'satiety cascade' described by Blundell and Halford (1994), but omit reference to external information concerning the environment (e.g. temperature or photoperiod) and the food (e.g. appearance or smell). Three levels of regulation of food intake have been categorised, and these are designated short, medium and long term. Short-term regulation refers to factors that influence the size of a single meal (or daily intake), medium-term regulation

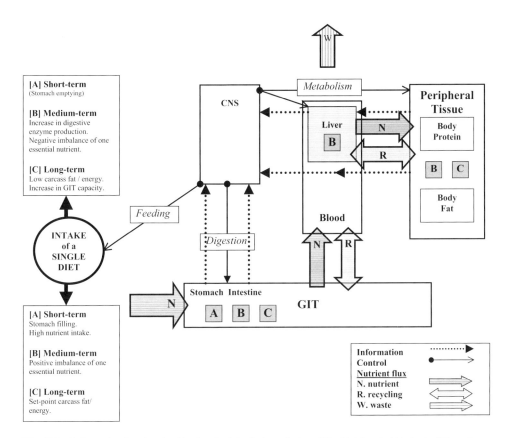

Fig. 13.9 Information flow involved in the physiological regulation of food intake via [A] short-, [B] medium- and [C] long-term factors acting from the GIT (gastrointestinal tract), blood, liver, peripheral tissues and the CNS. Nutrient flow (N) via food intake, nutrients absorbed from GIT and retained as growth; recycling of nutrients (R); waste following respiration or nitrogenous excretion (W).

involves food intake over a period of days, and long-term regulation refers to regulation over periods of weeks to years (or complete life cycles). A factor is viewed as anything that elicits a response by the fish. Factors will include the physical bulk of food stimulating mechano-receptors in the stomach wall (short-term), nutrient imbalances that induce changes in food intake (e.g. stimulate intake of a feed with a high content of the limiting nutrient) with links to circulating levels of nutrients and metabolites (medium-term), and whole-body energy balance (long-term). Specific factors may be involved at more than one level, but the main purpose is to emphasise the importance of an integrated multifactorial approach to information transfer, rather than to consider specific neurological or hormonal pathways (see Chapter 12). The first point of interest is in restricting the model to consider intake of a single feed of given composition. In the majority of experiments that investigate physiological effects a single feed is used, and a choice between two or more feeds is rarely given (e.g. Cuenca *et al.* 1993). This means that varying the quantity consumed is the only mechanism available for controlling nutrient intake. This differs from the natural situation where fish can select prey that may differ in nutrient concentrations (Forbes 1999).

In the gastrointestinal phase, several of the factors involved in short-term regulation relate to the capacity and filling of the stomach or intestinal-bulb in relation to its emptying and transfer of ingesta into the intestine. These factors may, in turn, be linked to feeding frequency and the physical and chemical properties of the food (Fänge & Grove 1979; Jobling 1987; dos Santos & Jobling 1991). In the medium term, feed composition may influence the production of digestive enzymes and the density of cell membrane nutrient transporters, resulting in changes in digestive and absorptive capacity (Kapoor *et al.* 1975; Buddington *et al.* 1987, 1997; Ferraris & Diamond 1989; Hirst 1993). Some of these factors may also act in the long term. For example, feeding fish with feeds of different bulk may result in the development of differences in the physical capacity of the GIT (Ruohonen & Grove 1996). The GIT is metabolically active (related to defence mechanisms as well as digestion and absorption), and has the potential to alter the composition of nutrients leaving it through both metabolism of dietary nutrients and via recycling (Reeds *et al.* 1999) (Fig. 13.9). It has also been suggested that blood-borne amino acids are metabolised preferentially by the GIT over dietary amino acids (Reeds *et al.* 1999). The consequences of this in the regulation of food intake relate principally to the short- and medium-term effects of circulating amino acids. However, little is known about the importance of these processes in fish, except that protein turnover in the GIT is high, as it is in the GIT of terrestrial animals (Houlihan *et al.* 1988).

When one feed is available, the relative importance of different factors in the metabolic phase is due to a balance between the negative (toxic) effects caused by the increased intake of some nutrients and the negative effects associated with low intake of others (Forbes 1999). These opposing forces will be mediated through circulating levels of metabolites or other physiological effects. For example, in mammals a high intake of dietary protein has negative effects associated with it (e.g. increased metabolism and body temperature; increased toxic nitrogenous waste in the blood), and under some circumstances food intake may be reduced to limit protein intake (Forbes 1999). Plasma concentrations of amino acids, glycerol and fatty acids and glucose have all been implicated in the regulation of food intake in mammals (Forbes 1992), but there is little information available concerning fish. Results of experiments run to determine protein and amino acid requirements indicate that food intake may be reduced when feeds contain either very low or high concentrations of the essential nutrient (see Chapter 11). This suggests that, like mammals and birds (Langhans 1999), fish are responsive to amino acids and avoid the metabolic consequences of an unbalanced amino acid intake. The poor ability of fish to regulate plasma glucose, as well as the low concentrations of glucose receptors in most tissues, argues against glucose being of major importance in the regulation of food intake under most circumstances (Peter 1979; Wilson 1994). However, it is probably more constructive to view the role of metabolites as part of a larger system in which their relative importance may change depending on a variety of endogenous and exogenous factors. Consequently, in the summary diagram, metabolites are depicted as being part of the flow of nutrients to the peripheral tissues via the blood and liver and also being subject to recycling (Fig. 13.9).

Responses based on circulating metabolites will act over the short to medium term. There are examples of experiments where feeds are used that contain extreme amounts of protein or specific amino acids (e.g. Rodehutscord *et al.* 1995), and fish seem to sense nutrient imbalance and regulate intake accordingly. Further, if fish are offered a choice of feeds they, like other animals, appear to have the ability to adjust the intake of the different feeds in order to maintain a balanced nutrient intake (Cuenca *et al.* 1993; see Chapter 11).

HIF reflects general metabolic activity following feeding, and it has been suggested that its magnitude may have an influence on food intake. As HIF increases, the scope for further metabolic activity decreases and the time-course of HIF should also reflect an increase in metabolites in the blood (Vahl 1979). Since dogfish (*Scyliorhinus canicula*) consumed food (approximately 17% of a full meal) at the time of maximum HIF, HIF was viewed as not having the major influence on food intake in this species (Sims & Davies 1994). Similarly, Atlantic cod may consume additional meals under conditions where metabolic rates are still elevated due to the effects of HIF (Soofiani & Hawkins 1982). Such experiments probably show the complexity of food intake control in that it is difficult to isolate the effect of a single factor when other factors are also exerting their effects.

That there is long-term regulation of body fat (or energy) to a set point (Kennedy 1953), thought to be primarily mediated via leptin and insulin, is generally accepted in mammals (Langhans 1999; see Chapter 12). Evidence for a similar long-term regulatory system in fish, although sparse, is becoming persuasive (Metcalfe & Thorpe 1992; Jobling & Miglavs 1993; Company *et al.* 1999; Silverstein *et al.* 1999). These studies show that attaining and maintaining a given energy status may be important for survival and reproduction. Both body protein and lipid should be considered as contributing to the 'energy status' (Fig. 13.9). Since fish species differ in their use of protein and lipid (and carbohydrate) reserves during feed deprivation (see Chapter 15), it will be of interest to see if these differences are reflected in long-term regulation.

Food intake has an influence on many physiological processes, and it is important to be able to attribute cause to effect. Our attempt to model the regulation of food intake in fish highlights the need to consider an integrated approach rather than place too much importance on one or other factor. It is likely that the relative importance of different factors will change according to conditions such as the species, environmental characteristics or nutritional history.

13.11 References

Abrahams, M.V. & Sutterlin, A. (1999) The foraging and antipredator behaviour of growth-enhanced Atlantic salmon. *Animal Behaviour*, **58**, 933–942.

Alanärä, A. (1992) The effect of time-restricted demand feeding on feeding activity, growth and feed conversion in rainbow trout (*Oncorhynchus mykiss*). *Aquaculture*, **108**, 357–368.

Ali, M., Przybylski, M. & Wootton, R.J. (1998) Do random fluctuations in daily ration affect the growth rate of juvenile three-spined sticklebacks? *Journal of Fish Biology*, **52**, 223–229.

Allen, J.R.M. & Wootton, R.J. (1982) The effect of ration and temperature on the growth of the three-spined stickleback, *Gasterosteus aculeatus* L. *Journal of Fish Biology*, **20**, 409–422.

Arzel, J., Metailler, R., Kerleguer, C., Delliou, H. & Guillaume, J. (1995) The protein requirement of brown trout (*Salmo trutta*) fry. *Aquaculture*, **130**, 67–78.

Aschoff, J. (1979) Circadian rhythms: general features and endocrinological aspects. In: *Endocrine Rhythms* (ed. D.T. Krieger), pp. 23–56. Raven Press, New York.

Bastrop, R., Jürss, K. & Wacke, R. (1992) Biochemical parameters as a measure of food availability and growth in immature rainbow trout (*Oncorhynchus mykiss*). *Comparative Biochemistry and Physiology*, **102A**, 151–161.

Beamish, F.W.H. (1974) Apparent specific dynamic action of largemouth bass, *Micropterus salmoides*. *Journal of the Fisheries Research Board of Canada*, **31**, 1763–1769.

Bergot, F. and Breque, J. (1983) Digestibility of starch by rainbow trout: effects of the physical state of starch and of the intake level. *Aquaculture*, **34**, 203–212.

Bence, J.R. (1986) Feeding rate and attack specialisation: the roles of predator experience and energetic tradeoffs. *Environmental Biology of Fishes*, **16**, 113–121

Benfey, T.J. (1992) Hepatic ornithine decarboxylase activity during short-term starvation and refeeding in brook trout, *Salvelinus fontinalis*. *Aquaculture*, **102**, 105–113.

Benfey, T.J., Saunders, R.L., Knox, D.E. & Harmon, P.R. (1994) Muscle ornithine decarboxylase activity as an indication of recent growth in pre-smolt Atlantic salmon, *Salmo salar*. *Aquaculture*, **121**, 125–135.

Björnsson, B.T. (1997) The biology of salmon growth hormone: from daylight to dominance. *Fish Physiology and Biochemistry*, **17**, 9–24.

Blundell, J.E. & Halford, J.C.G. (1994) Regulation of nutrient supply: the brain and appetite control. *Proceedings of the Nutrition Society*, **53**, 407–418.

Boujard, T. & Leatherland, J.F. (1992) Circadian rhythms and feeding time in fishes. *Environmental Biology of Fishes*, **35**, 109–131.

Boujard, T., Gelineau, A. & Corraze, G. (1995) Time of a single daily meal influences growth performance in rainbow trout, *Oncorhynchus mykiss* (Walbaum). *Aquaculture Research*, **26**, 341–349.

Boujard, T., Jourdan, M., Kentouri, M. & Divanach, P. (1996) Diel feeding activity and the effect of time-restricted self-feeding on growth and feed conversion in European sea bass. *Aquaculture*, **139**, 117–127.

Brafield, A.E. (1985) Laboratory studies of energy budgets. In: *Fish Energetics New Perspectives* (eds P. Tytler & P. Calow), pp. 257–281. Croom Helm, London, Sydney.

Brauge, C., Médale, F. & Corraze, G. (1994) Effects of dietary levels of carbohydrate levels on growth, body composition and glycaemia in rainbow trout, *Oncorhynchus mykiss*, reared in sea water. *Aquaculture*, **123**, 109–120.

Brauge, C., Corraze, G. & Médale, F. (1995) Effects of dietary levels of carbohydrate and lipid on glucose oxidation and lipogenesis from glucose in rainbow trout, *Oncorhynchus mykiss*, reared in freshwater or in seawater. *Comparative Biochemistry and Physiology*, **111A**, 117–124.

Brett, J.R. (1979) Environmental factors and growth. In: *Fish Physiology,* Vol. VIII (eds W.S. Hoar, D.J. Randall & J.R. Brett), pp. 599–675. Academic Press, New York.

Brett, J.R. (1995) Energetics. In: *Physiological Ecology of Pacific Salmon* (eds C. Groot, L. Margolis & W.C. Clarke), pp. 2–68. UBC Press, Vancouver.

Brett, J.R. & Groves, T. D. D. (1979) Physiological energetics. In: *Fish Physiology,* Vol. VIII (eds W.S. Hoar, D.J. Randall & J.R. Brett), pp. 279–352. Academic Press, New York.

Bromley, P.J. (1994) The role of gastric evacuation experiments in quantifying the feeding rates of predatory fish. *Reviews in Fish Biology and Fisheries*, **4**, 36–66.

Brown, C.R. & Cameron, J.N. (1991) The relationship between specific dynamic action (SDA) and protein synthesis rate in channel catfish. *Physiological Zoology*, **64**, 298–309.

Bryan, J.E. & Larkin, P.A. (1972) Food specialization by individual trout. *Journal of the Fisheries Research Board of Canada*, **29**, 1615–1624.

Buddington, R.K., Chen, J.W. & Diamond, J. (1987) Genetic and phenotypic adaptation of intestinal nutrient transport to diet in fish. *Journal of Physiology*, **393**, 261–281.

Buddington, R.K., Krogdahl, Å. & Bakke-McKellep, A.M. (1997) The intestines of carnivorous fish: structure and functions and the relations with diet. *Acta Physiologica Scandinavica*, **161** (Suppl. 638), 67–80.

Bulow, F.J. (1987) RNA-DNA ratios as indicators of growth in fish: a review. In: *The Age and Growth of Fish* (eds R.C. Summerfelt & G.E. Hall), pp.45–64. Iowa State University Press, Ames, Iowa.

Busacker, G.P. Adelman, I.R. & Goolish, E.M. (1990) Growth. In: *Methods for Fish Biology* (eds C.B. Schreck & P.B. Moyle), pp. 363–387. American Fisheries Society, Bethesda.

Carter, C.G. & Brafield, A.E. (1991) The bioenergetics of grass carp, *Ctenopharyngodon idella* (Val.): energy allocation at different planes of nutrition. *Journal of Fish Biology*, **39**, 873–887.

Carter, C.G. & Brafield, A.E. (1992) The relationship between specific dynamic action and growth in grass carp, *Ctenopharyngodon idella* (Val.). *Journal of Fish Biology*, **40**, 895–907.

Carter, C.G., Houlihan, D.F., McCarthy, I.D. & Brafield, A.E. (1992) Variation in food intake of grass carp, *Ctenopharyngodon idella* (Val.), fed singly or in groups. *Aquatic Living Resources*, **5**, 225–228.

Carter, C.G., Houlihan, D.F., Brechin, J. & McCarthy, I.D. (1993a) The relationships between protein intake and protein accretion, synthesis and retention efficiency for individual grass carp, *Ctenopharyngodon idella* (Val.). *Canadian Journal of Zoology*, **71**, 393–400.

Carter, C.G., Houlihan, D.F., Buchanan, B. & Mitchell, A.I. (1993b) Protein-nitrogen flux and protein growth efficiency of individual Atlantic salmon (*Salmo salar* L.). *Fish Physiology and Biochemistry*, **12**, 305–315.

Carter, C.G., McCarthy, I.D., Houlihan, D.F., Johnstone, R., Walsingham, M.V. & Mitchell, A.I. (1994) Food consumption, feeding behaviour and growth of triploid and diploid Atlantic salmon, *Salmo salar*, L. parr. *Canadian Journal of Zoology*, **72**, 609–617.

Carter, C.G., He, Z.Y., Houlihan, D.F., McCarthy, I.D. & Davidson, I. (1995a) Effect of feeding on tissue free amino acid concentrations in rainbow trout (*Oncorhynchus mykiss* Walbaum). *Fish Physiology and Biochemistry*, **14**, 153–164.

Carter, C.G., McCarthy, I.D., Houlihan, D.F., Fonseca, M., Perera, W.M.K. & Sillah, A.B.S. (1995b) The application of radiography to the study of fish nutrition. *Journal of Applied Ichthyology*, **11**, 231–239.

Carter, C.G., Houlihan, D.F. & Owen, S.F. (1998) Protein synthesis and nitrogen excretion and long-term growth of flounder *Pleuronectes flesus. Journal of Fish Biology*, **53**, 272–284.

Carter, C. G., Houlihan, D. F. & He, Z. Y. (2000) Changes in tissue free amino acid concentrations in Atlantic salmon, *Salmo salar* L., after consumption of a low ration. *Fish Physiology and Biochemistry* **23**, 295–306.

Chen, T.T., Lu, J.-K., Shamnlott, M.J., Cheng, C.M., *et al.* (1995) Transgenic fish: ideal models for basic research and biotechnological applications. *Zoological Studies*, **34**, 215–234.

Cho, C.Y., Slinger, S.J. & Bayley, H.S. (1982) Bioenergetics of salmonid fishes: energy intake, expenditure and productivity. *Comparative Biochemistry and Physiology*, **73B**, 25–41.

Christiansen, J.S. & Jobling, M. (1990) The behaviour and the relationship between food intake and growth of juvenile Arctic charr, *Salvelinus alpinus* L. subjected to sustained exercise. *Canadian Journal of Zoology*, **68**, 2185–2191.

Christiansen, J.S., Jørgensen, E.H. & Jobling, M. (1991) Oxygen consumption in relation to sustained exercise and social stress in Arctic charr (*Salvelinus alpinus* L.). *Journal of Experimental Zoology*, **260**, 149–156.

Company, R., Calduch-Giner, J.A., Kaushik, S. & Perez-Sanchez, J. (1999) Growth performance and adiposity in gilthead seabream (*Sparus aurata*): risks and benefits of high energy diets. *Aquaculture*, **171**, 279–292.

Conceicao, L., van der Meven, T., Verreth, J., Evjen, M.S., Houlihan, D.F. & Fyhn, H.J. (1997) Amino acid metabolism and protein turnover in larval turbot (*Scophthalmus maximus*) fed natural zooplankton or Artemia. *Marine Biology*, **129**, 255–265.

Cook, J.Y., McNiven, M.A., Richardson, G.F. & Sutterlin, A.M. (2000a) Growth rate, body composition and feed digestibility/conversion of growth-enhanced transgenic Atlantic salmon (*Salmo salar*). *Aquaculture*, **188**, 15–32.

Cook, J.Y., McNiven, M.A. & Sutterlin, A.M. (2000b) Metabolic rate of pre-smolt growth-enhanced transgenic Atlantic salmon (*Salmo salar*). *Aquaculture*, **188**, 33–45.

Cook, J.Y., Sutterlin, A.M. & McNiven, M.A. (2000c) Effect of food deprivation on oxygen consumption and body composition of growth-enhanced transgenic Atlantic salmon (*Salmo salar*). *Aquaculture*, **188**, 47–63.

Couture, P., Dutil, J.-D. & Guderley, H. (1998) Biochemical correlates of growth and condition in juvenile Atlantic cod (*Gadus morhua*) from Newfoundland. *Canadian Journal of Fisheries and Aquatic Sciences*, **55**, 1591–1598.

Cowey, C.B. (ed.) (1982) Special Issue on Fish Biochemistry. *Comparative Biochemistry and Physiology*, **73B**, 1–180.

Cowey, C.B., Mackie, A.M. & Bell, J.G. (eds) (1985) *Nutrition and Feeding in Fish*. Academic Press, London.

Cuenca, E.M., Diz, L.G. & de la Higuera, M. (1993) Self-selection of a diet covering zinc needs in the trout. In: *Fish Nutrition in Practice* (eds S.J. Kaushik & P. Luquet), pp. 413–418. INRA, Biarritz.

Cui, Y. & Wootton, R.J. (1988) Bioenergetics and growth of a cyprinid, *Phoxinus phoxinus* (L.): the effect of ration and temperature on growth rate and efficiency. *Journal of Fish Biology*, **33**, 763–773.

Cui, Y., Chen, S. & Wang, S. (1994) Effect of ration size on the growth and energy budget of the grass carp, *Ctenopharyngodon idella* Val. *Aquaculture*, **123**, 95–107.

Cui, Y., Hung, S.S.O. & Zhu, X. (1996) Effect of ration and body size on the energy budget of juvenile white sturgeon. *Journal of Fish Biology*, **49**, 863–876.

Damsgård, B., Arnesen, A.M. & Jobling, M. (1999) Seasonal patterns of feed intake and growth of Hammerfest and Svalbard Arctic charr maturing at different ages. *Aquaculture*, **171**, 149–160.

Devlin, R.H., Johnsson, J.I., Smailus, D.E., Biagi, C.A., Jonsson, E. & Björnsson, B.T. (1999) Increased ability to compete for food by growth hormone-transgenic coho salmon *Oncorhynchus kisutch* (Walbaum). *Aquaculture Research*, **30**, 479–482.

Delahunty, G., Olcese, J., Prack, M., Vodicnik, M.J., Schreck, C.B. & DeVlaming, V. (1978) Diurnal variations in the physiology of the goldfish *Carassius auratus*. *Journal of Interdisciplinary Cycle Research*, **9**, 73–88.

Donaldson, E.M. & Devlin, R.H. (1996) Uses of biotechnology to enhance production. In: *Principles of Salmonid Culture* (eds W. Pennell & B.A. Barton), pp. 969–1020. Elsevier, Amsterdam.

Donaldson, E.M., Fagerlund, U.H.M., Higgs, D.A. & McBride, J.R. (1979) Hormonal enhancement of growth. In: *Fish Physiology,* Vol. VIII (eds W.S. Hoar, D.J. Randall & J.R. Brett), pp. 455–597. Academic Press, London.

dos Santos, J. & Jobling, M. (1991) Factors affecting gastric evacuation in cod, *Gadus morhua* L. fed single meals of natural prey. *Journal of Fish Biology*, **38**, 697–713.

Dutil, J.-D., Lambert, Y., Guderley, H., Blier, P.U., Pelletier, D. & Desroches, M. (1998) Nucleic acids and enzymes in Atlantic cod (*Gadus morhua*) differing in condition and growth rate trajectories. *Canadian Journal of Fisheries and Aquatic Sciences*, **55**, 788–795.

Ehlinger, T.J. (1990) Habitat choice and phenotype-limited feeding efficiency in bluegill: individual differences in trophic polymorphism. *Ecology*, **71**, 886–896.

Elliott, J.M. (1976) Energy losses in the waste products of brown trout (*Salmo trutta* L.). *Journal of Animal Ecology*, **45**, 561–580.

Elliott, J.M. (1979) Energetics of freshwater teleosts. In: *Fish Phenology: Anabolic Adaptiveness in Teleosts* (ed. P.S. Miller), pp. 29–61. Academic Press, London.

Elliott, J.M. (1994) *Quantitative Ecology and the Brown Trout*. Oxford University Press, Oxford.

Espe, M., Lied, E. & Torrissen, K.R. (1993) Changes in plasma and muscle free amino acids in Atlantic salmon (*Salmo salar*) during absorption of diets containing different amounts of hydrolysed cod muscle protein. *Comparative Physiology and Biochemistry*, **105A**, 555–562.

Fänge, R. & Grove, D. (1979) Digestion. In: *Fish Physiology,* Vol. VIII (eds W.S. Hoar, D.J. Randall & J.R. Brett), pp. 162–260. Academic Press, New York.

Ferraris, R.P. & Diamond, J.M. (1989) Specific regulation of intestinal nutrient transporters by their dietary substrates. *Annual Review of Physiology*, **51**, 125–141.

Fletcher, D.J. (1984) The physiological control of appetite in fish. *Comparative Biochemistry and Physiology*, **78A**, 617–628.

Fletcher, D.J., Grove, D.J., Basimi, R.A. & Ghaddaf, A. (1984) Emptying rates of single and double meals of different food quality from the stomach of the dab, *Limanda limanda* (L.). *Journal of Fish Biology*, **25**, 435–444.

Forbes, J.M. (1988) Metabolic aspects of the regulation of voluntary food intake and appetite. *Nutrition Research Reviews*, **1**, 145–168.

Forbes, J.M. (1992) Metabolic aspects of satiety. *Proceedings of the Nutrition Society*, **51**, 13–19.

Forbes, J.M. (1999) Natural feeding behaviour and feed selection. In: *Regulation of Feed Intake* (eds D. van der Heide, E.A. Huisman, E. Kanis, J.W.M. Osse & M.W.A. Verstegen), pp. 3–12. CAB Publishing, Wallingford, UK.

Foster, A.R., Hall, S.J. & Houlihan, D.F. (1993a) The effects of temperature acclimation on organ/tissue mass and cytochrome c oxidase activity in juvenile cod (*Gadus morhua*). *Journal of Fish Biology*, **42**, 947–957.

Foster, A.R., Houlihan, D.F. & Hall, S.J. (1993b) Effects of nutritional regime on correlates of growth rate in juvenile Atlantic cod (*Gadus morhua*): comparison of morphological and biochemical measurements. *Canadian Journal of Fisheries and Aquatic Sciences*, **50**, 502–512.

Fuller, M.F. & Garlick, P.J. (1994) Human amino acid requirements: can the controversy be resolved? *Annual Review of Nutrition*, **14**, 217–241.

Galbreath, P.F. & Thorgaard, G.H. (1995) Saltwater performance of all-female triploid Atlantic salmon. *Aquaculture*, **138**, 77–85.

Galbreath, P.F., St Jean, W., Anderson, V. & Thorgaard, G.H. (1994) Freshwater performance of all-diploid and triploid Atlantic salmon. *Aquaculture*, **128**, 41–49.

Gelineau, A., Medale, F. & Boujard, T. (1998) Effect of feeding time on postprandial nitrogen excretion and energy expenditure in rainbow trout. *Journal of Fish Biology*, **52**, 655–664.

Gerking, S.D. (1994) *Feeding Ecology of Fish*. Academic Press, San Diego.

Gjedrem, T. (1983) Genetic variation in quantitative traits and selective breeding in fish and shellfish. *Aquaculture*, **33**, 51–72.

Grand, T.C. & Grant, J.W.A. (1994) Spatial predictability of food influences its monopolization and defence by juvenile convict cichlids. *Animal Behaviour*, **47**, 91–100.

Gregory, T.R. & Wood, C.M. (1999) The effects of chronic plasma cortisol elevation on the feeding behaviour, growth, competitive ability, and swimming performance of juvenile rainbow trout. *Physiological and Biochemical Zoology*, **72**, 286–295.

Grisdale-Helland, B. & Helland, S.J. (1997) Replacement of protein by fat and carbohydrate in diets for Atlantic salmon (*Salmo salar*) at the end of the freshwater stage. *Aquaculture*, **152**, 167–180.

Grove, D.J. (1986) Gastro-intestinal physiology: rates of food processing in fish. In: *Fish Physiology: Recent Advances* (eds S. Nilsson & S. Holmgren), pp. 140–152. Croom Helm, London.

Grove, D.J., Lozoides, L. & Nott, J. (1978) Satiation amount, frequency of feeding and gastric emptying rate in *Salmo gairdneri*. *Journal of Fish Biology*, **12**, 507–516.

Habicht, C., Seeb, J.E., Gates, R.B., Brock, I.R. & Olito, C.A. (1994) Triploid coho salmon outperform diploid and triploid hybrids between coho salmon and chinook salmon during their first year. *Canadian Journal of Fisheries and Aquatic Sciences*, **51** (Suppl.1), 31–37.

Halver, J.E. (1972) *Fish Nutrition*. Academic Press, London.

Halver, J.E. (1989) *Fish Nutrition*, 2nd edition. Academic Press, London.

Heinsbroek, L.T.N., Tijssen, P.A.T., Flach, R.B. & De Jong, G.D.C. (1993) Energy and nitrogen balance studies in fish. In: *Fish Nutrition in Practice* (eds S.J. Kaushik & P. Luquet), pp. 375–389. INRA, Biarritz.

Henken, A.M., Brunink, A.M. & Richter, C.J.J. (1987) Differences in growth rate and feed utilization between diploid and triploid African Catfish *Clarias gariepinus* (Burchell 1822). *Aquaculture*, **63**, 233–242.

Hemre, G.I. & Sandnes, K. (1999) Effect of dietary lipid level on muscle composition in Atlantic salmon, *Salmo salar* L. *Aquaculture Nutrition*, **5**, 9–16.

Hemre, G.I., Sandnes, K., Lie, Ø. & Waagbø, R. (1995) Blood chemistry and organ nutrient composition in Atlantic salmon, *Salmo salar* L. fed graded amounts of wheat starch. *Aquaculture Nutrition*, **1**, 37–42.

Hidalgo, M.C., Urea, E. & Sanz, A. (1999). Comparative study of digestive enzymes in fish with different nutritional habits. *Aquaculture*, **170**, 267–282.

Higgs, D.A., Macdonald, J.S., Levings, C.D. & Dosanjh, B.S. (1995) Nutrition and feeding habits in relation to life history stage. In: *Physiological Ecology of Pacific Salmon* (eds C. Groot, L. Margolis & W.C. Clarke) pp. 159-315. UBC Press, Vancouver.

Hill, J., Kiessling, A. & Devlin, R. (2000) Coho salmon (*Oncorhynchus kisutchi*) transgenic for a growth hormone gene construct exhibit increased rates of muscle hyperplasia and detectable levels of differential gene expression. *Canadian Journal of Fisheries and Aquatic Sciences*, **57**, 939–950.

Hilton, J.W., Atkinson, J.L. & Slinger, S.J. (1983) Effects of increased dietary fiber on the return of appetite. *Canadian Journal of Fisheries and Aquatic Sciences*, **40**, 81–85.

Hirst, B.H. (1993) Dietary regulation of intestinal nutrient carriers. *Proceedings of the Nutrition Society*, **52**, 315–324.

Hofer, R. (1979a) The adaptation of digestive enzymes to temperature, season and diet in roach, *Rutilus rutilus*, and rudd, *Scardinius erythrophthalmus* L. 1. Amylase. *Journal of Fish Biology*, **14**, 565–572.

Hofer, R. (1979b) The adaptation of digestive enzymes to temperature, season and diet in roach, *Rutilus rutilus*, and rudd, *Scardinius erythrophthalmus*; Proteases. *Journal of Fish Biology*, **15**, 373–379.

Holmgren, S., Jonsson, A.-C. & Holstein, B. (1986) Gastro-intestinal peptides in fish. In: *Fish Physiology: Recent Advances* (eds S. Nilsson & S. Holmgren), pp. 119–139. Croom Helm, London.

Houlihan, D.F. (1991) Protein turnover in ectotherms and its relationships to energetics. *Advances in Comparative and Environmental Physiology*, **7**, 1–43.

Houlihan, D.F., Hall, S.J., Gray, C. & Noble, B.S. (1988) Growth rates and protein turnover in Atlantic cod, *Gadus morhua*. *Canadian Journal of Fisheries and Aquatic Sciences*, **45**, 951–964.

Houlihan, D.F., Hall, S.J. & Gray, C. (1989) Effects of ration on protein turnover in cod. *Aquaculture*, **79**, 103–110.

Houlihan, D.F., Mathers, E. & Foster, A.R. (1993) Biochemical correlates of growth rate in fish. In: *Fish Ecophysiology* (eds J.C. Rankin & F.B. Jensen), pp. 45–71. Chapman & Hall, London.

Houlihan, D.F., Carter, C.G. & McCarthy, I.D. (1995a) Protein synthesis in fish. In: *Biochemistry and Molecular Biology of Fishes*, vol. 4. (eds P. Hochachka & P. Mommsen), pp. 191–219. Elsevier Science, Amsterdam.

Houlihan, D.F., Carter, C.G. & McCarthy, I.D. (1995b) Protein turnover in animals. In: *Nitrogen and Excretion* (eds P.J. Wright & P.A. Walsh), pp. 1–29. CRC Press, Boca Raton.

Houlihan, D.F., McCarthy, I.D., Carter, C.G. & Marttin, F. (1995c) Protein turnover and amino acid flux in fish larvae. *ICES Marine Science Symposium*, **201**, 87–99.

Hung, S.S.O. (1991) Carbohydrate utilization by white sturgeon as assessed by oral administration tests. *Journal of Nutrition*, **121**, 1600–1605.

Jobling, M. (1981) The influences of feeding on the metabolic rate of fishes: a short review. *Journal of Fish Biology*, **18**, 385–400.

Jobling, M. (1982) Some observations on the effects of feeding frequency on the food intake and growth of plaice, *Pleuronectes platessa* L. *Journal of Fish Biology*, **20**, 431–444.

Jobling, M. (1985) Growth. In: *Fish Energetics: New Perspectives* (eds P. Tytler & P. Calow), pp. 213–230. Croom Helm, London, Sydney.

Jobling, M. (1986) Mythical models of gastric emptying and implications for food consumption. *Environmental Biology of Fishes*, **16**, 35–50.

Jobling, M. (1987) Influences of food particle size and dietary energy content on patterns of gastric evacuation in fish: test of a physiological model of gastric emptying. *Journal of Fish Biology*, **30**, 299–314.

Jobling, M. (1993) Bioenergetics: feed intake and energy partitioning. In: *Fish Ecophysiology* (eds J.C. Rankin & F.B. Jensen), pp. 1–44. Chapman & Hall, London.

Jobling, M. (1994) *Fish Bioenergetics*. Chapman & Hall, London.

Jobling, M. (1995) Feeding of charr in relation to aquaculture. *Nordic Journal of Freshwater Research*, **71**, 102–112.

Jobling, M. (1997) Temperature and growth: modulation of growth rate via temperature change. In: *Global Warming: Implications for Freshwater and Marine Fish* (eds C.M. Wood & D.G. McDonald), pp. 225–253. Cambridge University Press, Cambridge.

Jobling, M. & Baardvik, B.M. (1994) The influence of environmental manipulations on inter- and intra-individual variation in food acquisition and growth performance of Arctic charr, *Salvelinus alpinus*. *Journal of Fish Biology*, **44**, 1069–1087.

Jobling, M. & Miglavs, I. (1993) The size of lipid depots – a factor contributing to the control of food intake in Arctic charr, *Salvelinus alpinus*. *Journal of Fish Biology*, **43**, 487–489.

Jobling, M., Baardvik, B.M. & Jørgensen, E.H. (1989) Investigation of food-growth relationships of Arctic charr, *Salvelinus alpinus* L. using radiography. *Aquaculture*, **81**, 367–372.

Jobling, M., Baardvik, B.M., Christiansen, J.S. & Jørgensen, E.H. (1990) Feeding behaviour and food intake of Arctic charr, *Salvelinus alpinus* L. studied by X-radiography. In: *The Current Status of Fish Nutrition in Aquaculture* (eds M. Takeda & T. Watanabe), pp. 461–469. Tokyo University of Fisheries, Tokyo.

Jobling, M., Arnesen, A. M., Baardvik, B. M., Christiansen, J. S. & Jørgensen, E. H. (1995) Monitoring voluntary food intake under practical conditions, methods and applications. *Journal of Applied Ichthyology* **11**, 248–262.

Jürss, J. & Bastrop, R. (1995) Amino acid metabolism in fish. In: *Biochemistry and Molecular Biology of Fishes*, vol. 4 (eds P. Hochachka & P. Mommsen), pp. 159–189. Elsevier Science, Amsterdam.

Kapoor, B.C., Smith, H. & Verighina, I.A. (1975) The alimentary canal and digestion in teleosts. *Advances in Marine Biology*, **13**, 109–239.

Kaushik, S. & Luquet, P. (1977) Study of free amino acids in rainbow trout in relation to salinity changes. II. Muscle free amino acids during starvation. *Annals of Hydrobiology*, **8**, 375–387.

Kennedy, G.C. (1953) Role of depot fat in the hypothalamic control of food intake in the rat. *Proceedings of the Royal Society*, **140B**, 578–592.

Kiessling, A., Kiessling, K.-H., Storebakken, T. & Åsgård, T. (1991a) Changes in the structure and function of the epaxial muscle of rainbow trout (*Oncorhynchus mykiss*) in relation to ration and age II. Activity of key enzymes in the energy metabolism. *Aquaculture*, **93**, 335–356.

Kiessling, A., Åsgård, T., Storebakken, T., Johansson, L. & Kiessling, K.-H. (1991b) Changes in the structure and function of the epaxial muscle of rainbow trout (*Oncorhynchus mykiss*) in relation to ration and age III: Chemical composition. *Aquaculture*, **93**, 373–387.

Kiessling, A., Higgs, D.A., Dosanjh, B.S. & Eales, J.G. (1994) Influence of sustained exercise at two ration levels on growth and thyroid function of all-female chinook salmon (*Oncorhynchus tshawytscha*) in seawater. *Canadian Journal of Fisheries and Aquatic Sciences*, **51**, 1975–1984.

Kiessling, A., Dosanjh, B., Higgs, D., Deacon, G. & Rowshandeli, N. (1995) Dorsal aorta cannulation: a method to monitor changes in blood levels of astaxanthin in voluntary feeding Atlantic salmon. *Aquaculture Nutrition*, **1**, 43–50.

Kiron, V., Watanabe, T., Fukuda, H., Okamoto, N. & Takeuchi, T. (1995) Protein nutrition and defence mechanisms in rainbow trout *Oncorhynchus mykiss*. *Comparative Biochemistry and Physiology*, **111A**, 351–359.

Kolok, A.S. (1989) The relationship between maintenance ration and growth rate in two strains of rainbow trout, *Salmo gairdneri* Richardson. *Journal of Fish Biology*, **34**, 807–809.

Koskela, J., Jobling, M. & Pirhonen, J. (1997) Influence of the length of the daily feeding period on feed intake and growth of whitefish, *Coregonus lavaretus*. *Aquaculture*, **156**, 35–44.

Laidley, C.W. & Leatherland J.F. (1988) Circadian studies of plasma cortisol, thyroid hormone, protein, glucose and ion concentration, liver glycogen concentration and liver and spleen weights in rainbow trout, *Salmo gairdneri* Richardson. *Comparative Biochemistry and Physiology*, **89A**, 495–503.

Langhans, W. (1999) Appetite regulation. In: *Protein Metabolism and Nutrition* (eds G.E. Lobley, A. White & J.C. MacRae), pp. 225–251. Wageningen Pers., Wageningen, the Netherlands.

Langhans, W. & Scharrer, E. (1992) Metabolic control over eating. *World Review of Nutrition and Dietetics*, **70**, 1–67.

Le Bail, P.-Y. & Boeuf, G. (1997) What hormones may regulate food intake in fish? *Aquatic Living Resources*, **10**, 371–379.

Lemieux, H., Blier, P. & Dutil, J.D. (1999) Do digestive enzymes set the physiological limit on growth rate and food conversion efficiency in the Atlantic cod (*Gadus morhua*)? *Fish Physiology and Biochemistry*, **20**, 293–303.

Love, R.M. (1988) *The Food Fishes: Their intrinsic variation and practical implications*. Farrand Press, London.

Lyndon, A.R., Houlihan, D.F. & Hall, S.J. (1992) The effect of short-term fasting and a single meal on protein synthesis and oxygen consumption in cod, *Gadus morhua*. *Journal of Comparative Physiology*, **162B**, 209–215.

Lyndon, A.R., Davidson, I. & Houlihan, D.F. (1993) Changes in tissue and plasma free amino acid concentrations after feeding in Atlantic cod. *Fish Physiology and Biochemistry*, **10**, 365–375.

Maclean, N. (1998) Genetic manipulation of farmed fish. In: *Biology of Farmed Fish* (eds K.D. Black & A.D. Pickering), pp. 327–354. Sheffield Academic Press, Sheffield.

Magnan, P. & Stevens, E.D. (1993) Pyloric caecal morphology of brook charr, *Salvelinus fontinalis*, in relation to diet. *Environmental Biology of Fishes*, **36**, 205–210.

Mathers, E.M., Houlihan, D.F. & Cunningham, M.J. (1992a) Nucleic acid concentrations and enzyme activities as correlates of growth rate of the saithe *Pollachius virens*: growth rate estimate of open-sea fish. *Marine Biology*, **112**, 363–369.

Mathers, E.M., Houlihan, D.F. & Cunningham, M.J. (1992b) Estimation of saithe *Pollachius virens* growth rates around the Beryl oil platforms in the North Sea: a comparison of methods. *Marine Ecology Progress Series*, **86**, 31–40.

McCarthy, I.D., Carter, C.G. & Houlihan, D.F. (1992) The effect of feeding hierarchy on individual variability in daily feeding of rainbow trout, *Oncorhynchus mykiss* (Walbaum). *Journal of Fish Biology*, **41**, 257–263.

McCarthy, I.D., Houlihan, D.F., Carter, C.G. & Moutou, K. (1993) Variation in individual food consumption rates of fish and its implications for the study of fish nutrition and physiology. *Proceedings of the Nutrition Society*, **52**, 411–420.

McCarthy, I.D., Houlihan, D.F. & Carter, C.G. (1994) Individual variation in protein turnover and growth efficiency in rainbow trout, *Oncorhynchus mykiss* (Walbaum). *Proceedings of the Royal Society of London B*, **257**, 141–147.

McCarthy, I.D., Gair, D.J. & Houlihan, D.F. (1999) Feeding rank and dominance in *Tilapia rendalli* under defensible and indefensible patterns of food distributions. *Journal of Fish Biology*, **55**, 854–867.

McMillan, D.N. & Houlihan, D.F. (1988) The effect of refeeding on tissue protein synthesis in rainbow trout. *Physiological Zoology*, **61**, 429–441.

McMillan, D.N. & Houlihan, D.F. (1992) Protein synthesis in trout liver is stimulated by both feeding and fasting. *Fish Physiology and Biochemistry*, **10**, 23–34.

McNamara, P.T., Caldarone, E.M. & Buckley, L.J. (1999) RNA/DNA ratio and expression of 18S ribosomal RNA, actin and myosin heavy chain messenger RNAs in starved and fed larval Atlantic cod (*Gadus morhua*). *Marine Biology*, **135**, 123–132.

Médale, F., Parent, J.P. & Vellas, F. (1987) Responses to prolonged hypoxia by rainbow trout (*Salmo gairdneri*) I. Free amino acid and proteins in plasma, liver and white muscle. *Fish Physiology and Biochemistry*, **3**, 183–189.

Médale, F., Brauge, C., Vallee, F. & Kaushik, S. J. (1995) Effects of dietary protein/energy ratio, ration size, dietary energy source and water temperature on nitrogen excretion in rainbow trout. *Water Science and Technology* **31**, 185–194.

Médale, F., Poli, J.M., Vallée, F. & Blanc, D. (1999) Comparison de l'utilisation digestive et metabolique d'un regime riche en glucides per la carpe à 18°C et 25°C. *Cybium*, **23**, 139–152.

Mélard, C., Kestemont, P. & Grignard, J.C. (1996) Intensive culture of juvenile and adult Eurasian perch (*P. fluviatilis*): effect of major biotic and abiotic factors on growth. *Journal of Applied Ichthyology*, **12**, 175–180.

Metcalfe, N.B. & Thorpe, J.E. (1992) Anorexia and defended energy levels in over-wintering juvenile salmon. *Journal of Animal Ecology*, **61**, 175–181.

Meton, I., Mediavilla, D., Caseras, A., Canto, E., Fernandez, F. & Baanante, I.V. (1999) Effect of diet composition and ration size on key enzyme activities of glycolysis-gluconeogenesis, the pentose phosphate pathway and amino acid metabolism in liver of gilthead sea bream (*Sparus aurata*). *British Journal of Nutrition*, **82**, 223–232.

Meyer, A. (1987) Phenotypic plasticity and heterochrony in *Cichlasoma managuese* (Pisces, Cichlidae) and their implications for speciation in cichlid fishes. *Evolution*, **41**, 1357–1369.

Meyer, A. (1990a) Ecological and evolutionary consequences of the trophic polymorphism in *Cichlasoma citrinellum* (Pisces, Cichlidae). *Biological Journal of the Linnean Society*, **39**, 279–299.

Meyer, A. (1990b) Morphometrics and allometry in the trophically polymorphic cichlid fish, *Cichlasoma citrinellum*: alternative adaptations and ontogenetic changes in shape. *Journal of Zoology, London*, **221**, 237–260.

Meyer-Burgdorff, K.H. & Rosenow, H. (1995) Protein turnover and energy metabolism in growing carp. 2. Influence of feeding level and protein energy ratio. *Journal of Animal Physiology and Animal Nutrition*, **73**, 123–133.

Miglavs, I. & Jobling, M. (1989) Effects of feeding regime on food consumption, growth rates and tissue nucleic acids in juvenile Arctic charr, *Salvelinus alpinus*, with particular reference to compensatory growth. *Journal of Fish Biology*, **34**, 947–957.

Millward, D.J. & Rivers, J. (1988) The nutritional role of indispensable amino acids and the metabolic basis for their requirements. *European Journal of Clinical Nutrition*, **42**, 367–393.

Mittelbach, G.G., Osenberg, C.W. & Wainwright, P.C. (1992) Variation in resource abundance affects diet and feeding morphology in the pumpkinseed sunfish (*Lepomis gibbosus*). *Oecologia*, **90**, 8–13.

Mittelbach, G.G., Osenberg, C.W. & Wainwright, P.C. (1999) Variation in feeding morphology between pumpkinseed populations: phenotypic plasticity or evolution? *Evolutionary Ecology Research*, **1**, 111–128.

Moutou, K.A., McCarthy, I.D. & Houlihan, D.F. (1998) The effect of ration level and social rank on the development of fin damage in juvenile rainbow trout. *Journal of Fish Biology*, **52**, 756–770.

Nose, T. (1972) Changes in the pattern of free plasma amino acids in rainbow trout after feeding. *Bulletin of the Freshwater Fisheries Laboratory Tokyo*, **22**, 137–144.

NRC (1993) *Nutrient Requirements of Fish*. National Academy Press, Washington, D.C.

Oddy, V.H. (1999) Protein metabolism and nutrition in farm animals: an overview. In: *Protein Metabolism and Nutrition* (eds G.E. Lobley, A. White & J.C. MacRae), pp. 7–24. Wageningen Pers., Wageningen, the Netherlands.

Ogata, H. (1986) Correlations of essential amino acid patterns between the dietary protein and the blood, hepatopancreas, or skeletal muscle in carp. *Bulletin of the Japanese Society of Scientific Fisheries*, **52**, 307–312.

Ojolick, E.J., Cusack, R., Benfey, T.J. & Kerr, S.R. (1995) Survival and growth of all-female diploid and triploid rainbow trout (*Oncorhynchus mykiss*) reared at chronic high temperature. *Aquaculture*, **131**, 177–187.

O'Keefe, R.A. & Benfey, T.J. (1999) Comparative growth and food consumption of diploid and triploid brook trout (*Salvelinus fontinalis*) monitored by radiography. *Aquaculture*, **175**, 111–120.

Overli, O., Winberg, S., Damsgard, B. & Jobling, M. (1998) Food intake and spontaneous swimming activity in Arctic charr (*Salvelinus alpinus*): role of brain serotonergic activity and social interactions. *Canadian Journal of Zoology*, **76**, 1366–1370.

Owen, S.F., Houlihan, D.F., Rennie, M.J. & van Weerd, J.H. (1998) Bioenergetics and nitrogen balance of the European eel (*Anguilla anguilla*) fed at high and low ration levels. *Canadian Journal of Fisheries and Aquatic Sciences*, **55**, 2365–2375.

Owen, S.F., McCarthy, I.D., Watt, P.W., *et al.* (1999) In vivo rates of protein synthesis in Atlantic salmon (*Salmo salar* L.) smolts determined using a stable isotope flooding dose technique. *Fish Physiology and Biochemistry*, **20**, 87–94.

Pelletier, D., Dutil, J.-D., Blier, P. & Guderley, H. (1994) Relation between growth rate and metabolic organisation of white muscle, liver and digestive tract in cod, *Gadus morhua. Journal of Comparative Physiology*, **164B**, 179–190.

Pelletier, D., Blier, P.U., Lambert, Y. & Dutil, J.-D. (1995) Deviation from the general relationship between RNA concentration and growth rate in fish. *Journal of Fish Biology*, **47**, 920–922.

Perera, W.M.K. (1995) *Growth performance, nitrogen balance and protein turnover of rainbow trout,* Oncorhynchus mykiss *(Walbaum), under different dietary regimens.* PhD Thesis, University of Aberdeen.

Peter, R.E. (1979) The brain and feeding behaviour. In: *Fish Physiology,* Vol. VIII (eds W.S. Hoar, D.J. Randall & J.R. Brett), pp. 121–159. Academic Press, New York.

Pickova, J. (1997) *Lipids in eggs and somatic tissues in cod and salmonids.* MSc Thesis, Swedish University of Agricultural Sciences, Uppsala, Sweden.

Plisetskaya, E.M., Buchelli-Narvaez, L.I., Hardy, R.W. & Dickhoff, W.W. (1991) Effects of injected and dietary arginine on plasma insulin levels and growth of Pacific salmon and rainbow trout. *Comparative Biochemistry and Physiology*, **98A**, 165–170.

Prieto, P.A., Kopchick, J.J. & Kelder, B. (1999) Transgenic animals and nutrition research. *Journal of Nutritional Biochemistry*, **10**, 682–695.

Ramnarine, I.W., Pirie, J.M., Johnstone, A.D.F. & Smith, G.W. (1987) The influence of ration size and feeding frequency on ammonia excretion by juvenile Atlantic cod, *Gadus morhua* L. *Journal of Fish Biology*, **31**, 545–559.

Reeds, P.J., Burrin, D.G., Stoll, B. & van Goudoever, J.B. (1999) Consequences and regulation of gut metabolism. In: *Protein Metabolism and Nutrition* (eds G.E. Lobley, A. White & J.C. MacRae), pp. 127–153. Wageningen Pers., Wageningen, the Netherlands.

Robinson, B.W., Wilson, D.S. & Shea, G.O. (1996) Trade-offs of ecological specialization: an intraspecific comparison of pumpkinseed sunfish phenotypes. *Ecology*, **77**, 170–178.

Rodehutscord, M., Jacobs, S., Pack, M. & Pfeffer, E. (1995) Response of rainbow trout (*Oncorhynchus mykiss*) growing from 50 to 150 g to supplements of DL-methionine in a semi-purified diet containing low or high levels of cystine. *Journal of Nutrition*, **125**, 964–969.

Rozin, P. & Mayer, J. (1961) Regulation of food intake in the goldfish. *American Journal of Physiology*, **201**, 968–974.

Rungruangsak-Torrissen, K., Carter, C.G., Sundby, A., Berg, A. & Houlihan, D.F. (1999) Maintenance ration, protein synthesis capacity, plasma insulin and growth of Atlantic salmon (*Salmo salar* L.) with genetically different trypsin isozymes. *Fish Physiology and Biochemistry*, **21**, 223–233.

Ruohonen, K. & Grove, D.J. (1996) Gastrointestinal responses of rainbow trout to dry pellet and low-fat herring diets. *Journal of Fish Biology*, **49**, 501–513.

Sæther, B.S., Johnsen, H.K. & Jobling, M. (1996) Seasonal changes in food consumption and growth of Arctic charr exposed to either simulated natural or a 12:12 LD photoperiod at constant water temperature. *Journal of Fish Biology*, **48**, 1113–1122.

Saunders, R.L. (1963) Respiration of the Atlantic cod. *Journal of the Fisheries Research Board of Canada*, **20**, 373–386.

Schluter, D. (1995) Adaptive radiation in sticklebacks: trade-offs in feeding performance and growth. *Ecology*, **76**, 82–90.

Schreck, C.B. & Moyle, P.B. (eds) (1990) *Methods for Fish Biology*. American Fisheries Society, Bethesda.

Shearer, K.D. (1994) Factors affecting the proximate composition of cultured fishes with emphasis on salmonids. *Aquaculture*, **119**, 63–88.

Shelverton, P.A. & Carter, C.G. (1998) The effect of ration on behaviour, food consumption and growth in juvenile greenback flounder (*Rhombosolea tapirina*: Teleostei). *Journal of the Marine Biological Association of the United Kingdom*, **78**, 1307–1320.

Silverstein, J.T., Shearer, K.D., Dickhoff, W.W. & Plisetskaya, E.M. (1999) Regulation of nutrient intake and energy balance in salmon. *Aquaculture*, **177**, 161–169.

Sin, F.Y.T. (1997) Transgenic fish. *Reviews in Fish Biology and Fisheries*, **7**, 417–441.

Sims, D.W. & Davies, S.J. (1994) Does specific dynamic action (SDA) regulate return of appetite in the lesser spotted dogfish, *Scyliorhinus canicula*? *Journal of Fish Biology*, **45**, 341–348.

Solomon, D.J. & Brafield, A.E. (1972) The energetics of feeding, metabolism and growth of perch (*Perca fluviatilis*). *Journal of Animal Ecology*, **41**, 699–718.

Soofiani, N.M. & Hawkins, A.D. (1982) Energetic costs at different levels of feeding in juvenile cod, *Gadus morhua* L. *Journal of Fish Biology*, **21**, 577–592.

Steele, N.C. & Pursel, V.G. (1990) Nutrient partitioning by transgenic animals. *Annual Review of Nutrition*, **10**, 213–232.

Stevens, E.D., Sutterlin, A.M. & Cook, T. (1998) Respiratory metabolism and swimming performance in growth hormone transgenic Atlantic salmon. *Canadian Journal of Fisheries and Aquatic Sciences*, **55**, 2028–2035.

Storebakken, T. & Austreng, E. (1987a) Ration level for salmonids. I. Growth, survival, body composition, and feed conversion in Atlantic salmon fry and fingerlings. *Aquaculture*, **60**, 189–206.

Storebakken, T. & Austreng, E. (1987b) Ration level for salmonids. II. Growth, feed intake, protein digestibility, body composition, and feed conversion in rainbow trout weighing 0.5–1.0 kg. *Aquaculture*, **61**, 207–221.

Sugden, P.H. & Fuller, S.J. (1991) Regulation of protein turnover in skeletal and cardiac muscle. *Biochemical Journal*, **273**, 21–37.

Talbot, C. & Higgins, P.J. (1983) A radiographic method for feeding studies of fish using metallic iron powder as a marker. *Journal of Fish Biology*, **23**, 211–220.

Thodesen, J., Grisdale-Helland, B., Helland, S.J. & Gjerde, B. (1999) Feed intake, growth and feed utilization of offspring from wild and selected Atlantic salmon (*Salmo salar*). *Aquaculture*, **180**, 237–246.

Thompson, I. (1993) *Nutrition and Disease Resistance in Fish*. PhD Thesis, University of Aberdeen, Aberdeen.

Tomas, F.M., Pym, R.A. & Johnson, R.J. (1991) Muscle protein turnover in chickens selected for increased growth rate, food consumption or efficiency of food utilisation: effects of genotype and relationship to plasma IGF-1 and growth hormone. *British Poultry Science*, **32**, 363–376.

Torrissen, K.R., Lied, E. & Espe, M. (1994) Differences in digestion and absorption of dietary protein in Atlantic salmon (*Salmo salar*) with genetically different trypsin isozymes. *Journal of Fish Biology*, **45**, 1087–1104.

Tveiten, H., Johnsen, H.K. & Jobling, M. (1996) Influence of maturity status on the annual cycles of feeding and growth in Arctic charr reared at constant temperature. *Journal of Fish Biology*, **48**, 910–924.

Vahl, O. (1979) An hypothesis on the control of food intake in fish. *Aquaculture*, **17**, 221–229.

Verbeeten, B., Carter, C.G. & Purser, G.J. (1999) The combined effect of feeding time and ration on growth performance and nitrogen metabolism of greenback flounder. *Journal of Fish Biology*, **55**, 1328–1343.

Wainwright, P.C., Osenberg, C.W. & Mittelbach, G.G. (1991) Trophic polymorphism in the pumpkinseed sunfish (*Lepomis gibbosus* Linnaeus): effects of environment on ontogeny. *Functional Ecology*, **5**, 40–55.

Walton, M. & Wilson, R. (1986) Postprandial changes in plasma and liver amino acids of rainbow trout fed complete diets containing casein. *Aquaculture*, **51**, 105–115.

Walzem, R.I., Storebakken, T., Hung, S.S.O. & Hansen, R.J. (1991) Relationship between growth and selected liver enzyme activities of individual rainbow trout. *Journal of Nutrition*, **121**, 1090–1098.

White, J.R. & Li, H.W. (1985) Determination of the energetic cost of swimming from the analysis of growth rate and body composition in juvenile chinook salmon, *Oncorhynchus tshawytscha*. *Comparative Biochemistry and Physiology*, **81A**, 25–33.

Wilson, R.P. (ed.) (1991) *Handbook of Nutrient Requirements of Finfish*. CRC Press, Boca Raton.

Wilson, R.P. (1994) Utilization of dietary carbohydrate by fish. *Aquaculture*, **124**, 67–80.

Winberg, S., Carter, C.G., McCarthy, I.D., He, Z.-Y., Nilsson, G.E. & Houlihan, D.F. (1993) Feeding rank and brain serotonergic activity in rainbow trout *Oncorhynchus mykiss*. *Journal of Experimental Biology*, **179**, 197–211.

Xie, S., Cui, Y., Yang, Y. & Liu, J. (1997) Energy balance of Nile tilapia (*Oreochromis niloticus*) in relation to ration size. *Aquaculture*, **154**, 57–68.

Chapter 14
Feeding Management

Anders Alanärä, Sunil Kadri and Mihalis Paspatis

14.1 Introduction

The efficiency with which fish utilise the food supply is a major factor determining the economic returns of a fish farm. In addition, improved feed utilisation can reduce environmental pollution. There would be few other businesses in which there is such potential for reducing environmental impact through measures designed to increase economic returns. Poor feed conversion is often the consequence of poor feed management, resulting from a lack of knowledge about energetic needs of the fish, inadequate distribution of feed and inferior feeding techniques. Feed management is multidisciplinary, requiring inputs from nutrition, physiology, behaviour and feeding techniques; this places a great demand on the fish farmer and manager.

This chapter is written with a practical approach in mind. There is a focus on how feed budgets are calculated, on how fish growth is modelled, and upon the feeding techniques that are available. Earlier chapters in this book have dealt with factors that influence the feed intake of fish in one way or another: considerations of environmental (see Chapters 6 and 7), behavioural (see Chapters 8, 9 and 10) and physiological (see Chapter 13) factors provided in other chapters will be used as the basis for the discussion of the standards for feed management given here.

14.2 Feed planning and production plans

In order to optimise production, a fish farmer has to feed the fish at a level that ensures good growth and minimal waste. Although estimation of feed requirements may be relatively easy in theory, the estimate will seldom match the need of the fish at any specific time because there are often large variations in feed intake both between days and over longer time periods. This variation is very difficult to predict, since it is related to a range of uncontrolled behavioural, physiological and environmental variables (see Chapter 7). As will be discussed later in this chapter (see feeding techniques), new feeding systems that take these variations into account are available (i.e. demand feeders where the fish themselves set the daily feed ration). However, there is always a risk in relying too much on advanced technical systems, and any mishaps may result in substantial feed losses. Estimates of daily feed requirements based on theoretical considerations can serve as a control of such systems. In addition, if traditional feeding systems (e.g. timer-controlled systems where daily feed rations are based on theoretical estimates) are combined with visual observations of feeding activity, there is the possibility to 'fine-tune' feed ration. Finally, models that allow estimation of feed requirements are important for production plans, i.e. the long-term planning of feed use.

Fish farmers often use feeding charts provided by feed companies when deciding upon daily feed rations. These charts are usually based on models describing feed requirements in relation to fish size and water temperature; they vary in their level of detail and possibilities to adjust the feeding regime to local conditions. A better approach for farmers would be to develop their own feed budgets. A feed budget can be written as:

$$FA = (n \times TER)/DE \tag{14.1}$$

where FA is feed allowance (kg/day), n is the number of fish, TER is the theoretical energy requirement (MJ DE/fish), and DE is the digestible energy content of the feed (MJ/kg). DE can be estimated from knowledge of the amounts of each energy-giving nutrient in the feed, their gross energy values and assumed digestibility coefficients (Hillestad *et al.* 1999), or digestibility trials can be carried out directly (see also Chapters 1 and 2).

The energy need of fish can be estimated by constructing an energy budget, a complete energy budget being a balance sheet of energy income set against energy expenditures (Brett 1979, 1995; Jobling 1994):

$$R = F + U + M + P \tag{14.2}$$

where R is the energy gained as food, F is the energy lost in the faeces, U is the energy lost in nitrogenous excretion products, M is the energy costs for bodily functions (metabolism), and P is the energy retained as production (body + gonads). Numerous studies of fish energetics have been performed since the seminal work of Winberg (1956), and many potential sources of error have been identified (see review by Brafield 1985). Methods have improved with the passage of time, but few models have been presented in a form that makes them useful for application under commercial conditions. One of these, presented by Rasmussen and From (1991), described the energy requirements of rainbow trout (*Oncorhynchus mykiss*) in relation to temperature and body size:

$$TER = 1.176 \times \exp(0.076 \times T) \times W^{0.674} \tag{14.3}$$

where T is temperature (°C) and W is body weight (kg).

Another way of estimating TER is via a route involving the calculation of the digestible energy (DEN) needed to produce 1 kg of weight increment (MJ DE/kg). This can be represented algebraically as:

$$DEN = FI \times DE/Wi \tag{14.4}$$

where FI is the feed intake (kg) and Wi is the weight increment (kg). The advantage of this method is that energy expenditures do not need to be quantified, but feed consumption and weight increment need to be recorded over a period of time (in single or groups of fish). The fish can be held in rearing facilities resembling those used under commercial conditions, so metabolic costs due to ingestion and digestion of feed, and to swimming activity, are likely to be more in accordance with those in culture than when using a complete energy budget approach (see Brafield 1985 for methods). There are, of course, limitations with this simplistic approach. The metabolic rates of feeding fish are temperature-dependent (Brett 1995), and metabolic costs increase with increasing temperature (Jobling 1997). Data for rainbow trout presented by Cho and Bureau (1998) indicate a positive relationship between temperature and DEN, but in comparison with the effect of body size on DEN, the effect of temperature seems relatively small. The balance between protein and fat in the feed may influence DEN (Cho 1982), as will the digestibility coefficients used for each energy-giving nutrient in the calculation of DE (Jobling 1983a). Other potentially important factors are genetic differences in growth efficiency between stocks (e.g. Thodesen *et al.* 1999), and the maturity status of the fish. Despite the limitations, DEN may be a useful measure because of the ease and relatively low costs of estimation, making it possible to evaluate the energy requirements of fish at the farm, or local stock level.

Due to size-related differences in metabolism and body composition of fish (Jobling 1994), the DEN required to produce small fish is less than that to produce large fish, e.g. ranging between 10 MJ for small juvenile rainbow trout to 19 MJ for a fish of 3 kg (Fig. 14.1) (Cho & Bureau 1998). The data given by Cho and Bureau (1998) for rainbow trout can be expressed as:

$$DEN = 10.22 + 1.13 \times Ln\,(W) \tag{14.5}$$

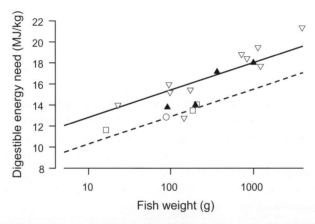

Fig. 14.1   The relationship between body weight and digestible energy need (DEN) to produce 1 kg of weight increment. The solid line is derived from data presented by Cho and Bureau (1998) (Equation 14.5), whereas the hatched line indicates the lower range of DEN (Equation 14.5, with an intercept of 7.7). Circles are data for rainbow trout (Alanärä 1994), open triangles Atlantic salmon (Kiessling *et al.* 1995; Paspatis & Boujard 1996; Grisdale-Helland & Helland 1997; Refstie *et al.* 1998; Helland & Grisdale-Helland 1998a; Sveier & Lied 1998; Sveier *et al.* 1999; Bjerkeng *et al.* 1999; Thodesen *et al.* 1999), squares Arctic charr (Jobling 1983b; Alanärä & Kiessling 1996; Larsson & Berglund 1998), and filled triangles Atlantic halibut (Aksnes & Mundheim 1997; Helland & Grisdale-Helland 1998b; Grisdale-Helland & Helland 1998). Values of DEN from different studies were calculated on the basis of DE contents of feeds, weight gain and feed intake. The following digestible energy values were used: protein 20.9 kJ/g, fat 35.1 kJ/g and carbohydrate 11.0 kJ/g (Hillestad *et al.* 1999). The best growing groups or individual fish have been selected.

where W is body weight (g).

This model also seems to fit data obtained for Atlantic salmon (*Salmo salar*) within the 20 to 4000-g size range. However, some data for Arctic charr (*Salvelinus alpinus*) and rainbow trout indicate that values of DEN might be lower than this model predicts (Fig. 14.1), an intercept of 7.7 (rather than 10.22) seeming to provide a better fit to these data. Atlantic halibut (*Hippoglossus hippoglossus*) is a cold-water species that seems to show a similar relationship between body size and DEN as the salmonids (Fig. 14.1). Results of studies on sea bass (*Dicentrarchus labrax*), of weight range 9 to 205 g, indicate that DEN is 18–25 MJ, with a mean of 22 MJ (Ballestrazzi *et al.* 1998; Dias *et al.* 1998; Kaushik 1998; Peres & Oliva-Teles 1999), whereas data for sea bream (*Sparus aurata*) (25 to 250 g) give a DEN of 18–19 MJ (Kaushik 1998; Company *et al.* 1999). In contrast to the findings for salmonids, the DEN values for sea bass and sea bream do not seem to correlate with body weight. Possible reasons for this may be related to differences in the way that body proximate composition changes occur among species as they grow, i.e. sea bass and sea bream invest less energy in fat storage than cold-water species like salmonids, or it could be a result of the rather limited size ranges of bass and bream studied hitherto.

Once values of DEN have been obtained, the TER can be estimated:

$$TER = DEN \times TWi \tag{14.6}$$

where TWi is the daily theoretical weight increment (kg/day) (see next section for details about growth predictions).

When TER is known, farmers can then create their own feed budget models or feeding charts (see example below). This procedure may be simplified by the use of computer spreadsheet software, when values of TER can be calculated automatically by combining equations and the specific farm data for temperature, fish size and DE content of feed.

Water temperature effects have a major influence in all models relating to energy requirements and growth. Starting at the lowest limit, an increase in temperature leads to increased feeding and growth; these peak at some intermediate temperature and then decline as the temperature continues to rise (Jobling 1997). Models describing the energy need (or growth) of fish are generally created under specific rearing conditions, using specific stocks of fish. The real energy need (or growth) may, therefore, differ from model predictions because of stock (genetic) differences, environmental conditions, and type of rearing facility, etc. Furthermore, requirements may differ at different times of the year; for example, salmonids often show reduced feed intake during autumn as compared with spring and summer (Smith *et al.* 1993; Tveiten *et al.* 1996).

To illustrate how energy requirements and daily feed allowance are calculated an example will be given based on data collected for sea bass held in groups of 20 (Ballestrazzi *et al.* 1998). Initial weight was 79.1 g and, after 195 days at 23.1°C, the weight was 291.9 g. Over the 195 days, the mean individual feed intake was 0.292 kg and the weight increment 0.213 kg. The gross energy content of the feed was 21.5 MJ/kg and the digestible energy content (DE) was estimated to be 18.3 MJ/kg (based on digestible energy values; protein 20.9 kJ/g, fat 35.1 kJ/g and carbohydrate 11.0 kJ/g).

According to Equation 14.4, the DEN can be calculated as:

DEN $= 0.292 \times 18.3/0.213 = 25.1$ MJ/kg

To calculate the TER using DEN, estimates of daily weight increment are required. In this example, weight increment will be calculated using the thermal-unit growth coefficient (TGC) (see section 14.3). (Values of TGC are calculated using Equation 14.9):

$$TGC = [(291.9^{0.3333} - 79.1^{0.3333})/(23.1 \times 195)] \times 1000 = 0.52$$

Based on TGC, the body weight of sea bass at 23.1°C after one day of growth can be calculated according to Equation 14.10:

$$FBW \text{ (final body weight)} = [79.1^{0.3333} + (0.52/1000 \times 23.1 \times 1)]^3 = 79.7 \text{ g}$$

The theoretical daily weight increment (TWi) in kg is then:

$$TWi = 0.0797 - 0.0791 = 0.0006 \text{ kg/day}.$$

When DEN and the daily weight increment are known, the TER can be calculated according to Equation 14.6:

$$TER = 25.1 \times 0.0006 = 0.016 \text{ MJ/day}.$$

Finally, the daily feed allowance can be calculated as (Equation 14.1):

$$FA = (20 \times 0.016)/18.5 = 0.017 \text{ kg/day (or 17 g/day)}.$$

Table 14.1 shows an example of a feeding chart where TER is based on the data given above. If some basic data on feed intake and growth are available, feeding charts like this are easily constructed by executing the series of calculations for different temperatures and body sizes.

Table 14.1 Theoretical energy requirements (TER, MJ DE/fish) of sea bass of various body sizes at different temperatures. Values are based on the data of Ballestrazzi *et al.* (1998), and calculations are summarised above.

Temperature	Body size (g)				
	10	50	100	250	500
14	0.002	0.007	0.011	0.018	0.027
16	0.003	0.008	0.012	0.021	0.032
18	0.003	0.009	0.014	0.024	0.037
20	0.004	0.010	0.016	0.028	0.041
22	0.004	0.011	0.017	0.031	0.046
24	0.004	0.012	0.019	0.034	0.051
26	0.005	0.013	0.021	0.037	0.056
28	0.005	0.014	0.022	0.040	0.061

14.3 Estimating growth

Body weight and weight gain are two important factors influencing feed intake of fish. Thus, one important step in estimating the feed requirements is to predict weight gain in a given period of time under given water temperature and rearing conditions.

The most commonly used expression of fish growth is the instantaneous growth rate, also known as the specific growth rate (SGR). SGR is calculated by taking natural logarithms of body weight, and expresses growth as %/day (Ricker 1979):

$$SGR = (\ln FBW - \ln IBW) \times 100/D \tag{14.7}$$

where FBW is final body weight (g), IBW is initial body weight (g) and D is the number of days between weighings. Weight change over a period of time can be estimated from initial weight and SGR:

$$FBW = IBW \times \exp(SGR \times D/100) \tag{14.8}$$

The problem with building models for SGR is that records are required for fish of different body sizes reared at different temperatures, so data collection is very time-consuming. Consequently, only a few models describing SGR of fish in culture are available (Table 14.2), but they may form an important base for prediction of growth in commercial farming.

An alternative way of growth prediction is based on the model originally presented by Iwama and Tautz (1981) and later modified by Cho (1990). The equation used to calculate TGC is written as:

$$TGC = [(FBW^{0.3333} - IBW^{0.3333})/(T \times D)] \times 1000 \tag{14.9}$$

where T is water temperature (°C) and D the number of days between weighings ($T \times D$ is the thermal sum in degree-days). Values of TGC vary with species, stock (genetics), feed, husbandry, environment and other factors (Cho 1990). However, the TGC for a given stock of fish may be relatively stable over a range of body weights (Kaushik 1998) and temperatures (Cho & Bureau 1998). Table 14.3 lists some examples of TGC values for different species. In practice, TGC is calculated for a given set of rearing conditions using past growth records for

Table 14.2 Specific growth rate (SGR, %/day) models for Atlantic salmon, rainbow trout, Arctic charr and gilthead sea bream. T is temperature and W is body weight (in grams). Ranges of fish weights and temperatures used in model development are given. Data for rainbow trout and Atlantic salmon were originally presented in tables, and the model was created using multiple regression analysis. The growth data for sea bream were originally expressed as g per fish per day, so recalculation was performed for model development.

Species	SGR model	W (g)	T (°C)	Reference
Rainbow trout and Atlantic salmon	$SGR = 0.665 \times (0.273 + 0.967 \times T) \times W^{-0.288}$	10–3000	2–16	Austreng *et al.* (1987)
Arctic charr	$SGR = 7.500 \times (0.022 + 0.073 \times T) \times W^{-0.351}$	10–600	1–14	Jørgensen *et al.* (1991)
Gilthead sea bream	$SGR = 0.509 \times (0.542 \times T - 0.522) \times W^{-0.374}$	1–400	19–26	Lupatsch & Kissil (1998)

Table 14.3 Examples of values for thermal-unit growth coefficient (TGC) calculated for different species of fish. For practical application, TGC should be calculated for the specific stock used, under the defined environmental and rearing conditions.

Species	TGC	Reference
Rainbow trout *Oncorhynchus mykiss*	2.98	Alanärä (1992a,b)
Atlantic salmon *Salmo salar*	3.05	Kiessling *et al.* (1995)
Arctic charr *Salvelinus alpinus*	1.65	Alanärä & Kiessling (1994)
Common carp *Cyprinus carpio*	1.40	Kaushik (1998)
Tilapia *Oreochromis aureus*	1.28	Kaushik (1998)
European catfish *Silurus glanis*	2.00	Kaushik (1998)
European sea bass *Dicentrarchus labrax*	0.64	Peres & Oliva-Teles (1999)
Gilthead sea bream *Sparus aurata*	0.62	Company *et al.* (1999)

the farm site, or records obtained from similar stocks and environmental conditions. Once the TGC is known, prediction of weight change is possible using the following equation:

$$FBW = [IBW^{0.3333} + (TGC/1000 \times T \times D)]^3 \tag{14.10}$$

The TGC is not able to handle the change in growth that occurs over the entire rate–temperature curve (Jobling 1997), and its application should be limited to the ascending limb of the curve. SGR, on the other hand, can be corrected for this by using the appropriate temperature function (see example in Elliott *et al.* 1995). Irrespective of the type of model used to predict growth, regular weighings are needed to allow periodic adjustment of calculations. In general, weighing at monthly intervals is recommended during periods of high production, but the time interval may be extended if it is seen that the model gives an accurate prediction of growth.

14.4 Variation in feed intake

Feed intake may vary considerably between days (Grove *et al.* 1978; Juell *et al.* 1993), and seasons (Rowe & Thorpe 1990; Jobling & Baardvik 1991). The causes for these variations are largely unknown, and it is most likely that several biological and environmental factors contribute (see Chapters 6, 7 and 8). Nevertheless, some influential factors are known and predictable, which means that they can be taken into account in farming conditions. This section will briefly highlight some of the most important factors known to influence variations in feed intake through time.

14.4.1 Temporal variation

One of the most obvious factors influencing feed intake is water temperature. Since temperature has a major influence on metabolism, it is obvious that feed intake is also affected (Brett 1979; Jobling 1994, 1997). Moderate changes in water temperature within the fish's range of tolerance generally lead to a corresponding change in feed intake. For example, Alanärä (1992b) showed that daily self-feeding activity of cage-reared rainbow trout tracked long-term changes in temperature quite closely. Abrupt and large changes in temperature may, however, lead to a depression in feed intake.

Other environmental factors known to influence feed intake from day to day are dissolved oxygen (Thetmeyer *et al.* 1999) and ammonia concentrations (Rasmussen & Korsgaard 1996; PersonLeRuyet *et al.* 1997) (see Chapters 6 and 7). Fish are dependent on a sufficiency of dissolved oxygen, and reductions in oxygen concentrations may lead to decreased feed intake. Husbandry practices may result in reduced dissolved oxygen in the culture environment. Figure 14.2 shows an example of an immediate increase in daily feed intake as a result of increasing water turnover in a cage by changing badly fouled nets for clean ones. Ammonia and ammonium are metabolic end-products excreted by fish, and they are known to influence feeding when present at moderate concentrations; they may also cause acute toxic effects at high concentrations. Both low oxygen and high ammonia/ammonium concentrations may arise as a consequence of low water turnover rates, within both tanks and sea cages.

Disease, in addition to having pathological effects, may stress fish and so indirectly cause markedly reduced feed intake or even complete anorexia (Bloch & Larsen 1993; Roberts & Shepherd 1997; Dawson *et al.* 1999). For example, Atlantic salmon suffering from a heavy sea lice infestation may reduce feeding, and feeding activity has been observed to increase by as much as 100% within 10 days of treatment for the infestation (S. Kadri, unpublished data).

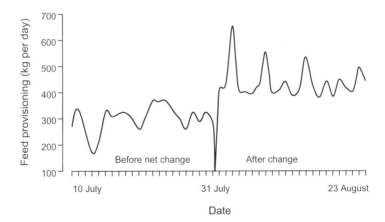

Fig. 14.2   Differences in feed provisioning to Atlantic salmon before and after changing a heavily fouled cage net. At the time of the net change, the cage contained about 14 000 fish with mean weight of about 2.8 kg. Water temperature was 12 ± 1°C during the week before and after net change. Feed intake was monitored using an Aquasmart AQ1 interactive feedback system (S. Kadri, unpublished data).

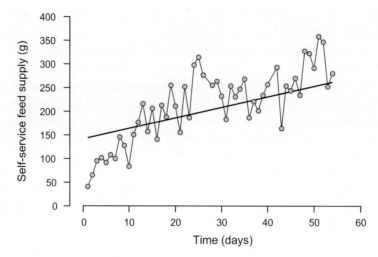

Fig. 14.3   Day-to-day variation in feeding activity (circles) in a group (n = 100) of rainbow trout (117 g) held under constant temperature (14°C) and light conditions (18L:6D) and fed using a self-feeder (unpublished data from Alanärä 1994). The solid line indicates feed requirements calculated according to Equation 14.1 (see p. 333). The DE content of the feed was estimated to be 19.4 MJ/kg, TGC to be 2.16 (Equation 14.10), and DEN to be 13.5 MJ/kg (Equation 14.6).

However, even under stable and favourable environmental conditions there may be considerable day-to-day variation in feeding activity (Fig. 14.3). One reason for this may be related to gastric evacuation time, because it has been claimed that return of appetite depends on stomach fullness (Grove *et al.* 1978; Bromley 1994), and that this in turn depends upon water temperature and body size. Daily fluctuations in feeding might therefore be induced by changes in stomach fullness and the time required for emptying. Whatever the cause, fish eating to satiation may show day-to-day variations in feed intake, and if these occur in synchrony, the day-to-day variations may be observed within the whole population.

14.4.2 Seasonal variation

Seasonal variations in feeding have been most widely studied in salmonid species, such as Atlantic salmon (Thorpe 1994), Arctic charr (Sæther *et al.* 1996) and chinook salmon (Clarke & Blackburn 1994). These fish typically increase feeding in spring, and there is a depression during the autumn (Smith *et al.* 1993; Tveiten *et al.* 1996). Fish that inhabit temperate regions appear to rely on seasonally changing environmental factors to time their annual cycles of feeding activity, growth and reproduction (see also Chapters 8 and 15). Thus, seasonal changes in feed intake will usually occur quite predictably, so it is possible to take this into account for long-term feed management.

The mechanisms behind these seasonal variations in feeding have been discussed in terms of endogenous circannual rhythms that are under the influence of the seasonal photoperiodic cycle (Eriksson & Lundqvist 1982; Eriksson & Alanärä 1992; see Chapter 8). In general, it is assumed that increasing day-length in spring and early summer is stimulatory while the decreasing autumnal day-length has the opposite effect (Higgins & Talbot 1985). For temperate species such as salmonids, equinoxes (when day-length is changing most rapidly) and solstices (when the change in day-length switches direction) are often important environmental cues which influence feeding activity (e.g. Kadri *et al.* 1996). A physiological trigger for reduced feeding during the autumn may be the level of fat depots, i.e. it has been proposed that fish may reduce feeding once they have acquired sufficient energy reserves to survive the winter (Tveiten *et al.* 1996). Further, there may be an inverse relationship between body fat content and feed intake, which would partly explain the large increase in feed intake after a long winter, when energy reserves are depleted (Metcalfe & Thorpe 1992; Jobling & Miglavs 1993; Shearer *et al.* 1997; Silverstein *et al.* 1999; see also Chapter 15).

Seasonal variations in feeding are known to be markedly influenced by life history events in some species. In Arctic charr and Atlantic salmon, for example, increased feeding during the early stages of maturation contrast with a depression during the later stages (Metcalfe *et al.* 1988; Kadri *et al.* 1996; Sæther *et al.* 1996; Tveiten *et al.* 1996). The level of depression varies, with maturing fish often undergoing a complete cessation of feeding and becoming anorectic (Thorpe 1994): this may occur during the summer under conditions of increasing temperatures and day-length when immature siblings are increasing feed intake (Kadri *et al.* 1996, 1997). The time of year when maturing fish cease feeding varies depending on species and age (size). For example, maturing adult Atlantic salmon held in sea cages, ceased feeding in late spring and early summer, several months prior to spawning (Kadri *et al.* 1995, 1996), whereas maturing Arctic charr decreased their feeding during late summer, and more or less stopped in September a few weeks prior to the spawning season (Tveiten *et al.* 1996). By contrast, some fish species, such as sea bass and bream cultured in the Mediterranean, mature during winter and spring: they continue to feed to some extent even though weight gain is reduced (G. Smart, personal observation), possibly as a result of the favouring of gonadal growth rather than flesh deposition.

Seasonal effects on feed intake are well documented for salmonid species, but systematic data for non-salmonids are sparse. Nevertheless, fish culturists are aware of the seasonal changes in feed intake that occur within their stocks as exemplified above for the Mediterranean.

14.5 Distribution of feed

14.5.1 Number of meals and feeding rate

A meal can be defined as the amount of food consumed in a single feeding bout, and in the case of fish, the duration of a meal is often from 30 min to 2 h. Traditionally, fish reared in large-scale aquaculture operations have been provided with feed at a relatively low intensity over prolonged periods. The choice of a 'little-and-often' feeding regime has partly been the result of technical limitations in some of the feeding systems used in farming operations, and partly because many farmers have thought that intense feeding was associated with feed waste. Today, there is evidence that larger fish of several species, such as salmonids, feed intensively in distinct periods or 'meals'. One to four meals per day may be sufficient for most salmonids to achieve maximum feed intake and growth (Elliott 1975; Grayton & Beamish 1977; Jobling 1983a,b; Juell *et al.* 1994), and the same may be true for some other species. The ideal number of meals is probably dependent on the water temperature and body size:

Cho (1990) suggested that rainbow trout larger than 200 g might be fed one to two meals per day, whereas fish smaller than 50 g should be provided with three to four meals per day. The inclusion of longer non-feeding periods may be beneficial for the growth of fish, since repeated feeding throughout long periods of the day increases swimming activity and energy expenditure (Alanärä 1992a; Johansen & Jobling 1998).

As the number of meals is decreased, the rate at which feed is supplied (i.e. number of pellets per fish per unit time) must be increased. This may influence feed waste, because the fish may not be able to capture all the pellets before they pass out of the culture system. This means that attempts should be made to adjust rates of feed supply to the feeding rate of the species, and the duration of meals should be adjusted according to the time required for all feeding individuals to become satiated. There is, however, a lack of knowledge about feeding rates of fish in culture. Juell *et al.* (1994) employed feed supply rates of about 0.4 and 1.8 pellets per fish per min, in a study on Atlantic salmon, and observed no significant effects on feed intake and growth; this indicated that the higher feed supply rate was too low to induce feed waste.

The time between consecutive feed deliveries is also an important factor. This should not be shorter than the total time the pellets are present in the rearing unit. For species that do not utilise the entire rearing volume, e.g. those that feed at the surface or are bottom feeders, the effective feeding depth may be less than the depth of the rearing unit. For species that feed in the water column, the sinking rate of the pellets will have a major influence on the feeding process and set a minimum time for consecutive feed deliveries. Thus, farmers should attempt to assess the part of the water column that the fish use during feeding, e.g. by underwater video camera.

In principle, the faster the feed supply rate the better, because low rates of feed supply may lead to feed monopolisation by the most competitive individuals (Grant 1993).

14.5.2 Diel timing of feeding

Many fish species show diel fluctuations in feeding activity, activities being concentrated at the times of day when the balance between food availability and predation risk is best (Helfman 1993; see Chapter 8). For example, salmonids – which are usually considered visual feeders – may feed diurnally with peaks in activity around dawn and dusk (Eriksson & Alanärä 1992). However, this activity pattern is seen mainly during spring, summer and autumn, but during winter, the juveniles of several salmonid species are active mainly during the night, e.g. Atlantic salmon (Fraser *et al.* 1995), Arctic charr (Jørgensen & Jobling 1989; Linnér *et al.* 1990) and rainbow trout (Landless 1976; Grove *et al.* 1978). Sea bass show a similar shift, being diurnal during summer and autumn and nocturnal during the winter (Sánchez-Vázquez *et al.* 1998). The benefits of a switch in diel activity are most likely related to seasonal shifts in the density of both prey and predators (Eriksson 1973; Fraser *et al.* 1995). For example, the nocturnal winter activity of young salmonids is synchronised with the drift activity of insects, which is a mainly nocturnal event in winter, at least in northern temperate zones (Müller 1978; Eriksson & Alanärä 1992). Alternatively, nocturnal activity during winter could be a strategy for reducing predator pressure (Fraser *et al.* 1995).

Most studies of diel activity patterns in fish have been performed under laboratory conditions using juvenile fish held in small groups. Data from large-scale rearing conditions, how-

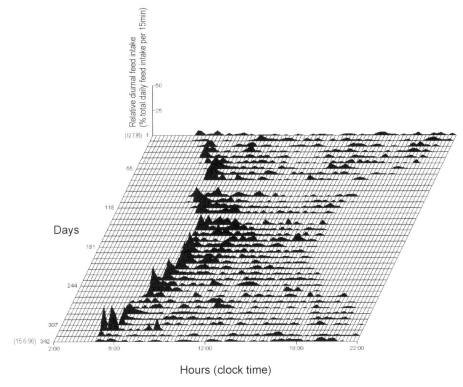

Fig. 14.4   Daily feeding patterns of Atlantic salmon in cages throughout a full production cycle. Data are presented as proportions of the daily ration provided through the day. Data were collected using the Aquasmart AQ1 interactive feedback system. (After Blyth *et al.* 1999.)

ever, indicate that generalisations drawn from laboratory conditions may be applicable in commercial culture (Kadri *et al.* 1991; Blyth *et al.* 1993). For example, Atlantic salmon seems to exhibit two peaks of feeding activity, one around dawn and another around dusk (Fig. 14.4). The feeding of salmonids in net cages at night should be avoided since they are inefficient at capturing pellets in the water column, thereby increasing the risk of feed wastage (Alanärä 1992b; Smith *et al.* 1993). Even under the most favourable night-time conditions (full moon and clear sky), the efficiency with which juvenile Atlantic salmon capture feed is low (35% of diurnal efficiency), and fish fed at less than 10% of the daytime rate when the moon was not full, skies were overcast or they were feeding in the shadows (Fraser & Metcalfe 1997). In contrast, sea bass may feed nocturnally in cages during periods of low water temperature and the efficiency of pellet capture appears to be similar to that seen during daytime feeding (Azzaydi *et al.* 2000).

14.5.3 Feed dispersal

The spatial distribution of feed may potentially influence feed intake and growth of fish in culture. Presentation of feed within a localised area may enable highly competitive individuals to monopolise the feeding area and suppress the feed intake of less competitive conspecifics (Metcalfe *et al.* 1992; Ryer & Olla 1995, 1996), and such observations have been made on Atlantic salmon in sea cages (Thorpe *et al.* 1990). Feed dispersal over a large proportion of the rearing unit volume should make feed indefensible and thereby reduce the problem of unequal feed distribution amongst individuals (Grant 1993; Jørgensen *et al.* 1996). However, under commercial farming conditions, a too-wide feed dispersal may lead to increased feed waste if the swimming of the fish creates strong centrifugal currents within the cage. For example, fish within groups of rainbow trout, Atlantic salmon and yellowtail (*Seriola* spp.) tend to swim uniformly in the same direction and when swimming at high speed, they will create an upwelling water current. Deeper water rises in the centre of the cage and is then forced sideways. Under such circumstances feed close to the net wall will be transported out of the cage in the current. In such cases, a more localised feed delivery close to the centre of the cage may be preferable, since the movements of the fish may serve to disperse the feed (S. Kadri & P.J. Blyth, unpublished data).

14.6 Feeding techniques

Several techniques for feeding fish have been developed during the past two decades or so, and some comparative studies have been conducted to test the influence of feeding technique on fish growth, feed efficiency and feed wastage (Pfeffer 1977; Thorpe *et al.* 1990; Alanärä 1992a; Kentouri *et al.* 1993; Azzaydi *et al.* 1998; Paspatis *et al.* 1999).

A modern feeding system should comprise three parts: a control unit; a feed dispenser; and a feed input regulating device. The control unit may range from a simple timer that activates the feed dispenser at fixed intervals to a much more complex data processing system. The feed dispenser is a container that stores and distributes the feed. Size may vary from small hoppers for hatchery use to the large silos used in large-scale production cages. Feeders developed for large-scale production conditions are often fitted with a feed spreading device. Feed may be spread either mechanically (e.g. rotating disc, Sterner AS, Sweden), or it can be carried by air or water which is pumped along pipes (e.g. Feeding Systems AS, Norway). In order to take variation in feeding activity into account, the feeding system should also contain a regulating device that matches feed input to demand. This may take the form of a sensor that registers feeding activity (fish biting on a trigger, video camera or hydroacoustic system registering feeding or swimming activity), or a detector that registers feed waste.

The following will focus on control functions and feed input regulation, because these are important for the adjustment and manipulation of feeding regimes.

14.6.1 Hand-feeding

Hand-feeding is the oldest and simplest feeding practice used in aquaculture. Provided that a farmer has experience of how the feed should be distributed, and is able to detect when the fish approach satiation, hand-feeding may be an efficient technique.

An advantage with hand-feeding is that the farmer has daily contact with the fish, and for fish species that are predominately surface active it is possible to make visual observations of feeding activity. In most cases, farmers take into consideration the feeding or swimming activity of fish in the upper part of the water column before they decide to continue, slow or stop feed delivery. In contrast, for species that avoid the surface or feed between surface and deeper water, reliable visual observations may be difficult to obtain. However, this problem may be solved by using an underwater camera or some kind of feed waste detecting system (i.e. waste pellet collection using an air-lift system, echosounders, pellet waste sensors) in combination with hand-feeding. From the point of view of large-scale commercial production, hand-feeding has not been considered as a viable option due to the high labour cost. Hand-held feed cannons may be used to increase the rate of feed supply, and this makes hand-feeding more cost-efficient. The working hours of farm staff may lead to the imposition of feeding times that diverge from the natural rhythms of the fish. It may therefore be beneficial to combine hand-feeding with some automatic system that could deliver a part of the daily ration outside normal working hours (see section 14.6.4).

14.6.2 Fixed feed ration systems

With the growth of intensive fish farming during the 1970s, the need for automation of feeding increased, and electronically operated timer-controlled feeding systems became popular. In such systems a timer unit sends out electric pulses that activate the feeder. The timer can be adjusted to regulate the periods of day for feed distribution as well as the frequency and duration of meals. Timer-controlled systems were, for the most part, replaced by computer-based control units by the end of the 1980s. The purchase cost of these feeding systems is high, especially for electric feeders, whereas operating costs are low. Computer-controlled systems offer more possibilities for the regulation and timing of feed delivery than do timer-controlled systems. Such systems may allow continuous recording of several abiotic factors (e.g. temperature, dissolved oxygen, salinity, light intensity, water flow direction and speed) that may influence feeding. Additional information relating to production management may be added and a feed distribution model designed. The recording of all these factors allows farm managers to regularly update information about the rearing units in order to organise future investments and fish production. However, computerised systems often need skilled staff both to ensure that the input and interpretation of data are correct and to check that the system is operating correctly. In addition, critical points are high purchase cost and fragility, particularly for environmental sensors and the central control unit.

When timer- or computer-controlled systems are used, the quantities of feed to be supplied are determined on the basis of either feed tables or mathematical models describing daily energy requirements (see p. 333). In other words, the daily feed ration is predetermined, and not regulated by the demand of the fish. This is a critical point because it means that possible short-term variations in feed intake are not taken into consideration in the farm operation.

14.6.3 Demand feeding systems

These types of feeding systems, which have been developed for use on commercial farms during the 1980s and 1990s, are based on the principle that the supply of feed is regulated by the demand of the fish. Demand feeders work in one of two ways: either by fish requesting feed themselves (self-feeders); or by automatic cessation or reduction of feed delivery when feeding activity declines (interactive feedback system).

Self-feeders

The principle of the self-feeder is that the fish regulate the level of feeding by activating (through presses or bites) a trigger positioned at, or slightly above the water surface, thereby causing feed to be released from a feed dispenser. In general, two types of self-feeder are available: those that manually deliver food and those that are electronically controlled. The manually controlled ones generally consist of a feed holder (hopper) with an aperture whose opening is controlled by a movable gate. A pendulum (trigger), the tip of which extends down into the water, is attached to the gate. Lateral movements of the pendulum cause the gate mechanism to open, allowing food to fall out of the food holder into the water (Shepherd & Bromage 1989). In electronically controlled systems, movement of the pendulum generates an electric pulse which is sent to a feed dispenser (Anthouard & Wolf 1988; Boujard *et al.* 1992; Cuenca & de la Higuera 1994; Sanchez-Vasquez *et al.* 1994). Alanärä (1994) described a trigger incorporating a string that ended in a plastic, pellet-shaped bead. An electronic pulse was generated when the fish bit at the bead and stretched the string. Thus, the fact that the bead had to be bitten by a fish in order for feed delivery made 'accidental feeding' unlikely. The big advantage of electronically controlled self-feeders over manual ones is that it is possible to connect them to most feed dispensers available on the market.

For self-feeders to work in a commercial setting some prerequisites need to be fulfilled. First, the fish must be able to learn to activate the trigger within a reasonable period of time, i.e. some days or weeks. Some salmonids (Alanärä 1996; Alanärä & Kiessling 1996; Paspatis & Boujard 1996), sea bass (Boujard *et al.* 1996; Azzaydi *et al.* 1998; Paspatis *et al.* 1999) and sparids (Divanach *et al.* 1993) seem to be able to do this, but there is little information relating to other commercially important species. Further, for self-feeders to function well it is important that a single trigger actuation provides food for more than one fish, i.e. one fish releases food for several others. The level of reward should correspond with the number of pellets that the fish can catch before they move out of reach and are lost (see discussion about feed delivery rate in the previous section). Thus, the reward must be adjusted according to the behaviour and feeding rate of the specific species. However, the only species for which there is a reasonable amount of knowledge relating to reward levels is rainbow trout (Alanärä 1996).

Social hierarchies and agonistic behaviour may become a problem when using self-feeders. For example, when salmonids are held at moderate densities in small tanks there may be monopolisation of the trigger by a small number of socially dominant fish (Alanärä 1996). No such information is available for large-scale rearing conditions, but the existence of such problems cannot be ruled out. Thus, knowledge about the social behaviour of the species is important if a self-feeding system is to be adjusted to function optimally, e.g. with respect to the size of reward level, spreading of feed, etc.

Interactive feedback systems

These can be defined as providing an indirect control of feed demand. The supply of feed is adjusted either by the monitoring of uneaten feed or evaluation of fish feeding activity. In the first case, specially designed echosounders may be used to detect uneaten feed (Juell 1991; Juell *et al.* 1993), or sensors placed under the feeding area can be used to monitor uneaten pellets as they sink through the water column (Blyth *et al.* 1993). Video technology with image analysis software has also been developed to detect uneaten feed pellets (Foster *et al.* 1995; Ang & Petrell 1997). Echosounders, which are sometimes used in conjunction with underwater video cameras, may be used to detect the position of fish. For example, if the fish are close to the water surface this may mean that they are hungry, whereas avoidance of the surface may mean that they are not eager to feed. Interpretation of the distribution data may then be used to control feed delivery.

The feedback received following detection of feed waste or reduced feeding activity is usually used to introduce an abrupt stop to feed delivery, but can also be used to gradually decrease the rate of delivery. The latter approach has been developed as proprietary software (Aquasmart International AS, Trondheim). For effective regulation of feed delivery, software must be developed or adjusted for each species because there are species-specific differences in how feeding activity declines and ceases as the fish approach satiation (e.g. abruptly or slowly).

Both self-feeders and interactive feedback system offer the possibility of collecting data from the continuous recordings of feed demand. These may give useful information about day-to-day or seasonal variations in feed intake, as well as about the diel timing of feeding at different times of the year.

14.6.4 The combined use of feeding techniques

In practice, the choice of feeding technique must be made following considerations of type and level of production, and the cost of the feeding system. Feeding procedures can be changed during the production cycle, and several techniques may be combined. The preferred methodology for rearing juveniles is often based on an automatic fixed ration system because of their high daily feed requirements. There can then be a change to combined feeding for larger fish, where much larger quantities of feed must be dispensed at much faster rates. Practical considerations, and not necessarily the control of fish satiation, may often determine the most appropriate combination. In small farms, with few skilled staff, hand-feeding in combination with timer-controlled feeders or self-feeders may be sufficient. A proportion of the daily feed ration can be distributed by hand, but feed will still be available to the fish when staff are not present, or when there is increased feed demand. Low investment and simplicity are often the priorities for small-scale farmers. On large farms, the effectiveness of a feeding system may be evaluated according to the degree of automation and control provided. Intensive large-scale fish rearing demands high rates of feed delivery with the least waste: this target may be met with demand feeding systems that are programmed to function optimally.

14.7 References

Aksnes, A. & Mundheim, H. (1997) The impact of raw material freshness and processing temperature for fish meal on growth, feed efficiency and chemical composition of Atlantic halibut (*Hippoglossus hippoglossus*). *Aquaculture*, **149**, 87–106.

Alanärä, A. (1992a) Demand feeding as a self-regulating feeding system for rainbow trout (*Oncorhynchus mykiss*) in net-pens. *Aquaculture*, **108**, 347–356.

Alanärä, A. (1992b) The effect of time-restricted demand feeding on feeding activity, growth and feed conversion in rainbow trout (*Oncorhynchus mykiss*). *Aquaculture*, **108**, 357–368.

Alanärä, A. (1994) The effect of temperature, dietary energy content and reward level on the demand feeding activity of rainbow trout (*Oncorhynchus mykiss*). *Aquaculture*, **126**, 349–359.

Alanärä, A. (1996) The use of self-feeders in rainbow trout (*Oncorhynchus mykiss*) production. *Aquaculture*, **145**, 1–20.

Alanärä, A. & Kiessling, A. (1996) Changes in demand feeding behaviour in Arctic charr, *Salvelinus alpinus* L., caused by differences in dietary energy content and reward level. *Aquaculture Research*, **27**, 479–486.

Ang, K.P. & Petrell, R.J. (1997) Control of feed dispensation in seacages using underwater video monitoring: effects on growth and food conversion. *Aquacultural Engineering*, **16**, 45–62.

Anthouard, M. & Wolf, V. (1988) A computerized surveillance method based on self-feeding measures in fish populations. *Aquaculture*, **71**, 151–158.

Austreng, E., Storebakken, T. & Åsgård, T. (1987) Growth rate estimates for cultured Atlantic salmon and rainbow trout. *Aquaculture*, **60**, 157–160.

Azzaydi, M., Madrid, J.A., Sánchez-Vázquez, F.J. & Martinez, F.J. (1998) Effect of feeding strategies (automatic, *ad libitum* demand-feeding and time-restricted demand-feeding) on feeding rhythms and growth in European sea bass (*Dicentrarchus labrax* L.). *Aquaculture*, **163**, 285–296.

Azzaydi, M., Martínez, F.J., Zamora, S., Sánchez-Vázquez, F.J. & Madrid, J.A. (2000) Nocturnal versus diurnal feeding of European sea bass in winter environmental conditions: influence on feeding rhythms and growth. *Aquaculture*, **182**, 329–338.

Ballestrazzi, R., Lanari, D. & D'Agaro, E. (1998) Performance, nutrient retention efficiency, total ammonia and reactive phosphorus excretion of growing European sea-bass (*Dicentrarchus labrax*, L.) as affected by diet processing and feeding level. *Aquaculture*, **161**, 55–65.

Bjerkeng, B., Hatlen, B. & Wathne, E. (1999) Deposition of astaxanthin in fillets of Atlantic salmon (*Salmo salar*) fed diets with herring, capelin, sandeel, or Peruvian high PUFA oils. *Aquaculture*, **180**, 307–319.

Bloch, B. & Larsen, J.L. (1993) An Iridovirus-like agent associated with systemic infection in cultured turbot *Scophthalmus maximus* fry in Denmark. *Diseases of Aquatic Organisms*, **15**, 235–240.

Blyth, P.J., Purser, G.J. & Russell, J.F. (1993) Detection of feeding rhythms in seacaged Atlantic salmon using new feeder technology. In: *Fish Farming Technology* (eds H. Reinertsen, L. A. Dahle, L. Jørgensen & K. Tvinnereim), pp. 209–216. Balkema, Rotterdam.

Blyth, P.J., Kadri, S., Valdmirsson, S.K., Mitchell, D.F. & Purser, G.J. (1999) Diurnal and seasonal variation in feeding patterns of Atlantic salmon, *Salmo salar* L., in sea cages. *Aquaculture Research*, **30**, 530–544.

Boujard, T., Dugy, X., Genner, D., Gosset, C. & Grig, G. (1992) Description of a modular, low cost, eater meter for the study of feeding behavior and food-preferences in fish. *Physiology and Behavior*, **52**, 1101–1106.

Boujard, T., Jourdan, M., Kentouri, M. & Divanach, P. (1996) Diel feeding activity and the effect of time-restricted self-feeding on growth and feed conversion in European sea bass. *Aquaculture*, **139**, 117–127.

Brafield, A.E. (1985) Laboratory studies of energy budgets. In: *Fish Energetics, New Perspectives* (eds P. Tytler & P. Calow), pp. 257–282. Croom Helm, Sydney, Australia.

Brett, J.R. (1979) Environmental factors and growth. In: *Fish Physiology.* Vol. VIII (eds W.S. Hoar, D.J. Randall & J.R. Brett), pp. 599–675. Academic Press, New York.

Brett, J.R. (1995) Energetics. In: *Physiological Ecology of Pacific Salmon* (eds C. Groot, L. Margolis & W.C. Clarke), pp. 1–68. UBC Press, Vancouver.

Bromley, P.J. (1994) The role of gastric evacuation experiments in quantifying the feeding rates of predatory fish. *Reviews in Fish Biology and Fisheries*, **4**, 36–66.

Cho, C.Y. (1982) Effects of dietary protein and lipid levels on energy metabolism of rainbow trout (*Salmo gairdneri*). In: *Energy Metabolism of Farm Animals* (eds A. Ekern & F. Sundstol), pp. 176–179. Proceedings of the 9th EAAP Symposium, Lillehammer, Norway.

Cho, C.Y. (1990) Fish nutrition, feeds, and feeding with special emphasis on salmonid aquaculture. *Food Reviews International*, **6**, 333–357.

Cho, C.Y. & Bureau, D.P. (1998) Development of bioenergetic models and the Fish-PrFEQ software to estimate production, feeding ration and waste output in aquaculture. *Aquatic Living Resources*, **11**, 199–210.

Clarke, W.C. & Blackburn, J. (1994) Effect of growth on early sexual maturation in stream-type chinook salmon (*Oncorhynchus tshawytscha*). *Aquaculture*, **121**, 95–103.

Company, R., Calduch-Giner, J.A., Kaushik, S. & Pérez-Sánchez, J. (1999) Growth performance and adiposity in gilthead sea bream (*Sparus aurata*): risks and benefits of high energy diets. *Aquaculture*, **171**, 279–292.

Cuenca, E.M. & de la Higuera, M. (1994) A microcomputer-controlled demand feeder for the study of feeding behavior in fish. *Physiology and Behavior*, **55**, 1135–1136.

Dawson, L.H.J., Pike, A.W., Houlihan, D.F. & McVicar, A.H. (1999) Changes in physiological parameters and feeding behaviour of Atlantic salmon *Salmo salar* infected with sea lice *Lepeophtheirus salmonis*. *Diseases of Aquatic Organisms*, **35**, 89–99.

Dias, J., Huelvan, C., Dinis, M.T. & Métailler, R. (1998) Influence of dietary bulk agents (silica, cellulose and a natural zeolite) on protein digestibility, growth, feed intake and feed transit time in European seabass (*Dicentrarchus labrax*) juveniles. *Aquatic Living Resources*, **11**, 219–226.

Divanach, P., Kentouri, M., Charalambakis, G., Pouget, F. & Sterioti, A. (1993) Comparison of growth performance of six Mediterranean fish species reared under intensive farming conditions in Crete (Greece), in raceways with the use of self feeders. In: *Production, Environment and Quality* (eds G. Barnabé & P. Kestemont), Bordeaux Aquaculture 1992. European Aquaculture Society. Sp. Publication No. 18, Ghent, Belgium.

Elliott, J.M. (1975) Number of meals in a day, maximum weight of food consumed in a day and maximum rate of feeding for brown trout, *Salmo trutta* L. *Freshwater Biology*, **5**, 287–303.

Elliott, J.M., Hurley, M.A. & Fryer, R.J. (1995) A new, improved growth model for brown trout, *Salmo trutta*. *Functional Ecology*, **9**, 290–298.

Eriksson, L.-O. (1973) Spring inversion of the diel rhythm of locomotor activity in young sea-going brown trout, *Salmo trutta* L. and Atlantic salmon, *Salmo salar* L. *Aquilo Series Zoologicae*, **14**, 68–79.

Eriksson, L.-O. & Alanärä, A. (1992) Timing of feeding behavior in salmonids. In: *The Importance of Feeding Behavior for the Efficient Culture of Salmonid Fishes* (eds J.E. Thorpe & F.A. Huntingford), pp. 41–48. World Aquaculture Society, Baton Rouge, LA, USA.

Eriksson, L.-O. & Lundqvist, H. (1982) Circannual rhythms and photoperiod regulation of growth and smolting in Baltic salmon (*Salmo salar*). *Aquaculture*, **28**, 113–121.

Foster, M., Petrell, R., Ito, M.R. & Ward, R. (1995) Detection and counting of uneaten food pellets in a sea cage using image analysis. *Aquacultural Engineering*, **14**, 251–269.

Fraser, N.H.C. & Metcalfe, N.B. (1997) The cost of becoming nocturnal: feeding efficiency in relation to light intensity in juvenile Atlantic salmon. *Functional Ecology*, **11**, 385–391.

Fraser, N.H.C., Heggenes, J., Metcalfe, N.B. & Thorpe, J.E. (1995) Low summer temperatures cause juvenile Atlantic salmon to become nocturnal. *Canadian Journal of Zoology*, **73**, 446–451.

Grant, J.W.A. (1993) Whether or not to defend? The influence of resource distribution. *Marine Behaviour and Physiology*, **23**, 137–153.

Grayton, B.D. & Beamish, F.W.H. (1977) Effects of feeding frequency on food intake, growth and body composition of rainbow trout (*Salmo gairdneri*). *Aquaculture*, **11**, 159–172.

Grisdale-Helland, B. & Helland, S.J. (1997) Replacement of protein by fat and carbohydrate in diets for Atlantic salmon (*Salmo salar*) at the end of the freshwater stage. *Aquaculture*, **152**, 167–180.

Grisdale-Helland, B. & Helland, S.J. (1998) Macronutrient utilization by Atlantic halibut (*Hippoglossus hippoglossus*): diet digestibility and growth of 1 kg fish. *Aquaculture*, **166**, 57–65.

Grove, D.J., Loizides, L.G. & Nott, J. (1978) Satiation amount, frequency of feeding and gastric emptying rate in *Salmo gairdneri*. *Journal of Fish Biology*, **5**, 507–516.

Helfman, G. (1993) Fish behaviour by day, night, and twilight. In: *The Behaviour of Teleost Fishes* (ed. T.J. Pitcher), pp. 366–387. Croom-Helm, London.

Helland, S.J. & Grisdale-Helland, B. (1998a) The influence of replacing fish meal in the diet with fish oil on growth, feed utilization and body composition of Atlantic salmon (*Salmo salar*) during the smoltification period. *Aquaculture*, **162**, 1–10.

Helland, S.J. & Grisdale-Helland, B. (1998b) Growth, feed utilization and body composition of juvenile Atlantic halibut (*Hippoglossus hippoglossus*) fed diets differing in the ratio between the macronutrients. *Aquaculture*, **166**, 49–56.

Higgins, P.J. & Talbot, C. (1985) Growth and feeding in juvenile Atlantic salmon (*Salmo salar* L.). In: *Nutrition and Feeding in Fish* (eds C.B. Cowey, A.M. Mackie & J.G. Bell), pp. 243–263. Academic Press, London.

Hillestad, M., Åsgård, T. & Berge, G.M. (1999) Determination of digestibility of commercial salmon feeds. *Aquaculture*, **179**, 81–94.

Iwama, G.K. & Tautz, A.F. (1981) A simple growth model for Salmonids in hatcheries. *Canadian Journal of Fisheries and Aquatic Sciences*, **38**, 649–656.

Jobling, M. (1983a) A short review and critique of methodology used in fish growth and nutrition studies. *Journal of Fish Biology*, **23**, 685–703.

Jobling, M. (1983b) Effects of feeding frequency on food intake and growth of Arctic charr, *Salvelinus alpinus* L. *Journal of Fish Biology*, **23**, 177–185.

Jobling, M. (1994) *Fish Bioenergetics*. Chapman & Hall, London.

Jobling, M. (1997) Temperature and growth: modulation of growth rate via temperature change. In: *Global Warming. Implications for Freshwater and Marine Fish* (eds C.M. Wood & D.G. McDonald), pp. 225–253. Cambridge University Press, Cambridge.

Jobling, M. & Baardvik, B.M. (1991) Patterns of growth of maturing and immature Arctic charr, *Salvelinus alpinus*, in a hatchery population. *Aquaculture*, **94**, 343–354.

Jobling, M. & Miglavs, I. (1993) The size of lipid depots – a factor contributing to the control of food intake in Arctic charr, *Salvelinus alpinus*? *Journal of Fish Biology*, **43**, 487–489.

Johansen, S.J.S. & Jobling, M. (1998) The influence of feeding regime on growth and slaughter traits of cage-reared Atlantic salmon. *Aquaculture International*, **6**, 1–17.

Juell, J.-E. (1991) Hydroacoustic detection of food waste – A method to estimate maximum food intake of fish populations in sea cages. *Aquacultural Engineering*, **10**, 207–217.

Juell, J.E., Furevik, D.M. & Bjordal, Å. (1993) Demand feeding in salmon farming by hydro acoustic food detection. *Aquacultural Engineering*, **12**, 155–167.

Juell, J.-E., Bjordal, Å., Fernö, A. & Huse, I. (1994) Effect of feeding intensity on food intake and growth of Atlantic salmon, *Salmo salar* L., in sea cages. *Aquaculture and Fisheries Management*, **25**, 453–464.

Jørgensen, E.H. & Jobling, M. (1989) Patterns of food-intake in arctic charr, *Salvelinus alpinus*, monitored by radiography. *Aquaculture*, **81**, 155–160.

Jørgensen, E.H., Jobling, M. & Christiansen, J.S. (1991) Metabolic requirements of Arctic charr, *Salvelinus alpinus* (L.), under hatchery conditions. *Aquaculture and Fisheries Management*, **22**, 377–378.

Jørgensen, E.H., Baardvik, B.M., Eliassen, R. & Jobling, M. (1996) Food acquisition and growth of juvenile Atlantic salmon (*Salmo salar*) in relation to spatial distribution of food. *Aquaculture*, **143**, 277–289.

Kadri, S., Metcalfe, N.B., Huntingford, F.A. & Thorpe, J.E. (1991) Daily feeding rhythms in Atlantic salmon in sea cages. *Aquaculture*, **92**, 219–224.

Kadri, S., Metcalfe, N.B., Huntingford, F.A. & Thorpe, J.E. (1995) What controls the onset of anorexia in maturing adult female Atlantic salmon? *Functional Ecology*, **9**, 790–797.

Kadri, S., Mitchell, D.F., Metcalfe, N.B., Huntingford, F.A. & Thorpe, J.E. (1996) Differential patterns of feeding and resource accumulation in maturing and immature Atlantic salmon, *Salmo salar*. *Aquaculture*, **142**, 245–257.

Kadri, S., Thorpe, J.E. & Metcalfe, N.B. (1997) Anorexia in one-sea-winter Atlantic salmon (*Salmo salar*) during summer, associated with sexual maturation. *Aquaculture*, **151**, 405–409.

Kaushik, S.J. (1998) Nutritional bioenergetics and estimation of waste production in non-salmonids. *Aquatic Living Resources*, **11**, 211–217.

Kentouri, M., Divanach, P. & Maignot, E. (1993) Comparaison de l'efficacité-coût de trois techniques de rationnement de la daurade *Sparus aurata*, en élevage intensif en bassins. In: *Production, Environment and Quality, Bordeaux Aquaculture '92* (eds G. Barnabé & P. Kestemont), European Aquaculture Society Special Publication No. 18, pp. 273–283. Ghent.

Kiessling, A., Dosanjh, B., Higgs, D., Deacon, G. & Rowshandeli, N. (1995) Dorsal aorta cannulation: a method to monitor changes in blood levels of astaxanthin in voluntarily feeding Atlantic salmon, *Salmo salar* L. *Aquaculture Nutrition*, **1**, 43–50.

Landless, P.J. (1976) Demand-feeding behaviour of rainbow trout. *Aquaculture*, **7**, 11–25.

Larsson, S. & Berglund, I. (1997) Growth and food consumption of 0+ Arctic charr fed pelleted or natural food at six different temperatures. *Journal of Fish Biology*, **52**, 230–242.

Linnér, J., Brännäs, E., Wiklund, B.-S., & Lundqvist H. (1990) Diel and seasonal locomotor activity patterns in Arctic charr, *Salvelinus alpinus* (L.). *Journal of Fish Biology*, **37**, 675–685.

Lupatsch, I. & Kissil, G.W. (1998) Predicting aquaculture waste from gilthead seabream (*Sparus aurata*) culture using a nutritional approach. *Aquatic Living Resources*, **11**, 265–268.

Metcalfe, N.B. & Thorpe, J.E. (1992) Anorexia and defended energy levels in over-wintering juvenile salmon. *Journal of Animal Ecology*, **61**, 175–181.

Metcalfe, N.B., Huntingford, F.A. & Thorpe, J.E. (1988) Feeding intensity, growth rates, and the establishment of life-history patterns in juvenile Atlantic salmon *Salmo salar*. *Journal of Animal Ecology*, **57**, 463–474.

Metcalfe, N.B., Wright, P.J. & Thorpe, J.E. (1992) Relationships between social status, otolith size at first feeding and subsequent growth in Atlantic salmon (*Salmo salar*). *Journal of Animal Ecology*, **61**, 585–589.

Müller, K. (1978) The flexibility of the circadian system of fish at different latitudes. In: *Rhythmic Activity of Fishes* (ed. J.E. Thorpe), pp. 91–104. Academic Press, London.

Paspatis, M. & Boujard, T. (1996) A comparative study of automatic feeding and self-feeding in juvenile Atlantic salmon (*Salmo salar*) fed diets of different energy levels. *Aquaculture*, **145**, 245–257.

Paspatis, M., Batarias, C., Tiangos, P. & Kentouri, M. (1999) Feeding and growth response of sea bass (*Dicentrarchus labrax*) reared by four feeding methods. *Aquaculture*, **175**, 293–305.

Peres, H. & Oliva-Teles, A. (1999) Influence of temperature on protein utilization in juvenile European seabass (*Dicentrarchus labrax*). *Aquaculture*, **170**, 337–348.

PersonLeRuyet, J., Galland, R., LeRoux, A., & Chartois, H. (1997) Chronic ammonia toxicity in juvenile turbot (*Scophthalmus maximus*). *Aquaculture*, **154**, 155–171.

Pfeffer, E. (1977) Studies on the utilization of dietary energy and protein by rainbow trout (*Salmo gairdneri*) fed either by hand or by an automatic self-feeder. *Aquaculture*, **10**, 97–107.

Rasmussen, G. & From, J. (1991) Improved estimates of a growth model and body composition of rainbow trout, *Oncorhynchus mykiss* (Walbaum, 1792) as a function of feeding level, temperature and body size. *Dana*, **9**, 15–30.

Rasmussen, R.S. & Korsgaard, B. (1996) The effect of external ammonia on growth and food utilization of juvenile turbot (*Scophthalmus maximus* L) *Journal of Experimental Marine Biology and Ecology*, **205**, 35–48.

Refstie, S., Storebakken, T. & Roem, A.J. (1998) Feed consumption and conversion in Atlantic salmon (*Salmo salar*) fed diets with fish meal, extracted soyabean meal or soyabean meal with reduced content of oligosaccharides, trypsin inhibitors, lectins and soya antigens. *Aquaculture*, **162**, 301–312.

Ricker, W.E. (1979) Growth rates and models. In: *Fish Physiology,* Vol. VIII (eds W.S. Hoar, D.J. Randall & J.R. Brett), pp. 677–743. Academic Press, New York.

Roberts, R.J. & Shepherd, C.J. (1997) *Handbook of Trout and Salmon Diseases*. Fishing News Books, Farnham.

Rowe, D.K. & Thorpe, J.E. (1990) Differences in growth between maturing and non-maturing male Atlantic salmon, *Salmo salar* L., parr. *Journal of Fish Biology*, **36**, 643–658.

Ryer, C.H. & Olla, B.L. (1995) The influence of food distribution upon the development of aggressive and competitive behaviour in juvenile chum salmon, *Oncorhynchus keta*. *Journal of Fish Biology*, **46**, 264–272.

Ryer, C.H. & Olla, B.L. (1996) Growth depensation and aggression in laboratory reared coho salmon: the effect of food distribution and ration size. *Journal of Fish Biology*, **48**, 686–694.

Sæther, B.-S., Johnsen, H.K. & Jobling, M. (1996) Seasonal changes in food consumption and growth of Arctic charr exposed to either simulated natural or a 12:12 LD photoperiod at constant water temperature. *Journal of Fish Biology*, **48**, 1113–1122.

Sanchez-Vasquez, F.J., Martinez, M., Zamora, S. & Madrid, J.A. (1994) Design and performance of an accurate demand feeder for the study of feeding behaviour in sea bass, *Dicentrarchus labrax* L. *Physiology and Behavior*, **56**, 789–794.

Sánchez-Vázquez, F. J., Azzaydi, M., Martinez, F.J., Zamora, S. & Madrid, J.A. (1998) Annual rhythms of demand-feeding activity in sea bass: evidence of a seasonal phase inversion of the diel feeding pattern. *Chronobiology International*, **15**, 607–622.

Shearer, K.D., Silverstein, J.T. & Plisetskaya, E.M. (1997) Role of adiposity in food intake control of juvenile chinook salmon (*Oncorhynchus tshawytscha*). *Comparative Biochemistry and Physiology*, **118**, 1209–1215.

Shepherd, J. & Bromage, N. (1989) *Intensive Fish Farming.* Blackwell Scientific Publications, Oxford.

Silverstein, J.T., Shearer, K.D., Dickhoff, W.W. & Plisetskaya, E.M. (1999) Regulation of nutrient intake and energy balance in salmon. *Aquaculture*, **177**, 161–169.

Smith, I.P., Metcalfe, N.B., Huntingford, F.A. & Kadri, S. (1993) Daily and seasonal patterns in the feeding behaviour of Atlantic salmon (*Salmo salar*) in a sea cage. *Aquaculture*, **117**, 165–178.

Sveier, H. & Lied, E. (1998) The effect of feeding regime on growth, feed utilisation and weight dispersion in large Atlantic salmon (*Salmo salar*) reared in seawater. *Aquaculture*, **165**, 333–345.

Sveier, H., Wathne, E. & Lied, E. (1999) Growth, feed and nutrient utilisation and gastrointestinal evacuation time in Atlantic salmon (*Salmo salar* L.): the effect of dietary fish meal particle size and protein concentration. *Aquaculture*, **180**, 265–282.

Thetmeyer, H., Waller, U., Black, K.D., Inselmann, S. & Rosenthal, H. (1999) Growth of European sea bass (*Dicentrarchus labrax* L.) under hypoxic and oscillating oxygen conditions. *Aquaculture*, **174**, 355–367.

Thodesen, J., Grisdale-Helland, B., Helland, S.J. & Gjerde, B. (1999) Feed intake, growth and feed utilization of offspring from wild and selected Atlantic salmon (*Salmo salar*). *Aquaculture*, **180**, 237–246.

Thorpe, J.E. (1994) Reproductive strategies in Atlantic salmon, *Salmo salar* L. *Aquaculture and Fisheries Management*, **25**, 77–87.

Thorpe, J.E., Talbot, C., Miles, M.S., Rawlings, C. & Keay, D.S. (1990) Food consumption in 24 hours by Atlantic salmon (*Salmo salar* L.) in sea cage. *Aquaculture*, **90**, 41–47.

Tveiten, H., Johnsen, H.K. & Jobling, M. (1996) Influence of maturity status on the annual cycles of feeding and growth in Arctic charr reared at constant temperature. *Journal of Fish Biology*, **48**, 910–924.

Winberg, G.G. (1956) *Rate of metabolism and food requirements of fishes.* Beloruss State University, Minsk (Fisheries Research Board Canada. Translation Series, **194**, 1960).

Chapter 15
Nutrient Partitioning and the Influence of Feed Composition on Body Composition

Malcolm Jobling

15.1 Introduction

Fish have been used as food by humans from ancient times, and fish currently form an important part of the diet in many countries (Love 1988; Haard 1992; Macrae *et al.* 1993). Fish are important as a protein source, but are also a source of lipid. At least as important as the lipid content *per se* is the nature of the lipid. For example, fish are generally low in cholesterol and, compared to meat from terrestrial animals, have a complex spectrum of lipids high in polyunsaturated fatty acids. Of particular importance is the high proportion of (n-3) fatty acids, contrasting with the (n-6) fatty acids of vegetable origin (see Chapter 1). Fish are also valuable sources of some vitamins and minerals. The muscle tissue (fillet) is generally the chief food source, but non-muscular tissues such as fins, gonads and livers of some species are also consumed; in instances where the bones are eaten, these may represent important dietary sources of calcium and phosphorus. Given the importance – and variety – of fish and fish products consumed by humans it is important to have knowledge about the chemical compositions of fish; the partitioning of nutrients among different organs and tissues, and how fish body composition may be influenced by the types of feed consumed.

As a crude approximation the animal body can be considered to comprise two compartments: 'lean body mass' (LBM) and lipid storage depots. The composition of LBM is similar across animal species and life history stages, being approximately 19 moisture:5 protein:1 mineral ash, with smaller amounts of carbohydrates and structural lipids. Thus, the bulk of the LBM comprises moisture and protein, but growing animals will also accumulate reserves of storage lipid. As a consequence, the relative changes in the chemical composition of a growing animal will be a reflection of differences in the rates of accretion of LBM and lipid reserves. Similarly, the changes that occur during times of feed shortage, or starvation, will reflect the differences in the rates at which the chemical components of the different body compartments are mobilised to meet energy demand.

The composition of farmed fish may be much more stable than that of their wild counterparts, because farmed fish are provided with food throughout the year and are not subject to the vagaries of changes in prey abundance (Love 1988; Haard 1992; Cowey 1993; Macrae *et al.* 1993). Nevertheless, there may be seasonal changes in the composition of farmed fish because fish of some species, such as salmonids, undergo periods of voluntary anorexia prior to spawning (Kadri *et al.* 1995; Tveiten *et al.* 1996). Thus, the final stages of gonadal maturation may be dependent upon the mobilisation of endogenous body components, and thereby represent a major drain on storage reserves of lipids and other nutrients.

The fact that farmed fish are usually provided with a good supply of nutrient-dense formulated feeds enables them to deposit large reserves of lipid. Where in the body the lipid is stored will influence several attributes of farmed fish: 'dressout' losses during processing, fillet texture and storage properties, and the nutritional value of the fillet. The focus of this chapter is the way in which the various nutritional attributes of fish are influenced by feed supply and feed composition.

15.2 Morphometrics and relationships among chemical components

Several biometric indices have been used for the estimation of energy status of fish, with a non-destructive assessment relating to 'condition' being most commonly used (Weatherley & Gill 1987; Sutton *et al.* 2000). The 'condition indices' are derived from length–weight relationships, which can be expressed by the equation:

$$W = cL^m$$

where W is weight, L is length, and c and m are constants. This equation indicates that length is the primary determinant of the weight of a fish, but there can be wide variations in weight between fish of the same length, both within and between populations. Further, weight can be both gained and lost. Thus, when fish are feeding and growing well, an individual may have a greater-than-usual weight at a particular length, but when feeding conditions are poor the fish may lose weight and be light for their length. This means that the length–weight relationship can be used to assess the 'well-being' of individual fish, and the equation can be re-arranged to give an index of condition, or condition factor (CF):

$$CF = [IW/IL^m] \times 100$$

where IW and IL are the weight and length of an individual fish, respectively, and m is the exponent in the length–weight relationship. Alternatively, the condition of an individual fish can be calculated as IW/EW, where EW is the 'expected weight' of the fish calculated from the length–weight relationship. In practice, because m in the length–weight relationship is often close to 3, the condition index is usually calculated as Fulton's Condition Factor (K):

$$K = [IW/IL^3] \times 100$$

In most cases, K will be satisfactory for analysing differences in condition related to sex or season when the fish used for comparison are of approximately the same length; however, if the length range is large, spurious results will be generated if m differs from 3.

Changes in condition are accompanied by changes in the biochemical composition of the body tissues, the most marked changes being in the percentages of lipids and moisture. In some studies the condition factor has been reported to be a reasonably good predictor of body lipid (Herbinger & Friars 1991), but in others it has fared less well (Simpson *et al.* 1992;

Adams *et al.* 1995; Sutton *et al.* 2000). Predictions of body lipid content could, however, be made from multiple regressions incorporating measurements of several body proportions (Simpson *et al.* 1992; Adams *et al.* 1995). Thus, the morphometrics approach may have potential as a non-destructive technique for assessment of energy status when repeated measures are made on the same fish over time.

The amounts of different chemical components within the fish body will usually increase as the body mass of the fish increases. This can create problems for the comparison of body compositions of fish that are growing at different rates. The problems arise because it may be inappropriate to make direct comparisons of the percentage, or relative, compositions of fish of different sizes (Sutton *et al.* 2000). The situation may be even more complex because fish that are growing at different rates may differ in body composition even when analyses are carried out on individuals that are matched for body size (Rasmussen & Ostenfeld 2000). For example, in the two salmonid species studied by Rasmussen and Ostenfeld (2000), the relative carcass yield was lower in fast-growing fish, presumably due to increased lipid deposition in the visceral cavity. Overall, the trend was for lipid deposition to be greater in the fish that were growing at the fastest rates (Table 15.1).

Allometric analysis has been used to compare the relative growth of body organs in animals, and this sort of analysis may have application in comparisons of changes in chemical composition because it circumvents some of the problems listed above (Shearer 1994; Sutton *et al.* 2000). In allometric analysis the logarithm of the mass of a chemical component (e.g. protein, lipid, a mineral element) is usually plotted against the logarithm of the fish body mass. These log-log plots most often produce relationships that are linear. Provided that certain conditions are met, treatment effects can then be tested for using analysis of covariance (ANCOVA). The allometric analysis also provides information about the way in which the proportions of different components change as the fish increase in mass: when the slope of the log-log regression is 1, both body mass and the mass of the component are increasing at the same rate, when the slope is <1, then body mass is increasing more rapidly than the mass of the component (i.e. the concentration of the chemical component is decreasing), and when the slope is >1 the specific component makes up an increasing proportion of the body mass as the fish increase in size.

Table 15.1 Influences of specific growth rate (SGR, % body weight/day) on body composition of rainbow trout, *Oncorhynchus mykiss*, and brook charr, *Salvelinus fontinalis*, when fast- and slow-growing fish were analysed at similar body size (rainbow trout ca. 260 g; brook charr ca. 230 g). (Data from Rasmussen & Ostenfeld 2000.)

	Rainbow trout *Oncorhynchus mykiss*		Brook charr *Salvelinus fontinalis*	
	SGR 1.26% bw/day	SGR 0.56% bw/day	SGR 1.00% bw/day	SGR 0.52% bw/day
Whole body				
Dry matter (%)	31.3 ± 1.2	27.3 ± 0.9	31.7 ± 0.4	30.0 ± 0.6
[Moisture (%)]	[68.7]	[72.7]	[68.3]	[70.0]
Lipid (%)	11.9 ± 1.2	8.2 ± 1.1	12.3 ± 0.9	10.2 ± 0.7
Protein (%)	16.8 ± 0.2	17.1 ± 0.4	17.7 ± 0.4	17.8 ± 0.6
Fillet				
Dry matter (%)	26.6 ± 1.0	25.7 ± 0.6	27.7 ± 0.6	27.8 ± 0.8
[Moisture (%)]	[73.4]	[74.3]	[72.3]	[72.2]
Lipid (%)	5.5 ± 1.2	4.4 ± 0.3	5.6 ± 0.6	5.2 ± 0.6
Protein (%)	19.4 ± 0.5	19.7 ± 0.3	20.4 ± 0.8	20.6 ± 0.3

The sum of the proportions of body lipid and moisture is often ca. 80%, and this seems to apply across body compartments, organs and tissues (Table 15.1) and species (Table 15.2) (e.g. Stirling 1976; Otwell & Rickards 1981; Manthey *et al.* 1988; Aursand *et al.* 1994; Lie *et al.* 1994; Hatlen 1997; Jobling *et al.* 1998; Koskela *et al.* 1998). As a consequence of this, there is a strong negative correlation between the percentages of body lipid and body moisture, and this relationship seems to hold under different conditions of feeding, growth and gonad development (Weatherley & Gill 1987; Love 1988; Shearer 1994). For example, the relative proportions of body lipid and moisture in juvenile walleye pollock, *Theragra chalcogramma*, are influenced by body size, season and feeding conditions (Sogard & Olla 2000), and there is a negative correlation between percent moisture (%M) and percent lipid (%L):

$$\%M = 85.823 - 1.367\%L$$

Similarly, there was a strong inverse relationship between percent moisture and percent lipid in rainbow trout, *Oncorhynchus mykiss*, fed either high- or low-protein feeds at different ration levels (Reinitz 1983), the relationship being described by the following equation:

$$\%M = 81.398 - 1.075\%L$$

As a final example, the body composition of gilthead sea bream, *Sparus aurata* (initial weight ca. 20 g), was found to be influenced by both feeding rate and the lipid concentration in

Table 15.2   Proximate percentage chemical composition of the fillet (muscle) of a range of fish species. (Data from Otwell & Rickards 1981; Manthey *et al.* 1988; Lie *et al.* 1994.)

Fish species	Percentage of chemical component		
	Moisture	Lipid	Protein
American eel	67.0	14.5	16.0
Anguilla rostrata			
European eel	46.0	32.5	17.5
Anguilla anguilla			
Wels (Sheatfish)	76.5	4.5	18.5
Silurus glanis			
Channel catfish	71.5	9.0	17.5
Ictalurus punctatus			
African catfish	75.0	3.0	20.0
Clarias gariepinus			
Rainbow trout	70.0	10.0	17.0
Oncorhynchus mykiss			
Atlantic salmon	69.0	10.0	18.5
Salmo salar			
Atlantic cod	80.5	0.5	18.0
Gadus morhua			
Wolf-fish	77.5	2.5	18.5
Anarhichas lupus			
Atlantic halibut	72.0	10.5	16.0
Hippoglossus hippoglossus			
Turbot	79.0	2.5	16.0
Scophthalmus maximus			

the feed. Percentage body lipid was lowest (4.4%) in fish fed a low-lipid feed (9% lipid) at the lowest ration level (1% body weight/day), and highest (10.9%) in those fed a high-lipid feed (17% lipid) to satiation each day (Company *et al.* 1999). The reverse trend was seen for percentage moisture, and there was a significant negative correlation between the percentages of body lipid and body moisture (Fig. 15.1):

$$\%M = 78.223 - 0.957\%L$$

Given the close interrelationships among the major chemical components – moisture, lipid and protein – it should be possible to derive predictive equations for the assessment of proximate composition. Estimates of body composition could then be made based upon the analysis of a single chemical component. It is easier to measure moisture than either lipid or protein, so moisture is most frequently used as the independent variable in the predictive equations. Such equations have been developed for several fish species, and some examples are presented in Table 15.3.

15.3 Patterns of lipid deposition and storage

Farmed fish will usually contain relatively more lipid than their wild counterparts (Stirling

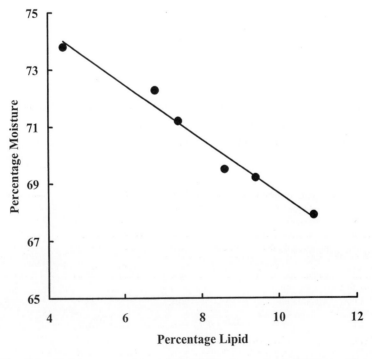

Fig. 15.1   The relationship between percentage moisture (%M) and percentage lipid (%L) in gilthead sea bream, *Sparus aurata,* fed high (17% lipid) or low (9% lipid) lipid feeds at three ration levels (1%, 1.75% or satiation). The regression is described by: %M = 78.223 – 0.957%L ($n = 6$; $R^2 = 0.973$). (Calculated from data given in Company *et al.* 1999).

Table 15.3 Interrelationships between the relative proportions of the major biochemical components (moisture, protein and lipid) in a range of fish species.

Species	Lipid (%L)–moisture (%M) relationship	Source
European eel *Anguilla anguilla*	%L = 95.3 – 1.22%M	Boëtius & Boëtius (1985)
Whitefish *Coregonus lavaretus*	%L = 55.85 – 0.648%M	Koskela *et al.* (1998)
Sockeye salmon *Oncorhynchus nerka*	%L = 50.95 – 0.591%M	Brett *et al.* (1969)
Rainbow trout *Oncorhynchus mykiss*	%L = 48.53 – 0.529%M	Jobling *et al.* (1998)
Atlantic salmon *Salmo salar*	%L = 61.99 – 0.764%M	Stangnes *et al.*, unpublished
Brown trout *Salmo trutta*	%L = 56.38 – 0.700%M % L = 51.63 – 0.613%M	Jonsson & Jonsson (1998) Elliott (1976)
Sea bass *Dicentrarchus labrax*	%L = 56.78 – 0.704%M	Stirling (1976)

Species	Protein (%P)–moisture (%M) relationship	Source
Sockeye salmon *Oncorhynchus nerka*	%P = 50.53 – 0.459%M	Brett *et al.* (1969)
Brown trout *Salmo trutta*	%P = 42.92 – 0.353%M %P = 47.06 – 0.380%M	Elliott (1976) Jonsson & Jonsson (1998)
Sea bass *Dicentrarchus labrax*	%P = 45.47 – 0.393%M	Stirling (1976)

1976; Otwell & Rickards 1981; Henderson & Tocher 1987; Haard 1992), and fish fed on lipid-rich feeds will usually be found to deposit more storage lipid than those fed on leaner diets (Table 15.4) (Manthey *et al.* 1988; Cowey 1993; dos Santos *et al.* 1993; Shearer 1994; Higgs *et al.* 1995; Bell 1998). The majority of the storage lipids comprise triacylglycerols (TAGs), the fatty acid composition of which can be manipulated by making modifications to the fatty acid composition of the feed. However, where in the body the lipids are stored differs between species, and fish of a given species may also have major lipid depots in several different organs or tissues.

Table 15.4 Influence of lipid concentration in the feed (12.6 versus 27.5%) on lipid concentrations in tissues of the whitefish, *Coregonus lavaretus*, and rainbow trout, *Oncorhynchus mykiss*. (Data from Koskela *et al.* 1998; Jobling *et al.* 1998.)

Body compartment or tissue	Whitefish *Coregonus lavaretus*		Rainbow trout *Oncorhynchus mykiss*	
	12.6% Lipid	27.5% Lipid	12.6% Lipid	27.5% Lipid
Whole fish				
% Lipid	11.5 ± 1.5	14.0 ± 1.9	10.5 ± 0.7	15.4 ± 1.6
% Moisture	67.2 ± 1.4	64.6 ± 3.4	69.4 ± 1.2	64.5 ± 2.0
Viscera				
% Lipid	22.1 ± 10.0	26.9 ± 9.4	23.3 ± 3.7	38.0 ± 6.7
% Moisture	60.1 ± 7.4	56.8 ± 6.3	59.8 ± 3.8	48.1 ± 5.2
Carcass				
% Lipid	10.5 ± 1.3	12.7 ± 1.9	9.3 ± 0.7	12.4 ± 1.0
% Moisture	67.8 ± 1.4	65.4 ± 3.7	70.5 ± 1.3	67.1 ± 1.8

The muscle is generally considered to be the site of the major lipid store in salmonids (Table 15.5) (Love 1988; Haard 1992; Macrae *et al.* 1993; Aursand *et al.* 1994). The majority of the adipocytes are located within the myosepta and in the connective tissue layers that lie between the muscle fibres. Lipid deposition in the muscle is dependent upon the lipid present in the feed. There is often a positive correlation between percentage muscle lipid and feed lipid concentration (Bell 1998). Further, the fact that the fatty acid composition of the feed influences the fatty acid composition of the muscle lipids has been demonstrated in many studies carried out on salmonids (e.g. Hardy *et al.* 1987; Henderson & Tocher 1987; Polvi & Ackman 1992; Aursand *et al.* 1994; Higgs *et al.* 1995).

Farmed salmonids may also deposit large amounts of lipid in the visceral cavity (Tables 15.4 and 15.5). These visceral stores, which comprise over 95% TAGs, may account for 10–12% of the total body lipid content (Aursand *et al.* 1994). Thus, the feeding of lipid-rich, high-energy feeds may lead to the deposition of excess visceral lipid that is lost during dressing (Jobling *et al.* 1998), resulting in a reduced yield following processing (Higgs *et al.* 1995).

The skeletal tissue of fish contains a higher concentration of lipid than does that of terrestrial animals, and oil-filled bones are common in fish (Phleger 1998). Some species have over 24% lipid in the skeleton; in the majority of fish species the bone lipid is mostly comprised of TAGs, but some species store wax esters in their bones. Various functions have been proposed for the bone lipids, including buoyancy, posture adjustment, and as an energy store that can be easily mobilised during periods of feed deprivation (Phleger 1998). Skeletal lipids appear to represent an important reserve in salmonids (Phleger *et al.* 1989, 1995; Aursand *et al.* 1994; Jørgensen *et al.* 1997). For example, the vertebral centra and neurocrania of prespawning Pacific salmon, *Oncorhynchus* spp., caught in the ocean have been found to contain up to 15% and 7% lipid, respectively. TAGs comprised over 90% of the bone lipids, and these seemed to be depleted during the spawning migration (Phleger *et al.* 1989, 1995). Similarly, the skeleton may be an important lipid depot in farmed Atlantic salmon, *Salmo salar*: the head and backbone have been found to contain about 20% of the total body lipid, and over 95% of the lipids present were TAGs (Aursand *et al.* 1994). The skin of the Atlantic salmon has also been reported to be relatively lipid-rich (18% lipid, whereof 97% TAGs), and this tissue may account for about 9% of the body lipid content (Aursand *et al.* 1994). Thus, the head, backbone and skin may account for ca. 30% of the body lipid in farmed Atlantic salmon, the muscle ca. 45%, and the visceral tissue ca. 12% (Table 15.5) (Aursand *et al.* 1994).

Table 15.5 Relative sizes (as % body weight) of various tissues in Atlantic salmon, *Salmo salar* (ca. 3 kg body weight), their relative (%) contents of moisture and lipid, and their relative contributions to whole body lipid depots. (Calculated from data in Aursand *et al.* 1994.)

Tissue	% Body weight	% Moisture	% Lipid	% Body lipid depot
White muscle	56.3	68.9	9.6	35.4
Red muscle	4.6	56.7	27.2	7.8
Belly flap	7.9	55.6	28.1	13.7
Viscera	6.9	60.0	26.6	11.7
Liver	1.3	73.7	4.9	0.4
Skin, head and backbone	22.6	58.0	19.7	29.4

Like salmonids, the halibut, *Hippoglossus hippoglossus*, is usually considered to be a 'medium-fat' fish, so the muscle tissue represents a major depot for lipid storage (Table 15.2). In this species, lipid concentrations tend to be highest in the liver (ca. 5–25%), in the 'notch' at the base of the anal fin (sometimes in excess of 40%), and in the red muscle (ca. 14%) (Haug *et al.* 1988; Berge & Storebakken 1991; Cowey 1993; Nortvedt & Tuene 1998), and there is differential deposition of lipid throughout the white muscle mass. Lipid accumulation in the area around the base of the anal fin appears to be common among flatfish, being attributed to fatty tissue filling the spaces between the muscles to the fin rays. Within the white muscle of the halibut the greatest relative amount of lipid is found in the abdominal area, with relative lipid content decreasing from the more anterior towards the posterior regions (Haug *et al.* 1988; Hemre *et al.* 1992; Nortvedt & Tuene 1998). The lipid fractions of the liver, red and white muscle are dominated by neutral lipids (ca. 75% of the total lipids), TAGs making up ca. 60–70% of the total lipids in each tissue (Haug *et al.* 1988). As with other species, the body composition of the halibut is influenced by the composition of the feed; proportions of dry matter and lipid in the carcass tend to increase with increasing dietary lipid concentrations (Aksnes *et al.* 1996; Helland & Grisdale-Helland 1998; Nortvedt & Tuene 1998). For example, Helland and Grisdale-Helland (1998) reported that an increase in dietary lipid from 20 to 27% resulted in a difference in carcass lipid of two percentage points (ca. 8.5% versus ca. 10.5%). Further, lipid deposition tends to be higher in rapidly growing fish, and also increases with increasing fish size (Nortvedt & Tuene 1998).

Unlike the salmonids and halibut, the cod, *Gadus morhua*, is a 'lean' fish in which the muscle lipids seldom exceed 1% of the fillet mass (Table 15.2). The structural phospholipids represent 65% or more of the muscle lipids, and there are only very small proportions of storage lipids, which are predominantly TAGs (Lie *et al.* 1986; Love 1988; dos Santos *et al.* 1993). The storage lipids of the cod are found in the liver, and in farmed cod the liver may account for as much as 15–20% of the body weight (Lie *et al.* 1986, 1988; Jobling *et al.* 1991, 1994). Thus, when cod are fed on lipid-rich feeds they tend to develop enlarged livers in which lipids, particularly TAGs, may represent over 65% of the liver weight (Lie *et al.* 1986, 1988; Jobling *et al.* 1991, 1994; dos Santos *et al.* 1993). The relative size of the liver seems to have a direct relationship with the dietary lipid concentration, or the quantity of lipid consumed.

The fatty acid composition of the feed is reflected in that of the liver TAGs, but also influences the fatty acid composition of the phospholipids of the liver and muscle. For example, when cod were given a feed containing peanut oil with high concentrations of 18:1(n-9) and 18:2(n-6) they deposited large proportions of these fatty acids in the liver TAGs. Cod provided with feeds containing marine oils (either cod-liver or Greenland halibut, *Reinhardtius hippoglossoides*) had much reduced concentrations of 18:1(n-9) and 18:2(n-6) in their liver TAGs. The same trends were seen in the fatty acid compositions of the phospholipids of both the liver and muscle, in which 18:2(n-6) made up ca. 13% of the phospholipid fatty acids in the cod fed peanut oil, but only ca. 2% in those fed the marine oils (Lie *et al.* 1986).

15.4 Temporal changes in body composition

Growth encompasses tissue elaboration, the storage and mobilisation of energy reserves, and reproductive growth, all of which may invoke changes in body composition, including large net losses at certain times of the year. Seasonal changes in body composition will usually be most extreme in fish that inhabit temperate zone or high-latitude environments, and there may also be changes in composition that are directly related to body size (Weatherley & Gill 1987; Love 1988; Shearer 1994; Jørgensen *et al.* 1997; Jonsson & Jonsson 1998; Sogard & Olla 2000; Sutton *et al.* 2000). In other words, the fish tend to accumulate storage lipids during the summer growth season, and the reserves are then mobilised to provide metabolic fuel during the winter, when food consumption is low. The reserves may also be mobilised to support gonadal development (Weatherley & Gill 1987; Love 1988; Jørgensen *et al.* 1997).

Seasonality in cycles of growth, reproduction, and depletion and repletion of energy reserves among temperate zone and high-latitude fish can be illustrated using the crucian carp, *Carassius carassius*, as an example. This species is common in freshwater ponds of northern Europe. The crucian carp tolerates temperatures from 0 to 38°C, and can survive several months of anoxia at low temperature. Populations at the northernmost extent of the species range experience continuous daylight during the summer months, and temperatures may rise 30°C: summer is a brief period of rapid growth and reproduction. During winter, however, the ponds may be covered by thick layers of ice and snow, and the fish live for many months in dark, hypoxic waters at temperatures close to freezing: during winter the fish are torpid and anorectic, surviving on reserves accumulated during the previous summer (Holopainen *et al.* 1997).

Winter torpor leads to a marked reduction in metabolic costs, but exposure to hypoxia or anoxia imposes a metabolic strain on the fish due to the inefficiency of anaerobic metabolic pathways in comparison with aerobic metabolism. The ability of the fish to survive several months of winter hypoxia is based upon three major adaptations: winter torpor leading to reduced metabolic costs; the accumulation of large glycogen reserves that enable the fish to survive a long period without feeding even though using inefficient anaerobic metabolic routes; and an anaerobiosis that involves the production of ethanol, CO_2 and ammonia as endproducts. The latter circumvents the problems that might arise due to accumulation of lactate. The shift in energy partitioning from growth and reproduction to accumulation of glycogen reserves occurs in July, even though waters are still warm and oxygen-rich. The shift seems to be triggered by changes in photoperiod, i.e. day-length is becoming shorter. Glycogen is deposited in the liver and muscle, and by late September–early October – when the fish become anorectic and torpid – the liver may represent 15% of body mass, and contain 35% glycogen. Glycogen may increase to 4% of the mass of the swimming muscle, and to ca. 8% of the mass of the heart. The onset of anorexia and torpidity may be induced by the fall in temperature that occurs during the autumn; fasting can be induced by exposure of fish to low temperature (<5°C) under conditions of constant photoperiod and high oxygen concentrations (Holopainen *et al.* 1997). Thus, the crucian carp may rely on at least two environmental cues – photoperiod and temperature – to regulate different phases of the seasonal cycles of growth, reproduction and energy partitioning.

Since fish of many species undergo natural periods of feed deprivation it is not surprising that fish are generally 'starvation-tolerant', and have considerable ability to endure and recover from prolonged periods of food shortages (Weatherley & Gill 1987; Love 1988; Sogard & Olla 2000). Fish may reduce their energy demands during periods of feed deprivation, and this is reflected in a reduction in oxygen consumption, but the fish will suffer gradual weight loss as the energy reserves are mobilised and depleted. Studies on the depletion of energy reserves during starvation indicate that the various chemical constituents may be mobilised at different rates, and that a given constituent may be depleted from different organs and tissues at different rates (e.g. Brett *et al.* 1969; Stirling 1976; Boëtius & Boëtius 1985; Black & Love 1986; Weatherley & Gill 1987; Love 1988; Jørgensen *et al.* 1997; Einen *et al.* 1998). However, the general tendency seems to be the conservation of body protein at the expense of lipid and glycogen. Some proteins are mobilised and depleted, with the myofibrillar proteins of the white muscle seeming to be most depleted during prolonged starvation (Black & Love 1986; Weatherley & Gill 1987).

Stirling (1976) noted that feed deprivation invoked a rapid initial decrease in glycogen followed by a progressive decline in lipid in sea bass, *Dicentrarchus labrax*, although there was also some mobilisation of protein. This resulted in progressive changes in the relative proportions of the chemical components making up the fish body: percentages of lipid, glycogen and protein declined, whereas the relative proportions of moisture and mineral ash increased. These effects on relative composition were observed in the white muscle and liver, as well as at the level of the whole fish. Conclusions about the mobilisation and utilisation of energy reserves drawn from examination of the relative changes in composition of organs and tissues should be treated with caution. A much clearer picture will emerge if absolute values are considered (Stirling 1976; Boëtius & Boëtius 1985; Jørgensen *et al.* 1997).

For example, Boëtius and Boëtius (1985) examined the effects of prolonged starvation (758 days) at ca. 14°C on changes in body weight (*W*) and a range of chemical components – moisture (M), lipid (L), protein (P) and 'residue' (R, mostly mineral ash) – in the European eel, *Anguilla anguilla* (male silver eels; initial weight ca. 90 g). They found that the decline in both body weight and the amounts of the different chemical components could be described by a series of negative exponential equations:

$$W = 89.6\exp^{-0.0012\,\text{Time}}$$

$$M = 47.9\exp^{-0.0012\,\text{Time}}$$

$$L = 25.6\exp^{-0.0010\,\text{Time}}$$

$$P = 12.7\exp^{-0.0017\,\text{Time}}$$

$$R = 3.33\exp^{-0.0007\,\text{Time}}$$

Most of the body weight loss was accounted for by a decline in the amounts of moisture and lipid, with body weight and these two components declining at similar rates (i.e. the slopes of the regressions were similar). The loss of 'residue' took place at the slowest rate, indicating little mobilisation of the mineral ash in the bones, and a conservation of skeletal tissue. It was the rate of decline in body protein that was most rapid, although protein contributed relatively little to the total loss of body mass (ca. 9 g of a total of ca. 50 g). Thus, the differences in the rates of mobilisation of lipid and protein (i.e. different slopes to the regressions) resulted in an increase in the ratio of lipid:protein from ca. 2 to ca. 3.5 during starvation, indicating a relative increase in body lipid relative to protein.

Farmed fish are routinely deprived of feed prior to slaughter to ensure that the gut is empty, and a longer period of feed deprivation may be applied if the fish have been treated with medicated feeds. How the imposition of periods of feed deprivation that differ in duration (days, weeks or months) influences the yield and composition of the final product (usually the fillet) has seldom been studied in detail, although some data are available for farmed Atlantic salmon (initial weight ca. 5 kg; deprived of feed for up to 80 days at 3–6°C) (Einen *et al.* 1998). During the early stages, up to 30 days, there was a decline in VSI (weight of viscera relative to body weight), indicating that there was a more rapid mobilisation of reserves in the visceral cavity than elsewhere in the body. Fillet yield, as a percentage of whole body weight, was fairly stable at ca. 58%, indicating that there was no preferential mobilisation of fillet reserves at this stage. In absolute terms, however, there was a substantial loss of fillet mass during the first 30 days of feed deprivation (Table 15.6).

Beyond 30 days, fillet yield gradually declined, to ca. 55% of whole body weight after 80 days of feed deprivation. This gradual decline is indicative of increased mobilisation of fillet reserves with the passage of time, leading to a lower proportion of muscle mass relative to the head, skin and skeletal tissues. Thus, during the 80 days of feed deprivation the fish lost 11–11.5% of their body mass, and most of this was accounted for by a reduction in the weight of the fillet (Table 15.6). Both lipid and protein were mobilised, and there was little change in the lipid:protein ratio. Perhaps more significant from a production point of view was the fact that the prolonged period of feed deprivation did not result in there being any marked reduction in the relative proportion (percentage) of lipid in the fillet. Thus, the imposition of a prolonged period of feed deprivation resulted in a substantial loss of fillet mass without having any marked effect on the percentage of lipid.

The body composition of farmed fish is usually more stable than that of their wild counterparts (Love 1988; Haard 1992; Cowey 1993; Macrae *et al.* 1993), but there may be seasonal changes in chemical composition. For example, Bell (1998) reported seasonal changes in fillet lipid concentrations of farmed Atlantic salmon: the highest concentrations were observed during spring and early summer, and the lowest values were recorded during the late summer and autumn. The changes probably reflect the accumulation, and later mobilisation and redistribution of lipids in fish that are undergoing sexual maturation; it is known that salmonids of several species undergo periods of voluntary anorexia prior to spawning (Kadri *et al.* 1995; Tveiten *et al.* 1996; Hatlen 1997).

Under conditions prevalent in the wild the growth of fish will often be food-limited, and the cycles of depletion and repletion of energy reserves may reflect seasonal availability of

Table 15.6   The influence of feed deprivation on whole body and fillet weight changes in Atlantic salmon, *Salmo salar*. Initial body weight was ca. 5 kg, and fish were held at 3–6°C. (Calculated from data given in Einen *et al.* 1998.)

Day	Round weight (g)	Fillet components (g)			Fillet 'loss' (g) since day 0			Fillet % composition	
		Weight	Lipid	Protein	Weight	Lipid	Protein	Lipid	Protein
0	5000	2900	545	565				18.8	19.4
14	4675	2715	495	550	185	50	15	18.2	20.3
30	4600	2670	485	540	230	60	25	18.1	20.3
58	4590	2575	465	525	325	75	40	18.0	20.3
80	4435	2440	450	470	460	95	95	18.4	19.3

prey organisms. Fish appear to be able to adapt to seasonal feast-and-famine conditions by showing marked growth spurts, known as catch-up or compensatory growth, when food supplies are increased following a period of undernutrition. The responses seem to be common across vertebrate species: increased food intake (i.e. hyperphagia), rapid growth and the repletion of energy reserves (Broekhuizen *et al.* 1994). It is usually the animals that are in the poorest condition that show the greatest response (Jobling *et al.* 1991, 1994). Further, among fish that have been held on restricted rations the catch-up growth response appears to be inversely related to the severity of the feed restriction (Jobling *et al.* 1999; Sæther & Jobling 1999).

Although the underlying mechanisms remain to be elucidated, it is possible that a shift in the balance between the relative sizes of the LBM and energy storage depots induced by a period of undernutrition may be involved in triggering the changes in feed intake (i.e. hyperphagia) and growth that constitute the catch-up response (Broekhuizen *et al.* 1994). For example, when cod were deprived of food for a prolonged period they depleted their reserves of liver lipids, liver and muscle glycogen, and then there was increased mobilisation of white muscle proteins. The integrity of the red muscle, used for continuous swimming rather than burst performance, seemed to be maintained at the expense of the white muscle (Black & Love 1986). When the cod were refed, both muscle types were repleted (predominantly the white muscle, as red muscle is conserved under starvation). This involved an increase in weight, a relative increase in lipid and glycogen, and a decrease in the percentage of moisture in the muscle. There was little repletion of the liver lipid stores before the percentage of moisture in the muscle fell below 82% (Black & Love 1986). Thus, during the initial stages of recovery from a period of food deprivation cod seem to give priority to the repletion of muscle tissue, and this results in a marked increase in the fillet yield of cod that are undergoing catch-up growth (Jobling *et al.* 1991, 1994).

15.5 Muscle (fillet) composition and factors that influence 'quality'

Although the composition of the muscle of different animal species is similar, comprising similar proteins, there are some differences between fish muscle and the muscle of terrestrial animals. These differences are mainly associated with the structural requirements for swimming. Also, the fact that fish live in aquatic environments means that their muscle requires less structural support than that of terrestrial animals. As a consequence of the latter, fish muscle tends to have less connective tissue than the muscle of terrestrial animals, and this results in fish muscle having a more tender texture.

The yield of muscular tissue from fish is generally relatively larger than from terrestrial vertebrates. Depending upon species, the locomotor muscle makes up 30–70% of the fish body mass, e.g. ca. 30–40% in 'catfishes' such as *Silurus glanis*, *Ictalurus punctatus* and *Clarias gariepinus* (Manthey *et al.* 1988), and ca. 50–65% in salmonids (Aursand *et al.* 1994; Hatlen 1997; Einen *et al.* 1998; Rasmussen & Ostenfeld 2000). The muscle is divided into a lateral red component, used during low-speed cruising, and a much larger white muscle mass which is active during short bursts of high-speed swimming. The two types of muscle have a different enzyme spectrum and metabolism, differ in the degree of vascularisation, and also differ in lipid content. The red muscle contains much more lipid than does white (Table 15.5)

(Porter *et al.* 1992; Macrae *et al.* 1993; Aursand *et al.* 1994). The red-brown colour of the lateral muscle is due to the presence of the iron-containing haem pigment, myoglobin. The differences between the red and white muscle influence storage properties, with the fillets containing a high proportion of red muscle being more susceptible to oxidative damage due to the content of highly unsaturated fatty acids (HUFAs) (Freeman & Hearnsberger 1994). The differences in proportions of red and white muscle among species seem, however, to have little influence upon essential amino acid patterns, which appear to be similar across a wide range of fish and other vertebrates (Table 15.7) (Haard 1992; Cowey 1993; Lie *et al.* 1994).

The muscle tissue comprises muscle fibres that are bound together by connective tissue, and fish muscle usually contains 15–20% protein by weight (Tables 15.1 and 15.2). The muscle comprises a limited number of abundant proteins – actin, myosin and collagen – and the muscle proteins are often categorised based upon their solubility: water-soluble or sarcoplasmic proteins, salt-soluble or myofibril proteins, and the insoluble stroma proteins, primarily collagen (Love 1988; Haard 1992; Macrae *et al.* 1993). Fish muscle tends to contain a relatively high concentration of myofibril protein (65–75%), and lower concentrations of sarcoplasmic (20–30%) and stroma (2–5%) proteins. The myofibril complexes contain myosin and actin, the main components of the thick and thin muscle filaments, respectively. Myosin comprises 50–60% of the myofibrillar contractile proteins, and actin 15–30%. Thus, myosin is the most abundant of the muscle proteins, making up about 35–40% of the total. Myosin molecules are connected to actin molecules to form the actomyosin complex that is responsible for muscle contraction and relaxation. Actomyosin plays a major role in determining the textural properties of the fish fillet because it is quite labile and is influenced by processing and storage conditions. For example, actomyosin becomes progressively less soluble during frozen storage, and this results in the fillet becoming increasingly tough.

Depending upon species and season, the lipid content of fish muscle can range from as little as 0.5% to over 20% (Table 15.2). From a human nutritional standpoint fish can be classified into groups according to their muscle lipid content (Haard 1992; Cowey 1993; Macrae *et al.* 1993): lean (<2% lipid), low-fat (2–4% lipid), medium-fat (4–8% lipid) and high-fat (>8% lipid). The muscle lipids of the lean species are structural phospholipids that contain high levels of HUFAs, whereas the muscle lipids of the 'fattier' species are dominated by storage TAGs. TAGs usually contain lower proportions of HUFAs than do phospholipids.

Salmonids are usually considered to be medium-fat fish, with muscle lipid contents within the range 2–7% (Henderson & Tocher 1987; Hatlen 1997; Rasmussen & Ostenfeld 2000), but it is not uncommon for the muscle of farmed Atlantic salmon to have a lipid content of 15% or more (Aursand *et al.* 1994; Wold *et al.* 1996; Bell 1998; Einen *et al.* 1998). Further, the lipid is not uniformly distributed throughout the muscle tissue of salmonids and other 'fattier' fish species. The tissue of the body wall surrounding the visceral cavity (known as the 'belly flap') has the highest concentrations of lipid, and there may also be a decrease in muscle lipid moving in a caudad direction (Henderson & Tocher 1987; Aursand *et al.* 1994; Zhou *et al.* 1996; Hatlen 1997; Einen *et al.* 1998). The lipid concentration of the belly flap tissue is usually greater than that of the red muscle, and may be several times that of the dorsal white muscle (Table 15.5). Thus, the belly flaps may represent one of the major lipid depots in the salmon. Belly flap lipid may account for 12–14% of the total body lipid in the salmon (Aursand *et al.* 1994), and the fact that the belly flap lipids may be depleted during starvation points to this being a labile store (Zhou *et al.* 1996).

Table 15.7 The essential (indispensable) amino acid (EAA or IAA) compositions (expressed as grams AA per kg protein) of the fillet (muscle) of some fish species. (Adapted from Manthey *et al.* 1988; Lie *et al.* 1994.)

	Atlantic salmon *Salmo salar*	Atlantic cod *Gadus morhua*	Wolf-fish *Anarhichas lupus*	Atlantic halibut *Hippoglossus hippoglossus*	Turbot *Scophthalmus maximus*	Wels *Silurus glanis*	Channel catfish *Ictalurus punctatus*	African catfish *Clarias gariepinus*
Arginine	71	55	54	62	57	53	55	55
Histidine	43	22	22	25	19	21	20	22
Isoleucine	60	39	38	43	38	30	29	30
Leucine	92	72	70	80	69	73	70	74
Lysine	92	88	86	105	82	82	81	84
Methionine	33	28	27	31	25	25	27	27
Phenylalanine	54	39	43	37	44	36	35	35
Threonine	49	39	43	43	44	50	43	42
Tryptophan	11	11	10	19	13	14	13	15
Valine	60	39	38	49	38	35	36	33

Among those species which are usually considered to be 'lean' or 'low-fat', farmed fish will often be found to contain higher proportions of muscle lipid than their wild counterparts. This is especially the case when the farmed fish have been fed on lipid-rich feeds. For example, farmed turbot, *Scophthalmus maximus*, were found to have a higher percentage of intramuscular lipid than wild turbot (1.06 versus 0.64%), the difference being largely the result of an increased deposition of neutral lipids in the muscle of the farmed fish (0.52 versus 0.24%) (Sérot *et al.* 1998). There were also differences between the farmed and wild fish in fatty acid compositions of both the phospholipids and the neutral lipids, although differences were most pronounced for the neutral lipids. The lipids of the farmed turbot had higher proportions of 18:2(n-6), 20:1 and 22:1, and lower proportions of 20:5 and 22:6(n-3) than those of the wild turbot. This was clearly a reflection of the composition of the feed provided to the farmed fish: the feed contained 18:2(n-6) derived from cereals and plant oils, and 20:1 and 22:1 from the fish meal and capelin, *Mallotus villosus*, oil components (Sérot *et al.* 1998).

Similarly, when cod were fed on either prawn, *Pandalus borealis*, or herring, *Clupea harengus*, the fatty acid compositions of the prey were reflected in the fatty acid profiles of the muscle lipids, ca. 75% of which were phospholipids (Table 15.8) (dos Santos *et al.* 1993). Thus, the muscle lipids of the cod fed on prawn had higher levels of 18:1 fatty acids than did those of cod fed on herring, whereas the muscle lipids of the fish fed on herring had higher levels of 20:1 and 22:1. However, since the majority of the muscle lipids were phospholipids, the fatty acid profiles were dominated by HUFAs [20:5(n-3) and 22:6(n-3)], irrespective of which type of prey the cod had consumed.

Whilst prey type had some influence upon the fatty acid profiles of the muscle lipids, the effects on liver lipids – most of which were TAGs – were much more pronounced. Cod fed on prawn, which contained relatively high levels of 16:0, 18:1, 20:5 and 22:6 fatty acids, had

Table 15.8 Lipid compositions of two types of prey (Prawn, *Pandalus borealis*, or Herring, *Clupea harengus*) fed to cod, *Gadus morhua*, and the effect of prey type on the lipid compositions of the liver and muscle of cod. (Data from dos Santos *et al.* 1993.) TL = Total lipid; TAG = Triacylglycerol; PL = Phospholipid; TFA = Total fatty acids.

			Composition of cod tissues			
	Prey		Liver		Muscle	
	Prawn	Herring	Prawn	Herring	Prawn	Herring
Lipid (% wet weight)	3.5	12.0	36.4	66.8	1.0	1.0
Lipid classes (%TL)						
TAG (%)	44	88	76	94	3	5
PL (%)	28	1	4	1	77	78
Fatty acids (% TFA)						
16:0	17.1	14.6	13.7	10.5	17.1	15.7
18:1 (n−9)	17.0	9.9	22.6	12.1	9.6	7.1
18:1 (n−7)	9.4	2.9	11.3	5.3	5.6	3.0
20:1 (n−9)	2.0	13.2	2.9	12.4	1.4	4.4
20:5 (n−3)	16.5	5.8	12.4	9.4	22.2	17.1
22:1 (n−11)	1.3	20.0	1.8	8.7	0.8	2.5
22:6 (n−3)	12.5	5.3	9.6	11.9	24.9	29.8

liver lipids dominated by 16:0, 18:1 and 20:5, whereas the liver lipids of cod fed on herring had higher concentrations of 20:1 and 22:1 fatty acids, both of which were present in large quantities in the feed.

Thus, the fatty acid composition of the TAGs most usually reflects the fatty acid composition of the feed (Haard 1992; Cowey 1993; dos Santos *et al.* 1993; Sérot *et al.* 1998), so it is possible to manipulate the fatty acid composition of the muscle TAGs via dietary means. However, TAGs usually contain lower proportions of HUFAs than do phospholipids, and the incorporation of HUFAs into the TAGs will usually be dependent upon them being provided preformed in the feed. Further, the HUFAs, which contain numerous double bonds within the molecule, are susceptible to peroxidative damage so must be protected by antioxidants to prevent the fillet becoming rancid during storage (Hsieh & Kinsella 1989; Haard 1992; Freeman & Hearnsberger 1994).

Tissue antioxidant systems comprise a variety of components, including glutathione and melatonin, that are highly efficient scavengers of hydroxyl and peroxyl radicals, enzymes, such as superoxidase dismutase, catalase and glutathione peroxidase, which catalyse radical and peroxide quenching reactions, and metal-binding proteins which sequester free iron and copper ions that may induce lipid peroxidation (Hsieh & Kinsella 1989; Chan 1993; Jacob 1995). Ascorbic acid (vitamin C) is an important water-soluble antioxidant, and the tocopherols (vitamin E and related compounds) provide antioxidant protection in the lipid phase, e.g. for the HUFAs of cell membranes. The antioxidant properties of ascorbic acid and tocopherols appear to be synergistic, with the ascorbic acid seeming to play a role in the regeneration of tocopherols rather than acting directly to trap hydroperoxide radicals. Carotenoids are thought to provide antioxidant protection to lipid-rich tissues, but this protection may also result from interaction with the tocopherols rather than directly. Thus, there is evidence of a series of interactions between carotenoids, tocopherols and ascorbic acid, and the fish must be provided with a sufficient dietary supply of these antioxidants to mitigate the risk of peroxidation of the tissue HUFAs. However, the requirement for these contributors to natural antioxidant activity will obviously depend upon tissue levels of HUFAs at risk from peroxidation; the greater the concentration of HUFAs the greater the need for antioxidant protection.

In addition to their possible role as tissue antioxidants the carotenoids may contribute to the appearance of fishery products, and thereby have an influence on market acceptability. For example, flesh pigmentation is particularly important in the marketing of salmonid species that normally obtain carotenoids from the crustaceans that form part of their natural diet (Torrissen *et al.* 1989). In wild salmonids the carotenoids present in the flesh are dominated by astaxanthin and its esters: the pigment picture of farmed salmonids may be more complex because a wide range of ingredients, containing a variety of pigments, may be used in formulating feeds (e.g. Hatlen 1997). Further, the carotenoids may be affected by processing, frozen storage and exposure to fluorescent lights at sales outlets, leading to colour changes and fading. Compensation must be made for this by producing farmed salmonids with flesh carotenoid concentrations elevated above the minimum acceptable level of ca. 4 mg/kg. However, as it is virtually impossible to distinguish flesh carotenoid concentrations above 7–8 mg/kg visually, it should not be necessary to provide a finished product with higher pigment levels because the eye will be unable to distinguish it from a product that is less well pigmented.

Although the carbohydrate (predominantly glycogen) content of fish muscle is normally <1%, the amount of glycogen present at the time of slaughter is an important determinant of the quality of the final product. This is because the glycogen content of the muscle influences both the timing of onset, and the duration of, rigor mortis. Rigor mortis, or death-stiffening, refers to the physical changes that occur in the muscle as a result of the complex reactions that occur post mortem (Love 1988; Haard 1992; Macrae *et al*. 1993; Robb *et al*. 2000). Rigor mortis lasts longer when the fish has exerted little muscular activity immediately prior to death, and rigor is also prolonged if the fish is chilled after slaughter.

Hydrolysis of muscle glycogen to glucose and lactic acid occurs when the fish struggles prior to death, and the hydrolysis of glycogen continues after death. Concomitant with the production of lactic acid, adenosine triphosphate (ATP) is produced. As reserves of glycogen are consumed, and lactic acid accumulates, there is a fall in pH. The post-mortem changes in the pH of fish muscle may be influenced by amino acid and dipeptide buffering agents, such as histidine, carnosine and anserine. The levels of these organic buffering agents tend to be highest in fish that have a high proportion of red muscle, and in those that are in 'poor' nutritional condition (Haard 1992).

As time progresses, levels of ATP fall as the ATP is degraded via dephosphorylation and deamination to form inosine monophosphate (IMP). Within a few hours, ATP concentrations decline to a critical level, and the muscles start to contract. The fish stiffens (having entered rigor mortis), and remains in this condition for several hours or days. Eventually the muscles go limp again, and rigor is said to be resolved. The softening of the muscle that occurs is not a reversal of the process of rigor mortis, but arises due to autolysis. The post-mortem pH in the muscle has a major influence on texture, and fish with a low post-mortem pH usually have a firm-textured flesh. However, a rapid pH decline results in an early onset of rigor, a soft-textured flesh with poor water-holding properties, and a fillet that is prone to 'gaping' (Love 1988).

High levels of activity immediately prior to slaughter result in a rapid drop in muscle pH, reduce the time to the onset of rigor, and may also lead to an increase in the intensity of rigor. This may affect the strength of the muscle, and the myosepta of the muscles may break under strain, resulting in 'gaping' of the fillet (Love 1988; Robb *et al*. 2000). 'Gaping' occurs when the connective tissue myosepta break, and the myotomes (muscle blocks) become separated. This creates problems because it makes the processing of the fish more difficult. In salmonids, the flesh of fish that have struggled prior to slaughter may be lighter and less red than that of fish displaying lower levels of activity (Robb *et al*. 2000). These colour differences may be a reflection of changes in muscle proteins. In animals that have struggled prior to death the rapid drop in pH post mortem may lead to protein denaturation and increased insolubility, and to a loss of moisture from the muscle. These changes may affect the reflection of light from the muscle surface, hence changing the perception of colour.

The flavour of the fillet develops as small peptides and free amino acids are liberated due to hydrolysis of the muscle proteins, there is oxidation of fatty acids, and there is accumulation of trimethylamine. Trimethylamine is thought to arise due to a bacteria-initiated reduction of trimethylamine oxide coupled to the oxidation of lactic acid. The IMP formed from the degradation of ATP has flavour-enhancing properties, but the IMP is slowly degraded to inosine, which, in turn, is cleaved to ribose and hypoxanthine. Measurement of the latter is sometimes used as a 'spoilage indicator'. As spoilage continues there is an accumulation of the acidic products of bacterial degradation, including those derived from the sulphur-containing amino acids, and there are increases in concentrations of biogenic amines and ammonia.

15.6 Concluding comments

Fish is an excellent source of protein, and fish muscle (fillet) has a well-balanced amino acid composition, making fish an important component of the human diet in many countries. The proportions of protein (Tables 15.1 and 15.2) and essential amino acids (Table 15.7) in the fillet are similar across species, but there are considerable inter-specific differences in the percentages of fillet moisture and lipid (Tables 15.1 and 15.2). These differences depend upon the relative amounts of red and white muscle in the fillet, and the extent to which a particular species deposits storage lipid in the muscle. The extent to which there is lipid storage in the muscle will also influence the balance between TAGs and the structural phospholipids. Since the fatty acid composition of the TAGs usually reflects that of the feed, it is possible to manipulate the fatty acid composition of the storage lipids by dietary means.

The beneficial effects of the (n-3) HUFAs on human health have been extensively reported (Connor *et al.* 1992; Drevon 1992; Lands 1992): they may protect against cardiovascular disease and some chronic infections, they are required for normal fetal development, and they may also be involved in protection against diabetes and some forms of cancer. The natural abundance of the (n-3) HUFAs in the lipids of fish has led to an increased consumer awareness of fish as a 'health food'. However, storage TAGs usually contain lower proportions of HUFAs than do the structural phospholipids, and the extent to which HUFAs are incorporated into TAGs will be dependent upon the dietary supply. Thus, the extent to which the flesh of farmed fish contains high levels of HUFAs will be governed by the relative proportions of red and white locomotor muscle in the fillet, by muscle lipid content (i.e. whether the fish is one of the 'lean' or 'fatty' species), and by the fatty acid composition of the feed consumed. Nevertheless, the fillet of farmed fish is envisaged to have considerable potential as a vehicle for delivering large quantities of HUFAs to human consumers.

The HUFAs, which contain several double bonds within the molecule, are susceptible to oxidation and so must be protected by antioxidants to prevent the fillet becoming rancid during storage. Consequently, attention needs to be paid to the potential risk of peroxidation of the HUFAs in the fillets of farmed fish, and precautions must be taken to prevent the development of flesh rancidity during transport and storage.

15.7 References

Adams, C.E., Huntingford, F.A. & Jobling, M. (1995) A non-destructive morphometric technique for estimation of body and mesenteric lipid in Arctic charr: a case study of its application. *Journal of Fish Biology*, **47**, 82–90.

Aksnes, A., Hjertnes, T. & Opstvedt, J. (1996) Effect of dietary protein level on growth and carcass composition in Atlantic halibut (*Hippoglossus hippoglossus* L.). *Aquaculture*, **145**, 225–233.

Aursand, M., Bleivik, B., Rainuzzo, J.R., Jørgensen, L. & Mohr, V. (1994) Lipid distribution and composition of commercially farmed Atlantic salmon (*Salmo salar*). *Journal of the Science of Food and Agriculture*, **64**, 239–248.

Bell, J.G. (1998) Current aspects of lipid nutrition in fish farming. In: *Biology of Farmed Fish* (eds K.D. Black & A.D. Pickering), pp.114–145. Sheffield Academic Press, Sheffield.

Berge, G.M. & Storebakken, T. (1991) Effect of dietary fat level on weight gain, digestibility and fillet composition of Atlantic halibut. *Aquaculture*, **99**, 331–338.

Black, D. & Love, R.M. (1986) The sequential mobilisation and restoration of energy reserves in tissues of Atlantic cod during starvation and refeeding. *Journal of Comparative Physiology*, **156B**, 469–479.

Boëtius, I. & Boëtius, J. (1985) Lipid and protein content in *Anguilla anguilla* during growth and starvation. *Dana*, **4**, 1–17.

Brett, J.R., Shelbourn, J.E. & Shoop, C.T. (1969) Growth rate and body composition of fingerling sockeye salmon, *Oncorhynchus nerka*, in relation to temperature and ration size. *Journal of the Fisheries Research Board of Canada*, **26**, 2363–2394.

Broekhuizen, N., Gurney, W.S.C., Jones, A. & Bryant, A.D. (1994) Modelling compensatory growth. *Functional Ecology*, **8**, 770–782.

Chan, A.C. (1993) Partners in defense, vitamin E and vitamin C. *Canadian Journal of Physiology and Pharmacology*, **71**, 725–731.

Company, R., Calduch-Giner, J.A., Kaushik, S. & Pérez-Sánchez, J. (1999) Growth performance and adiposity in gilthead sea bream (*Sparus aurata*): risks and benefits of high energy diets. *Aquaculture*, **171**, 279–292.

Connor, W.E., Neuringer, M. & Reisbick, S. (1992) Essential fatty acids: the importance of n-3 fatty acids in the retina and brain. *Nutrition Reviews*, **50**, 21–29.

Cowey, C.B. (1993) Some effects of nutrition on flesh quality of cultured fish. In: *Fish Nutrition in Practice* (eds S.J. Kaushik & P. Luquet), pp. 227–236. INRA, Paris.

dos Santos, J., Burkow, I.C. & Jobling, M. (1993) Patterns of growth and lipid deposition in cod (*Gadus morhua* L.) fed natural prey and fish-based diets. *Aquaculture*, **110**, 173–189.

Drevon, C.A. (1992) Marine oils and their effects. *Nutrition Reviews*, **50**, 38–45.

Einen, O., Waagan, B. & Thomassen, M.S. (1998) Starvation prior to slaughter in Atlantic salmon (*Salmo salar*). I. Effects on weight loss, body shape, slaughter- and fillet-yield, proximate and fatty acid composition. *Aquaculture*, **166**, 85–104.

Elliott, J.M. (1976) Body composition of brown trout (*Salmo trutta* L.) in relation to temperature and ration size. *Journal of Animal Ecology*, **45**, 273–289.

Freeman, D.W. & Hearnsberger, J.O. (1994) Rancidity in selected sites of frozen catfish fillets. *Journal of Food Science*, **59**, 60–63, 84.

Haard, N.F. (1992) Control of chemical composition and food quality attributes of cultured fish. *Food Research International*, **25**, 289–307.

Hardy, R.W., Scott, T.M. & Harrell, L.W. (1987) Replacement of herring oil with menhaden oil, soybean oil, or tallow in the diets of Atlantic salmon raised in marine net pens. *Aquaculture*, **65**, 267–277.

Hatlen, B. (1997) *Muscle pigmentation of Arctic charr,* Salvelinus alpinus *(L.)*. Doctoral thesis, University of Tromsø.

Haug, T., Ringø, E. & Pettersen, G.W. (1988) Total lipid and fatty acid composition of polar and neutral lipids in different tissues of Atlantic halibut, *Hippoglossus hippoglossus* (L.). *Sarsia*, **73**, 163–168.

Helland, S.J. & Grisdale-Helland, B. (1998) Growth, feed utilization and body composition of juvenile Atlantic halibut (*Hippoglossus hippoglossus*) fed diets differing in the ratio between macronutrients. *Aquaculture*, **166**, 49–56.

Hemre, G.I., Bjørnsson, B. & Lie, Ø. (1992) Haematological values and chemical composition of halibut (*Hippoglossus hippoglossus* L.) fed six different diets. *Fiskeri Direktoratets Skrifter, Serie Ernæring*, **5**, 89–98.

Henderson, R.J. & Tocher, D.R. (1987) The lipid composition and biochemistry of freshwater fish. *Progress in Lipid Research*, **26**, 281–347.

Herbinger, C.M. & Friars, G.W. (1991) Correlation between condition factor and total lipid content in Atlantic salmon, *Salmo salar* L., parr. *Aquaculture and Fisheries Management*, **22**, 527–529.

Higgs, D.A., Macdonald, J.S., Levings, C.D. & Dosanjh, B.S. (1995) Nutrition and feeding habits in relation to life history stage. In: *Physiological Ecology of Pacific Salmon* (eds C. Groot, L. Margolis & W.C. Clarke), pp.160–315. UBC Press, Vancouver.

Holopainen, I.J., Tonn, W.M. & Paszkowski, C.A. (1997) Tales of two fish: the dichotomous biology of crucian carp (*Carassius carassius* (L.)) in northern Europe. *Annales Zoologia Fennici*, **34**, 1–22.

Hsieh, R.J. & Kinsella, J.E. (1989) Oxidation of polyunsaturated fatty acids: mechanisms, products and inhibition with emphasis on fish. *Advances in Food and Nutrition Research*, **33**, 233–341.

Jacob, R.A. (1995) The integrated antioxidant system. *Nutrition Research*, **15**, 755–766.

Jobling, M., Knudsen, R., Pedersen, P.S. & dos Santos, J. (1991) Effects of dietary composition and energy content on the nutritional energetics of cod, *Gadus morhua*. *Aquaculture*, **92**, 243–257.

Jobling, M., Meløy, O.H., dos Santos, J. & Christiansen, B. (1994) The compensatory growth response of the Atlantic cod: effects of nutritional history. *Aquaculture International*, **2**, 75–90.

Jobling, M., Koskela, J. & Savolainen, R. (1998) Influence of dietary fat level and increased adiposity on growth and fat deposition in rainbow trout, *Oncorhynchus mykiss* (Walbaum). *Aquaculture Research*, **29**, 601–607.

Jobling, M., Koskela, J. & Winberg, S. (1999) Feeding and growth of whitefish fed restricted and abundant rations: influences on growth heterogeneity and brain serotonergic activity. *Journal of Fish Biology*, **54**, 437–449.

Jonsson, N. & Jonsson, B. (1998) Body composition and energy allocation in life-history stages of brown trout. *Journal of Fish Biology*, **53**, 1306–1316.

Jørgensen, E.H., Johansen, S.J.S. & Jobling, M. (1997) Seasonal patterns of growth, lipid deposition and lipid depletion in anadromous Arctic charr. *Journal of Fish Biology*, **51**, 312–326.

Kadri, S., Metcalfe, N.B., Huntingford, F.A. & Thorpe, J.E. (1995) What controls the onset of anorexia in maturing adult female Atlantic salmon? *Functional Ecology*, **9**, 790–797.

Koskela, J., Jobling, M. & Savolainen, R. (1998) Influence of dietary fat level on feed intake, growth and fat deposition in the whitefish, *Coregonus lavaretus*. *Aquaculture International*, **6**, 95–102.

Lands, W.E.M. (1992) Biochemistry and physiology of n-3 fatty acids. *FASEB Journal*, **6**, 2530–2536.

Lie, Ø., Lied, E. & Lambertsen, G. (1986) Liver retention of fat and fatty acids in cod (*Gadus morhua*) fed different oils. *Aquaculture*, **59**, 187–196.

Lie, Ø., Lied, E. & Lambertsen, G. (1988) Feed optimization in Atlantic cod (*Gadus morhua*): fat versus protein content in the feed. *Aquaculture*, **69**, 333–341.

Lie, Ø., Lied, E., Maage, A., Njaa, L.R. & Sandnes, K. (1994) Nutrient content in fish and shellfish. *Fiskeri Direktoratets Skrifter, Serie Ernæring*, **6**, 83–105.

Love, R.M. (1988) *The Food Fishes: Their intrinsic variation and practical implications.* Farrand Press, London.

Macrae, R., Robinson, R.K. & Sadler, M.J. (eds) (1993) *Encyclopaedia of Food Science, Food Technology and Nutrition.* Academic Press, London.

Manthey, M., Hilge, V. & Rehbein, H. (1988) Sensory and chemical evaluation of three catfish species (*Silurus glanis, Ictalurus punctatus, Clarias gariepinus*) from intensive culture. *Arkiv für Fischerei-wissenschaft*, **38**, 215–227.

Nortvedt, R. & Tuene, S. (1998) Body composition and sensory assessment of three weight groups of Atlantic halibut (*Hippoglossus hippoglossus*) fed three pellet sizes and three dietary fat levels. *Aquaculture*, **161**, 295–313.

Otwell, W.S. & Rickards, W.L. (1981) Cultured and wild American eels, *Anguilla rostrata*: fat content and fatty acid composition. *Aquaculture*, **26**, 67–76.

Phleger, C.F. (1998) Buoyancy in marine fishes: direct and indirect role of lipids. *American Zoologist*, **38**, 321–330.

Phleger, C.F., Laub, R.J. & Benson, A.A. (1989) Skeletal lipid depletion in spawning salmon. *Lipids*, **24**, 286–289.

Phleger, C.F., Laub, R.J. & Wambeke, S.R. (1995) Selective skeletal fatty acid depletion in spawning Pacific pink salmon, *Oncorhynchus gorbuscha. Comparative Biochemistry and Physiology*, **111B**, 435–439.

Polvi, S.M. & Ackman, R.G. (1992) Atlantic salmon (*Salmo salar*) muscle lipids and their response to alternative dietary fatty acid sources. *Journal of Agricultural and Food Chemistry*, **40**, 1001–1007.

Porter, P.J., Kramer, D.E. & Kennish, J.M. (1992) Lipid composition of light and dark flesh from sock-eye salmon. *International Journal of Food Science and Technology*, **27**, 365–369.

Rasmussen, R.S. & Ostenfeld, T.H. (2000) Effect of growth rate on quality traits and feed utilisation of rainbow trout (*Oncorhynchus mykiss*) and brook trout (*Salvelinus fontinalis*). *Aquaculture*, **184**, 327–337.

Reinitz, G. (1983) Relative effect of age, diet and feeding rate on the body composition of young rainbow trout (*Salmo gairdneri*). *Aquaculture*, **35**, 19–27.

Robb, D.H.F., Kestin, S.C. & Warriss, P.D. (2000) Muscle activity at slaughter: I. Changes in flesh colour and gaping in rainbow trout. *Aquaculture*, **182**, 261–269.

Sæther, B-S. & Jobling, M. (1999) The effects of ration level on feed intake and growth, and compensatory growth after restricted feeding, in turbot *Scophthalmus maximus* L. *Aquaculture Research*, **30**, 647–653.

Sérot, T., Gandemer, G. & Demaimay, M. (1998) Lipid and fatty acid compositions of muscle from farmed and wild turbot. *Aquaculture International*, **6**, 331–343.

Shearer, K.D. (1994) Factors affecting the proximate composition of cultured fishes with emphasis on salmonids. *Aquaculture*, **119**, 63–88.

Simpson, A.L., Metcalfe, N.B. & Thorpe, J.E. (1992) A simple non-destructive biometric method for estimating fat levels in Atlantic salmon, *Salmo salar. Aquaculture and Fisheries Management*, **23**, 23–29.

Sogard, S.M. & Olla, B.L. (2000) Endurance of simulated winter conditions by age-0 walleye pollock: effects of body size, water temperature and energy stores. *Journal of Fish Biology*, **56**, 1–21.

Stirling, H.P. (1976) Effects of experimental feeding and starvation on the proximate composition of the European bass *Dicentrarchus labrax. Marine Biology*, **34**, 85–91.

Sutton, S.G., Bult, T.P. & Haedrich, R.L. (2000) Relationships among fat weight, body weight, water weight, and condition factors in wild Atlantic salmon parr. *Transactions of the American Fisheries Society*, **129**, 527–538.

Torrissen, O.J., Hardy, R.W. & Shearer, K.D. (1989) Pigmentation of salmonids – Carotenoid deposition and metabolism. *CRC Critical Reviews in Aquatic Sciences*, **1**, 209–225.

Tveiten, H., Johnsen, H.K. & Jobling, M. (1996) Influence of maturity status on the annual cycles of feeding and growth in Arctic charr reared at constant temperature. *Journal of Fish Biology*, **48**, 910–924.

Weatherley, A.H. & Gill, H.S. (1987) *The Biology of Fish Growth.* Academic Press, London.

Wold, J.P., Jakobsen, T. & Krane, L. (1996) Atlantic salmon average fat content estimated by Near-Infrared Transmittance Spectroscopy. *Journal of Food Science*, **61**, 74–77.

Zhou, S., Ackman, R.G. & Morrison, C. (1996) Adipocytes and lipid distribution in the muscle tissue of Atlantic salmon (*Salmo salar*). *Canadian Journal of Fisheries and Aquatic Sciences*, **53**, 326–332.

Glossary of Terms

Compiled by Malcolm Jobling

Abiotic Non-living substances or factors; the abiotic environment is that part of an organism's environment consisting of non-biological factors such as topography, climate and other physical factors.

Absorption Passage of nutrients across the gut and into the blood; uptake of fluid and solutes by cells and tissues.

Absorption efficiency *See* Digestibility.

Acclimation Process by which an organism adjusts to a new circumstance, e.g. to exposure to water of different temperature or salinity.

Acclimatisation The sum of the set of processes by which an organism adapts to a new set of environmental conditions.

Accuracy A measure of how close an estimate is to the true value. *cf.* Precision.

Actin Protein that forms the thin filaments of muscle fibres, and combines with myosin to give the contractile actomyosin complex.

Actomyosin Complex of myosin and actin in the muscle filaments; integral part of the contractile threads of the muscles.

Ad libitum Feeding at will until no more food is accepted.

Adaptation The process of adjustment of an organism to a new set of environmental conditions. *Alt.* A morphological, physiological, developmental or behavioural character that enhances survival and reproductive success of an organism.

Adenohypophysis (anterior pituitary) *See* Pituitary gland.

ADF (acid detergent fibre) The ADF of a feed reflects the carbohydrates not solubilised by acid detergent; considered to be those plant constituents that are not readily available, and are poorly utilised by animals.

Adipocyte (or adipose cell)   An animal cell specialised for lipid storage, and which contains large fat globules in the cytoplasm.

Adipose tissue   Connective tissue containing large numbers of adipocytes; tissue used for deposition of fat stores in animals.

Adrenaline (epinephrine)   *See* Catecholamines.

Aflatoxins   Mycotoxins, frequently carcinogenic, that are produced by certain species of moulds, e.g. *Aspergillus* spp., that grow on feedstuffs that are stored under inadequate conditions. The feedstuffs that are particularly susceptible to attack by moulds are cottonseed, peanut, maize (corn) and cereals.

Agar   A gelatinous polysaccharide extracted from red algae, that forms thick gels at low concentrations.

Aggression   Patterns of behaviour which either intimidate or damage another organism, but excluding the predatory behaviours used for obtaining food; refers to a hostile act or threat made to protect territory, the family group or offspring, or to establish dominance.

Agonist   A substance responsible for triggering a response in a cell, such as a hormone, neurotransmitter, etc.; a drug that mimics the action of a hormone, neurotransmitter, etc.

Agonistic behaviour   The complex of aggression, threat, appeasement and avoidance behaviours that may occur during encounters between members of the same species; social interaction between members of a species, involving aggression or threat and conciliation or retreat.

Alevin   Stage in the life of a young fish encompassing the time from hatching until the end of dependence upon endogenous nutrition (yolk); usually used with reference to young, newly hatched salmonids.

Alginate   A carbohydrate polymer derived from alginic acid, a gel-like polysaccharide found in the cell walls of brown algae. Used as a binding and texturising agent in feeds.

Alimentary canal   The tube running from the mouth to the anus in which food is digested, and from which nutrients are absorbed; the gut or gastrointestinal tract; in vertebrates, typically comprising the oesophagus, stomach, intestines and rectum.

Alkalinity   The capacity of water to neutralise acids; usually expressed as ppm (parts per million) or mg l^{-1} calcium carbonate equivalents. In natural water bodies, alkalinity is usually due to dissolved-mineral carbonates, bicarbonates and hydroxides, and to a lesser extent borates, silicates and phosphates. *cf.* Water hardness.

Alkaloids   A large group (*ca.* 6000 are known) of toxic, basic nitrogenous compounds found in plants. The term alkaloid has not been precisely defined, but the group encompasses heterocyclic nitrogen bases derived from the amino acids ornithine, lysine, phenylalanine, tyrosine and tryptophan.

Allergen   A substance to which an individual is hypersensitive; causes an immune response usually characterised by local inflammatory reactions, but sometimes induces severe shock symptoms.

Allometry   The study of size and its consequences; examination of relative growth and the changes in proportions with increase in size.

Amino acid   A member of the class of compounds having the general formula $RCH(NH_2)COOH$, where R is a distinctive side chain. Around 20 different amino acids are present in proteins; of these 10 are considered indispensable (or essential) dietary factors for animals – Arg, His, Ile, Leu, Lys, Met, Phe, Thr, Trp and Val. Amino acids are also precursors for many important biological molecules, e.g. histamine, thyroxine, catecholamines, and serotonin.

Ammonia toxicity   Ammonia is toxic to aquatic organisms, but toxicity depends both upon absolute quantity and the extent of ionic dissociation (i.e. relative proportions of NH_3 and the ammonium ion, NH_4^+). Ionic dissociation is influenced by factors such as pH, temperature and salinity.

Ammonotelic   Excreting nitrogen mainly as 'ammonia' (i.e. NH_3 or the ammonium ion, NH_4^+), as in most teleost fish and aquatic invertebrates. Total ammonia nitrogen (TAN) is the sum of the ammonia-nitrogen found in the unionised (NH_3) and ionised (NH_4^+) form.

Amphihaline   Term used to describe a species showing broad tolerance to salinity, and being capable of surviving in both fresh water and the sea. *cf.* Euryhaline.

Anabolism   *See* Metabolism.

Anadromous   Fish that spawn in fresh water, but live much of their lives in the sea. *See* Diadromous.

Anorexia   Reduced, or lack of, appetite; administration of anorectic agents causes a loss of appetite and reduces feeding.

Anoxic   Used to describe a habitat devoid of molecular oxygen; *anoxia* refers to a lack of oxygen. *cf.* Hypoxic.

Antagonist   A substance that counteracts the effects of another chemical substance, e.g. a drug that reduces or prevents the action of a hormone, neurotransmitter, etc.

Antibiotics   Originally defined as chemicals produced by micro-organisms that have the capacity of inhibiting the growth of, or killing, other micro-organisms, but now also encompasses a range of synthetic antimicrobial compounds; examples include penicillin, oxytetracycline, streptomycin and erythromycin; used to control specific pathogenic organisms, but several antibiotics are known to have growth-promoting properties when given to animals in low doses.

Antifungal agents   Chemicals, including certain antibiotics, which either kill or inhibit the growth of fungi; agents with fungicidal or fungistatic properties used to treat fungal diseases, or as preservatives in feeds.

Antinutritional factor (ANF)   A substance that reduces the value of a feed or feedstuff, primarily via interference with the digestion, absorption or metabolism of nutrients.

Antioxidant   A substance that inhibits the oxidation of other compounds. Important natural antioxidants include vitamins C and E, carotenoids, glutathione and melatonin.

Appetite   Physical craving or desire for food. The term *hunger* may be used to denote both a craving for nourishment and the corresponding physiological need.

Appetitive behaviour   Purposeful feeding behaviours resulting in the identification and location of specific food items.

Ascorbic acid (vitamin C)   A water-soluble vitamin that acts as a co-factor in oxidation-reduction reactions, and is involved in the synthesis of cartilage and bone.

Ash   The mineral residues remaining after combustion of feeds, feedstuffs, or plant and animal tissues at 450–500°C.

ASR (aquatic surface respiration)   Ventilation of the gills with water from the thin (a few mm thick), oxygen-rich layer found close to the water surface; a behavioural adaptation adopted by several fish species when exposed to hypoxic conditions.

Astaxanthin   A carotenoid pigment which is found naturally in many crustaceans; astaxanthin and its esters impart the distinctive pink-red colour to the flesh of salmonids (the fish obtain the carotenoid pigments by consuming crustacean prey, such as copepods, krill, shrimps and prawns).

ATP (adenosine triphosphate)   Nucleotide made up of the purine base adenine linked to the pentose sugar ribose, which carries 3 phosphate groups linked in series; an energy-rich molecule important in the metabolic reactions occurring in cells.

Autolysis   Self-digestion, or dissolution, of the cells and tissues of an organism by endogenous hydrolytic enzymes, as occurs after death, and in some pathological conditions.

Automatic feeder   An automated, usually electrically operated, device used for feed distribution; often programmed to dispense feed at preselected times in predetermined amounts.

Autoradiography   Technique by which macromolecules, cell components or tissues are radioactively labelled (e.g. with a radioisotope or radiolabelled hormone) and their image recorded on a photographic plate producing an autoradiograph.

Balance trial   A technique used for determination of the quantitative nutrient requirements of animals, in which total intake and loss of a nutrient is measured over time.

Balanced diet   *See* Complete diet.

Barbel   A fleshy outgrowth on the lower jaw of some species of fish, having tactile (mechanoreceptor) and/or gustatory (taste) functions.

Belly flap   Thin tissue layer surrounding the visceral cavity of fish; portion of the muscle tissue which may be removed when preparing the fillets of some fish species, e.g. salmonids, for sale; tissue may be used as a lipid storage depot in salmonids, and some other fish species.

Benthic   Descriptive of an organism that lives on the bottom, or buried in bottom sediments.

Betaine   A tertiary amine (nonprotein nitrogenous compound) formed by oxidation of choline; in the body betaine may be transformed into trimethylamine.

Binder   The component(s) of a compound mixture, such as a formulated feed, that serve(s) to hold together the other constituents.

Bioavailability (biological availability)   Describes the proportion of a nutrient in a feed or feedstuff that may be utilised by an animal. Bioavailability of a nutrient can be subdivided into three constituent phases: availability in the intestine for absorption, absorption and/or retention in the body, and utilisation. Bioavailability is studied to evaluate the 'quality' of feeds and feedstuffs, and to provide data for establishing nutritional requirements.

Bioenergetics   The study of the flow and the transformation of energy that occur in living organisms.

Biogenic amines   Amines that are formed following the decarboxylation of amino acids, e.g. arginine is decarboxylated to form putrescine, lysine gives cadavarine, and decarboxylation of histidine yields histamine. Accumulation of biogenic amines, following the degradation of amino acids by microbial action, is used to assess the deterioration of fish or feed that occurs during storage.

Biological clock   Name given to the mechanisms responsible for time-keeping in living organisms. Many of the biological rhythms exhibited by animals are maintained by clock mechanisms that are endogenous in the sense that the rhythm persists when the animal is isolated from environmental time cues. However, it is often the case that an external *Zeitgeber*, or time-setter, is responsible for the maintenance of synchrony between the clock rhythm and the rhythm of environmental events, so that when animals are isolated from the exogenous *Zeitgeber* their biological rhythms drift out of step with those in the environment.

Biological rhythm   A self-sustained cyclic change in a behavioural or physiological process that repeats at regular intervals.

Biological value (BV) of a protein   The proportion of the absorbed nitrogen that is retained in the body; a measure of the biological usefulness of that part of the dietary protein that has been digested and absorbed.

Biomass   The amount of living matter in a specified habitat or part thereof; mass of living matter per unit of water surface or volume.

Biotic factors   Biological factors such as availability of food, numbers of conspecifics and competitors, and predators, etc., that affect the abundance, distribution and life histories of individuals and species.

Biuret reaction   Simple test for the presence of proteins and peptides, which is based upon a reaction with the peptide bond. Solutions of copper sulphate and sodium hydroxide are added to the test sample, and a purple copper complex is formed if a compound containing a peptide bond is present.

BOD (biochemical oxygen demand, sometimes termed biological oxygen demand)
The amount of oxygen required by micro-organisms to stabilise organic matter under aerobic conditions. A measure of the amount of organic material present in water; measured as the amount of oxygen taken up from a water sample containing a known amount of dissolved oxygen kept at 20°C for 5 days. A low BOD indicates little organic material is present, whereas a high BOD indicates an increased activity of heterotrophic micro-organisms due to the presence of large amounts of organic matter.

Bomb calorimeter   The equipment used to measure the heat of combustion resulting from the complete oxidation of a feed, feedstuff, or dried tissue sample; *See* Gross energy.

Brine shrimp   The common name for *Artemia salina*. *Artemia* is an anostracan, brachiopod crustacean; probably the single most important live food source used in aquaculture for the feeding of larval fish and crustaceans.

Broodstock   Sexually mature individuals of both sexes held for the purposes of reproduction, or younger (immature) individuals destined to be used for breeding.

Brown algae (Phaeophyta) A class of over 1500 species of algae (mostly marine seaweeds) which are greenish yellow to deep brown in colour; the colour is imparted by the predominance of orangish/brown pigments (carotene, fucoxanthin) over the chlorophylls. Brown algae, which includes the kelps, are most abundant in the cooler waters of the world.

Browning *See* Maillard reaction.

Buccal cavity The mouth cavity.

Canola A variety of rape which has been selectively bred to contain low levels of several antinutritional factors.

Carbohydrates Compounds containing carbon, oxygen and hydrogen, and having the general formula $C_X(H_2O)_Y$ (although some compounds classifed as carbohydrates do not comply to this general definition). The group includes simple sugars (monosaccharides) and their derivatives, oligosaccharides and polysaccharides, such as starch and cellulose.

Carboxymethylcellulose (CMC) A derivative of cellulose formed when cellulose is treated with chloracetic acid under alkaline conditions; used as a thickening and binding agent.

Carnivore Generally applied to animals that feed upon other animals, although there are also carnivorous (flesh-eating) plants that trap and digest insects and other small animals.

Carotenoids A group of *ca.* 600 widely-distributed orange, yellow, red and brown lipid-soluble pigments that can be synthesised by plants. Some carotenoids are effective antioxidants, and carotene is a provitamin, or precursor, for vitamin A.

Carrageenan An algal polysaccharide from red algae; used as a thickening and binding agent.

Catabolism *See* Metabolism.

Catadromous Fish that spawn in salt water but live much of their lives in fresh water. *See* Diadromous.

Catalysis The acceleration of a reaction due to the presence of a substance (catalyst) which can be recovered unchanged after the reaction.

Catch-up growth (compensatory growth) The rapid growth observed when undernourished, malnourished or feed-deprived animals are returned to adequate feeding conditions; usually accompanied by hyperphagia and efficient feed utilisation.

Catecholamines Catecholamines, and other monoamines, function widely as neurotransmitters. The amino acid tyrosine can be converted to the catecholamines noradrenaline (*alt.* norepinephrine) and adrenaline (*alt.* epinephrine) via dopamine. Neurones that use catecholamines as transmitters are called adrenergic, and the receptors for the catechol-amines may be of two types, α and β.

Cellulose   A linear polysaccharide made up of glucose residues joined by β-1,4 linkages; the most abundant organic compound in the biosphere, comprising the bulk of the cell walls in plants and algae.

Chitin   A long chain polymer of N-acetylglucosamine units; the main polysaccharide in fungal cell walls, and also found in the exoskeletons of insects and crustaceans.

Cholecystokinin (CCK)   Gastrointestinal peptide released from cells in the duodenal mucosa in response to the presence of components of gastric chyme, but also present in neurones of the peripheral and central nervous system; CCK inhibits gastric emptying and the secretion of gastric acid, stimulates gallbladder contraction, stimulates pancreatic enzyme secretion, and is implicated in the central regulation of food intake via actions as a 'satiety hormone'.

Cholesterol   A steroid alcohol ($C_{27}H_{45}OH$) that is the precursor for the steroid hormones and bile acids.

Chromatography   A physical process by which separation of components of a mixture is achieved due to differential partitioning between a stationary phase and a fluid phase; may be divided into gas chromatography and liquid chromatography on the basis of the fluid phase involved.

Chronobiology   The study of timekeeping systems in living organisms, including the study of biological rhythms.

Circadian pacemaker   An internal timekeeping mechanism capable of generating or co-ordinating circadian rhythms. *See* circadian rhythm, pineal organ.

Circadian rhythm   A biological rhythm having a periodicity of 24 ± 4 hours, or a frequency of *ca.* 1 cycle per 24 hours. *cf.* Diel, Diurnal rhythm.

Circannual rhythm   A rhythm or cycle having a periodicity of approximately one year.

COD (chemical oxygen demand)   A measure of the amount of organic matter in a water sample that can be oxidised by a strong chemical oxidising agent, such as potassium di-chromate. Used to assess the amount of organic matter in industrial and domestic wastes. COD will be higher than BOD, because chemical oxidation will also lead to the oxidation of biologically resistant organic matter present in the water sample.

Coefficient of variation (CV)   An expression of relative variability that relates sample vari-ability to the mean of the sample: $CV = (s.d./mean) \times 100$. Used to compare the variations of populations independent of the magnitude of their means.

Cofactor   A non-protein substance required by a protein for biological activity, especially in enzyme-catalysed reactions, where coenzymes are cofactors.

Collagen   Insoluble fibrous proteins of high tensile strength; abundant proteins found in skin, cartilage, the walls of blood vessels, etc.; are rich in the amino acid glycine, and contain hydroxyproline and hydroxylysine, but lack cysteine and tryptophan.

Compensatory growth   *See* Catch-up growth.

Competition   Occurs when two or more individuals are using resources that are in short supply. Food, space and mates are the commonest resources for which animals compete. Competition between members of the same species is termed *intraspecific*, and may take the form of direct interference and aggression between individuals. *Interspecific* competition is that which occurs between individuals of different species; it commonly takes the form of exploitation of a resource by members of one species, thus denying the use of the resource to members of other species by reducing availability.

Complete (or balanced) diet   One that fulfils all the known nutritional requirements of an animal.

Compound feed   A feed composed of several ingredients.

Compressed pellets   Traditional type of animal feed formed by forcing steam-conditioned ingredients through a die under pressure. There is incomplete gelatinisation of starch, and pellets are less durable than expanded and extruded pellets.

Condition factor (K)   An index describing the relationship between the weight (W) and length (L) of a fish: $K = (W/L^b) \times 100$, where the exponent b is usually assumed to be 3. Low K values indicate fish that have a low body weight for their length, and may indicate that fish are in 'poor condition', i.e. have been deprived of food or underfed for a prolonged period.

Consummatory behaviour   Feeding behaviours that include taking a potential food item into the oropharynx, determining its palatability, and then either ingesting or rejecting it.

Corticotropin-releasing factor/hormone (CRF or CRH)   A 41-amino acid peptide that is synthesised in the hypothalamus; has neuromodulatory actions and acts on the anterior pituitary leading to release of adrenocorticotropic hormone (ACTH). Implicated in the regulation of energy balance via suppression of food intake.

Cortisol (hydrocortisone)   A glucocorticosteroid hormone produced by the interrenal tissue of fish, and the cortex of the adrenal gland in higher vertebrates; secretion is stimulated in response to many stressful stimuli; influences carbohydrate metabolism and osmoregulatory functions, and is an immunosuppressant.

Crepuscular   Pertaining to the twilight hours around dusk and dawn; *cf.* Diurnal, Nocturnal.

Crowding   The situation in which the movements, or other activities, of individuals in a group are restricted by the physical presence of others.

Crude fibre   The insoluble carbohydrates remaining after boiling a feed or feedstuff in acid and alkali; this fraction represents the carbohydrates that are not readily available to, and are poorly utilised by, animals.

Crude protein   The content of nitrogen in a feed, feedstuff, plant or animal tissue multiplied by a factor (usually 6.25, since most proteins contain about 16% N) to provide an estimate of the protein content; this only gives a crude estimate of protein content because variable amounts of nonprotein nitrogen will be present in analysed samples.

Daily ration   Amount of food consumed in a 24-h period.

Deamination   The removal of an amino group $(-NH_2)$ from a molecule, as occurs during the metabolism of amino acids.

Decarboxylation   The removal of a carboxyl group $(-COOH)$ from a molecule.

Deficient diet   *See* Incomplete diet.

Detritus   Nonliving organic material that is in a partial state of decomposition. Can be derived from either plant or animal remains. Usually in small particles suspended in the water or mixed in bottom sediments.

Dextrin   Small, soluble polysaccharides formed on partial hydrolysis of starch.

Diadromous   A term used to describe fish which migrate between fresh water and the sea, usually for the purposes of feeding and spawning. *Anadromous* refers to those species which move from the sea to fresh water to spawn (e.g. most salmonids, sturgeons and some members of the herring family), whereas *catadromous* species (e.g. some eels) spend a large portion of their life in fresh water and migrate to the sea to spawn.

Diel   Pertaining to 24 hours; occurring at 24-h intervals.

Diencephalon   Part of the forebrain underlying the cerebral hemispheres, and containing the inferior lobes and ventro-posterior regions of the hypothalamus.

Diet   A selection or mixture of feeds, or feedstuffs, provided on a continuous or prescribed schedule; a balanced, or complete, diet supplies all nutrients needed for normal health and growth.

Digestibility (or absorption efficiency) The proportion of a feed nutrient that is digested and absorbed from the gastrointestinal tract, as indicated by intake minus faecal output of the nutrient; apparent digestibility differs from true digestibility in that the former takes no account of any endogenously derived nutrient that may appear in the faeces.

Digestion Process by which nutrients are rendered soluble and capable of being absorbed; the actions of the various hydrolytic enzymes that break down the proteins, carbohydrates and lipids into smaller units.

Diploid An organism whose cells (apart from the gametes) have two sets of chromosomes, i.e. having a double set of homologous chromosomes and, therefore, two copies of the basic genetic complement of the species.

Diurnal Active during daylight hours; *cf.* Nocturnal, Crepuscular. *Alt.* occurring every day.

Diurnal rhythm A biological rhythm having a periodicity of about 1 day (24 hours) in length. *cf.* Circadian rhythm.

DNA (Deoxyribonucleic acid) The genetic material of the cell found within the nucleus; provides codes for protein synthesis, and serves as a blueprint for cell replication.

Dominance systems Social systems in which physical domination of one individual over others is initiated and sustained by aggression or other behavioural patterns. *Despotism* describes the case in which one individual dominates all others, and there are no intermediate ranks. In dominance, or social, hierarchies there are distinct ranks, with individuals of any rank dominating those below them and submitting to those above them.

Dominant The highest ranking individual in a dominance hierarchy; the alpha (α) individual.

Dopamine A biogenic amine, formed from tyrosine, which acts as a neurotransmitter in the central nervous system.

Down-regulation Decrease in number, as of receptors on cell membranes, or a decrease in the rate of production of a particular substance; regulation which reduces the number and/or sensitivity of receptors.

Dress-out percentage Usually refers to the percentage weight of the fish following removal of the viscera, head and tail; may also be used to refer to the gutted carcass, including the head and tail.

Drug A chemical that modifies a physiological process in an animal; a substance used to prevent, cure or treat a disease condition; usually encompasses those substances with sufficient potential to cause harm that manufacture, sale and use are controlled through regulatory restrictions.

Dry matter   The portion of a feed, feedstuff or tissue remaining after water (moisture) is removed.

Dyestuff   Indigestible pigment that may be added to a feed for the purpose of monitoring consumption of that feed by examination of stomach and intestinal contents or faeces.

Ectotherm   An animal whose body temperature is determined primarily by the temperature of the environment; also called poikilotherms, and colloquially 'cold-blooded' animals. *cf.* Endotherm.

Effluent   The wastewater flowing out of a fish farm (or rearing unit), power station, or industrial enterprise. Aquacultural effluent can impact on the environment because of the dissolved and suspended matter it contains, including phosphorus, nitrogen and organic matter.

Egestion   The process of ridding the body of waste material by defaecation; the voiding of unabsorbed material from the gut.

Electrophysiology   The study of electrochemical properties of living tissue.

Endocrine   *See* Hormone.

Endogenous   Originating from within the organism, i.e. produced within or caused by factors within the body.

Endorphins   A family of peptides produced in the gut, brain and pituitary that can mimic the effects of morphine, by binding to opiate receptors. Endorphins, together with the enkephalins and dynorphins, are known as the endogenous opioids or opiates.

Endotherm   An animal capable of maintaining body temperature relatively independent of the temperature of the environment; animal that generates its own body heat; also called homeotherms, and colloquially 'warm-blooded' animals. *cf.* Ectotherm.

Energy budget   The balance of energy input and use in a biological system, expressed in terms of consumption (intake or ingestion), production (growth), respiration, faecal losses and excretory losses.

Enkephalin   A pentapeptide, found in the brain and some other tissues, that binds to opiate receptors and mimics the effects of morphine.

Entraining agent   A factor that synchronises an organism's biological rhythm to the outside world; e.g. the light-dark cycle may be an entraining agent for circadian rhythms. *See* Zeitgeber.

Erucic acid   A 22C monounsaturated fatty acid – 22:1(n–9) – that is a normal constituent of rapeseed oil. Considered to be cardiotoxic due to the production of lesions via focal necrosis of cardiac muscle fibres.

Essential amino acid (EAA)   *See* Indispensable amino acid.

Essential fatty acid (EFA)   Fatty acids that cannot be synthesised by animals, and which must be present in the diet. Fatty acids of the (n–3) and (n–6) series are the EFAs.

Essential nutrient   A nutrient that cannot be synthesised by an animal, and which must be present in the diet.

Ester   A compound formed by the condensation of an acid with an alcohol; includes acylglycerols, which are esters of fatty acids and glycerol, wax esters, and phospholipids.

Ether extract   The portion of a feed, feedstuff or tissue that is soluble in ethyl ether and is removed by extraction in this solvent; represents the majority of the lipid fraction of a feed.

Euryhaline   Term used to describe organisms that are tolerant of a wide range of salinity. *cf.* Amphihaline, Stenohaline.

Eurythermal   Term used to describe organisms that are able to tolerate a wide range of temperature. *cf.* Stenothermal.

Eutrophic   Pertaining to water bodies that contain high levels of nutrients, and may therefore support high plant productivity.

Excretion   The elimination of waste materials from the body; specifically the elimination of waste materials produced by metabolism, e.g. nitrogenous excretion refers to the elimination of nitrogenous compounds (ammonia, urea, uric acid, etc.) resulting, primarily, from the metabolism of proteins and nucleic acids.

Exogenous   Originating from, or caused by factors, outside the body.

Expanded feed (or pellet)   Type of dry, relatively low-density pelleted feed with a low sinking rate.

Extraoral   Located outside the oropharynx.

Extruded feed (or pellet)   Pelleted feed produced by the extrusion process.

Extrusion   The process by which feeds are prepared by passing the ingredients through a die under high temperature (110–150°C) and pressure. When the feed leaves the extruder die it cools rapidly, takes up air, and increases in volume to give a pellet of low density. The high temperatures and pressures used under extrusion lead to gelatinisation of starch, but increase the risk of degradation of some fatty acids, vitamins, and amino acids.

Facial lobe   An enlargement in the dorsal medulla oblongata of the brain, associated with connections of the facial nerve (nervus facialis; cranial nerve VII).

Facial nerve   A cranial nerve innervating taste buds on the lips, barbels and in parts of the mouth; cranial nerve VII.

Faeces   The undigested feed residues, plus some endogenous digestive secretions, sloughed cells of the intestinal lining, and bile metabolites that are expelled from the gut via the anus.

Fats   Simple lipids that are esters of fatty acids with the trihydric alcohol glycerol. Important energy sources in feeds, and the primary form of energy storage in animals.

Fatty acid   A long chain organic acid of the general formula $CH_3(C_XH_Y)COOH$, where the hydrocarbon chain is either saturated ($Y = 2X$) or there are double bonds between some of the adjacent carbon atoms (unsaturated fatty acid); a constituent of lipids.

Feed   Food for animals, usually manufactured from a range of ingredients, or feedstuffs. *Alt.* The actions involved in obtaining and ingesting food.

Feed additives   Non-nutritive components of feed formulations that either augment the nutritional value of feeds or otherwise result in increased fish production. Some additives (e.g. binders, antimicrobial compounds, antioxidants, feeding stimulants) affect feed quality (e.g. stabilise feeds and prevent deterioration during storage), whereas others (e.g. enzymes, pigments, hormones, chemotherapeutants) have more direct effects on fish health and product quality.

Feed consumption (or intake)   The amount ingested within a specified time interval.

Feed conversion efficiency (FCE)   *See* Feed efficiency.

Feed deprivation   Withdrawal or withholding of food for a protracted time period; a period of feed deprivation is usually introduced prior to the slaughter of farmed fish to ensure that feed residues have been voided from the gut.

Feed efficiency   Wet weight gain per unit feed consumed; often calculated as gain divided by the amount of feed provided, thereby including an error relating to unconsumed feed waste; may also be termed feed conversion efficiency (FCE) or feed utilisation. *cf.* Feed:gain ratio.

Feed gain ratio   The term used in commercial aquaculture to express the quantity of feed required to produce a given weight gain; the reciprocal of feed efficiency.

Feed selection (or preference)   The choice(s) made when several feeds are offered simultaneously; feeds may be presented simultaneously and choice examined *a posteriori* via examination of gut content, or choice may be assessed by examination of the number of visits made (trigger activation) to self-feeders providing different feeds.

Feedback control   Descriptive of a control system linked by reciprocal influences; operative in many mechanical, biological and biochemical processes in which the result of an action either inhibits (*negative feedback*) or promotes (*positive feedback*) further action.

Feedback feeder   *See* Interactive feeding system; On-demand feeder.

Feeding   Behaviour that includes all the activities that are involved in obtaining, handling and ingesting food.

Feeding deterrents   Chemical stimuli that prevent the continuation of feeding and hasten the termination of feeding.

Feeding rate (or level)   Term used in commercial aquaculture to describe the amount of feed provided over a given time interval (usually 1 day). *cf.* Ration.

Feeding stimulants   Chemical stimuli that promote ingestion and continuation of feeding.

Feedstuff   Material suitable for the manufacture of animal feed; several feedstuffs are usually combined to produce an animal feed, and several feeds may be provided to the animal to give a balanced diet.

Fillet   Strips of fish flesh cut parallel to the main body axis, and from which fins, major bones and, sometimes, skin and belly flap have been removed.

Fish farm   A generic term used to describe an aquaculture production unit; the term encompasses the rearing facilities (tanks, raceways, ponds, cages), farm buildings, service equipment and the stock of fish.

Fish meal   Ground product of dried, defatted fish or fish-processing wastes; 'quality' or grade depends upon the type of raw material used and the processing conditions employed during manufacture; an important source of protein in fish feeds.

Fish oil   Lipids extracted from whole fish or fish waste that are liquid at room temperature.

Fish pond   Any pond, either natural or man-made, containing fresh-, brackish- or seawater, in which fish are held or reared.

Fish protein concentrate (FPC)   Products of high protein content prepared from whole fish or fish processing waste using a variety of techniques for the selective separation of proteins from other constituents.

Fjord   Narrow, deep, steep-walled inlet of the sea formed either by the submergence of a mountainous coast or by the entrance of the sea into a deeply excavated glacial trough after the glacier melts away; a fjord may be several hundred metres deep, and often has a relatively shallow entrance sill.

Flavour enhancer   Also known as taste enhancers, these are compounds that exert a synergistic effect on other flavour components; examples include monosodium glutamate and the 5'-nucleotides, 5'-inosine monophosphate (IMP) and 5'-guanosine monophosphate (GMP).

Food   Material of plant or animal origin that contains nutrients, and which, following digestion and absorption, provides nourishment to the body; the material an animal feeds upon.

Foraging   Behaviours associated with searching for, capturing, subduing and consuming food.

Formulated feed   A compound feed containing ingredients that together provide specified levels of nutrients.

Fortification   The addition of nutrients to a feed so that total amounts exceed those present in the basic ingredients. Feeds are usually fortified with vitamins and minerals to ensure that the amounts present are more than adequate to meet requirements.

Fouling organisms   The assemblage of aquatic organisms that attach to and grow on underwater objects, such as the hulls of ships, and fish rearing enclosures (net pens and cages).

Free-running rhythm   A biological rhythm operating in the absence of environmental cues.

Frequency of occurrence   A numerical index used in stomach contents analysis; the percentage of stomachs within a sample containing a given type of prey, giving an estimate of the proportion of fish within a population that has fed on a particular prey type.

Galanin   A 29-amino acid peptide, named for terminal glycine and amidated alanine residues, produced by cells in both the gastrointestinal tract and central nervous system; central administration stimulates feeding.

Gaping (of fillet)   Rupture of the connective tissue myosepta in the fillet of fish that results in a separation of the myotomes (muscle blocks); makes processing of the fillet more difficult and also reduces the market value.

Gelatinisation   The rupturing of starch granules resulting from treatment with combinations of moisture, heat, pressure and mechanical shear; aids in the binding of feed ingredients.

General linear model (GLM)   A statistical model relating a dependent variable to one or more independent variables. The underlying relation is linear such that a unit change in the independent variable(s) produces a unit change in the dependent variable. Examples include simple linear regression, multiple regression, and log-linear models.

Glossopharyngeal nerve   A cranial nerve innervating taste buds located in the posterior part of the mouth and on the gill arches; cranial nerve IX.

Glucagon   A pancreatic polypeptide hormone involved in the regulation of blood glucose and fatty acid concentrations; stimulates the breakdown of liver glycogen to glucose.

Glucosinolates   A class (over 90 exist in dicotyledonous plants) of sulphur-containg compounds with a common structure consisting of a central thiocyanate with a β-glucose residue attached to the sulphur atom, and a sulphate ion attached to the nitrogen atom. The nature of the group attached to the thiocyanate carbon atom gives the various glucosinolates their individuality. The glucosinolates are not particularly harmful themselves, but upon enzymatic hydrolysis by myrosinase they form products that may impair thyroid function and have other deleterious effects.

Gluten   A complex of proteins – gliadins and glutenins – together with the other components of the endosperm cells of grains (cereals). As a protein source in feeds gluten is deficient in lysine, methionine and tryptophan.

Glycogen   A branched chain polysaccharide made up of glucose units joined by α-1,4 and α-1,6 linkages; the main storage carbohydrate in animals, found in the liver and muscles.

Glycolysis   The anaerobic breakdown of glucose to pyruvate in living cells, with the production of ATP; the pyruvate formed may either enter the tricarboxylic acid cycle, or be converted anaerobically to lactic acid, or ethanol and carbon dioxide, or other organic products.

Glycoprotein   A conjugated protein with one or more heterosaccharides as a prosthetic group; most glycoproteins contain either glucosamine or galactosamine, or both, and galactose and mannose may also be present. Glycoproteins have lubricant properties, and are components of mucous secretions.

Goitrogens   Agents that interfere with thyroid function, either by acting directly on the thyroid gland or by altering its regulatory mechanisms and thyroid hormone homeostasis. There are several antithyroid sulphur-containing organic compounds present in plant feedstuffs, e.g. thiocyanate and thiocyanate-derived compounds, and goitrin derived from glucosinolates.

Gossypol   A toxic, yellow, phenolic pigment found in the pigment glands of cottonseed.

Green algae (Chlorophyta)   Freshwater and marine algae that have chlorophylls and carotenoids as their pigments. Their main storage product is starch, and the cell walls are of
cellulose.

Gross energy   The heat of combustion of a material generated on complete oxidation; determined by burning samples in a bomb calorimeter.

Ground water   Water that is contained in subsurface geological formations, and has characteristics (e.g. pH, dissolved minerals) related to the rocks with which it has contact; often
has low levels of dissolved oxygen and stable temperature; often brought to the surface by
pumping or by use of artesian wells.

Growth hormone (GH; somatotropin)   A peptide hormone produced in the anterior pituitary; implicated in the regulation of several physiological processes in fish, including
growth and metabolism, and ionoregulation.

Gums   Materials, composed largely of polysaccharides, that result from the breakdown of
plant cell walls.

Gustation   Detection of chemical stimuli via the sense of taste.

Gut (or gastrointestinal tract)   *See* Alimentary canal.

Habituation   A form of learning in which repeated applications of a stimulus result in decreased responsiveness.

Haemagglutins   *See* Lectins.

Hand feeding   Manual feeding, as opposed to the use of automatic feeders.

Heat increment of feeding (HIF)   The increase in heat production (metabolism) that is
associated with the digestion, absorption and biochemical processing of a meal; is often
measured as the increase in oxygen consumption that follows the ingestion of a meal;
also known as Specific Dynamic Action, or Effect (SDA or SDE), the calorigenic or ther-
mogenic effect of feeding.

Heat of combustion   The heat produced when organic material is completely oxidised in
an oxygen atmosphere to yield carbon dioxide, water, and oxides of nitrogen and sulphur;
usually measured using a bomb calorimeter, and referred to as the gross energy of a feed
or feedstuff.

Hemicellulose   The plant cell wall polysaccharides that are extractable by treatment with
dilute sodium hydroxide solutions. Hemicellulose comprises 20–30% of plant cell walls,
and it includes all the cell wall polysaccharides except cellulose and pectin.

Hierarchy   A principle of organisation in which the elements are ordered in such a way that the higher-ranking elements control lower ones. Dominance hierarchies often occur in the social organisation of animals; in a linear dominance hierarchy an α animal is dominant over all others, a β animal dominates all but the α and so on, whereas in a branched hierarchy there may be several animals of equal rank.

High-energy feeds   Feeds that are formulated to contain adequate levels of protein, and high levels (sometimes in excess of 30%) of lipid (oil); extruded pellets which are given a 'top dressing' of oil in order to increase the lipid content.

Histamine   A biogenic amine, produced by decarboxylation of the amino acid histidine, with several physiological effects, e.g. stimulation of secretion of gastric acid, stimulation of blood vessels to dilate and become leaky (vasodilation).

Homology   Resemblance by virtue of common descent; affinity of structure and origin apart from form or use.

Hormone (or endocrine)   A chemical messenger that is released into the blood and usually reaches all tissues, but may be excluded by the blood-brain barrier. Specificity is the property of the target tissue, due to the presence of receptors. Intracellular events mean that a given hormone can have different effects on different target tissues.

HUFA (highly unsaturated fatty acid)   A fatty acid with more than 4 double bonds in the carbon chain.

Hydrolysis   A chemical process whereby a compound is broken down into simpler units via the uptake of water; the digestive enzymes, proteases, carbohydrases and lipases hydrolyse proteins, carbohydrates and lipids, respectively, and they can be commonly termed hydrolases.

Hyper-   A prefix indicating that a quantity or factor exceeds, or is over or above normal levels. *cf.* Hypo-.

Hyperphagia   An excessive, or abnormally high, food intake; increased ingestion often observed when malnourished or undernourished animals are returned to adequate feeding conditions.

Hypo-   A prefix indicating below or beneath; also used to denote a deficiency or lack. *cf.* Hyper-.

Hypophysis   *See* Pituitary gland.

Hypothalamus   The brain region located beneath the thalamus in the diencephalon; serves as an important link between the autonomic nervous system and the endocrine system, and is involved in the regulation of many physiological functions.

Hypoxic   Deficient or lacking in oxygen; *hypoxia* refers to low levels of oxygen. *cf.* Anoxic.

Immunoassay   Method used to assay a substance using its reaction with specific antibodies, e.g. radioimmunoassay (RIA) is a very sensitive method for the detection and measurement of substances using radioactively labelled antibodies or antigens.

Immunohistochemistry   A microscopic technique in which cells are stained with antibodies labelled with dyes or electron-dense material in order to detect or highlight specific structures; technique in which one or more antibodies are used to visualise the distribution of the corresponding antigens in a histological preparation.

Immunoreactive   Reacting to particular antigens or haptens (a small molecule capable of eliciting an immune response when attached to a larger macromolecule).

Incomplete (or deficient) diet   One that does not fulfil the nutritional requirements of an animal.

Index of relative importance (IRI)   A numerical index used in stomach contents analysis, that combines frequency of occurrence, percentage by number and volumetric data into a single index: IRI = (% by number + % by volume)(% frequency of occurrence).

Indispensable (essential) amino acids (EAAs)   Amino acids that either cannot be synthesised by the body, or are synthesised in insufficient amounts to allow maintenance of normal health and growth. EAAs must, therefore, be obtained via the diet; they are Arg, His, Ile, Leu, Lys, Met, Phe, Thr, Trp and Val.

Infradian rhythm   A biological rhythm having a periodicity longer than 28 h or a frequency of less than one cycle per 24 h; examples include lunar and semi-lunar rhythms, and seasonal rhythms.

Ingestion   The process of consuming food.

Inlet   A short narrow waterway connecting a bay or lagoon with the sea.

Inosine monophosphate (IMP)   A nucleotide which is the precursor to adenine monophosphate and guanosine monophosphate; IMP has flavour enhancing properties.

Insulin   A pancreatic polypeptide hormone that plays an important role in the regulation of blood glucose concentrations; reduces blood levels of glucose, fatty acids, and amino acids, and promotes their storage; its action is antagonistic to glucagon, glucocorticoids and adrenaline.

Insulin-like growth factors (IGF) (or somatomedins) Polypeptide hormones secreted by the liver, and some other tissues, in response to a growth hormone stimulus. IGFs act directly on their target cells to promote cellular multiplication, differentiation and growth.

Interactive feeding system An on-demand feeding system in which feed release is controlled either via waste feed detection or by monitoring the behavioural responses of the fish. Automatic feeders are timed to release feed at regular intervals, a sensor system is used to monitor behaviour or detect any uneaten feed, and feedback signals from the sensor determine whether or not feed release should be terminated.

Intracerebroventricular (ICV) Within the ventricular system of the vertebrate brain.

Kelp Common name for brown algae of the order Laminariales. Seaweeds with a broad-bladed thallus attached to the substratum by a tough stalk and holdfast.

Kjeldahl method A technique used to estimate the amount of 'crude' protein in a sample; following the measurement of the quantity of nitrogen present, crude protein is estimated by multiplication by a factor (usually 6.25, because most proteins contain about 16% N).

Krill Common name for euphausiids, a group of planktonic marine crustaceans.

Lagoon A shallow sound, pond or lake separated from the open sea by one or more inlets.

Landlocked A body of water enclosed, or almost enclosed, by land.

Lateralis tuberis nuclei A group of related periventricular cells in the ventral hypothalamus of the brain.

Leaching Process by which chemicals, such as nutrients, may be dissolved, washed out of, and lost from materials.

Lean body mass (LBM) The body weight less the weight of fat; represents the body mass devoid of storage fat, comprising primarily moisture, protein and ash.

Learning Process in which the behaviour of an animal becomes consistently modified as a result of experience; may include conditioning, habituation and imprinting.

Lecithin The traditional term for the phosphatidylcholine phospholipids.

Lectins (haemagglutins) Proteins which interact with carbohydrates via binding to sugar residues in polysaccharides, glycoproteins or glycolipids. The ability to bind to sugar residues on cell membranes gives these proteins their agglutination properties.

Legume A member, or the fruit, of the large family of dicotyledonous plants that includes peas, beans, clover, lupin, etc.; the Leguminosae.

Leptin (*ob* protein)   A 167-amino acid (16-kDa) protein that is synthesised in, and secreted by, cells of the white adipose tissue; implicated in the regulation of food intake, energy expenditure and energy balance.

Lignin   A polymer, comprising phenylpropylene monomers, that is an important constituent of higher plants due to its glue-like properties that hold cell walls together. The phenylpropylene monomers are produced by deamination of the aromatic amino acids tyrosine and phenylalanine. Lignin is nondigestible, and is a component of the crude fibre fraction of feeds and feedstuffs.

Limiting nutrient   An essential nutrient that is present in a feed, or feedstuff, at low concentration, and that is likely to result in a deficiency, e.g. a limiting amino acid.

Lipids   A diverse class of organic compounds that are insoluble in water but soluble in organic solvents such as ether, acetone, ethanol and chloroform; include acylglycerols, phospholipids, glycolipids and steroids. Some lipids are essential components of biological membranes, and others act as energy stores.

Lux   Unit of illumination; used for the quantitative expression of light intensity.

Macronutrients   Substances required in relatively large amounts for normal growth and development; the macronutrients present in animal feeds are the proteins, lipids and carbohydrates.

Macrophyte   Rooted aquatic plant with vascular tissue, or a large, erect, branching alga.

Maillard reaction (nonenzymatic browning)   The complex series of reactions that may occur between reducing sugars and the free amino groups of amino acids and proteins. In feedstuffs it is usually glucose, fructose, maltose, lactose and reducing pentoses that react with the amino acids and proteins. In proteins, it is, predominantly, the primary amino group of the lysine side-chains that reacts.

Maintenance ration   The amount of food needed to meet metabolic requirements, without any gain or loss of body weight; the level of intake that provides for the continuation of life processes but allows for no increase in biomass.

Marker   An inert, indigestible substance added to a feed for the purpose of assessing digestibility or consumption; particulate, radio-opaque (X-ray dense) markers must be added to the feed for estimation of intake using the X-radiographic technique.

Masking (of a rhythm)   Alteration of the usual shape and/or parameters of a rhythm due to the influence of random or non-random environmental stimuli. Masking persists for the duration of the stimulus without inducing alteration of the endogenous rhythm components.

Meal   Food consumed in a single feeding bout; sequence of feeding events demarcated by refractory periods during which no feeding occurs. *Alt.* Dried, processed and ground product of animal or plant origin, e.g. fish meal, blood meal, soybean meal, rapeseed meal.

Medicated feed   A feed to which drugs or antibiotics have been added.

Melanocyte stimulating hormone (MSH)   A peptide of the hypothalamus and pituitary having both neuromodulatory and hormonal actions; implicated in the regulation of energy balance via suppression of food intake and increasing energy expenditure.

Melatonin   A hormone derived from the amino acid tryptophan; produced in the pineal gland and several other tissues, including the eye. Melatonin secretion influences biological rhythms, affects skin pigmentation, and it is also a natural antioxidant in cells.

Metabolic rate   A measure of the rate of metabolic activity within an organism; the rate at which an organism uses energy to sustain life processes. It is most usually measured as heat production or oxygen consumption (sometimes together with carbon dioxide production), but the latter only gives an estimate of the contribution of aerobic metabolism.

Metabolism   An integrated network of biochemical reactions that occurs in living organisms; the biochemical processes by which the absorbed nutrients are transformed and stored (*anabolism*), broken down and energy made available for the performance of work (*catabolism*).

Microbound feeds   Small (50–700 μm) particulate, formulated feeds that are typically fed to larval fish and crustaceans; characterised by being held together by an internal binder, usually a complex carbohydrate or protein having adhesive and absorptive properties.

Microencapsulated feeds   Small, particulate, formulated feeds that are typically fed to larval fish and crustaceans; characterised by having a distinct wall or capsule usually made up of material that differs from the central nutrient core; the wall may be of carbohydrate, protein or lipid (triacylglycerols or phospholipids), sometimes combined with other materials.

Micronutrients   Substances required in small amounts for normal growth and development, e.g. trace elements and vitamins.

Mineral   Inorganic salt; some minerals are required by animals, in moderate or trace amounts, for correct functioning of physiological activities.

Moist feed (or pellet)   A feed containing approx. 30% moisture; prepared from a mixture of wet ingredients, such as minced fish, and dry ingredients, such as fish meal, cereals, vitamin and mineral premixes, and alginate binder.

Monoamine   An amine in which one of the hydrogen atoms has been replaced by an organic side chain (group).

Monogastric   A simple stomach; used to describe nonruminant animals.

Multivariate   More than one variable. Used in reference to statistical techniques that analyse more than one dependent or independent variable simultaneously. Examples include analysis of variance, multiple regression, principal components analysis and discriminant function analysis.

Mycotoxins   A group of structurally diverse toxic substances produced by moulds (microfungi). The term mycotoxin is confined to describing the toxic metabolites produced by microfungi, as opposed to the toxins produced by some mushrooms and toadstools (macrofungi). The moulds that produce the mycotoxins of greatest concern are representatives of the genera *Aspergillus, Penicillium* and *Fusarium*.

Myofibril   Contractile protein fibril of muscle cells, composed of many myofilaments; has the actomyosin complex as the major component.

Myoseptum (*pl.* myosepta) (*Alt.* myocomma)   Connective tissue lying between successive myotomes (myomeres).

Myosin   Protein which interacts with actin to form the contractile complex of muscle cells.

Myotome   A muscular segment of the 'lower' vertebrates, such as fish, and the segmented invertebrates; segment of muscle separated from the next by a thin layer of connective tissue.

Naloxone   An analogue of morphine that blocks opiate receptors; an antagonist of opioid action.

Nares   The nostrils; openings of the olfactory organ to the exterior.

NDF (neutral detergent fibre)   The proportion of a feed or feedstuff remaining after extraction in a neutral detergent; mostly comprises plant cell wall constituents of low biological availability.

Neap tide   Tide of decreased range that occurs about every 2 weeks when the moon is in quadrature.

Net pen   Large enclosures used for rearing fish; usually placed in protected bays, inlets or fjords; netting forms the sides and bottom of the pen (to allow water exchange), there is an anchorage and weighting system (to reduce deformation), and the pen is supported by a series of floats at, or close to, the water surface.

Neurocrines (and neurotransmitters) Chemical messengers that are located in nerves, and are usually released very close to the target tissue; they may act directly on target tissue, and may also stimulate, or inhibit, the release of endocrines or paracrines.

Neuroendocrine Pertaining to both the neuronal and endocrine systems, structurally and functionally.

Neurohypophysis (posterior pituitary) *See* Pituitary gland.

Neuromodulator A neuropeptide or neurotransmitter that modifies the response of a synapse to another neurotransmitter.

Neuropeptide Any of the many small peptides produced by cells within the nervous system; some act as neurotransmitters and others as neuromodulatory hormones.

Neuropeptide Y (NPY) A 36-amino acid peptide that is implicated in the regulation of food intake and energy expenditure; central administration results in increased food intake.

Neurotransmitter A chemical that mediates the transport of nerve impulses across a synapse; released by the presynaptic nerve ending and interacts with receptors in the post-synaptic membrane.

NIRS and NIT *See* Spectroscopy.

Nitrification The oxidation of ammonia to nitrite, followed by the oxidation of nitrite to nitrate; process carried out chiefly by a few groups of bacteria (nitrifiers) of the genera *Nitrosomonas* (oxidise ammonia to nitrite) and *Nitrobacter* (convert nitrite to nitrate), and by a few species of fungi.

Nitrogen balance method Parallels the energy budget, but is based upon protein-N as the key material for growth.

Nitrogen-free extractives (NFE) One of the six fractions in the system of proximate analysis: calculated by difference when the sum of the percentages of moisture, crude protein, ether extract, ash and crude fibre is subtracted from 100. NFE is a complex mixture of compounds that may include cellulose, hemicelluloses, lignin, sugars, starch and pectins, organic acids, resins and tannins, pigments and water-soluble vitamins.

Nociception The sensing of painful or injurious stimuli.

Nocturnal Active during the hours of darkness; *cf.* Diurnal, Crepuscular.

Nonenzymatic browning *See* Maillard reaction.

Noradrenaline (norepinephrine) *See* Catecholamines.

NPU (net protein utilisation) or PPV (protein productive value)   The proportion of the ingested nitrogen (or protein) that is retained in the body.

Nucleic acids (DNA and RNA)   High molecular weight compounds which, on hydrolysis, yield a mixture of basic nitrogenous compounds (purines and pyrimidines), a pentose (ribose or deoxyribose) and phosphoric acid. They play a fundamental role in living organisms as a store of genetic information, and they are the means by which this information is utilised in the synthesis of proteins.

Nucleoside   A nitrogenous compound formed by the linking of a pyrimidine (e.g. cytosine, thymine, uracil) or purine (e.g. adenine, guanine) to a pentose (five-carbon sugar).

Nucleotide   A derivative of purine and pyrimidine bases, a five-carbon sugar (pentose), and phosphoric acid; product of esterification of a nucleoside with phosphoric acid; hydrolytic product of nucleic acids.

Nutrient   A substance used by an organism for maintenance, growth and reproduction; designated as macronutrients (proteins, lipids and carbohydrates) and micronutrients (vitamins and trace elements) depending upon the amounts required.

Nutrient requirement   Minimum amount of a nutrient that is needed for the maintenance of growth, health and reproduction, with no safety margin.

Oestrogen   *See* Steroids.

Off-flavours   Undesirable flavours and odours that develop in fish and fish products, and that negatively affect the acceptability of the product to consumers. Off-flavour problems may be associated with oxidative rancidity of lipids, bacterial spoilage of the fillet, with accumulation of pollutants (e.g. hydrocarbons) in the flesh, and with chemicals produced during the blooming of certain species of blue-green algae in fish ponds.

Oil   A term used to describe lipid sources that are liquid at room temperature, e.g. fish oil, cod liver oil, soya oil, linseed oil. The primary constituents are glycerol-based simple lipids, and oils are liquid at room temperature because of the high proportions of unsaturated fatty acids they contain.

Oilseed meal   Ground, dried residues remaining after the oil has been extracted from soybeans, rape/canola, cottonseed, sunflower, peanut, linseed, etc.; used as a protein source in fish feeds.

Olfaction   Detection of chemical stimuli via the sense of smell.

Olfactory nerve   Nerve that conducts odour stimuli from the olfactory organ to the brain; cranial nerve I.

Oligotrophic   Refers to a water body that contains low levels of nutrients, and therefore supports limited plant productivity.

Omnivore   An animal that feeds upon a mixed diet comprising both plants and animals.

On-demand feeders   Feeding systems in which both the timing and quantities of feed released are controlled by the fish: may be either self-feeders or interactive feeding systems.

On-growing   Colloquial term for the rearing of organisms from the time they are small juveniles until they reach marketable size.

Opiate   Any of a class of compounds mimicking the effects of opium (morphine) in the brain; several peptides found in the brain have an opiate action, e.g. endorphins and enkephalins.

Organoleptic   Properties of materials, usually foodstuffs, perceived by sensory organs, usually those involved in the senses of taste (gustation) and smell (olfaction).

Oropharynx   The region of the alimentary canal of fishes located caudal to the mouth and rostral to the oesophagus, including the continuous oral and pharyngeal cavities.

Palatability   The degree of acceptability of a feed or feedstuff, as affected by a range of physical (e.g. appearance, texture) and chemical (e.g. odour, taste) properties.

Paracrine   Chemical messengers that are released from endocrine cells and diffuse to target tissues. Paracrine effects are 'distance limited'. Paracrine agents can act on endocrine tissue to stimulate, or inhibit, the release of endocrines.

Parr   A juvenile life stage of salmonids, which extends from the time of the start of exogenous feeding until the fish transform into smolt and migrate to the sea.

Passive integrated transponder (PIT tag)   A small transmitter attached to an animal that transmits an identification signal only when activated by an external electronic stimulus.

Pectin   A high-molecular-weight polymeric carbohydrate that, along with cellulose, imparts firmness and rigidity to plant cell walls. The molecule consists of a chain of α-1,4,-linked, galacturonic acid units (100–1000 units in length) with varying amounts of neutral sugars forming side chains.

Pellet   Feed produced when ingredients are mixed and compacted by forcing them through a die. Binding of ingredients is usually achieved by the gelatinisation of starch in the mixture resulting from steam and heat treatment.

Peptide   A chain of a small number (up to *ca.* 20) of amino acids linked by peptide bonds.

Peptide bond   The covalent bond joining the α-amino group of one amino acid to the carboxyl group of another amino acid with the loss of a water molecule; the type of bond that links the amino acids of peptide and protein chains.

PER (protein efficiency ratio)   The gain in wet weight per unit weight protein consumed.

Percentage composition by number   A numerical index used in stomach contents analysis, that provides an estimate of the relative abundance of a particular prey type in the diet of an animal, or population of animals. The number of items of each prey type found in each stomach sampled is counted; the metric is the number of items of a given prey type expressed as a percentage of the total number of prey counted. Similar indices can be calculated from volumetric and gravimetric data, these being percentage composition by volume and percentage composition by weight, respectively.

Peroxidation   Destruction of unsaturated fatty acids due to oxidative processes, resulting in the production of toxic by-products; processes by which fats and oils become rancid.

pH   The negative $\log_{10}$ of the hydrogen ion concentration: the pH of a neutral solution is 7, that of acid solutions less than 7, and of alkaline solutions greater than 7.

Phospholipids   Lipids consisting of glycerol, two fatty acids and a phosphorylated alcohol; form the lipid bilayer in biological membranes.

Photoperiod   Durations of the daily exposure of an organism to light and dark conditions, e.g. a photoperiod of 12L:12D describes an environment where there are 12 h of light and 12 h of darkness.

Phototaxis   Directed movement in response to light; positive phototaxis is a movement towards a light source, whereas negative phototaxis involves movement away from the light source.

Phytate   Hexaphosphoinositol, the phosphate derivative of the sugar alcohol inositol, is mostly found in seeds; the phosphorus in phytate is not readily available to fish, and phytate also chelates with di- and trivalent metal ions thereby reducing their biological availability.

Phytoestrogens   Compounds that occur naturally in plants, and that can mimic the effects of oestrogens. Most of these compounds, e.g. isoflavones and coumestans, are only very weakly oestrogenic.

Pineal gland (pineal organ)   A light-sensitive organ that lies on the uppermost part of the diencephalon of the brain; produces the endocrine factor *melatonin,* and is involved in the regulation of physiological activities that vary in relation to the light cycle, either on a daily or seasonal basis.

Piscivore   An animal that preys upon, and derives a major part of its food from, fish.

Pituitary gland (hypophysis)   An endocrine gland that is closely allied to the hypothalamus; consists of relatively distinct anterior (adenohypophysis) and posterior (neurohypophysis) portions. The hormones of the pituitary gland consist of a range of peptides that have a variety of effects upon other endocrine organs and body tissues, e.g. growth hormone, gonadotropins, MSH and TSH.

Plankton   Organisms, either plant (phytoplankton) or animal (zooplankton), that are suspended in the water column and generally have limited swimming ability. Their major movements are influenced by wave action and water currents.

Pollution   Harmful or undesirable change in the environment as a result of the release of chemicals (e.g. industrial wastes and pesticides), heat, radioactivity, large amounts of organic matter (e.g. sewage), etc.; usually applied to deleterious environmental changes arising from human activities.

Polyculture   The rearing of two or more species in the same culture system (e.g. multispecies culture of cyprinids in freshwater ponds, as practised in China and several other countries in south-east Asia); usually involves the rearing of compatible species that do not compete for food and that have commercial value.

Polymorphism   The existence of different forms of individuals within a species or population; occurrence of different forms of, or different forms of organs in, the same individual at different periods of life; showing a marked degree of variation in body form during the life cycle, or within the species.

Polypeptide   A long chain of amino acids linked together by peptide bonds; a structural unit of a protein, some proteins consisting of one, and others of several.

Polysaccharide   A high molecular weight carbohydrate formed by the linking together of a large number of monomer units; classified as homopolysaccharides composed of a single type of monomer unit, and heteropolysaccharides which are made up of a mixture of different monomers. Polysaccharides are found as storage products (e.g. starch and glycogen) and as structural components of cell walls (e.g. cellulose).

Polyunsaturated fatty acids (PUFAs)   Fatty acids with more than one double bond (usually 2–4) in the hydrocarbon chain, e.g. linolenic [18:3(n–3)] and linoleic [18:2(n–6)] acids.

Post-prandial   Following a meal.

PPV (protein productive value)   *See* NPU.

Precision   A measure of repeatability, or of how close repeated measures are to one another. *cf.* Accuracy.

Premix   A mixture of micronutrients (e.g. vitamins or trace elements) with a bulk 'carrier'. The carrier is added to increase dispersion, and more uniform mixing, of the micronutrients in the finished feed.

Preoptic nuclei   A group of neurones in the preoptic area of the diencephalon of the brain; consists of the pars parvocellularis (small cells located anteroventally) and the pars magnocellularis (large cells located posterodorsally).

Preservative   A chemical added to a feed to prevent oxidation, fermentation or other forms of deterioration, primarily due to bacterial or fungal attack.

Prosthetic group   A nonprotein chemical group that is bound to a protein; the prosthetic group of enzymes usually forms part of the active site, and is, therefore, essential for the expression of biological activity.

Protease inhibitor   Molecules, often proteins (such as the trypsin inhibitors), that inhibit the actions of proteolytic enzymes.

Protein   An organic molecule consisting of chains of amino acids, and containing chiefly C, H, O, N and S. One or more polypeptide chain may comprise a protein, and such chains may also be associated with nonprotein components (prosthetic groups). Essential in living organisms as enzymes and structural constituents of tissues.

Protein quality   The nutritional value of a protein, determined primarily by its amino acid composition (i.e. relative content of essential amino acids) and its digestibility.

Protein synthesis   The synthesis of a protein at the ribosomes using messenger RNA as a template.

Proximate analysis   A combination of analytical procedures used to quantify the moisture and dry matter contents of feeds, feedstuffs or plant and animal tissues. Six fractions are recognised: moisture, crude protein, ether extract, ash, crude fibre and nitrogen-free extractives.

Quaternary ammonium compounds   Term used to describe compounds that contain a nitrogen atom that is bonded to four organic groups, e.g. glycine-betaine or trimethylglycine.

Raceway   A long, narrow rearing unit, usually constructed of concrete; the water inlet and outlet are at opposite ends so water flows along the length of the unit.

Radioimmunoassay (RIA)   *See* Immunoassay.

Radioreceptor studies   Use of a technique in which radiolabelled substances, e.g. hormones, are used to quantify receptor densities in tissues.

Raffinose   A trisaccharide that gives glucose, fructose and galactose on hydrolysis.

Rape (rapeseed)   An oilseed plant of the *Brassica* family grown in temperate regions for the production of its oils; rapeseed meal has a high protein content and may be included in animal feeds, but use may be restricted due to the presence of several antinutritional factors (efforts have been made to reduce ANFs by selective breeding).

Ration   A fixed portion of feed, usually expressed as the 'daily allowance' or the amount of feed consumed by an animal each day.

Recirculation system   A closed, or partially-closed, rearing system used in aquaculture in which the effluent water from the rearing units is treated to enable its reuse; *cf.* Single-pass system.

Red algae (Rhodophyta)   A diverse assemblage of eukaryotic algae in which the red colour is imparted by the predominance of phycobilins over the other pigments. The storage carbohydrate is floridean starch, which resembles amylopectin. The red algae are largely marine, they are distributed worldwide, and there are many species in both tropical and temperate waters.

Refractory   Unresponsive; a period during which a stimulus fails to induce a response.

Reinforcer   An environmental change which increases the likelihood that an animal will make a particular response, e.g. a reward acting as a positive reinforcer.

Rendered animal products   Products made from the offal or 'wastes' arising from the slaughter and processing of domestic animals for human consumption; the offal may include blood, viscera and skeletal tissues, skin and feathers, etc.; the offal may be dried and processed into meals for inclusion in animal feeds.

Reward   That which is given in return for performance of a given act; a positive reinforcement. A food reward can be given as a given weight or number of food items.

Reward (reinforcement) schedules   There are 4 simple types of reward schedule: *fixed-* and *variable-ratio schedules*, and *fixed-* and *variable-interval schedules.* With a self-feeder operating on a fixed-ratio *N* schedule, feed delivery would follow immediately after every *N*th trigger actuation, independently of its time of occurrence; thus there is a fixed ratio between the number of actuations made and the number of rewards obtained. On a fixed-interval *T* seconds schedule, feed would be delivered immediately following the first trigger actuation made *T* or more seconds after the previous delivery; thus on an interval schedule, reward is given at intervals of time determined by the experimenter, and there is a time-restriction imposed upon the way in which feed is delivered. The number of trigger actuations (responses) required for each feed delivery on ratio schedules, or the time that must elapse before a response is effective on interval schedules need not be fixed, but can be varied in some fashion. Such schedules are termed variable-ratio and variable-interval, respectively.

Rhythms   Wave-like variations that are characterised by given periods, frequencies and amplitudes. The period of a rhythm is the interval of time between successive peaks, the frequency is the number of peaks in a given period of time, and the amplitude is defined as half the magnitude of the change that occurs between a peak and a trough. Biological rhythms may be driven by a clock which is endogenous in the sense that the rhythmic mechanism is independent of external events, they may be entrained by external time-setters, *Zeitgebers*, responsible for the maintenance of synchrony between the clock rhythm and the rhythm of environmental events, or they may be completely under exogenous influence.

Rigor mortis   Stiffening of the body after death due to contraction and temporary rigidity of the muscles.

RNA (ribonucleic acid)   Large linear molecule made up of ribonucleotide subunits containing the bases uracil, guanine, cytosine and adenine; found in cells as transfer (tRNA), ribosomal (rRNA) and messenger (mRNA) RNA, these RNAs being synthesised by transcription of chromosomal DNA acting as a template.

RQ (respiratory quotient)   The ratio of the volume of carbon dioxide produced to the volume of oxygen used in respiration; provides an indication of the type of substrate used in respiration, e.g. an RQ close to 0.7 suggests that lipid is the primary fuel, whereas an RQ close to 1 suggests that the major respiratory substrate is carbohydrate.

Ruminant   Herbivorous ungulate (Artiodactyla) that possesses a 4-chambered stomach and chews the cud (ruminates); the stomach contains micro-organisms that break down the cellulose in plant material; examples include cattle, sheep, goats, deer and antelope.

Saponins   A heterogeneous group of plant glycosides, in which the structure consists of an aglycone unit linked to one or more carbohydrate chains. When saponins are agitated in water they form a soapy lather. Other properties generally ascribed to saponins are haemolytic effects on red blood cells, cholesterol-binding properties, and a bitter taste. They are widely distributed in plants of agricultural importance, particularly legumes.

Sarcoplasmic proteins   Proteins found within the cytoplasm between the fibrils of muscle tissue.

Satiate   To gratify fully, or glut, as in being fed to satiation. Satiety may be considered as the satisfaction of appetite that results in the cessation of feeding, i.e. when feeding undisturbed from a source of freely available food an animal will eventually stop, and it is then said to be satiated. During the process of satiation the rate of feeding will usually gradually decline until it reaches zero, although an abrupt cessation of feeding is sometimes observed.

Saturated fatty acid   A fatty acid that is completely hydrogenated, and therefore lacks double bonds between adjacent carbon atoms.

Sea cage   A floating enclosure, comprising a wire or netting pen suspended from a floating framework; used to hold and rear fish in open water. *See* Net pen.

Secondary metabolites (compounds)   Compounds produced by microbes, plants, and some animals (e.g. antibiotics, alkaloids and tannins) that may not be essential to the growth of the organism, but may have other functions e.g. in chemical defence.

Self-feeder   An on-demand feeding system in which feed release is controlled by a trigger-release mechanism operated by the fish. Feed demand may be monitored via recording the timing and numbers of trigger actuations, but the amount of feed consumed is unknown unless feed wastage can also be recorded.

Serotonin (5-hydroxytryptamine; 5-HT)   A biogenic amine, derived from tryptophan, that acts as a neurotransmitter within the central nervous system.

SGR (specific growth rate)   Term describing an exponential change in body mass with time; $SGR = [(\ln W_2 - \ln W_1)/t] \times 100$, where W_1 and W_2 are initial and final weights, and t is the time (in days) between weighings.

Silage   Material that has undergone anaerobic fermentation, usually due to the actions of lactic acid bacteria, e.g. fish silage.

Single-cell protein (SCP)   The generic term used to describe the crude or refined protein which originates from bacteria, yeasts, moulds or unicellular algae.

Single-pass system   A system in which water is passed through rearing units only once, and is discharged without being recycled. *cf.* Recirculation system.

Slack water   Interval when the speed of the tidal current is very weak or zero; usually refers to the period of reversal between ebb and flood currents, so in most places slack water occurs near times of high and low water.

Smolt   A juvenile life stage in most species of salmonids; the fish at the time it is physiologically and behaviourally prepared to undertake migration from fresh water to the sea.

Social facilitation   Behaviour that is initiated, or increased in rate or frequency, by the presence of another animal carrying out the same behaviour.

Social interactions   Encounters between individuals in which the behaviour of each individual influences the behaviour of the other.

Solvent-extracted   Materials from which the fats and oils have been removed by treatment with organic solvents, e.g. solvent-extracted soybean meal.

Somatomedins   *See* Insulin-like growth factors.

Somatotropin   *See* Growth hormone.

Specific dynamic action (SDA)   *See* Heat increment of feeding (HIF).

Spectroscopy   Encompasses a range of techniques for acquiring information about atomic and molecular structure via the study of patterns of absorption or emission of electromagnetic radiation. Near-infrared spectroscopy is a technique that is used for the rapid analysis of the chemical composition of feeds and feedstuffs, and instruments that measure either reflectance (NIRS) or transmission (NIT) have been developed.

Spectrum   A range of radiations given in order of wavelength.

Spring tide   Tide of increased range, which occurs about every 2 weeks when the moon is new or full.

Stachyose   A tetrasaccharide comprising a glucose, a fructose and two galactose residues.

Starch   A polysaccharide made up of a long chain of glucose units joined by α-1,4 linkages, that may either be unbranched (amylose) or branched at an α-1,6 linkage (amylopectin); an important storage carbohydrate in plants, occurring as granules.

Starter feed   Feed provided to larval and juvenile fish immediately following, and for a short period of time after, the transition from endogenous to exogenous nutrition.

Steam conditioning   The treatment of feed ingredients with steam to improve the bioavailability of the nutrients they contain, and to cause partial gelatinisation of starch prior to the production of feed pellets.

Stenohaline   Term used to describe organisms which tolerate only a narrow range of salinity. *cf.* Euryhaline.

Stenothermal   Term used to describe organisms which have a limited range of tolerance to temperature. *cf.* Eurythermal.

Steroid (hormones)   Hormones that have cholesterol (27C) as their chemical precursor; include the 21C steroids of the interrenal (e.g. cortisol), the male sex hormones (androgens; 19C), and the female sex hormones (oestrogens; 18C).

Stomach contents analysis   A method used for determining the types of prey consumed by animals, that may also be used to give information about prey selection. Various indices may be calculated based upon the collection of numerical, volumetric or gravimetric data.

Stress   The response of an animal to a demand that causes a physiological disturbance to the point at which performance may be compromised and the chances of survival reduced; stressors in aquaculture are typically physical disturbances (e.g. those caused by handling, grading or transport) that invoke acute stress, or those that are chronically stressful (e.g. poor water quality, overcrowding). Stress responses have been broadly classified as *primary* (neuroendocrine and endocrine responses involving stimulation of the hypothalamic-pituitary-interrenal axis), *secondary* (includes changes in blood chemistry and other haematological factors that relate to physiological functions such as metabolism and hydromineral balance) and *tertiary* (refer to whole-animal performance, such as growth, disease resistance and survival).

Synchronisation   State of a system when two or more variables exhibit periodicity with the same frequency and stable phase relation.

Tachycardia   A rapid heart rate, or an increase in heart rate to above normal levels.

TAN (total ammonia nitrogen)   *See* Ammonotelic.

Tannins   These are complex, soluble, phenolic secondary metabolites that play a prominent role in the general defence strategies of plants. They are believed to be by-products of the metabolism of the aromatic amino acid phenylalanine.

Taste bud   A sense organ (gustatory) consisting of a small flask-shaped group of cells; found chiefly within the oropharynx, but also extraorally on barbels or free rays of the pectoral fins of some fish species.

Taxis   Directed movement of an organism towards (positive) or away from (negative) a source of stimulation, such as light (phototaxis) or chemicals (chemotaxis); behavioural orientation to a a directional stimulus.

Tegmentum mesencephali   A mass of white fibres, with grey matter, in the mesencephalon (middle portion of the vertebrate brain); part of the ventral midbrain above the substantia nigra.

Telencephalon   Anterior part of the forebrain, including the cerebral hemispheres, lateral ventricles, optic part of the hypothalamus and the anterior portion of the third ventricle.

Terpene   Member of a class of volatile aromatic compounds built up from 5C isoprene units; components of plant essential oils, and including menthol, limonene and geraniol.

Territory   An area which an animal defends by fighting or by demarcation and behavioural signals; the marks or signals act as a deterrent to entry.

TGC (thermal growth coefficient)   An expression of growth that attempts to take account of body size and temperature influences; $TGC = (^3\sqrt{W_2} - ^3\sqrt{W_1})(\Sigma T)^{-1}$, where ΣT is the sum day-degrees Celsius, and W_1 and W_2 are initial and final weights.

Thermocline   The zone of rapid temperature change in a thermally-stratified body of water; horizontal temperature discontinuity layer between a warm upper layer (epilimnion) and a cooler bottom layer of water (hypolimnion) in a freshwater lake; a thermocline may arise seasonally due to the heating of surface water during the summer, or may be a permanent feature of a water body.

Thyroid hormones   Iodine-containing hormones that are derivatives of the amino acid tyrosine; hormones secreted by the thyroid follicular cells; thyroxine (tetraiodothyronine, T_4) and triiodothyronine (T_3); hormones involved in the regulation of metabolism.

Thyroxine   An iodine-containing hormone that is produced by the thyroid gland; it is a derivative of the amino acid tyrosine, and has the chemical name tetraiodothyronine; involved in the regulation of body metabolism.

TOBEC   Total body electrical conductivity, a non-invasive method for the estimation of the composition of biological materials that relies on the fact that different chemical constituents differ in electrical properties.

Tocopherols   *See* Vitamin E.

Tomography (computerised tomography; CT)   A non-invasive technique used for the estimation of the composition of body tissues, and the distributions of various chemical constituents; the technique relies upon the fact that there are differences in X-ray transmission and reflection among the various constituents making up the different organs and tissues of the body.

Torpor   State of inactivity which is usually accompanied by a greatly reduced metabolic rate; may occur on a daily basis, or seasonally.

Trace element   An inorganic nutrient that is essential for normal growth and development of an organism, but which is required in minute amounts.

Transgenic   Animals and plants into which genes from another species have been deliberately introduced by genetic engineering, and in which there is qualitative or quantitative modification as a result of the expression of the introduced genetic constructs or transgenes.

Trigger actuation (activation)   The prerequisite for feed release from the hopper when self-feeders are being used. Self-feeders are usually equipped with triggering devices that must be pulled, pushed or bitten to cause the release of feed.

Triploid   An organism with three sets of chromosomes per somatic cell. Triploidy can be induced in fish to create sterile individuals.

Trypsin   A proteolytic enzyme (serine esterase) originating in the pancreas. Attacks peptide linkages in which lysine and arginine are present.

Trypsin inhibitors   The most common inhibitors present in legumes, such as soya bean, act on serine proteases, a group of proteolytic enzymes of which trypsin and chymotrypsin are representatives. Serine protease inhibitors are proteins that form stable complexes with these digestive enzymes, thereby reducing their activity. The two main protease inhibitors found in legumes are soya bean trypsin inhibitor (Kunitz) and soya bean protease inhibitor (Bowman–Birk).

Turbidity   Opacity resulting from the presence of suspended matter; the reduction in light transmittance in water resulting from suspended or colloidal matter, or the presence of planktonic organisms.

Turbulence   Motion in fluids in which velocities and pressures fluctuate irregularly.

TVN (total volatile nitrogen)   Volatile nitrogenous compounds, such as ammonia and volatile acids and bases; they are present in fish and fish products, and increase as a result of bacterial action; measurement of TVN provides an indication of 'freshness', and the degree of spoilage resulting from microbial activity.

Ultradian rhythm   A biological rhythm having a periodicity shorter than 20 h, or a frequency of more than one cycle per 24 h; examples include heart rhythms, and some behavioural rhythms linked to tidal cycles.

Vagal lobe   An enlargement in the dorsal medulla oblongata of the brain associated with connections of the vagus nerve (nervus vagus; cranial nerve X), and, in some species, the glossopharyngeal nerve (nervus glossopharyngeus; cranial nerve IX).

Viscera   A collective term for the internal organs, such as the gut, liver, pancreas, spleen and associated organs.

Vitamins   These are organic compounds required by animals in small amounts for the maintenance of a variety of metabolic functions; classified as lipid-soluble (vitamins A, D, E and K) or water-soluble (ascorbic acid, myo-inositol, choline and the vitamin B complex). Vitamins are essential nutrients that must be obtained via the diet, and a lack of a vitamin will lead to the development of a deficiency disease.

Vitamin C   *See* Ascorbic acid.

Vitamin E (Tocopherols)   A group of lipid-soluble compounds with strong antioxidant properties; have important functions in the stabilisation of biological membranes and in the prevention of oxidation of cell components.

Water hardness   The concentration of divalent ions that results when mineral salts are leached into the water from limestone and other rock formations; due primarily to the carbonate and bicarbonate salts of calcium and magnesium, although other metallic cations also contribute. Aquatic animals and plants require calcium and magnesium for normal growth and development, so the hardness of water has been used as a measure of its potential to support aquatic productivity: water low in hardness (soft water) is usually acidic, whereas harder water tends to be alkaline. Hardness is expressed as ppm (parts per million) or mg l^{-1} of calcium carbonate equivalents.

Water quality   In aquaculture, a description of how well the characteristics of the water body match the requirements of the farmed species in question; water can be of 'good' or 'poor' quality. Good-quality water is defined as water capable of supporting the survival and growth of the species being farmed, and of maintaining sanitary conditions of a standard required for harvesting and marketing the species as intended (e.g. as food for humans).

Wax esters   Esters of fatty acids and long chain monohydric alcohols that are insoluble in water and are difficult to hydrolyse; found as protective waterproof coatings on leaves and fruits, and on the exoskeletons of insects.

Wean   To accustom a young animal to a novel form of nourishment; traditionally used to describe the transition from the reliance of young mammals on mother's milk, but in aquaculture used to refer to the process of transition from the consumption of live prey to the acceptance of formulated feeds by larval and juvenile fish.

Xenobiotic   A substance foreign to a living organism; often used to describe drugs, or environmental pollutants, such as pesticides or industrial chemicals that enter water bodies in run-off from the land.

Zeitgeber   An external stimulus that serves to trigger or phase a biological rhythm. A *Zeitgeber* does not induce a rhythm or 'give time', but determines the arrangement of a rhythm in time.

Index